Powell's   Kinematics & Design of Planar

$35.00 / 9.98   PC

Mathematics & Physics   117163

Kinematics and Design of Planar Mechanisms

C. H. Chiang

KRIEGER PUBLISHING COMPANY
MALABAR, FLORIDA
2000

Original Chinese Edition 1997
1st English Edition 2000

Printed and Published by
**KRIEGER PUBLISHING COMPANY
KRIEGER DRIVE
MALABAR, FLORIDA 32950**

Copyright © 1997 Chinese Edition by McGraw-Hill, Inc. (Taiwan)
Copyright © 2000 English Edition by Krieger Publishing Company

This book was produced from camera-ready art provided by the author.

All rights reserved. No part of this book may be reproduced in any form or by any means, electronic or mechanical, including information storage and retrieval systems without permission in writing from the publisher.
No liability is assumed with respect to the use of the information contained herein.
Printed in the United States of America.

FROM A DECLARATION OF PRINCIPLES JOINTLY ADOPTED BY A COMMITTEE OF THE AMERICAN BAR ASSOCIATION AND A COMMITTEE OF PUBLISHERS:
This publication is designed to provide accurate and authoritative information in regard to the subject matter covered. It is sold with the understanding that the publisher is not engaged in rendering legal, accounting, or other professional service. If legal advice or other expert assistance is required, the services of a competent professional person should be sought.

Library of Congress Cataloging-in-Publication Data

Chiang, C. H.
    Kinematics and design of planar mechanisms / C.H. Chiang.
      p. cm.
    Includes bibliographical references and index.
    ISBN 1-57524-154-4 (alk. paper)
    1. Machinery, Kinematics of. 2. Links and link-motion. I. Title.

TJ175 .C43 2000
621.8'11—dc21

                                                  00-039066

PREFACE

Since 1950, as I began to give lectures on elementary kinematics of mechanisms, I was very much impressed by the two papers "What Is Wrong With Kinematics and Mechanisms" and "A Brief Account of Modern Kinematics" by A.E.R. de Jonge (de Jonge, 1942, 1943). To my knowledge the only lecture on Advanced Kinematics given in China before World War-II was that given by Professor R. Kraus at the Tung-Chi University, Woosung (Shanghai). I was able to teach Advanced Kinematics of Mechanisms from 1973 until my retirement in 1990. The materials gathered in this book are largely based on my lectures during that period. The contents of this book are comparable with those of the Beyer's book: Kinematische Getriebesynthese, Springer-Verlag, 1953, Berlin, but are more systematically organized and this book also encompasses some new techniques and concepts that have been developed in the following past forty years.

Research works in planar kinematics, both in theoretical treatment and practical applications, remain the main thrust in the kinematics field. Investigators dealing with spherical and spatial kinematics have to be equipped with sound basic knowledge of planar kinematics. This book introduces first some fundamental concepts of instantaneous planar kinematics to accustom the reader in applying these concepts in synthesis, or design problems. The following sections, apart from those dealing with some special techniques of mechanism analysis, are devoted to dimensional synthesis of mechanisms, and the last chapter introduces harmonic analysis of four-link mechanisms. Three types of dimensional synthesis problem are considered: body guidance, path generation and function generation. For each sort of problem, a geometrical as well as an algebraic method is presented. The algebraic method used in this book is the matrix method, to facilitate computer programming. As non-linear equations are not involved, the numerical method is excluded from this book, although a minimal numerical operation is needed in optimizing the transmission angle of a crank-rocker.

As a text it is not possible to cite all reference materials from the bulk of papers published within the past decades, therefore only articles that are closely related to certain subjects are cited. This book serves as a graduate-level text, as well as a self-teaching handbook for the professional designer of mechanisms.

Due to the shortage of available nomenclature, use of some notation representing more than one definition is unavoidable, particularly because some letter symbols are already standardized. Every effort has been made to avoid any possible ambiguities.

Some exercises, excluding the verification exercises, may not have a unique solution. In such cases, a unique answer cannot be given. Teachers may also assign existing examples as exercises.

The first edition of this book was published in 1997 in Chinese by McGraw-Hill International Enterprises Inc. (Taiwan). The pressent English edition is a complete translation of the Chinese version.

As can be observed from the numerous papers published in journals and conferences in recent years, the subject planar kinematics has not yet been exhausted. It is hoped that new ideas and techniques may be developed in the future in this branch of kinematics.

I am very much obliged to Drs. Jen-San Chen and Ching-Kuo Lin for their assistance in devising numerical examples and answers to some exercises, and in computer programming; to Dr. Hon-Cheung Yu for reviewing the English translation; and to Mr Jin-Wang Lin and Ms Ming-Ling Shih for drawing illustrations with the aid of computer graphics. In particular I am indebted to Dr. Samuel Molian for his encouragement in publishing the English version of this book. I would also like to thank my wife Danhsia and my daughter Gisela for their continuous support without whom I would not have been able to complete this book. Finally I would like to extend my gratitude to Krieger Publishing Company for their warm collaboration in presenting this book accurate in all details.

Taipei
June, 1999

C. H. Chiang

CONTENTS

PART I 1

1 Fundamental concepts of planar instantaneous kinematics 3
1.1 Pole, polodes and pole changing velocity 3
1.2 Hartmann construction 5
1.3 The inflection circle 7
1.4 The tangential circle 9
1.5 Relations between δ and u, and a_P and u 12
1.6 Euler-Savary equation 13
1.7 The Bobillier construction 17
1.8 Quadratic transformation 23
1.9 Kinematic inversion and the return circle 27
1.10 The generating curve and its envelope 28
1.11 Equation of the envelope of a coupler straight line of a four-bar linkage 30
Exercises 33

2 On some special techniques of mechanism analysis 39
2.1 The Freudenstein equation 39
2.2 Velocity analysis 40
2.3 Simplified acceleration analysis of a four-bar linkage 43
2.4 Simplified acceleration analysis of complex mechanisms 47
Exercises 49

PART II Synthesis of mechanisms 51

3 Dimensional synthesis of four-bar linkages — body guidance problems 53
3.1 Guiding a body through two finitely separated positions 53
3.2 Guiding a body through two infinitesimally separated positions 55
3.2.1 General concepts 55
3.2.2 Equations of fixed and moving polodes of a four-bar linkage 58
3.3 Guiding a body through three finitely separated positions --- geometrical method 62
3.3.1 General case 62
3.3.2 Three homologous points on a straight line 69
3.3.3 Three homologous lines passing through a fixed point 73
3.3.4 The R_M - and R^1-curves 78
3.4 Guiding a body through three finitely separated positions --- algebraic method 82

3.4.1 Angle of rotation of a body displacement.. 82
3.4.2 The displacement matrix.. 83
3.4.3 The point A_0 corresponding to A_1, A_2, A_3 84
3.4.4 The point A_1 corresponding to A_0 85
3.4.5 The R_M- and R^1-curves 86
3.4.6 The displacement matrix based on coordinates of the pole 87
3.4.7 Selection of coordinate system 88
3.5 Guiding a body through three infinitesimally separated positions 89
3.5.1 Geometrical method 89
3.5.2 Algebraic method 93
3.5.3 The curve of equal radius of curvature of the first kind --- the ρ- and ρ_M –curves 94
3.5.4 The curve of equal radius of curvature of the second kind --- the q_1- and $q_M(q_{M1})$-curves 100
3.6 Guiding a body through four finitely separated positions --- the centre-point curve and circle-point curve 113
3.6.1 The centre-point curve... 113
3.6.2 Construction methods of centre-point curve.. 115
3.6.3 The circle-point curve.. 119
3.6.4 The break-ups of centre-point curve and of circle-point curve 121
3.6.5 Four homologous lines passing through a fixed point and four homologous points on a line.. 124
3.6.6 Circle-point curve and centre-point curve by algebraic method...................... 126
3.6.7 Four homologous points on a line by algebraic method................................... 130
3.7 Guiding a body through four infinitesimally separated positions --- the circling-point curve and centering-point curve.. 133
3.7.1 The circling-point curve 133
3.7.2 The centering-point curve... 137
3.7.3 Construction of circling-point curve and of centering-point curve 139
3.7.4 The velocity pole as a moving point... 144
3.7.5 Circling-point curve and centering-point curve of a given four-bar linkage by algebraic method... 149
3.7.6 The break-ups of circling-point curve and of centering-point curve 149
3.7.7 The Ball point (four infinitesimally separated homologous points on a line) and four infinitesimally separated homologous lines passing through a fixed point 156
3.7.8 The Ball curve 159
3.8 Guiding a body through five finitely separated positions --- the Burmester points.. 160
3.9 Guiding a body through five infinitesimally separated positions 163
3.9.1 The general case 163
3.9.2 Special cases 164
3.9.3 To find the Burmester points of a given four-bar linkage 165
3.10 Intermediate cases... 166

3.10.1	Cases $P_1P_2-P_3$ and $P_1-P_2P_3$	167
3.10.2	Intermediate cases of four positions	168
3.10.3	Intermediate cases of five positions	172
3.11	Closing address	172
	Exercises	172

4 Two other methods of guiding an instantaneous motion of a body 185

4.1	The principle of instantaneous invariants	185
4.2	The polode method	194
4.2.1	The Grübler-Hall equation	194
4.2.2	Relative polodes between input link and output link of a four-bar linkage	195
4.2.3	The Sieker-Beyer equation	199
4.2.4	Maximum and minimum values of the radii of curvature of the polodes	202
4.2.5	Synthesis to match given radii of curvature of polodes and their rates of change	207
4.3	Comparison of three methods of guiding an instantaneous motion	210
	Exercises	212

5 Balance of number of coordinates in synthesis problems 217

5.1	Kraus's concept of valence	217
5.2	Geometrical constructions for finding a certain unknown joint	219
5.3	Examples of balancing the number of coordinates	220
5.4	The method of point-position reduction	225
	Exercises	229

6 Dimensional synthesis of linkages --- path generation problems 231

6.1	Coordination of two coupler point positions with one crank angle	231
6.2	Coordinations of two coupler point positions with one pair of crank angles	232
6.3	Algebraic method	234
6.3.1	Coordinations of three coupler point positions with two crank rotation angles	234
6.3.2	Coordinations of four coupler point positions with three crank rotation angles --- the k_{A0}-curve	236
6.4	Vector method	242
6.4.1	Coordinations of four coupler point positions with three crank rotations	243
6.4.2	Coordinations of five coupler point positions with four crank rotations	245
6.5	Algebraic equation of the coupler (point) curve	246
6.5.1	Derivation of the coupler curve equation	246
6.5.2	Nodes of a coupler point curve	249
6.5.3	Singular foci of a coupler point curve	252
6.6	Roberts-Chebyshev theorem	252
6.6.1	The content of Roberts-Chebyshev theorem	252
6.6.2	A simple proof of Roberts-Chebyshev theorem	254
6.6.3	Special cases	255

- 6.6.4 Cognate linkages of multiple-bar linkages ... 259
- 6.7 R_M-curve as a special case of coupler point curve ... 260
- 6.8 Transition curve ... 262
- 6.9 Generation of ellipses ... 268
- 6.9.1 Basic geometrical concept ... 269
- 6.9.2 Six finitely separated positions of a body --- the conic section point curve ... 269
- 6.9.3 A special case of conic section point curve ... 271
- 6.9.4 An ellipse tangent to a given curve in four points ... 275
- 6.9.5 Dwell mechanism by means of osculating ellipse ... 276
- 6.9.6 An ellipse tangent to a given curve in five points ... 279
- 6.10 Generating coupler curves with cusps ... 280
- 6.10.1 Generating coupler curves with two cusps ... 280
- 6.10.2 Generating coupler curves with three cusps ... 284
- 6.11 Generating symmetrical coupler curves ... 285
- 6.11.1 Proof of the principle ... 285
- 6.11.2 Simple proof of the condition (6.58) ... 287
- 6.11.3 Antuma's triangular nomogram ... 287
- 6.11.4 Generating symmetrical coupler curves by six-bar linkages ... 292
- 6.12 Higher order path curvature ... 295
- Exercises ... 304

7 Synthesis of function generators ... 309
- 7.1 Function generator and its applications ... 309
- 7.2 Coordination of finitely separated angular displacements --- geometrical method ... 313
- 7.2.1 Coordination of a single angle-pair $\Delta\phi_{12} : \Delta\psi_{12}$ $(P_1 - P_2)$ and the relative pole ... 313
- 7.2.2 Coordination of a pair of angular and linear displacements $\Delta\phi_{12} : \Delta s_{12}$---synthesis of a slider-crank function generator ... 316
- 7.2.3 Coordinations of two pairs of crank rotations $\Delta\phi_{12} : \Delta\psi_{12}$; $\Delta\phi_{13} : \Delta\psi_{13}$ $(P_1-P_2-P_3)$ ... 317
- 7.2.4 Coordinations of three pairs of crank rotations $\phi_2 : \psi_2 ; \phi_3 : \psi_3; \phi_4 : \psi_4$ $(P_1-P_2-P_3-P_4)$ ... 318
- 7.3 Order type synthesis --- geometrical method ... 319
- 7.3.1 Conversion of differential coefficients in an order type synthesis ... 319
- 7.3.2 Matching a single angular velocity ratio (P_1P_2) ... 320
- 7.3.3 Matching ψ' and ψ'' $(P_1P_2P_3)$ ... 322
- 7.3.4 Matching prescribed ψ', ψ'' and ψ''' $(P_1P_2P_3P_4)$ --- the Carter-Hall circle ... 323
- 7.4 Error of a function generator ... 325
- 7.5 Transmission angle ... 325
- 7.6 Simple geometric method for higher order synthesis ... 328
- 7.6.1 Matching ψ' and ψ'', third order synthesis $(P_1P_2P_3)$ ... 329
- 7.6.2 Matching ψ', ψ'' and ψ''', fourth order synthesis $(P_1P_2P_3P_4)$ ... 330
- 7.6.3 Matching ψ', ψ'', ψ''' and ψ'''', fifth order synthesis $(P_1P_2P_3P_4P_5)$ ... 333

CONTENTS

7.7 Algebraic methods 337
7.7.1 Basic equations 337
7.7.2 Coordinations of three angular displacement pairs --- four finitely separated relative positions (P_1–P_2–P_3–P_4) 339
7.7.3 Coordinations of four angle-pairs --- five finitely separated relative positions (P_1–P_2–P_3–P_4–P_5) 341
7.7.4 Intermediate cases of four relative positions 343
7.7.5 Synthesis of crank-rockers 346
7.7.6 Synthesis of double-rockers 354
7.7.7 Synthesis of double cranks (drag-links) 357
7.8 Spacing of precision points 365
7.9 Synthesis of geared five-bar function generators 368
Exercises 374

PART III 381

8 Harmonic analysis of four link mechanisms 383
8.1 General concepts 383
8.2 Harmonic analysis of central slider-crank mechanism (Biezeno & Grammel, 1953) 384
8.3 Harmonic analysis of the output angle of a four-bar linkage (Freudenstein, 1959b) 386
8.4 Harmonic analysis of inverted slider-cranks 393
8.5 Harmonic analysis of the rotation energy of the coupler of an offset slider-crank (Meyer zur Capellen, 1959) 396
8.6 Harmonic analysis of the kinetic energy of the inverted slider-crank (Meyer zur Capellen & Thünker, 1975) 400

Appendix 1 : Homogeneous coordinates and circular points 405
A1.1 Homogeneous coordinates 405
A1.2 Circular points 405
Appendix 2 : On some topics regarding plane algebraic curves 407
A2.1 Tangent of an algebraic curve 407
A2.2 Double points 408
A2.3 Asymptotes 410
A2.4 Formulae of a curve in homogeneous coordinates 411
A2.5 Foci and singular foci 416
A2.6 Line coordinates 416
A2.7 Duality 419
Appendix 3 : Equation of q_1-curve (3.76) in rectangular coordinates 420
Appendix 4 : Expressions of the 9 terms in equation (3.96) 421
Appendix 5 : Coefficients in equations (3.97), (3.99) 423
Appendix 6 : Coefficients in equation (3.102) 424
Appendix 7 : Coefficients in equations (3.103), (3.104) 426

Appendix 8 : Coefficients in equation (3.143) 427
Appendix 9 : The Frost equation of radius of curvature in bipolar coordinates 429
   A9.1 Derivation of Frost (1880-81) equation 429
Appendix 10 : Coefficients in equation (6.9) of k_{A0}-curve 433
Appendix 11 : Selection of γ_2 in equation (6.15) 434
Appendix 12 : Equations of the major and minor axes of an ellipse 437
   A12.1 Equation of an ellipse 437
   A12.2 Major and minor axes of the ellipse 437
Appendix 13 : Derivation of equation (7.8) --- diameter of Carter-Hall circle .. 440
Appendix 14 : Program for fourth order synthesis of four-bar function generators 443
Appendix 15 : Program for fourth order synthesis of slider-crank function generators 445
Appendix 16 : Algebraic equations for synthesis of function generators 447
   A16.1 Four precision points (P_1–P_2–P_3–P_4) 447
   A16.2 Five precision points (P_1–P_2–P_3–P_4–P_5) 449
Appendix 17 : Relations between displacement matrix, rotation angle and pole coordinates 451
Appendix 18 : Expansion of equation (7.97) 452

References 454
Name index 461
Subject index 463

PART I

This part consists of Chapters 1 and 2. In general, textbooks on elementary kinematics of planar mechanisms deal almost exclusively with analysis, and specifically, analysis of instantaneous kinematics of mechanisms. For this reason, Chapter 1 of this book introduces some fundamental concepts of instantaneous kinematics of planar motion. This paves the way to advanced kinematics, and serves as a basis for handling of planar synthesis in instantaneous kinematics. Chapter 2 deals with some special techniques of analysis that are not found in general textbooks. Apart from the velocity analysis explained in Section 2.2, Chapter 2 is often related to later sections.

1
Fundamental concepts of planar instantaneous kinematics

1.1 Pole, polodes and pole changing velocity

From elementary kinematics, it is well-known that for a body performing an instantaneous plane motion, there exists an *instantaneous centre of velocity,* and for two relatively moving bodies there exists a *centro.* Instantaneous centre of velocity is just the centro between a moving body and the fixed body. We shall call all of these *poles,* in accordance with internationally recognized terminology. What was used to be called instantaneous centre of velocity is just one kind of poles.

Fig. 1.1 shows a four-bar linkage A_0ABB_0. The four bars A_0A, B_0B, AB, A_0B_0 are represented by a, b, c and f respectively. c is called *coupler,* for it connects a and b. The coupler c, having no fixed centre of rotation, must therefore have an instantaneous centre of rotation, i.e. the pole between c and the fixed body f. Generally, this pole is denoted by P_{cf}. For simplicity, it is denoted for the time being by P. The location of P is at the intersection of the extensions of A_0A and

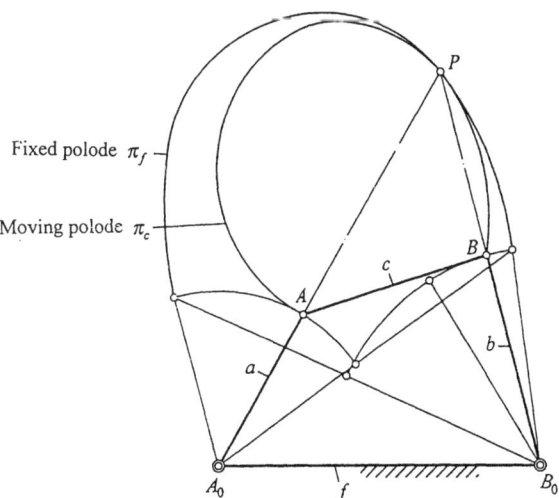

Fig. 1.1. Fixed polode π_f and moving polode π_c.

B_0B. As the four-bar linkage changes its position, P changes its position correspondingly. With respect to the body c, the new point P is not the original point P, and with respect to the fixed link f, the new point P is also not the old point P. Consider that c is extended into a large sheet of paper. If we puncture for each position of the four-bar linkage the point P with a pencil, we get then a point P on the paper c and also a point P on the fixed plane f. The locus of the P's on the fixed plane f becomes a curve which is called the *fixed polode* (fixed centrode), and is denoted by π_f; while the locus of the P's on the moving plane c also becomes a curve which is called the *moving polode* (moving centrode), and is denoted by π_c. It can be seen from Fig. 1.1 that the two polodes are always tangent to each other at P, and the fixed polode π_f is rigidly connected to the fixed link f, while the moving polode π_c is rigidly connected to the coupler c. In the position as shown in the figure, if we remove the two links A_0A and B_0B, then the original coupler motion of c can be replaced by π_c rolling without slip on the fixed polode π_f. (Please note that the two polodes shown in Fig. 1.1 are not complete.)

However, in the case of the so-called *kinematic inversion*, where AB ($= c$) is fixed and A_0B_0 ($= f$) is set to move, the original fixed polode becomes the moving polode, and the original moving polode the fixed polode. The motion of A_0B_0 can also be obtained by fixing π_c with π_f rolling without slip on π_c. In Section 1.9, we shall further discuss kinematic inversion. For equations of the polodes, please refer to Section 3.2.2.

Let us now return to the original four-bar linkage, i.e. with f fixed and c in planar motion. Since during the motion of c, the position of the pole P keeps changing, the question we want to ask is: what is the changing velocity of P if the driving crank a rotates with an angular velocity ω_a? Note that, if P is regarded as a point on the moving body c, then the instantaneous velocity of P is zero and therefore should not be called the *velocity of the pole*. Instead, it should be called the *pole changing velocity*. Moreover, it should not be denoted by the symbol $\mathbf{v}_P$ (since $\mathbf{v}_P = 0$, as has just been mentioned). It is denoted in this book by the symbol $\mathbf{u}$.

To find $\mathbf{u}$, we can proceed as in Fig. 1.2. Assume first $\omega_a = 1$. Imagine that the extensions of A_0A and B_0B were rigid rods; and that a pair of sliders were pinned together at their intersection point P, the sliders being able to slide on rods a and b respectively and to rotate with respect to each other. Now at the point P there are three coinciding points: a point P_a on rod a, a point P_b on rod b, and the centre point P_{cf} of the pin. What we want to find is the velocity $\mathbf{u}$ of the point P_{cf}. As shown in Fig. 1.1, for $\omega_a = 1$, we can find $\mathbf{v}_{Pa}$ from $\mathbf{v}_A$. Since the motion of the pin relative to P_a is along the rod a, the arrowhead of $\mathbf{u}$ lies on a line L_a drawn from the arrowhead of $\mathbf{v}_{Pa}$ parallel to rod a. Similarly we can find $\mathbf{v}_{Pb}$ from $\mathbf{v}_B$. The arrowhead of $\mathbf{u}$ should also lie on a line L_b drawn from the arrowhead of $\mathbf{v}_{Pb}$ parallel to the rod b. Therefore the arrowhead of $\mathbf{u}$ is at the intersection of the two

Fundamental concepts of planar instantaneous kinematics

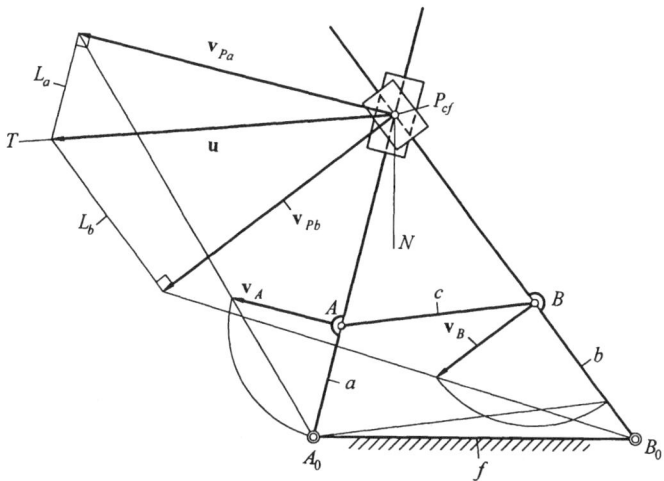

Fig. 1.2. Determination of pole changing velocity **u**.

lines L_a and L_b. Hence **u** can be found.

Combining Figs. 1.1 and 1.2, we can see that the direction of **u** is just the direction of the common tangent of the two polodes at P. This common tangent is denoted by PT, and is called *poletangent*. The line perpendicular to PT at P is denoted by PN, and is called *polenormal*.

The above construction of **u** from ω_a serves to highlight the geometrical characteristics of **u**. In general, however, it is not advisable to use this method. The reason is that the magnitude of **u**, though determined by ω_a, depends also on the dimensions and the position of the four-bar linkage.

1.2 Hartmann construction

We can carry out the construction mentioned in Section 1.1 in a reverse order. In Fig. 1.2, let **u** and the velocity $\mathbf{v}_A$ of A be known, but A_0 unknown. We can decompose from **u** a component $\mathbf{v}_{Pa}$ parallel to $\mathbf{v}_A$. The line joining the arrowheads of $\mathbf{v}_{Pa}$ and $\mathbf{v}_A$ cuts the line PA in A_0. Similarly, suppose there is another point E on the moving body as shown in Fig. 1.3. To find the centre of curvature E_0 of the path of E, we find first $\mathbf{v}_E$ from $\mathbf{v}_A$. Since **u** is known, we can decompose from **u** a component $\mathbf{u}_E$ parallel to $\mathbf{v}_E$. The line joining the arrowheads of $\mathbf{u}_E$ and $\mathbf{v}_E$ then cuts the line PE in E_0. This method is called *Hartmann construction* (Hartmann, 1893).

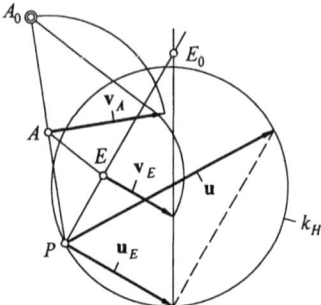

Fig. 1.3. Hartmann construction.

It should be noted that, all linear velocities are kinematic quantities, and the points A_0, E_0 .... are determined by geometrical relations. Linear velocities are indeed dependent variables of time. However, all linear velocities are linearly dependent. For example, if the magnitude of **u** is doubled, then the magnitudes of $\mathbf{u}_E$, $\mathbf{v}_E$ are also doubled. The line joining the arrowheads of the doubled $\mathbf{u}_E$ and $\mathbf{v}_E$ will cut the line PE in the same point E_0. In other words, E_0 remains unchanged.

With known **u**, if we want to find the centres of curvature of the paths of a number of points on the same moving body, we shall have to resolve each time the vector **u** into two mutually perpendicular components. One of the components such as $\mathbf{u}_E$ has its arrowhead lying on a circle k_H with diameter **u**, as shown in Fig. 1.3. The circle k_H is called *Hartmann circle*.

Fig. 1.4 shows an example of practical application of Hartmann construction. Fig. 1.4(a) shows a mincing machine. The mincing cutter is in the form of a wheel,

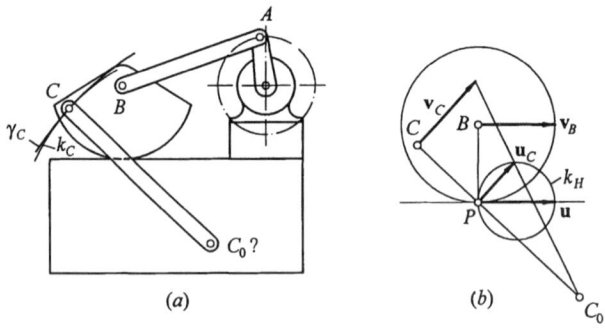

Fig. 1.4. Application of Hartmann construction. (*a*) A mincing machine. (*b*) Determination of the centre of curveature C_0 of the path of point C.

Fundamental concepts of planar instantaneous kinematics

which rolls reciprocately on a flat plate to mince. Given the location of a point C on the wheel. It is required to find C_0. In Fig. 1.4(b), the tangent point between the mincing wheel and the plate is its pole P. The pole changing velocity $\mathbf{u}$ is equal to the velocity $\mathbf{v}_B$ of the centre B of the wheel. The velocity $\mathbf{v}_C$ of the point C can easily be obtained from $\mathbf{v}_B$ and P. Construct the Hartmann circle with $\mathbf{u}$ as the diameter. Draw a line from P parallel to $\mathbf{v}_C$ to cut k_H to obtain the component $\mathbf{u}_C$. The line joining the arrowheads of $\mathbf{v}_C$ and $\mathbf{u}_C$ cuts the line CP in C_0. A link C_0C can be made to connect the two points C and C_0. Although the motion of the mincing wheel now is no longer the original pure rolling, within a small range it is sufficiently close to be regarded as such. In other words, the circle of curvature k_C of the path of C is quite close to the original path γ_C of the point C.

1.3 The inflection circle

On a body performing planar motion, there are points the normal accelerations of which vanish identically. The locus of such points is a circle called *inflection circle*, and is denoted by k_W, as shown in Fig. 1.5.

Proof: Please refer to Fig. 1.6. For a body in planar motion, let the velocity pole P and the pole changing velocity $\mathbf{u}$ be known. As before, the direction of $\mathbf{u}$ is denoted by PT, the poletangent. The line PN perpendicular to PT at P is the polenormal. Let E be a point on the body. The acceleration of E can be written as

$$\mathbf{a}_E = \mathbf{a}_P + \mathbf{a}_{EP}$$
$$= \mathbf{a}_P + \mathbf{a}'_{EP} + \mathbf{a}''_{EP} \qquad (1.1)$$

In equation (1.1), $\mathbf{a}_P$ is the acceleration of P, and $\mathbf{a}'_{EP}$ and $\mathbf{a}''_{EP}$ are the respective tangential and normal accelerations of E relative to P. As shown in Fig. 1.6, let $\mathbf{a}_P$ be moved to the point E, and be resolved into two mutually perpendicular components. One of the components $a_P \sin\theta$ is along the direction PE, where θ

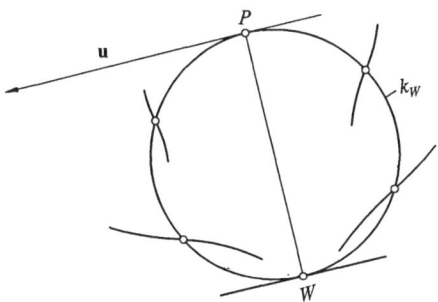

Fig. 1.5.   Inflection circle k_W.

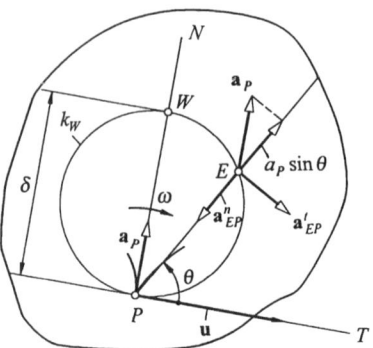

Fig. 1.6. Proof of the inflection circle.

$= \angle TPE$, and a_P is the magnitude of $\mathbf{a}_P$. The value of the normal acceleration of the point E is then

$$a_E^n = a_P \sin\theta - \overline{PE}\,\omega^2 \qquad (1.2)$$

In equation (1.2) ω is the angular velocity of the body. To find the locus of E for $a_E^n = 0$, we simply equate equation (1.2) to zero, i.e.,

$$a_P \sin\theta - \overline{PE}\,\omega^2 = 0$$

or

$$\overline{PE} = \frac{a_P}{\omega^2} \sin\theta \qquad (1.3)$$

In equation (1.3), a_P/ω^2 is a constant, which means that at a certain instant, with respect to the whole body, a_P/ω^2 remains unchanged regardless of the location of E. The dimension of a_P/ω^2 is a length, which will be denoted by δ, or

$$\delta = \frac{a_P}{\omega^2} \qquad (1.4)$$

hence

$$\overline{PE} = \delta \sin\theta \qquad (1.5)$$

Equation (1.5) shows that the locus of E is a circle, denoted by k_W in Fig. 1.6, with diameter δ. The point W on k_W opposite to P is called *inflection pole* (inflection centre). However, this terminology does not have any geometrical meaning.

From the expression $a_E^n = v_E^2 / \rho_E$, where v_E is the magnitude of the

Fundamental concepts of planar instantaneous kinematics 9

velocity of E, ρ_E the radius of curvature of the path of E, it can be seen that if a_E^n equals zero unless v_E vanishes, ρ_E becomes infinite. However, there is only one point on the body whose velocity vanishes, i.e. the pole P. Since E is not P, v_E does not vanish. Hence ρ_E must become infinite. This means that if the point E is not moving on a straight line, it must be passing through an inflection point on its path. As can clearly be seen from Fig. 1.5, every point on the inflection circle is passing through an inflection point of its path, hence the terminology *inflection circle*, short for *inflection point circle*. The inflection circle is discovered in 1706 by De La Hire.

In practical applications, for sufficiently small movement of the body, the path of any point on the inflection circle in the vicinity of the inflection point is close to a straight line, and hence can be regarded as a small straight line motion. The tangents of all paths of the points on the inflection circle pass through the point W. To grasp the geometry further, let there be an inflection point on a curve. Then any arbitrary line passing through the inflection point cuts the curve in three points. As the three points approach the inflection point indefinitely, the line approaches the tangent at the inflection point. We therefore deduce that the tangent at an inflection point is tangent to the curve in three points.

It should be noted that, although a_P and ω are kinematic quantities, i.e. dependent of time, the quantity $\delta = a_P / \omega^2$, as mentioned before, is a length, i.e. independent of the speed of the motion. For example, the inflection circle of a rolling wheel at any instant is determined regardless of its speed.

1.4 The tangential circle

On a body performing planar motion, there are points the tangential accelerations of which vanish identically. The locus of such points is also a circle that is called the *tangential circle*, and is denoted by k_T.

Proof: Please refer to Fig. 1.7. As before, let both P and $\mathbf{u}$ be known. Let F be a point on the moving body. The acceleration of F can be written as

$$\mathbf{a}_F = \mathbf{a}_P + \mathbf{a}'_{FP} + \mathbf{a}''_{FP} \tag{1.6}$$

As shown in Fig. 1.7, let $\mathbf{a}_P$ be moved to the point F, and be resolved into two mutually perpendicular components with one component $a_P \cos\theta$ perpendicular to PF. Then the tangential acceleration of F is

$$\begin{aligned} a_F^t &= a_{FP}^t - a_P \cos\theta \\ &= \overline{PF}\,\dot\omega - a_P \cos\theta \end{aligned} \tag{1.7}$$

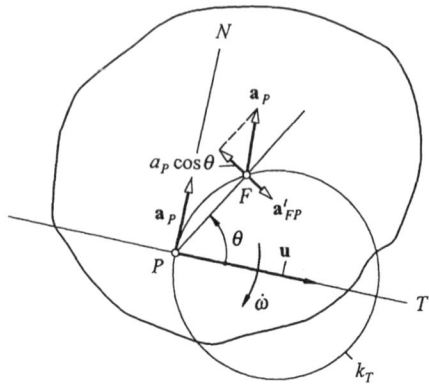

Fig. 1.7. Proof of the tangential circle.

In equation (1.7) $\dot{\omega}$ is the angular acceleration of the body. To find the locus of F for $a_F{}' = 0$, simply equate equation (1.7) to zero, i.e.,

$$\overline{PF}\,\dot{\omega} - a_P \cos\theta = 0$$

or
$$\overline{PF} = \frac{a_P}{\dot{\omega}}\cos\theta \qquad (1.8)$$

In equation (1.8), $a_P / \dot{\omega}$ is a constant, being a length. This equation shows that the locus of F is a circle, denoted by k_T, with diameter $a_P / \dot{\omega}$. As can be seen in Fig. 1.7, although the point P lies on k_T, P is a point with tangential acceleration $\mathbf{a}_P$. Hence P is a singular point on k_T. The tangential circle k_T is not simply a geometrical quantity, but varies with respect to $\dot{\omega}$. Inflection circle and tangential

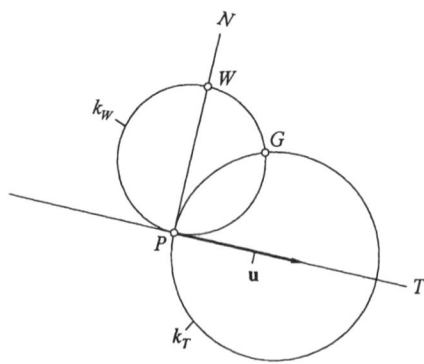

Fig. 1.8. Combination of inflection circle k_W and tangential circle k_T.

Fundamental concepts of planar instantaneous kinematics

circle are called *Bresse's circles*.

Fig. 1.8 superimposes the two circles k_W and k_T of Figs. 1.6 and 1.7 respectively. k_W and k_T intersect each other in 2 points: beside the point P also in another point G. As k_W is the locus of points with vanishing normal accelerations, and k_T is the locus of points with vanishing tangential accelerations, the total acceleration of G is zero. Please note that, as mentioned before, the acceleration of P is not zero. Fig. 1.9 shows the distribution of linear velocities of a moving body. The velocity of any point on the body is normal to the line joining that point and P, and the magnitude of this velocity is proportional to its distance from P. For this reason P is called the *velocity pole*. Similarly, Fig. 1.10 shows the distribution of linear accelerations on a body. For any point on the body, the angle between the

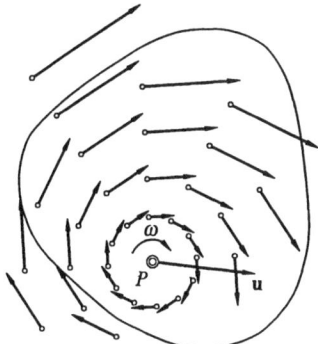

Fig. 1.9. Distribution of linear velocities.

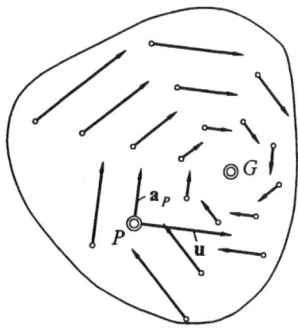

Fig. 1.10. Distribution of linear accelerations.

acceleration of that point and the line joining that point to G is always the same, and the magnitude of the acceleration of that point is proportional to its distance from G. For this reason G may be called the *acceleration pole*.

1.5 Relations between δ and u, and a_P and u

In Fig. 1.3, if E is an inflection point, then E_0 should lie at infinity, as shown in Fig. 1.11. In other words, $\mathbf{u}_E = \mathbf{v}_E$, and the line joining the arrowheads of $\mathbf{u}_E$ and $\mathbf{v}_E$ is parallel to PE, or E_0 lies at infinity in the direction of PE. We can now write

$$v_E = u_E = u \cos \alpha = \overline{PE}(-\omega)$$

ω in counterclockwise direction is considered as positive. In Fig. 1.11 ω is clockwise, hence it is negative. Thus we have

$$\overline{PE} = \left(-\frac{u}{\omega}\right) \cos \alpha \tag{1.9}$$

Comparing equation (1.9) with equation (1.5), we have

$$\delta = -\frac{u}{\omega} \tag{1.10}$$

Equation (1.10) can also be written as

$$u = -\delta \omega \tag{1.11}$$

or

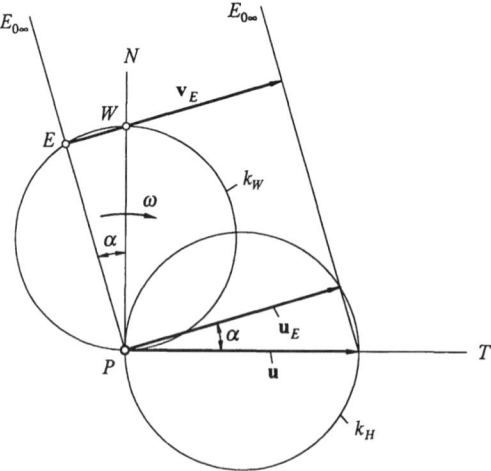

Fig. 1.11. A proof of $\delta = -u/\omega$.

Fundamental concepts of planar instantaneous kinematics

$$v_W = u \qquad (1.12)$$

Equation (1.12) means that the velocity of the inflection pole is equal to u. By means of the above equations, if u and ω are known, we can find the point W, and hence the inflection circle k_W.

On the other hand, we get from equation (1.4) $\delta = a_P/\omega^2$ and equation (1.11)

$$u = -\frac{a_P}{\omega}$$

or

$$a_P = -u\omega \qquad (1.13)$$

This is the acceleration of the pole P.

In Fig. 1.6, a small portion of the path in a cuspidal shape is drawn in the vicinity of P. This means that if P is regarded as a point on the moving body, its path will become such a shape just before it reaches and after it leaves the position as shown in the figure; and the tangent of the path at P is normal to PT, or along PN. The velocity of this point before it reaches the position P shown in the figure is not zero, and becomes zero at the position shown, and again is not zero after just leaving the position P shown. This means that the point P as an instantaneous velocity pole must have an acceleration, and this acceleration is a tangential acceleration, or a_P in equation (1.13).

Please refer in advance to Fig. 1.27(a). The acceleration of the point P is $a_P = r\omega^2 - v^2/r$, where r, ω, and v are respectively the radius, angular velocity and the velocity of the centre W of the wheel. Note that although this is a general form of normal acceleration, it is not the normal acceleration of P. It is indeed the relative normal acceleration of P relative to W, but is the absolute tangential acceleration of P.

1.6 Euler-Savary equation

Suppose in Fig. 1.12, P, $\mathbf{u}$, ω and δ for a moving body at an instant are known. A is a moving point. Let $\angle TPA$ be denoted by θ_A. We know from Hartmann construction that if a component $\mathbf{u}_A$ parallel to $\mathbf{v}_A$ is decomposed from $\mathbf{u}$, i.e. $u_A = u\sin\theta_A$, then the line joining the arrowheads of $\mathbf{u}_A$ and $\mathbf{v}_A$ will cut PA in A_0, where A_0 is the centre of curvature of the path of A. It should be noted that A is on the moving body, while A_0 is on the fixed body. We get from similar triangles

$$\frac{v_A}{u\sin\theta_A} = \frac{\overline{AA_0}}{\overline{PA_0}}$$

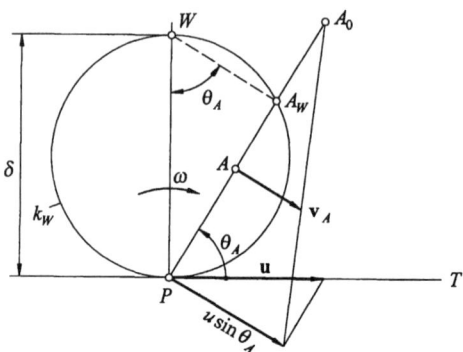

Fig. 1.12. For derivation of Euler-Savary equation.

or

$$\frac{\omega \overline{PA}}{\delta \omega \sin \theta_A} = \frac{\overline{AA_0}}{\overline{PA_0}} = \frac{\overline{PA_0} - \overline{PA}}{\overline{PA_0}}$$

or

$$\left(\frac{1}{\overline{PA}} - \frac{1}{\overline{PA_0}}\right) \sin \theta_A = \frac{1}{\delta} \tag{1.14}$$

Equation (1.14) is called *Euler-Savary equation*. We have then
Theorem 1: The Euler-Savary theorem, i.e. equation (1.14).

By means of Euler-Savary equation, if δ and θ_A are known, we can find A_0 from A, or A from A_0. This equation establishes a so-called 1:1 correspondence. Note that $\overline{PA}$ and $\overline{PA_0}$ are directed lengths, i.e. the direction of a ray along θ_A is (+), otherwise it is (−). In other words, $\overline{PA} = -\overline{AP}$. After a kinematic inversion, i.e. with the moving body fixed and the fixed body moving, the point A is the centre of curvature of the path of A_0.

In a special case as that shown in Fig. 1.13, if the point A is located at the centre of curvature M of the moving polode, then A_0 is at the centre of curvature M_0 of the fixed polode. In this case $\theta_A = 90°$, and the left hand side of equation (1.14) becomes $1/\overline{PM} - 1/\overline{PM_0}$, and the equation becomes

$$\frac{1}{\overline{PM}} - \frac{1}{\overline{PM_0}} = \frac{1}{\delta} \tag{1.15}$$

or in the form

Fundamental concepts of planar instantaneous kinematics

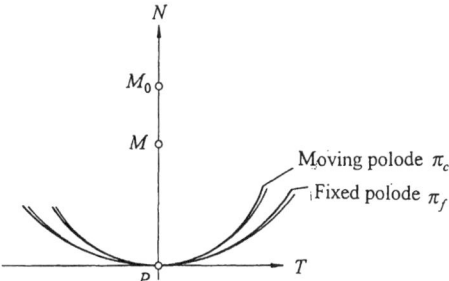

Fig. 1.13. A special case of Euler-Savary equation, M and M_0.

$$\frac{1}{\rho_c} - \frac{1}{\rho_f} = \frac{1}{\delta} \quad (1.15a)$$

In equation (1.15a) ρ_c, ρ_f are the respective radii of curvature of the moving polode π_c and fixed polode π_f.

We can now derive the equation of the inflection circle from Euler-Savary equation (1.14). Let $\overline{PA_0} \to \infty$, then equation (1.14) becomes

$$\overline{PA} = \delta \sin \theta_A \quad (1.16)$$

Equation (1.16) is the same as equation (1.5), i.e. the locus of A is a circle, the inflection circle, with diameter δ and tangent PT.

We rewrite equation (1.14) in the following form

$$\frac{\overline{PA_0} - \overline{PA}}{\overline{PA_0}} = \frac{\overline{PA}}{\delta \sin \theta_A} = \frac{\overline{PA}}{\overline{PA_W}}$$

where A_W is the intersection point of PA and the inflection circle, as shown in Fig. 1.12. We rewrite the above equation further in the following form

$$\frac{\overline{AA_0}}{\overline{PA_0}} = \frac{\overline{PA}}{\overline{PA_W}} \quad (1.17)$$

Equation (1.17) represents the four line segments in proportional relationship. Hence if the point P and inflection circle k_W are known, for an arbitrary point A, the line PA cuts k_W in A_W. The point A_0 can then be constructed as follows (please refer to Fig. 1.14) :

(a) draw an arbitrary line g passing through A;
(b) choose an arbitrary point X on g;

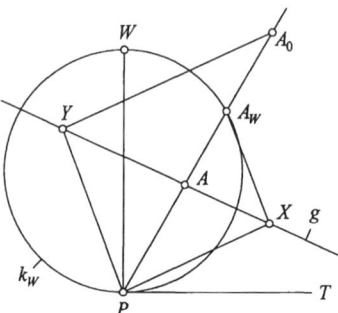

Fig. 1.14. Parallel line construction from A to A_0.

(c) join $A_W \rightarrow X \rightarrow P$, and draw $PY \| XA_W$, $YA_0 \| PX$, thus obtaining A_0.

Conversely, if A_0 is given, then two lines parallel to each other can be drawn respectively from A_0 and P, and two lines parallel to each other can be drawn respectively from A_W and P. These four lines intersect respectively in two points, X and Y. A is the intersection point of XY and PA_0.

It can easily be seen from Fig. 1.14 that if the inflection circle k_W and the point A_W are unknown, but A and A_0 are known, we can then follow the sequence $A_0 \rightarrow Y \rightarrow P \rightarrow X \rightarrow A_W$ drawing parallel lines and then obtaining A_W. This principle enables us to find the inflection circle of the coupler AB for any given four-bar linkage A_0ABB_0. Please refer to Fig. 1.15, and compare it with Fig. 1.14, where both line g and point Y were arbitrarily chosen. Now in Fig. 1.15, we have to choose g and Y such that they can be used for both A and B. Therefore the line AB is chosen as g, and the intersection point of AB and A_0B_0 is chosen as Y. Thus we join $Y \rightarrow P$, and draw $PX \| A_0 B_0$, $XB_W A_W \| PY$ to get B_W and A_W. Both A_W and B_W are points on the inflection circle k_W. Draw a line from A_W perpendicular to PA_W and a line from B_W perpendiaular to PB_W. The intersection of these two lines is the inflection pole W. The circle drawn with $\overline{PW}$ as diameter is k_W.

From similar triangles in Fig. 1.14 we get

$$\overline{AA_W} \cdot \overline{AA_0} = \overline{PA}^2 \tag{1.18}$$

Please note that the line segments in equation (1.18) are all directed lines; in other words, $\overline{AA_0} = -\overline{A_0A}$. It can be seen from equation (1.18) that the product of $\overline{AA_W}$ and $\overline{AA_0}$ is always positive. In other words, $\overline{AA_0}$ and $\overline{AA_W}$ are always of the same sense. This means that A_0 and A_W always lie on the same side of A. Similarly, B_0 and B_W also always lie on the same side of B. Equation (1.18) can be considered as another form of Euler-Savary equation.

Fundamental concepts of planar instantaneous kinematics

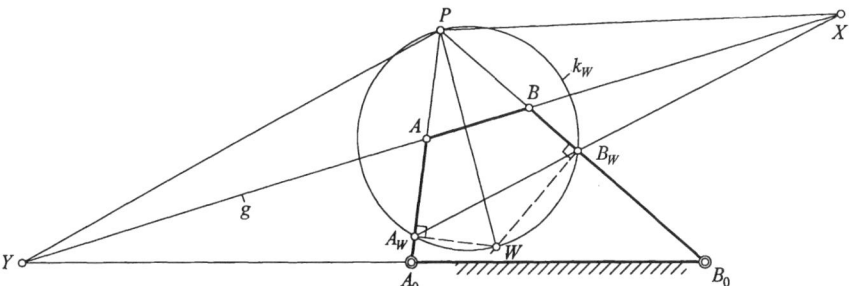

Fig. 1.15. A construction for finding the inflection circle of the coupler AB of a four-bar linkage A_0ABB_0.

The geometrical construction described above and shown in Fig.1.15 for finding the inflection circle of the coupler may not always be convenient or sufficiently accurate. To overcome this shortcoming, we can easily use equation (1.18) to compute A_W from the known points P, A and A_0, and to compute B_W from the known points P, B and B_0. With A_W, B_W, and P known, we can establish the circle k_W.

1.7 The Bobillier construction

In Fig. 1.16, let the pole P, poletangent PT and inflection circle k_W of the coupler AB of a four-bar linkage A_0ABB_0 be known. In $\triangle PDA$, we have from the law of sine

$$\frac{\overline{PD}}{\overline{PA}} = \frac{\sin(\lambda + \beta)}{\sin \lambda} = \cos \beta + \cot \lambda \sin \beta \tag{1.19}$$

also in $\triangle PDA_0$,

$$\frac{\overline{PD}}{\overline{PA_0}} = \frac{\sin(\alpha + \lambda + \beta)}{\sin(\alpha + \lambda)} = \cos \beta + \cot(\alpha + \lambda) \sin \beta \tag{1.20}$$

Subtracting equation (1.20) from equation (1.19), we have

$$\overline{PD}\left(\frac{1}{\overline{PA}} - \frac{1}{\overline{PA_0}}\right) = \sin \beta [\cot \lambda - \cot(\alpha + \lambda)] \tag{1.21}$$

However, we know from Euler-Savary equation

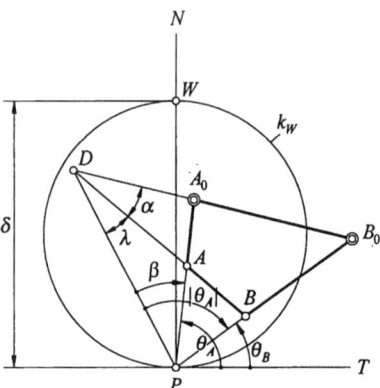

Fig. 1.16. A proof of Bobillier construction.

$$\left(\frac{1}{\overline{PA}} - \frac{1}{\overline{PA_0}}\right) \sin\theta = \frac{1}{\delta} \qquad [(1.14)]$$

Therefore equation (1.21) can be written as

$$\overline{PD} = \delta \sin\theta_A \sin\beta [\cot\lambda - \cot(\alpha+\lambda)] \qquad (1.22)$$

Similarly, from $\triangle PDB$ and $\triangle PDB_0$ we get

$$\overline{PD} = \delta \sin\theta_B \sin(\beta+\theta_A-\theta_B)[\cot\lambda - \cot(\alpha+\lambda)] \qquad (1.23)$$

Comparing equations (1.22) and (1.23), we get

$$\beta = \theta_B \qquad (1.24)$$

This is *Bobillier Theorem*. For a four-bar linkage A_0ABB_0, PD is called the *collineation axis*. Bobillier theorem states that

Theorem 2: For a four-bar linkage, the angle measured in one direction between one of the rotating links (e.g. A_0A) and the collineation axis (i.e. PD) is equal to the angle measured in opposite sense between another rotating link (e.g. B_0B) and the poletangent PT.

Obviously, the angle θ_A between A_0A and PT is also equal to the angle between B_0B and PD, both angles being measured in opposite senses. Bobillier theorem can also be stated as follows:

The collineation axis PD and the poletangent PT are symmetrically disposed with respect to the bisector (not shown in the figure) of $\angle APB$.

What Bobillier proved in his book (Bobillier, 1870) was not quite the same as

Fundamental concepts of planar instantaneous kinematics

the Bobillier theorem we have today. The Bobillier theorem in its present form may be considered as a corollary derived from the original Bobillier theory. With Bobillier theorem we can easily determine the direction of the poletangent *PT*. In Fig. 1.17, let A_0ABB_0 be a given four-bar linkage. We locate first the intersection D_{AB} of *AB* and A_0B_0; and $\angle APD_{AB} = \beta$. To obtain *PT*, we simply construct $\angle BPT = \beta$ from *BP* in the opposite sense. So, unlike in the case of Fig. 1.2, we do not have to find **u** in order to obtain *PT*. The main application of Bobillier construction is to find the centre of curvature say, B_0 of the path γ_B of a point *B* on a moving body. Let the points *A*, A_0, *P*, *PT*, and *B* be known. The construction is as follows:

(a) make $\angle APD = \angle BPT$, both angles being measured in opposite senses;

(b) points *A*, *B* and *D* being collinear, locate point *D*;

(c) points A_0, B_0 and *D* being collinear, locate point B_0.

As shown in Fig. 1.16, the point *D* is the intersection of *AB* and A_0B_0. However, as shown in Fig. 1.17 there can be a number of *D*'s. To avoid confusion, we use the symbol D_{AB} to denote the intersection of *AB* and A_0B_0, etc. Suppose we wish to find the centre of curvature E_0 of the path γ_E of a point *E* on the coupler *c* (a point on the coupler can simply be called a *coupler point*). We modify the above procedure as follows:

(a) make $\angle D_{AB}PD_{BE} = \angle APE$, both angles being measured in the same sense;

(b) points *B*, *E* and D_{BE} being collinear, locate D_{BE};

(c) points B_0, E_0, D_{BE} being collinear, locate E_0.

In this construction there is no need to find *PT*. $\angle D_{AB}PD_{BE} = \angle APE$ because

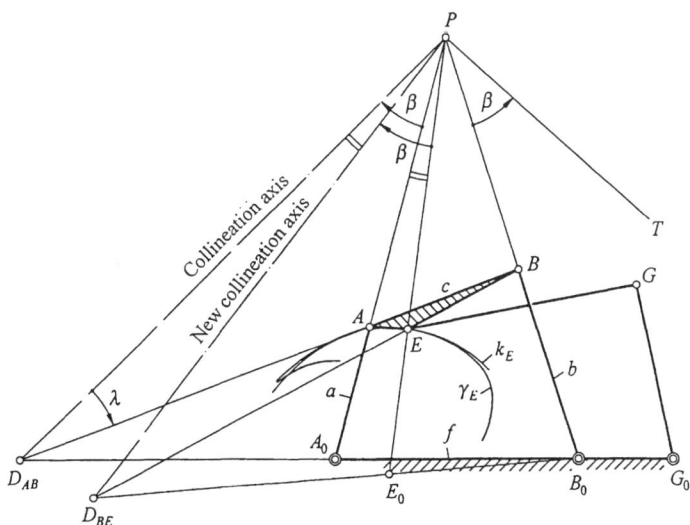

Fig. 1.17. Application of Bobillier construction.

$\sphericalangle EP\,D_{BE} = \sphericalangle AP\,D_{AB} = \beta$. The line PD_{AB} is the collineation axis of the original four-bar linkage A_0ABB_0, while PD_{BE} is the new collineation axis of the new four-bar linkage E_0EBB_0. We remark that in each construction we only have to identify a set of four points, for instance, the points A_0, A, B, B_0, or A_0, A, E, E_0, or A_0, A, W, W_0, etc. Furthermore, knowing any three of the four points, we can locate the fourth point.

It can be seen from Exercises 1.8 and 1.9 that, as a means of finding the centre of curvature A_0 of the path γ_A of a point A on the rolling wheel, Bobillier construction is even more straightforward than Hartmann construction.

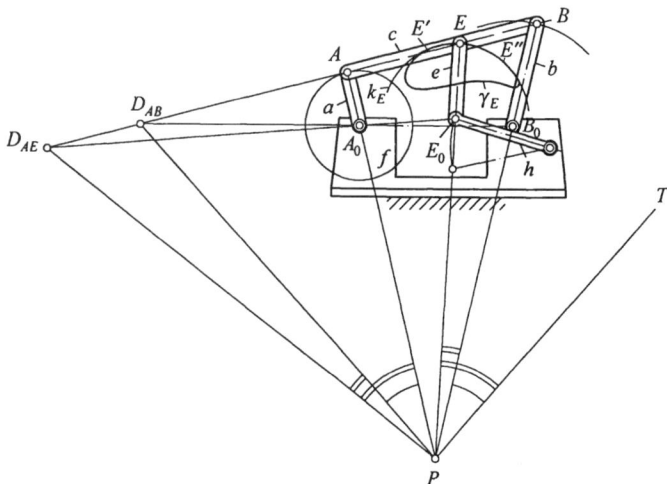

Fig. 1.18. An application example of Bobillier construction: dwell mechanism.

In Fig. 1.18 we apply Bobillier construction to find the centre of curvature E_0 of the path of a point E on the coupler c ($=AB$). We make $\sphericalangle D_{AE}PD_{AB} = \sphericalangle EPB$, both angles being measured in the same sense. Having found E_0, we connect links e and h. Thus, as the point E passes through its path segment $E'E''$, the point E_0 on the link e remains almost stationary, as does the link h. Mechanisms like this are called *dwell mechanisms*. Fig. 1.19 shows another dwell mechanism. The four-bar linkage and link e are taken over from Fig. 1.18. We replace h of Fig.1.18 by a sliding rod h to make it almost stationary. Dwell mechanisms that make use of the centre of curvature of the path of a coupler point are called *coupler-dwell mechanisms*. A comprehensive discussion on coupler-dwell mechanisms can be found in (Alt, 1932a). The two links e and h so connected is specifically called a *dyad*.

Fundamental concepts of planar instantaneous kinematics

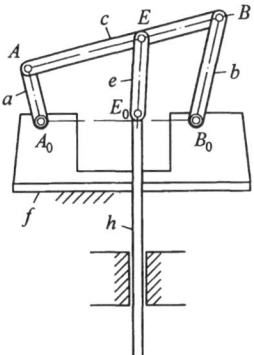

Fig. 1.19. Another type of dwell mechanism similar to that in Fig. 1.18.

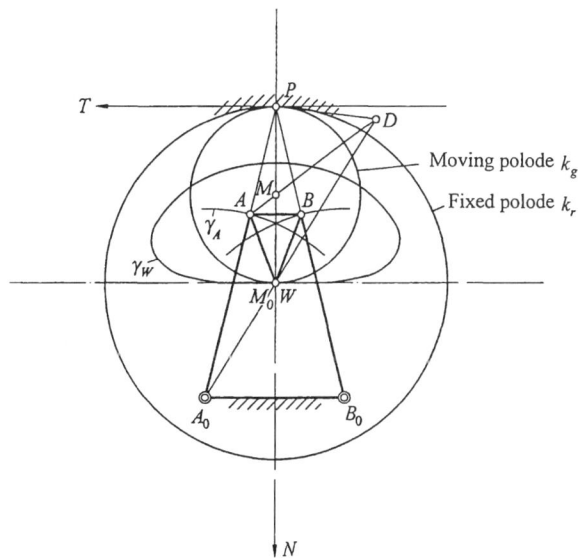

Fig. 1.20. Cardan circles k_g, k_r; determination of A_0, B_0 by Bobillier construction.

Fig. 1.20 shows a Cardan circle-pair k_g and k_r. These are a pair of polodes. The diameter of k_g is half that of k_r. k_g rolls without slip on the inside of k_r. In the configuration, P, PT, M and M_0 are all known. We wish to find the centre of curvature A_0 of the path γ_A of the moving point A. In this case the four points in question are M_0, M, A and A_0. Using Bobillier construction, we find the point A_0. From symmetry, we find the points B and B_0. The path of the point W on the coupler AB of the so constructed four-bar linkage A_0ABB_0 is approximately a straight line.

We shall return to this subject in the discussion of *Ball point* in Section 3.7.7.

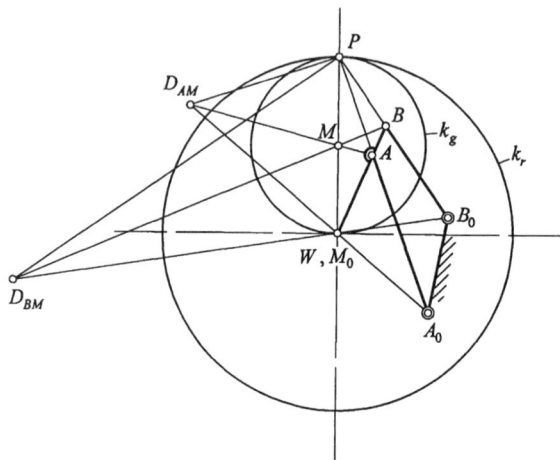

Fig. 1.21. Cardan circles k_g, k_r; determination of A_0, B_0 by Bobillier construction.

Fig. 1.21 shows the same Cardan circle-pair as that in Fig. 1.20. By means of Bobillier construction, and with respect to the two sets of four points, M_0, M, A, A_0 and M_0, M, B, B_0, we find the respective centres of curvature A_0 and B_0 of the paths of A and B. Having found A_0 and B_0, we construct a four-bar linkage A_0ABB_0. The path of the coupler point W is now approximately a straight line. This mechanism can be used for a crane.

Fig. 1.22 shows a special position of a four-bar linkage A_0ABB_0. A_0A is parallel to B_0B. Now the point P lies at infinity in the direction of A_0A, or of B_0B. Suppose we wish to find the centre of curvature E_0 of the path of a certain coupler point E. We draw first at E a line $g \parallel BB_0 \parallel AA_0$. Let the intersection of g and A_0B_0 be denoted by E_f. In the usual Bobillier construction, we should make $\sphericalangle D_{AE}PD_{AB} = \sphericalangle EPB$. However, now we mark on A_0B_0 the line segment $\overline{D_{AB}D_f}$, such that $\overline{D_{AB}D_f} = \overline{B_0E_f}$, both being measured in the same sense, to obtain D_f. We then draw a line l at D_f parallel to g. It can be seen that $D_{AB}P_\infty$ is the collineation axis of the original four-bar linkage A_0ABB_0, while l is the new collineation axis of the new four-bar linkage A_0AEE_0. The rest of the construction remains the same as before. If there is another point F on the same line g, obviously, we can apply the same procedure to find the centre of curvature F_0 of the path γ_F of the point F. It can be seen from Fig. 1.22 that $\overline{E_0E} = \overline{F_0F}$. In other words, the

Fundamental concepts of planar instantaneous kinematics

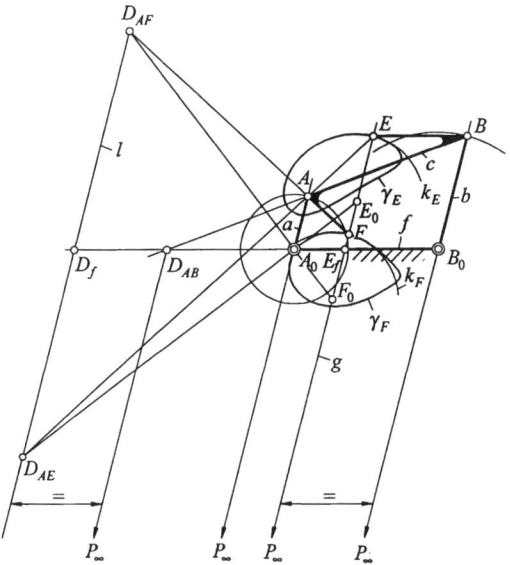

Fig. 1.22. Bobillier construction in the case of $A_0A \parallel B_0B$.

radii of curvature of the paths of all coupler points on the line g are equal.

1.8 Quadratic transformation

We denote as before the point on the moving body by A, and the centre of curvature of the path of A by A_0. There is a so called *quadratic transformation* between the locus of A and that of A_0. In Fig.1.23, let the poletangent PT be taken as the x- or ξ- axis, and the polenormal PN be taken as the y- or η- axis; let the

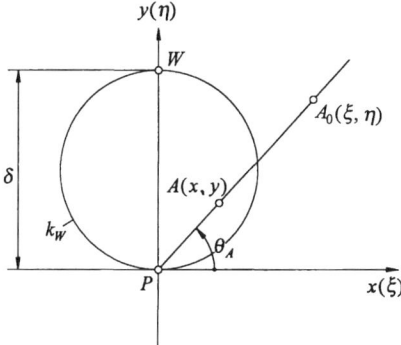

Fig. 1.23. Rectangular coordinates of A, A_0.

Cartesian coordinates of A be $A(x,y)$, and those of A_0 be $A_0(\xi,\eta)$. Then from equation (1.14), Euler-Savary equation

$$\left(\frac{1}{\overline{PA}} - \frac{1}{\overline{PA_0}}\right) \sin\theta = \frac{1}{\delta} \qquad [(1.14)]$$

we have

$$\left.\begin{array}{l} \xi = -\dfrac{\delta xy}{x^2 + y^2 - \delta y} \\[2mm] \eta = -\dfrac{\delta y^2}{x^2 + y^2 - \delta y} \end{array}\right\} \qquad (1.25a,b)$$

and

$$\left.\begin{array}{l} x = \dfrac{\delta \xi \eta}{\xi^2 + \eta^2 + \delta \eta} \\[2mm] y = \dfrac{\delta \eta^2}{\xi^2 + \eta^2 + \delta \eta} \end{array}\right\} \qquad (1.26a,b)$$

Equations (1.25a,b) and (1.26a,b) represent a *quadratic transformation*. In other words, in the general case, if the locus $\eta = \eta(\xi)$ of the point A_0 is an nth curve

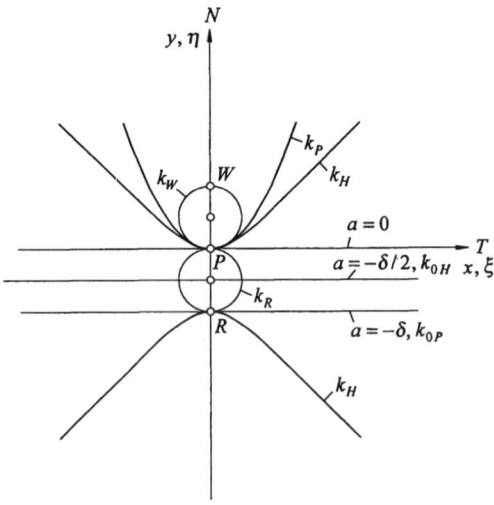

Fig. 1.24. Relations in a quadratic transformation.

Fundamental concepts of planar instantaneous kinematics

without any node, cusp or multiple points, then the locus $y = y(x)$ of the point A is a $2n^{th}$ curve, and vice versa (Primrose, 1955 Chap. IV). For example, in Fig. 1.24, assuming that the locus of A_0 is a line $\eta = a$ (a is a constant) parallel to the ξ-axis, we have then from equation (1.25b)

$$(x^2 + y^2 - \delta y)a + \delta y^2 = 0 \tag{1.27}$$

Equation (1.27) represents a pencil of conics. We consider the following cases:

(a) As $a \to \pm\infty$, the locus of A_0 is the line at infinity. As expected, the locus of A is the inflection circle k_W.

$$k_W: x^2 + y^2 - \delta y = 0$$

(b) As $a = 0$, the locus of A becomes the poletangent $y = 0$.

(c) As $a = -\delta$, i.e. the locus of A_0 is a line k_{0P} passing through R and parallel to PT, the locus of A is the parabola k_P.

$$k_P: x^2 - \delta y = 0$$

(d) As $a = -\delta/2$, i.e. the locus of A_0 is a line k_{0H}, the locus of A is the hyperbola k_H.

$$k_H: x^2 - y^2 - \delta y = 0$$

In this case any line drawn from P cuts k_{0H} and k_H respectively in A_{0H} and A_H.

(e) Similarly, if the locus of A is $y = \delta/2$, i.e. a line parallel to and at a distance $\delta/2$ above PT (not shown in Fig. 1.24), then the locus of A_0 is also a hyperbola (not shown in the figure):

$$\xi^2 - \eta^2 + \delta\eta = 0$$

(f) In general, if the locus of A is a conic k tangent to PT at P,

$$k: a_3 x^2 + a_2 xy + a_1 y^2 + a_0 y = 0$$

then the corresponding locus of A_0 is also a conic tangent to PT at P, as shown in Fig. 1.25:

$$k_0: (a_3 \xi^2 + a_2 \xi\eta + a_1 \eta^2 + a_0 \eta)\delta + a_0(\xi^2 + \eta^2) = 0$$

In a special case in which k degenerates into two lines k_1 and k_2, where k_2 is PT, as shown in Fig. 1.26, the locus k_0 of A_0 does not degenerate. If the points $A_1, A_2, \ldots$ are known, we can find the points $A_{10}, A_{20}, \ldots$ corresponding to $A_1, A_2, \ldots$ by Bobillier construction (Section

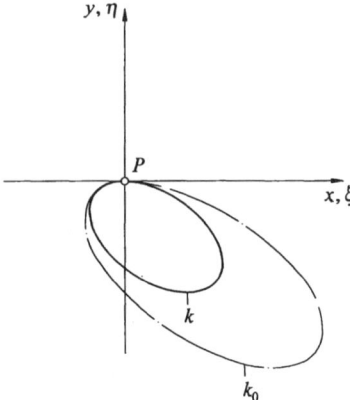

Fig. 1.25. Quadratic transformation: when the locus of point A is a conic section k tangent to the poletangent PT at P, the locus of A_0 is another conic section k_0 tangent to the poletangent PT at P.

1.7). The points $A_0, A_{10}, A_{20},...$ then lie on a conic k_0. Alternatively this conic can also be computed by means of the later equation (1.28). Furthermore, the point Q_0 corresponding to any point Q on k_2 is at P.

It should be pointed out here that, as shown in Fig. 1.24, the locus corresponding to the line k_{0P} is the quadratic k_P. This means that the locus corresponding to k_P is k_{0P}, being a straight line but not a quartic curve. This is because the quartic has been reduced to a straight line.

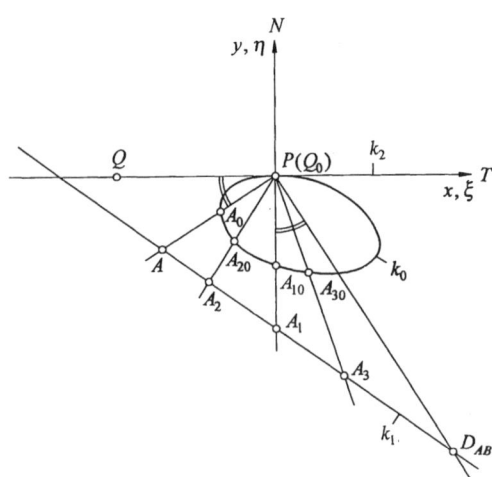

Fig. 1.26. A special case of Fig. 1.25; k degenerates into lines k_1 and k_2.

Fundamental concepts of planar instantaneous kinematics

Moreover, as shown in Fig. 1.25, k and k_0 being mutually correspondent, there exists no quartic curve.

(g) In case the locus of A is a general straight line $y = mx + n$, then the locus of its corresponding point A_0 is a conic

$$k_0: (\xi^2 + \eta^2 + \delta\eta)n - (\eta - m\xi)\eta\delta = 0 \tag{1.28}$$

1.9 Kinematic inversion and the return circle

Suppose that a moving body c is performing a planar motion relative to a fixed plane f. Then a person sitting on c would feel that the body f were moving. If a point E on c generates on f a path γ_E, then for that person γ_E would appear to be sliding through E all the time. If the point E is moving with a velocity $\mathbf{v}_E$, it seems then, relative to the body c, as if the coincident point on f were moving with a velocity $-\mathbf{v}_E$.

As mentioned in Section 1.5, if A is an inflection point, the location of A_0 is at infinity. What then is the location of the centre of curvature A_0 of the path of A if A lies at infinity? To find out, let $\overline{PA} \to \infty$ in equation (1.14), or

$$\overline{PA_0} = -\delta \sin \theta_A \tag{1.29}$$

Comparing equation (1.29) with equation (1.16), we see that the locus of A_0 represented by equation (1.29) is a circle of the same size as the inflection circle k_W, but on the opposite side of the poletangent PT. This circle is called the *return circle*, and is denoted by k_R. It is also called *cusp circle* (Hunt, 1978, Section 5.4; Dijksman, 1976, Section 2.2.6). The reason of this terminology will be explained in Section 3.5.1. The point R on k_R opposite to P is called the *return pole*, which has

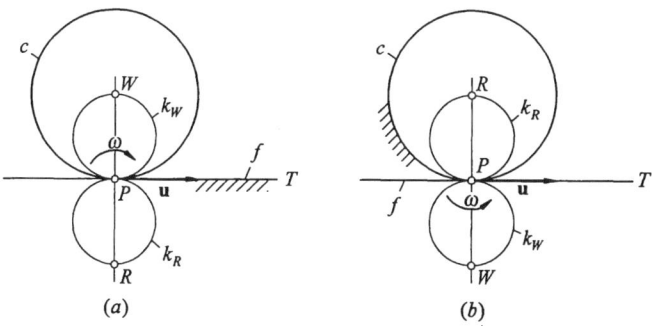

Fig. 1.27. *(a)* k_W and k_R when a wheel c rolls on a fixed flat plate f.
(b) k_W and k_R when a flat plate f rolls on the outside of a fixed wheel c.

no geometrical meaning.

The phrase *a moving point at infinity* seems to defy common sense. However, we may explain it by the concept of kinematic inversion. After a kinematic inversion, A_0 becomes the moving point, and A becomes the centre of curvature of the path of A_0. Therefore A_0 is an inflection point after inversion, and its locus is the inflection circle after inversion. In other words, the inflection circle and return circle exchange their rôles after inversion. This can easily be seen from Fig.1.27 (a)(b). In Fig. 1.27(a), the flat plate f is fixed, and the wheel c rolls on f. The inflection circle k_W is just equal to one half of the circle c. The return circle k_R is equal to k_W, but on the lower side of PT. Fig.1.27(b) shows the situation after inversion. The wheel c is fixed, while the flat plat f rolls without slip on c. The half circle of c on the upper side becomes the return circle, and the circle on the lower side of PT and of the same size as k_R becomes now the inflection circle k_W.

1.10 The generating curve and its envelope

Fig.1.28 shows a four-bar linkage A_0ABB_0 with a point C on its coupler c. The centre of curvature C_0 of the path of C can be obtained as before by Bobillier construction. Now let C be the centre of curvature of another curve h on c at the point H, i.e. $C = H_0$, and let the three points C_0, C and H be collinear. As the curve h moves together with c, it assumes a series of different positions that has a common

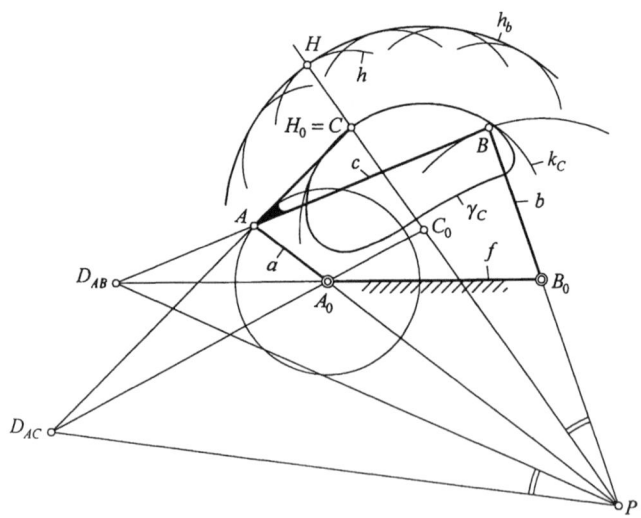

Fig. 1.28. The centre of curvature of the envelope h_b of the generating curve h at H is the centre of curvature C_0 of the path γ_C of $H_0(=C)$ at C.

Fundamental concepts of planar instantaneous kinematics

envelope h_b. It can be shown that the centre of curvature of the envelope h_b at H is just C_0. For if h is a circle with centre at C, and h is rigidly connected to c and moves together with c, then the distance between the envelope h_b and the path γ_C of C is constant along the normal direction. Hence we arrive at the following theorem:

Theorem 3: The centre of curvature of the envelope h_b of the generating curve h at H is the centre of curvature C_0 of the path γ_C of H_0 $(= C)$ at C.

Fig. 1.29 shows a special case of the envelope. The generating curve h here is a straight line. Draw a line PH from P normal to h. We may say that the centre of curvature $H_0 (= C)$ of h at H lies at infinity in the direction HP. According to Section 1.9, C_0 should lie on the return circle k_R. Hence C_0 can be found, as soon as k_R is found. Draw a circular arc k_{Hb} with C_0 as centre and $\overline{C_0 H}$ as its radius. If h_b is replaced by k_{Hb}, then within small range of the motion, h remains

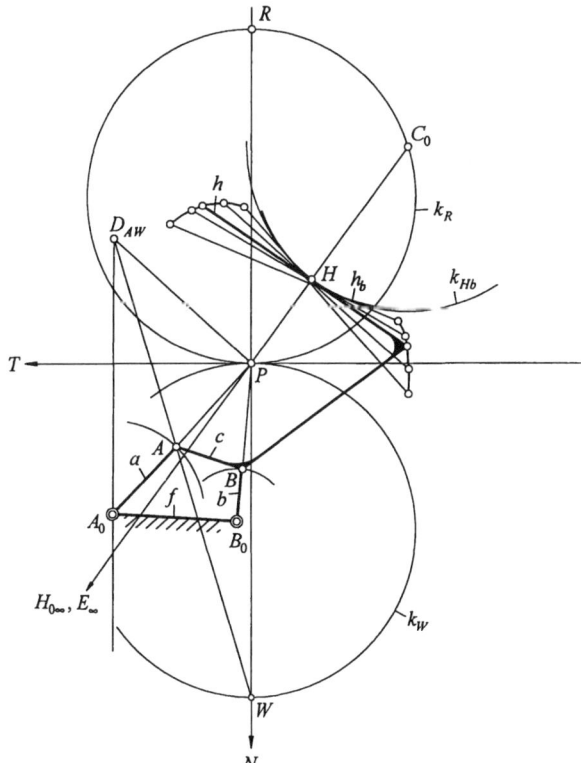

Fig. 1.29. The position of the centre of curvature C_0 of its envelope h_b at the point H, when the generating curve h is a straight line.

osculatory with k_{Hb}. We can exploit this property to construct a mechanism similar to a cam mechanism by making the coupler c as the driving member, and by making a member that contains k_{Hb} as the follower. During the motion of the mechanism, the follower will have a certain period of standstill; thus a dwell mechanism is constructed.

1.11 Equation of the envelope of a coupler straight line of a four-bar linkage

Appendix A2.6 outlines the concept of the equation of an envelope of an algebraic curve. The reader will find the present section easier to understand by first familiarizing himself with that concept.

Fig. 1.30 shows a four-bar linkage A_0ABB_0, with a straight line g rigidly connected to its coupler c. During the motion of the mechanism, g forms a family of straight lines. Our objective is to establish the equation of the envelope of this family of straight lines. Such an equation first appeared in (Hunt & Fichter, 1981). Its derivation is slightly more involved than that of the coupler point curve equation. We denote as before by the following symbols: $f = \overline{A_0B_0}$, $a = \overline{A_0A}$, $c = \overline{AB}$, $b = \overline{B_0B}$. Let A_0 be the origin of the fixed rectangular coordinate system (x, y), A_0B_0 the x-axis. The origin of the moving coordinate system (ξ, η) is taken at A, and AB is taken as the ξ-axis. The location of the straight line g on AB is determined by its normal distance e from A and the inclination β of this normal with respect to the ξ- axis. Both e and β are constants. The position of g with respect to the fixed coordinate system is determined by its normal distance d from A_0 and the inclination α of this normal with respect to the x-axis. Both d and α are variables.

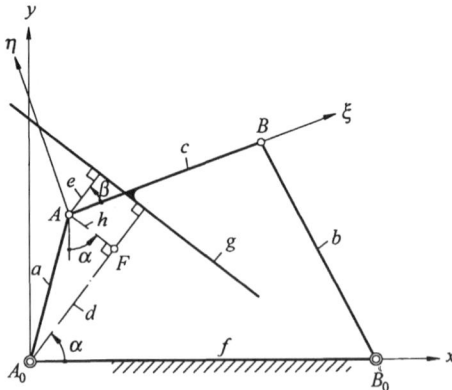

Fig. 1.30. Coordinates for the equation of the envelope of a coupler straight line.

Fundamental concepts of planar instantaneous kinematics

We shall now derive an equation $F(d, \alpha) = 0$ which contains, of course, the six constants a, c, b, f, e and β. First draw a line AF from A perpendicular to d. Let $\overline{AF} = h$. We have then

$$h^2 = a^2 - (d-e)^2 \tag{1.30}$$

The coordinates of A are

$$\left.\begin{array}{l} x_A = (d-e)\cos\alpha - h\sin\alpha \\ y_A = (d-e)\sin\alpha + h\cos\alpha \end{array}\right\}$$

and the coordinates of B are

$$\left.\begin{array}{l} x_B = x_A + c\cos(\alpha-\beta) = (d-e)\cos\alpha - h\sin\alpha + c\cos(\alpha-\beta) \\ y_B = y_A + c\sin(\alpha-\beta) = (d-e)\sin\alpha + h\cos\alpha + c\sin(\alpha-\beta) \end{array}\right\} \tag{1.31}$$

The equation of the circle on which B lies is

$$(x_B - f)^2 + y_B^2 = b^2$$

or

$$x_B^2 + y_B^2 - 2fx_B + (f^2 - b^2) = 0 \tag{1.32}$$

Substituting equation (1.31) into equation (1.32) and taking $x_A^2 + y_A^2 = a^2$ into consideration, we get after rearranging

$$\frac{a^2 + c^2 + f^2 - b^2}{2} + (d-e)(c\cos\beta - f\cos\alpha) - cf\cos(\alpha-\beta)$$
$$= h(c\sin\beta - f\sin\alpha) \tag{1.33}$$

In the above equation, $(a^2 + c^2 + f^2 - b^2)/2$ is a constant. Let it be denoted by z, i.e.

$$(a^2 + c^2 + f^2 - b^2)/2 = z$$

Squaring both sides of equation (1.33), and substituting equation (1.30) into it, we get

$$z^2 + (d-e)^2[c^2 + f^2 - 2cf\cos(\alpha-\beta)] + c^2f^2\cos^2(\alpha-\beta)$$
$$+ 2z(d-e)(c\cos\beta - f\cos\alpha) - 2zcf\cos(\alpha-\beta)$$
$$- 2cf(d-e)\cos(\alpha-\beta)(c\cos\beta - f\cos\alpha) - a^2(c\sin\beta - f\sin\alpha)^2 = 0 \tag{1.34}$$

Equation (1.34) is then the required equation $F(d, \alpha) = 0$. However, to obtain the envelope of the straight lines g, this equation has to be re-expressed as a *line equation*. Now in Fig.1.30, if the homogeneous coordinate equation of the line g with respect to the coordinate system xy is[§]

[§] Please refer to Appendix A2.6. However, care should be taken not to confuse the present symbols l, m with those l, m in Sections 3.7.1.and 3.7.2.

we should have
$$lx + my + nz = 0$$

$$\sin\alpha = \frac{m}{L}, \quad \cos\alpha = \frac{l}{L}, \text{ where } L = \sqrt{l^2 + m^2} \tag{1.35}$$

and
$$d = -n/L$$

Similarly, if the homogeneous coordinate equation of the line g with respect to the coordinate system $\xi\eta$ is

$$\lambda\xi + \mu\eta + \nu\zeta = 0$$

we should also have

$$\sin\beta = \frac{\mu}{\Lambda}, \quad \cos\beta = \frac{\lambda}{\Lambda}, \text{ where } \Lambda = \sqrt{\lambda^2 + \mu^2} \tag{1.36}$$

and
$$e = -\nu/\Lambda$$

Therefore
$$d - e = (L\nu - \Lambda n)/(L\Lambda) \tag{1.37}$$

$$\cos(\alpha - \beta) = (\lambda l + \mu m)/(L\Lambda) \tag{1.38}$$

Substituting equations (1.35) - (1.38) into equation (1.34) and rearranging, we get

$$z^2 + \frac{1}{\Lambda^2}[(c^2 + f^2)\nu^2 + 2z\lambda c\nu - a^2c^2\mu^2] + \frac{1}{L^2}[(c^2 + f^2)n^2 + 2zl fn - a^2f^2m^2]$$
$$+ \frac{cf(\lambda l + \mu m)}{L^2\Lambda^2}[4\nu n + 2cn\lambda + 2fl\nu + cf(\lambda l + \mu m)]$$
$$- \frac{2}{L\Lambda}[(c^2 + f^2)\nu n + zl f\nu + z\lambda cn + zcf(\lambda l + \mu m) - a^2cfm\mu]$$
$$- \frac{2cfn(\lambda l + \mu m)}{L^3\Lambda}(n + fl) - \frac{2cf\nu(\lambda l + \mu m)}{L\Lambda^3}(\nu + c\lambda) = 0 \tag{1.39}$$

Multiplying equation (1.39) by $L^3\Lambda^3$ and transferring terms, and squaring both sides, we get

$$[z^2L^2\Lambda^2 + [(c^2 + f^2)\nu^2 + 2z\lambda c\nu - a^2c^2\mu^2]L^2 + [(c^2 + f^2)n^2 + 2zfln - a^2f^2m^2]\Lambda^2$$
$$+ cf(\lambda l + \mu m)[cf(\lambda l + \mu m) + 2\nu(n + fl) + 2n(\nu + c\lambda)]^2 L^2\Lambda^2$$
$$= 4\{[c^2 + f^2)\nu n + z[cf(\lambda l + \mu m) + f\nu l + cn\lambda] - a^2cf\mu m]L^2\Lambda^2$$
$$+ cf\nu(\lambda l + \mu m)(\nu + c\lambda)L^2 + cfn(n + fl)(\lambda l + \mu m)\Lambda^2\}^2 \tag{1.40}$$

Equation (1.40) is then the required homogeneous coordinate equation in (l, m, n) of the envelope of the straight line g (please refer to Appendix A2.6).

Fundamental concepts of planar instantaneous kinematics

We shall here outline a few of the properties of the equation of the envelope. It can be seen from equation (1.40) that it is a homogeneous equation of 6th degree in the variables (l, m, n). If this equation is transformed into a point equation of the form $f(x, y, z) = 0$, its degree will be 12. Nevertheless, the coupler point equation (6.25) presented later is not the *dual* (please refer to Section A2.7) of equation (1.40). For example, in the Cardan motion (please refer to Fig. 3.67) the coupler curve is an ellipse, while the envelope of a coupler straight line is a complicated curve.

The envelope equation has at infinity a double root at each of the imaginary circular points I and J. Moreover, because $l^2 = 0$, $m^2 = 0$ satisfy equation (1.40), and $l = 0, m = 0$ represent the line at infinity (please refer to Section A2.6), this line is the bitangent of the envelope at infinity. Hence the envelope has 8 points at infinity. But the degree of the envelope is 12, it has 4 more points at infinity (real or imaginary). The tangents at these points are the asymptotes. In other words, the envelope has at most 4 real asymptotes.

Bitangent is the *dual* of a double point. According to Section A2.4(*b*), for a curve of *class* m_1, its maximum number of bitangents is $(m_1 - 1)(m_1 - 2)/2$. Therefore the maximum number of bitangents of a non-degenerate envelope of a coupler straight line is $(6-1)(6-2)/2 = 10$.

Also, to show that the line at infinity is a bitangent of the envelope, we put the lowest term of equation (1.40) (considered as a non-homogeneous equation in l, m) equal to zero, thus turning it into a quadratic equation (please refer to Section A2.2(*b*)).

Exercises

1.1 Given a four-bar linkage A_0ABB_0, in which $f = \overline{A_0B_0} = 81$, $a = \overline{A_0A} = 22$, $c = \overline{AB} = 37$, $b = \overline{B_0B} = 48$. Find the fixed polode and the moving polode for the motion of c relative f.

1.2 Explain how the pole changing velocity of the coupler AB of a four-bar linkage A_0ABB_0 (e.g. Fig. 2.2(*a*)) can be found from the velocities $\mathbf{v}_A$, $\mathbf{v}_B$ of the points A and B by means of Hartmann construction.

1.3 The figure shows two bodies a, b in sliding contact, rotating about their respective fixed centres A_0, B_0. The bodies a, b are in contact at a point E. Let this point on a, b be denoted respectively by E_a, E_b. The centre of curvature of the contacting curve on a is A, and that of the contacting curve on b is B. It is required to find the centre of curvature E_{a0} of the path of E_a relative to the body b.

(Hint: The equivalent four-bar linkage for the relative motion between a and b is A_0ABB_0, and the relative pole between a, b is P_{ab}. Imagine that B_0B ($= b$) were fixed, and assume an arbitrary velocity $\mathbf{v}_{A0}$ of A_0, to find the velocity $\mathbf{v}_A$ of the point A. The pole changing velocity $\mathbf{u}$ can then be found from $\mathbf{v}_{A0}$ and $\mathbf{v}_A$ according to Exercise 1.2. From $\mathbf{u}$ the centre of

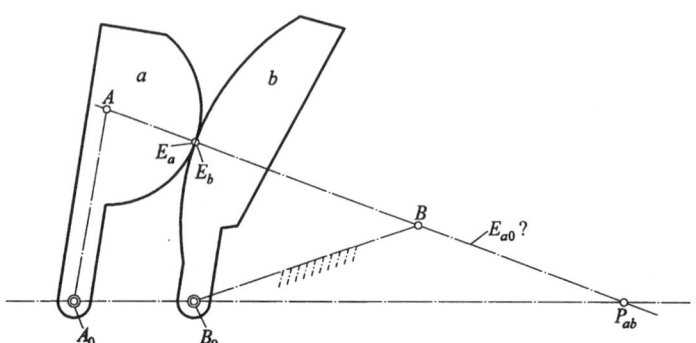

curvature E_{a0} of the path of E_a on b can be found by means of Hartmann construction.)

1.4 A small wheel rolls on the inside of a large wheel, the diameters of both wheels being 40 and 160.
(a) find the inflection circle and the inflection pole W_1 of the small wheel.
(b) The path of W_1 is an approximate straight line. As the centre of the small wheel moves from Q_1 to Q_2 ($\sphericalangle Q_1OQ_2 = 22.5°$), W_1 moves to W_2. Find the vertical deviation e of W_2 from the horizontal line passing through W_1.

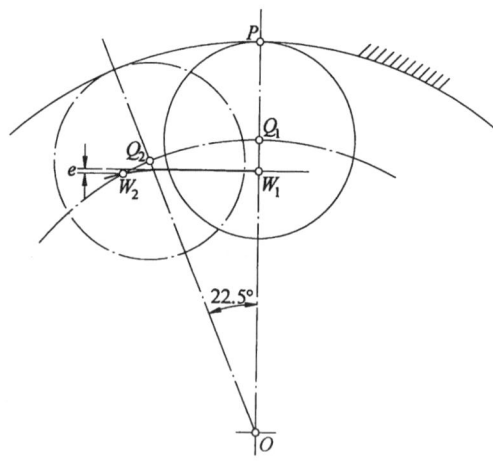

1.5 Using the relation $u = \delta\omega$ find:
the inflection circle of the wheel c in Fig. 1.27(a), and
the inflection circle of the flat plate f in Fig. 1.27(b).

Fundamental concepts of planar instantaneous kinematics

1.6 A small wheel k_g rolls without slip on the inside of a large wheel k_r. The radius of k_g is $R_g = 20$, and that of k_r is $R_r = 36$. Find the inflection circle k_W of the small wheel. Choose a point E on k_W, and find the path of E as k_g rotates through $\pm 30°$.

1.7 A small wheel rolls without slip on the outside of a lerge wheel. The radius of the small wheel is 15, and that of the large wheel is 30. Find the inflection circle k_W of the small wheel, and choose a point E on k_W. Let the joining line OC oscillate through $\pm 15°$ from its vertical position as shown in the figure. Find the path of E and draw the tangent of this path at E.

Hint: Only three positions of E are sufficient for drawing the path of E. The small wheel rotates through an angle 45° as OC oscillates through 15°. Do not use an awkward method to locate the positions of E.

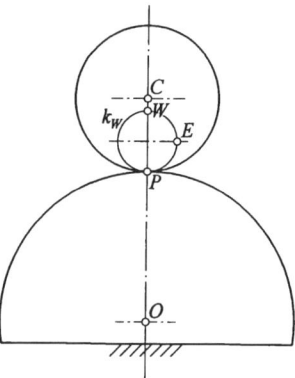

1.8 A small circle of diameter 30 rolls without slip on the outside of a large, fixed circle of diameter 60. Use Hartmann construction to find the centre of

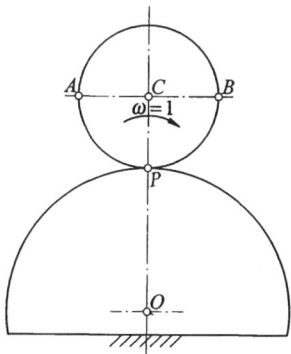

curvature A_0 of the path of the point A on the small circle. Similarly find B_0. Construct a four-bar linkage A_0ABB_0 to approximate the two original wheels.

1.9 Same as Exercise 1.8, but find A_0 and B_0 by Bobillier construction.

1.10 In Exercise 1.8, if P is considered as a point on the small circle,
(a) what is the acceleration of P if the angular velocity $\omega = 1$ of the small circle is a constant?
(b) what is the acceleration of P if the angular velocity ω of the small circle is not a constant, but with an angular acceleration α?

1.11 In the four-bar linkage shown, $f = \overline{A_0B_0} = 47$, $a = \overline{A_0A} = 55$, $c = \overline{AB} = 16$, $b = \overline{B_0B} = 37$. In the position shown in the figure, $\sphericalangle B_0A_0A = 45°$. If the method of Fig. 1.15 is applied here, the intersection of AB and A_0B_0 will fall beyond the paper. Try to apply the method of Fig. 1.14 to the two points A and B to find A_W and B_W, and then determine the inflection circle of AB. Choose any point E (except P) on this inflection circle, and let A_0A oscillate through $\pm 10°$. Draw the path of E.

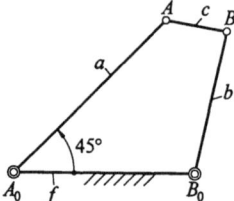

1.12 Suppose a certain position of a known four-bar linkage A_0ABB_0 is given. Show how to find the poletangent and diameter δ of the inflection circle of the coupler by means of Euler-Savary equation and Fig. 1.23.

1.13 Show that

$$\frac{1}{\overline{PM}} - \frac{1}{\overline{PM_0}} = \frac{1}{\delta}$$

in Fig. 1.13 by means of Hartmann construction and the relation $u = \delta\omega$.

1.14 Given the lengths of the links of a four-bar linkage: $\overline{A_0B_0} = 65$, $\overline{A_0A} = 44$, $\overline{AB} = 32$, $\overline{B_0B} = 33$. Find, in the position $\sphericalangle B_0A_0A = 55°$, the inflection circle of the coupler AB by parallel lines method. Choose any point E on this circle, and draw the path of E as the crank A_0A oscillates through $\pm 10°$.

1.15 Please refer to Fig. 1.29, and note that in a certain position, the centre of curvature at a certain point of the envelope of a straight line bound to the coupler lies on the return circle k_R. Apply this principle to Exercise 1.14, and draw the return circle k_R of the coupler AB. Show that we can construct a dwell mechanism by binding a round plate with A_0B_0, and binding a straight line with

AB, and by fixing AB and setting A_0B_0 to move.

1.16 The figure shows a construction method of finding the inflection circle of the coupler AB of four-bar linkage A_0ABB_0. Mark the letters at all unlabelled points and explain the construction procedure.

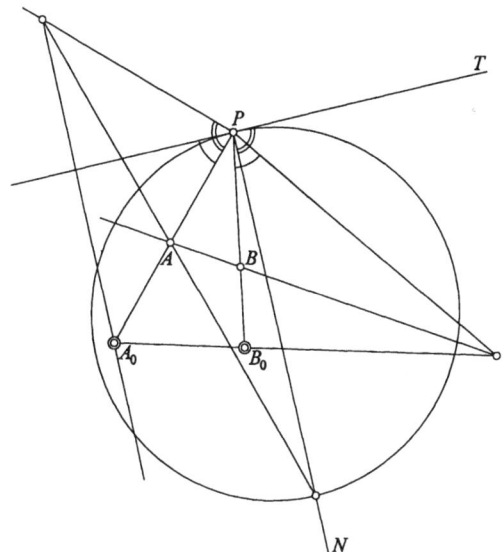

1.17 Find the centre of curvature C_0 of the path of C in Fig. 1.4(b) by means of Bobillier construction.

1.18 Extend the line A_0B_0 in Fig. 1.22 to the right to a point H, such that $\overline{B_0H} = \overline{D_{AB}A_0}$. Draw a line HT passing through H parallel to A_0A (parallel to B_0B as well).
 (a) Show that $\overline{E_0E} \cdot \overline{E_fH} = \overline{A_0A} \cdot \overline{A_0H}$ = constant.
 (b) Deduce from the above relation the location of E_0 if E lies on the line HT. What is the inflection circle of the coupler AB?

1.19 The figure shows a Grashof double rocker, in which $a = 50$, $c = 23$, $b = 41$ and $f = 34$. The coordinates of a line bound to the coupler c with respect to A are: $\beta = 81.5°$, $e = 1.8$ (please refer to Fig. 1.30). Find the equation of the envelope of this straight line, and draw all real bitangents.

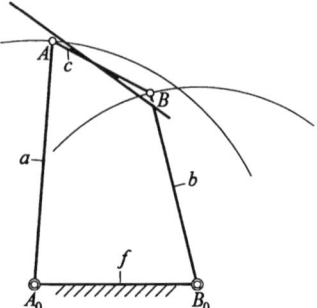

2
On some special techniques of mechanism analysis

2.1 The Freudenstein equation

Fig. 2.1 shows a four-bar linkage A_0ABB_0. The lengths of the four bars are represented respectively by $\overline{A_0A} = a$, $\overline{B_0B} = b$, $\overline{AB} = c$, and $\overline{A_0B_0} = f$. The input angle is ϕ, and the output angle is ψ. We shall assume from now on that all

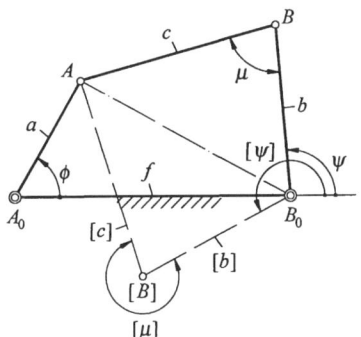

Fig. 2.1. Relation between ψ ([ψ]) and ϕ.

counterclockwise angular displacements are positive. a, b, c and f are constants, and ϕ, ψ are variables. In order to get a functional relationship between ϕ and ψ, the most concise form is the *Freudenstein equation* (Freudenstein, 1955) :

$$R_1 \cos\phi - R_2 \cos\psi - R_3 = -\cos(\phi - \psi) \tag{2.1}$$

where

$$R_1 = f/b \tag{2.2}$$

$$R_2 = f/a \tag{2.3}$$

$$R_3 = \frac{a^2 + b^2 - c^2 + f^2}{2ab} \tag{2.4}$$

Although the number of links of the four-bar linkage is 4, this relation between ϕ and ψ depends only on the three ratios R_1, R_2 and R_3. If a, b, c and f are known, then equation (2.1) can be considered as an implicit function of ψ with respect to ϕ. Solving this equation gives

$$\tan\frac{\psi}{2} = \frac{-V \pm (V^2 - U^2 + W^2)^{1/2}}{W - U} \tag{2.5}$$

where

$$U = a^2 + b^2 - c^2 + f^2 - 2af\cos\phi$$
$$V = 2ab\sin\phi$$
$$W = 2b(f - a\cos\phi)$$

The ($\pm$) sign in equation (2.5) results from the fact that for a certain set of the four lengths a, b, c, f, two different values of ψ and $[\psi]$ can be obtained for a single ϕ. The angle ψ corresponds to the *open* four-bar linkage, or A_0ABB_0 shown in Fig. 2.1; while $[\psi]$ corresponds to the *crossed* four-bar linkage, or $A_0A[B]B_0$. Let a known four-bar linkage be given as the *open* one as shown in Fig. 2.1. The rules are:

(a) If the four-bar linkage is a crank-rocker or a drag-link (double-crank), use the (+) sign to calculate ψ;

(b) if the four-bar linkage is a double-rocker, use the (+) sign to calculate ψ, and the (−) sign to calculate $[\psi]$.

This is because in case (a), a four-bar linkage cannot move from the open position shown in Fig. 2.1 to the crossed position shown by the chain line, unless it is disassembled by removing the pin at B and then reassembled. We shall assume from now on that for a crank-rocker or double-crank, the open configuration is always formed on the upper half of the line of centres A_0B_0.

2.2 Velocity analysis

We would like to raise here two points:

(a) In analyzing velocity for a four-bar linkage such as A_0ABB_0 shown in Fig. 2.2(a), most elementary text books on mechanisms teach the reader to calculate first the velocity v_A of the point A, then draw a velocity polygon to a certain scale with $\mathbf{o}_v\mathbf{a} = \mathbf{v}_A$ as that shown in Fig. 2.2(b). Such construction is not necessary. We can start by ignoring the actual value of ω_a, simply assuming $\omega_a = 1$ without dimension, and carrying out the whole velocity analysis. This is because all linear velocities and angular velocities of a mechanism are mutually linearly dependent. In other words, they are in proportion. The actual linear velocities and angular velocities can be obtained by multiplying those obtained for $\omega_a = 1$ by the actual value of ω_a. Thus for example, if $\omega_a = 3$ rad/s, all we have to do is to draw a velocity polygon

On some special techniques of mechanism analysis

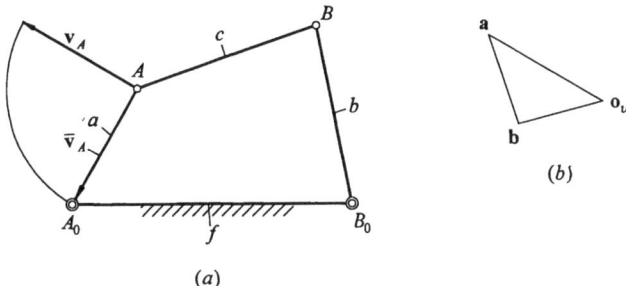

Fig. 2.2. (a) Four-bar linkage. (b) Velocity polygon.

like that shown in Fig. 2.2(b) with $\omega_a = 1$, and then multiply all velocities thus obtained by 3 and ascertain their units.

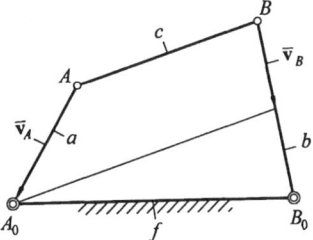

Fig. 2.3. Orthogonal velocities $\bar{v}_A$, $\bar{v}_B$.

(b) In most cases, the velocity polygon may be saved. Since we have assumed $\omega_a = 1$, the length of $\overline{AA_0}$ represents v_A, but rotated through 90°, as shown in Fig. 2.3. Let the rotated velocity of A be represented by $\bar{v}_A$ which is called an

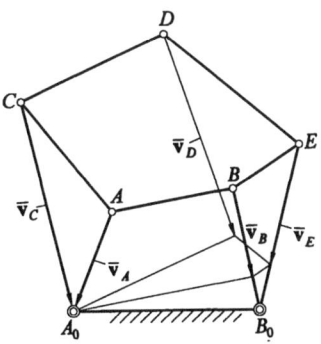

Fig. 2.4. Velocity analysis by orthogonal velocities.

orthogonal velocity. We only have to draw a line from the arrowhead of $\bar{v}_A$ parallel to AB to intersect B_0B to get the arrowhead of $\bar{v}_B$. In this way the actual velocities of all points can be obtained by rotating the orthogonal velocities about 90° in the sense of $-\omega_a$. In Fig. 2.4, the point A_0 is the common point of the arrowheads of $\bar{v}_C$ and $\bar{v}_A$. Draw a line from A_0 parallel to AB, to intersect B_0B gives $\bar{v}_B$; then draw a line parallel to BE to intersect B_0E to yield $\bar{v}_E$. Finally draw lines respectively from A_0 and the arrowhead of $\bar{v}_E$ parallel to CD and ED. Their intersection is $\bar{v}_D$.

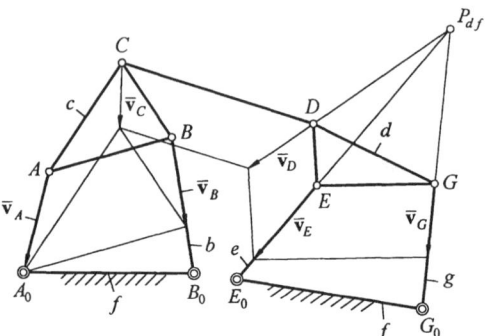

Fig. 2.5. Velocity analysis by orthogonal velocities.

In Fig.2.5, $\bar{v}_C$ is obtained by drawing lines from A_0 and the arrowhead of $\bar{v}_B$ parallel respectively to AC and BC to get their intersection. In order to find $\bar{v}_D$, it is necessary to find first the pole P_{df} of the link d which is the intersection of E_0E and G_0G. Draw a line from the arrowhead of $\bar{v}_C$ parallel to CD to intersect $P_{df}D$ will yield $\bar{v}_D$. A line drawn from the arrowhead of $\bar{v}_D$ parallel to DE will locate $\bar{v}_E$, and a line drawn further parallel to EG will locate $\bar{v}_G$.

Fig. 2.6 shows an eight-bar linkage in which all loops are five-bar loops. There are no four-bar loops. f is the fixed link, and a is the driving link. Assume $\omega_a = 1$. It is required to find the velocities of all joints in this mechanism. The procedure of solution is as follows (see Rosenauer, 1938). Extend AD, to meet C_0C in M. Consider M as a point on g, or a point on $\triangle CDF$. Since the orthogonal velocity $\bar{v}_C$ of C must be along the direction of C_0C, the orthogonal velocity $\bar{v}_M$ of M must also be along the direction of C_0C. Hence a line drawn from the arrowhead of $\bar{v}_A$, or the point A_0, parallel to ADM to cut C_0C will yield $\bar{v}_M$. Similarly, extend $A'E$ to meet B_0B in N. Consider the point N as a point on h, or a point on $\triangle BEF$. A line drawn from A_0 parallel to $A'E$ cutting B_0B will yield $\bar{v}_N$. The lines drawn from the arrowheads of $\bar{v}_M$ and $\bar{v}_N$ parallel respectively to MF and NF meet in

On some special techniques of mechanism analysis

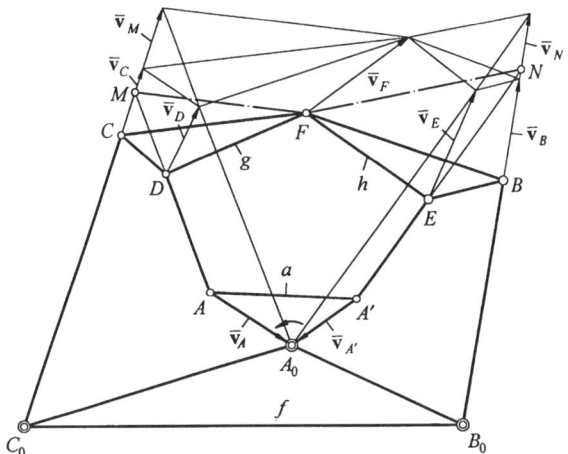

Fig. 2.6. Velocity analysis of an eight-bar linkage without four-bar loops.

$\bar{v}_F$. Finally draw lines from the arrowhead of $\bar{v}_F$ parallel to FD and FE, cutting respectively the lines drawn before from A_0 in $\bar{v}_D$ and $\bar{v}_E$. The lines drawn from the arrowhead of $\bar{v}_F$ parallel respectively to FC and FB, will cut C_0C and B_0B to give $\bar{v}_C$ and $\bar{v}_B$. The problem is thus solved.

2.3 Simplified acceleration analysis of a four-bar linkage

Fig. 2.7(a) shows a simplified technique to analyze the accelerations of a four-bar linkage A_0ABB_0. Given a constant angular velocity ω_a of the driving crank a, it is required to find the angular accelerations of the follower crank b and coupler c, and the tangential acceleration of the point B. As is well-known in elementary kinematics of mechanisms, the traditional method of analyzing accelerations of a four-bar linkage requires the construction of a velocity polygon and an acceleration polygon, and the calculation of a number of normal and relative normal accelerations. Such a procedure is lengthy and therefore discourages students from carrying out the analysis. What we are going to introduce here, is a technique that omits not only the velocity polygon but also the acceleration polygon (Chiang, 1970). The procedure is as follows: In the position of the four-bar linkage as shown in the figure,

(a) draw $A_0CP\|AB$, $PG\|BCD\|AA_0$, $GKF\|DP$;

(b) draw $KJ \perp B_0B$, $FJ \perp A_0C$; KJ and FJ intersect in J.

The construction is thus completed. From the construction we have:

$$a_B^t = \overline{JK}\omega_a^2 \tag{2.6}$$

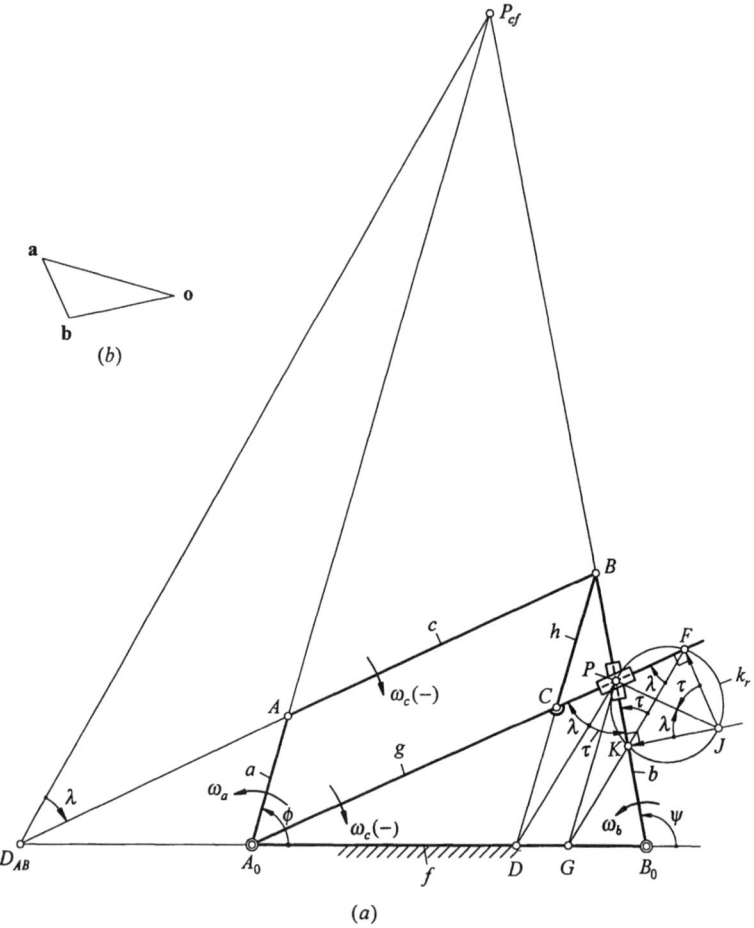

Fig. 2.7. (a) Simplified acceleration analysis of a four-bar linkage.
(b) Velocity polygon of the coupler AB.

$$a_{BA}^t = \overline{JF}\omega_a^2 \tag{2.7}$$

$$\alpha_b = a_B^t / \overline{B_0B} = \omega_a^2 \overline{JK} / \overline{B_0B} \tag{2.8}$$

$$\alpha_c = a_{BA}^t / \overline{AB} = \omega_a^2 \overline{JF} / \overline{AB} \tag{2.9}$$

where a_B^t represents the tangential acceleration of B, and a_{BA}^t the tangential acceleration of B relative to A, and α_b, α_c are the angular accelerations of b and c respectively.

On some special techniques of mechanism analysis

Proof: In Fig. 2.7(a), since we have assumed $\omega_a = 1$, therefore $\overline{\mathbf{v}}_A$ is represented by $\overline{A_0 A}$. $\triangle BPC$ is the velocity polygon rotated through 90°. Although ω_a in Fig. 2.7(a) is drawn counterclockwise, this velocity polygon is independent of the sense of ω_a: $\overline{A_0 A}$ is the $\mathbf{v}_A$ rotated through 90° in the sense of ω_a, or $\overline{\mathbf{v}}_A$, no matter whether ω_a is clockwise or counterclockwise; and $\triangle BPC$ is the velocity polygon rotated through 90° in the sense of ω_a. The three sides, $\overline{BC}$, $\overline{BP}$, $\overline{CP}$, of this triangle represent respectively $\overline{\mathbf{v}}_A$, $\overline{\mathbf{v}}_B$, and $\overline{\mathbf{v}}_{BA}$ or $\overline{\mathbf{v}}_C$. For clarity purposes, the unrotated velocity polygon of the coupler AB is shown in Fig. 2.7(b) for comparison. Since $\overline{\mathbf{v}}_B$ is represented by $\overline{PB}$, $\dot{\overline{v}}_B$ is represented by $\dot{\overline{PB}}$, which is the tangential acceleration a_B^t of the point B. In order to find $\dot{\overline{PB}}$, imagine that two crossed sliders are placed at P. These two sliders can slide on rods g and b respectively, but are joined together at P to rotate relative to each other. Comparing Fig.2.7(a) with Fig.1.2, we see that the four-bar linkage $A_0 CBB_0$ in Fig. 2.7(a) corresponds to the four-bar linkage $A_0 ABB_0$ in Fig. 1.2; and that the quadrilateral $PKJF$ in Fig. 2.7(a) corresponds to the quadrilateral in Fig.1.2 with $\mathbf{u}$ as a diagonal but rotated through 90° in the sense of ω_a. If we rotate $PKJF$ backwards through 90°, in the sense of $-\omega_a$, we can see that the relative sliding velocity of the slider on rod b is represented by $\overline{JK}$. In other words, $\overline{JK}$ represents $\dot{\overline{PB}}$, or $\dot{v}_B$. However, as

$$v_B = \overline{PB}\,\omega_a$$

$$\dot{v}_B = \dot{\overline{PB}}\,\omega_a$$

therefore

$$a_B^t = \dot{v}_B = \dot{\overline{PB}}\,\omega_a = \overline{JK}\,\omega_a^2 \qquad (2.10)$$

In equation (2.10) $\overline{JK}$ is a directed quantity, being directed from J to K. For this reason an arrowhead is put on $\overline{JK}$ in Fig. 2.7(a). If $\overline{JK}$ is translated to the point B, it represents the vector a_B^t.

Similarly, $\overline{CP}$ represents $\overline{\mathbf{v}}_{BA}$. It can be proved that $\overline{JF}$ represents $\dot{\overline{CP}}$, and

$$a_{BA}^t = a_C^t = \overline{JF}\,\omega_a^2 \qquad (2.11)$$

The arrowhead of $\overline{JF}$ in Fig. 2.7(a) points from J to F, indicating the direction of a_C^t, or of a_{BA}^t. If $\overline{JF}$ is translated to point B, it represents the vector a_{BA}^t.

The angular acceleration of link b is

and the angular acceleration of link b is

$$\alpha_b = \frac{a'_B}{B_0 B} = \frac{\overline{JK}}{B_0 B} \omega_a^2 \tag{2.12}$$

and the angular acceleration of link c is

$$\alpha_c = \frac{a'_{BA}}{AB} = \frac{\overline{JF}}{AB} \omega_a^2 \tag{2.13}$$

In Fig. 2.7(a), ω_a is drawn counterclockwise, hence ω_b is counterclockwise, while ω_c is clockwise. On the contrary, if ω_a is clockwise, then ω_b will also be clockwise, while ω_c will be counterclockwise. As to the senses of α_b and α_c, because both equations (2.12) and (2.13) contain ω_a^2, both α_b and α_c are counterclockwise, regardless of ω_a being clockwise or counterclockwise.

Comparing Fig. 2.7(a) with Fig. 1.2, we can see that P is the pole between the two bodies h and f. $\overline{PJ}$ represents just the changing velocity of this pole P, its magnitude being $\overline{PJ}\omega_a$, and its direction being that rotated therough 90° in the sense of ω_a. Comparing again Fig. 2.7(a) with Fig. 1.15, we can see that if the points C, A_0, B_0, B are replaced respectively by the points A_0, A, B, B_0 in Fig. 1.15, the circumscribed circle of the quadrilateral $PKJF$ becomes just the circumscribed circle of the quadrilateral PA_WWB_W in Fig. 1.15. In other words, the circle in Fig. 2.7(a) is the inflection circle of the relative motion of f relative to h (assuming h being fixed and f being moving), or the return circle of the relative motion of h relative to f, hence may be denoted by k_r.

In Fig. 2.7(a), let $\sphericalangle KJP = \lambda$, $\sphericalangle FJP = \tau$, then the velocity ratios i_{ba} and i_{ca} are

$$i_{ba} = \omega_b / \omega_a = \overline{KP} / \overline{PB} = \overline{PB} / \overline{B_0 B}$$

$$i_{ca} = \omega_c / \omega_a = \overline{PC} / \overline{AB} = \overline{FP} / \overline{A_0 P}$$

where i_{ba} is (+), while i_{ca} is (–). It can easily be seen that

$$\alpha_b = -i_{ba}(1 - i_{ba}) \cot \lambda \, \omega_a^2 \tag{2.14}$$

$$\alpha_c = -i_{ca}(1 - i_{ca}) \cot \tau \, \omega_a^2 \tag{2.15}$$

Note that, in Fig. 2.7(a), λ is (–), while τ is (+). The above two equations can also be written as

$$\cot \lambda = -\frac{di_{ba}/d\phi}{i_{ba}(1 - i_{ba})} \tag{2.16}$$

$$\cot \tau = -\frac{di_{ca}/d\phi}{i_{ca}(1-i_{ca})} \qquad (2.17)$$

The angles λ in Figs. 2.7(a) and 1.17 are equal, since it can easily be proved that in Fig. 2.7(a) the lines DP and $D_{AB}P_{cf}$ are parallel.

It can be seen from equation (2.14) that, when $\lambda = 90°$, $\alpha_b = 0$, the velocity ratio $i_{ba} = \omega_b/\omega_a$ becomes a maximum or minimum. We have therefore

Theorem 4: As the collineation axis is perpendicular to the centre line of the coupler, the ratio of the velocity of output link to that of the input link ($= \omega_b/\omega_a$ now) is a maximum or minimum.

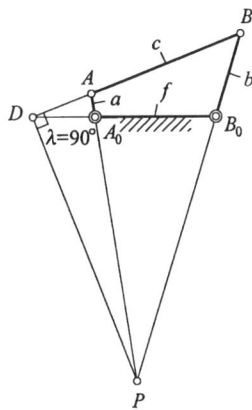

Fig. 2.8. ω_b/ω_a becomes an extreme value when collineation axis is perpendicular to AB.

This is the position of the four-bar linkage as that shown in Fig. 2.8. Moreover, τ is the angle between the collineation axis and the centre line of b. Similarly, it can be seen from equation (2.15) that as $\tau = 90°$, $\alpha_c = 0$, the velocity ratio $i_{ca} = \omega_c/\omega_a$ becomes a maximum or minimum, hence we have

Theorem 5: As the collineation axis is perpendicular to the centre line of one of the rotating cranks (e.g. b), the ratio of the angular velocity of the coupler (now ω_c) to that of the other rotating crank (now ω_a) becomes a maximum or minimum.

It can be proved in a similar manner that

Theorem 6: As the collineation axis is perpendicular to the line of centres (now A_0B_0), the relative angular velocity ratio ω_{cb}/ω_{ca} becomes a maximum or minimum.

2.4 Simplified acceleration analysis of complex mechanisms

Fig. 2.9 shows a six-bar linkage, the so-called Stephenson-II Mechanism. a is

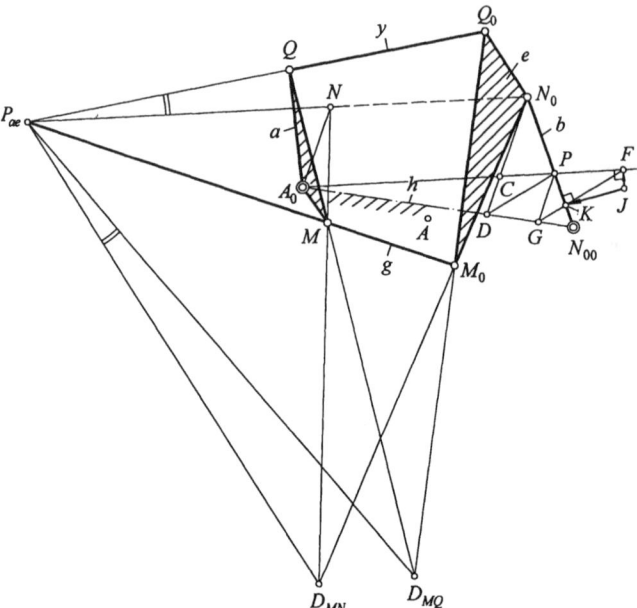

Fig. 2.9. Simplified acceleration analysis of a complex linkage.

the driving link, and b is the driven link. Suppose a rotates with a uniform angular velocity. It is required to find the angular acceleration of b. With the traditional velocity polygon and acceleration polygon method, the task would be quite tedious. It can be largely simplified by the following approach. As far as the four-bar loop is concerned, we can find first by means of Bobillier construction the centre of curvature N on link a of the relative path of the point N_0 on link e for the relative motion between e, a. Then $A_0 N N_0 N_{00}$ is considered as an instantaneous four-bar linkage. Applying the method of Section 2.3 to find $\overline{JK}$, then

$$\alpha_b = \frac{\overline{JK}}{\overline{N_{00}N_0}}\omega_a^2$$

Nowadays traditional graphical method is largely replaced by computer. However, the technique shown in Fig. 2.9 is not only a graphical method, but also a concise geometrical concept. This concept enables us to eliminate rigid computation, thereby greatly simplifing the computer programs and offering straightforward and more accurate results. Further application of this concept in higher order synthesis of function generators will be shown in Sections 7.6.1 and 7.6.2.

Exercises

2.1 Prove equation (2.5), i.e.
$$\tan\frac{\psi}{2} = \frac{-V \pm (V^2 - U^2 + W^2)^{1/2}}{W - U}$$

2.2 Derive from Freudenstein equation the expressions $\psi' = d\psi/d\phi$, $\psi'' = d^2\psi/d\phi^2$, and $\psi''' = /d^3\psi/d\phi^3$.
(Please note that the expression of ψ'' may contain ψ', and that of ψ''' may also contain ψ' and ψ''.)

2.3 In the mechanism shown, $\overline{A_0B_0} = 122$, $\overline{A_0A} = 43$, $\overline{AB} = 103$, $\overline{B_0B} = 69$, $\angle B_0A_0A = 60°$. A_0ABC is a parallelogram, and P is the velocity pole between links BC and A_0B_0. Let the velocities $\mathbf{v}_A$ and $\mathbf{v}_B$ be given as shown. Find the changing velocity of the pole P.

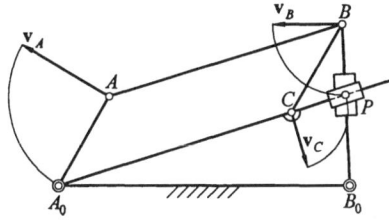

2.4 Find, in Exercies 2.3, the linear acceleration components a_B^t and a_{BA}^t, in terms of ω_a^2, where ω_a is the angular velocity of A_0A.

2.5 In the four-bar linkage shown, $\overline{A_0B_0} = 43$, $\overline{A_0A} = 69$, $\overline{AB} = 122$, $\overline{B_0B} = 103$.

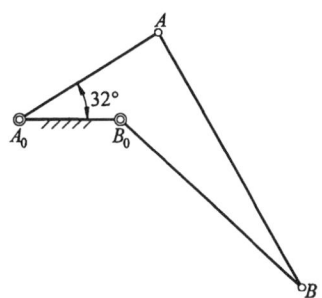

Find by means of Bobillier construction the inflection circle of AB.

2.6 Repeat Exercise 2.5, but by means of method of parallel lines to find the inflection circle of AB. Compare the constructed figure with Fig. 2.7 and with the result of Exercise 2.3.

2.7 In Fig. 2.9, in the relative motion between a and e, the centre of curvature of the relative path of a point A_0 on e is A, which belongs to e. Explain why the equivalent instantaneous four-bar linkage is $A_0AN_0N_{00}$, but not $A_0AN_0N_{00}$?

PART II
Synthesis of mechanisms

By synthesis of mechanisms it means in general three aspects, namely, *type synthesis, number synthesis,* and *dimensional synthesis.* A brief introduction of their definitions is given as follows.

(*a*) Type synthesis

It is also called *Reuleaux synthesis,* because this branch of synthesis is developed on the basis of the principle of pair inversion according to Reuleaux and the principle of mechanism inversion. In brief, in designing a mechanism transmission, it has to be decided what sort of mechanism is to be adopted; for example, whether it is a screw mechanism, link mechanism, gear mechanism, belt or chain mechanism, cam mechanism, intermittent mechanism, or even hydraulic mechanism. A decision has to be made so as to select the most appropriate mechanism for the specific purpose. Next to this, proper control arrangement for the linkage by electric, pneumatic or hydraulic devices has to be considered.

(*b*) Number synthesis

This may also be called *Grübler synthesis.* Basically it is a kind of synthesis based on the Grübler criterion to judge a linkage mechanism whether it is a constrained kinematic chain from its number of links and number of joints. For a four-bar linkage, the number of links is 4, and the number of joints is also 4; and that is all one can say. However, for multi-bar linkages, the problems can be quite complicated. For example, the number of *non-isomophos* eight-bar linkages is 16, as shown in Fig. II.1 (cited from an unpublished paper of Alt, according to Hain, 1955). In recent years, substantial progress has been made in this branch by applying *graph theory* and *topology.* We shall not go further into it. Readers who are interested may refer to (Yan & Harary, 1987), (Yan & Hwang, 1990) and (Yan, 1993).

(*c*) Dimensional synthesis

This branch includes investigations on mechanism synthesis problems based on geometrical relations and algebraic concepts. And it is this kind of synthesis that this book deals mainly with. The arrangement of materials follows substantially that of (Beyer, 1953).

Dimensional synthesis concerns to a large extent practical problems. According

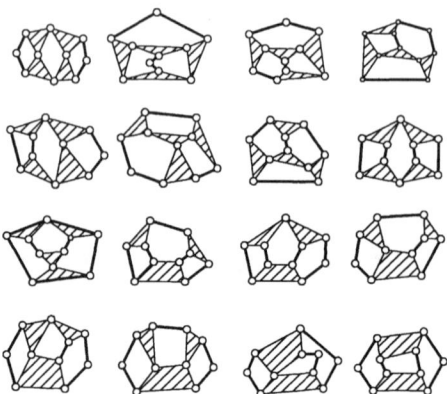

Fig. II.1. The 16 non-isomorphous eight-bar linkages.

to the nature of the problems, there are three categories:

(c1) body guidance problems;
(c2) path generation problems;
(c3) function generation problems.

In the following sections, we shall deal mainly with syntheses of four-bar linkages. In general, a problem can be said to have been solved once the four points A_0, A, B, B_0 have been found.

In the three definitions (a), (b), (c) mentioned above, the terms (a), (b) are often used alternatively by some authors. The definitions introduced here are due to (Beyer, 1931).

3
Dimensional synthesis of four-bar linkages --- body guidance problems

3.1 Guiding a body through two finitely separated positions

In this chapter we shall consider how to synthesize a four-bar linkage to guide a body through given positions. Since the body to be guided has no fixed centre of rotation, it is the coupler.

In a plane, the position of a body can be defined by two points on the body. Therefore a body can in general be represented by two points on the body. In Fig. 3.1,

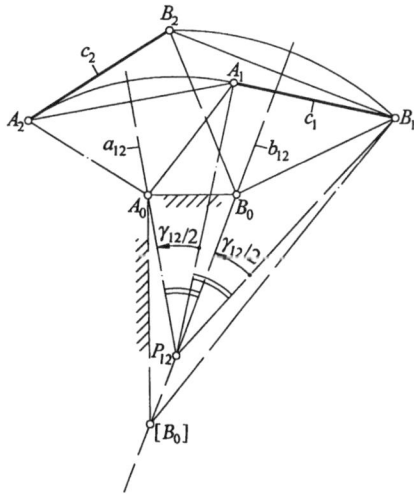

Fig. 3.1. Guiding a body through two finitely separated positions.

suppose the first position $c_1 (= A_1 B_1$) and the second position $c_2 (= A_2 B_2$) of a body c are given. It is required to find a four-bar linkage to guide the body c from its first position to the second position. The construction procedure is as follows: Join $\overline{A_1 A_2}$, and draw its perpendicular bisector a_{12}; join $\overline{B_1 B_2}$, and draw its perpendicular bisector b_{12}. The intersection point P_{12} of a_{12} and b_{12} represents the pole between the 1st and 2nd positions of the body c (please do not confuse this pole with the instantaneous pole between two bodies 1 and 2). Choose any point A_0 on a_{12}, and any point B_0 on b_{12}. Joining the four points $A_0A_1B_1B_0$ yields the required four-bar linkage. This is because $\overline{A_0A_1} = \overline{A_0A_2}$, and $\overline{B_0B_1} = \overline{B_0B_2}$. It can be seen from Fig. 3.1 that

$$\sphericalangle A_0P_{12}A_1 = \sphericalangle B_0P_{12}B_1 = \gamma_{12}/2 \tag{3.1}$$

In equation (3.1) the angle γ_{12} is $\sphericalangle A_1P_{12}A_2 = \sphericalangle B_1P_{12}B_2$, being the angle of rotation of the body c from c_1 to c_2. Consequently we have

$$\sphericalangle B_0P_{12}A_1 + \sphericalangle A_1P_{12}A_0 = \sphericalangle B_1P_{12}B_0 + \sphericalangle B_0P_{12}A_1$$

or

$$\sphericalangle B_0P_{12}A_0 = \sphericalangle B_1P_{12}A_1 \tag{3.2}$$

Since any point on b_{12} can be chosen as B_0, so if B_0 is chosen at $[B_0]$, it can be proved in a similar manner that

$$\sphericalangle [B_0]P_{12}B_1 = 180° - \sphericalangle A_0P_{12}A_1 \tag{3.3}$$

$$\sphericalangle [B_0]P_{12}A_0 = 180° - \sphericalangle B_1P_{12}A_1 \tag{3.4}$$

From the above four equations (3.1)--(3.4), we have the following theorem:

Theorem 7: Two rotating cranks (A_0A_1 and B_0B_1) subtend equal or supplementary angles at the pole. The coupler (A_1B_1) and the line of centres (A_0B_0) subtend also equal or supplementary angles at the pole. If the senses of reading around the pole are the same (e.g. A_0A_1 and B_0B_1, and A_0B_0 and A_1B_1), the two angles are equal, but if the senses of reading around the pole are opposite, the two angles are supplementary (e.g. A_0A_1 and $[B_0]B_1$, and $A_0[B_0]$ and A_1B_1).

The sense of reading is of course limited to an angle less than 180°.

Theorem 7, despite its simplicity, is of fundamental importance, and will frequently be referred to later on.

Since any point on a_{12} can be chosen as A_0, and any point on b_{12} can be chosen as B_0, it is of course permissible to choose both A_0 and B_0 at P_{12}. The four-bar linkage then becomes a triangle $\Delta A_1P_{12}B_1$, i.e., a *locked chain*. We have only to rotate the triangle with P_{12} as the centre of rotation, through an angle γ_{12}, to guide the body from A_1B_1 to A_2B_2. This is also the source of the definition of the terminology *pole*.

As long as two positions of a body are given, any point on the plane in its 1st position can be chosen as A_1, hence there are ∞^2 ways of choosing A_1. Similarly, any point can be chosen as B_1, and there are also ∞^2 ways of choosing B_1. After

Dimensional synthesis of four-bar linkages --- body guidance problems 55

having chosen A_1 and B_1, the points A_2 and B_2 are determined, hence a_{12} and b_{12} are also determined. There are ∞^1 ways of choosing A_0 on a_{12}, and ∞^1 ways of choosing B_0 on b_{12}. Altogether there are ∞^6 solutions. Alternatively it can be stated as follows. After the two positions of a body have been given, the pole P_{12} is determined uniquely, but there are still ∞^6 ways in choosing the four points A_1, B_1, A_0, B_0; this means that ∞^6 different four-bar linkages can be constructed to guide the coupler, or the body c.

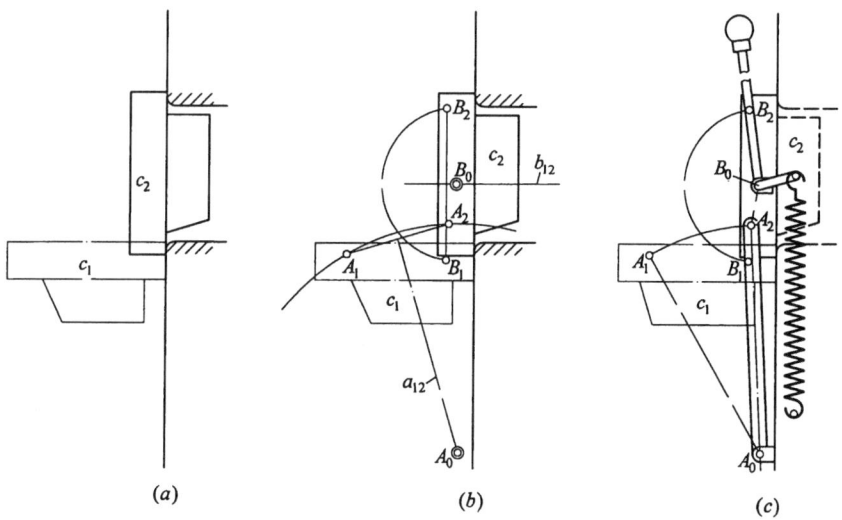

Fig. 3.2. A practical example of guiding a body through two finitely separated positions. (a) Given positions c_1, c_2. (b) Construction of a_{12}, b_{12}, and selection of A_0, B_0. (c) Synthesis of four-bar linkage $A_0A_1B_1B_0$.

Fig. 3.2 shows a practical example, taken from (Kraemer, 1959, Example 1). The open position c_1 and the closed position c_2 of an electric oven door are given, as shown in Fig. 3.2(a). The problem can be solved, provided that proper positions of A_1, B_1, A_0, B_0 are chosen.

3.2 Guiding a body through two infinitesimally separated positions

3.2.1 General concepts

In Fig. 3.1, suppose the two points A_1, A_2 approach to each other indefinitely to become a point A, and the points B_1, B_2 also approach to each other indefinitely to become a point B, as shown in Fig.3.3. Although the distance between A_1 and

56　　　　　　　　　　*Kinematics and Design of Planar Mechanisms*

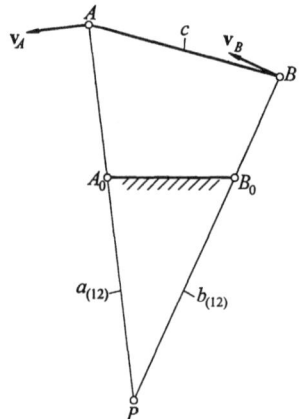

Fig. 3.3. Guiding a body through two infinitesimally separated positions.

A_2 disappears, its direction A_1A_2 still exists, being the direction of the moving velocity $\mathbf{v}_A$ of the point A. Similarly the direction of the velocity $\mathbf{v}_B$ of the point B is also determined. Thus a_{12} becomes the line passing through A and perpendicular to $\mathbf{v}_A$, written as $a_{(12)}$, and b_{12} becomes the line passing through B and perpendicular to $\mathbf{v}_B$, written as $b_{(12)}$. Choose any point on $a_{(12)}$ as A_0, and any point on $b_{(12)}$ as B_0. The four-bar linkage A_0ABB_0 is thus completed. The intersection point P of $a_{(12)}$ and $b_{(12)}$ is the original P_{12}, or the well-known instantaneous velocity pole of c. As long as the two infinitesimally separated positions of a body are given, P is uniquely determined. In other words, if A, B are given, and the directions of $\mathbf{v}_A$ and $\mathbf{v}_B$ are also given, P is uniquely determined. In this case there are ∞^2 ways of choosing either A or B, and there are ∞^1 ways of choosing either A_0 or B_0. There are altogether ∞^6 solutions.

Fig. 3.4 shows a practical example, being also taken from (Kraemer, 1959). A curved stiff pipe, with the directions of the velocities $\mathbf{v}_A$ and $\mathbf{v}_B$ of its two points A and B given. From these the pole P of the pipe can be determined. Rotating the pipe about P will enable the assembling and dismantling of the pipe.

If two points A, B of of a body c are given, and the velocity $\mathbf{v}_A$ of A is also given, we can either (1) assign the direction of the velocity $\mathbf{v}_B$ of B, or (2) assign the magnitude v_B of the velocity $\mathbf{v}_B$. Suppose the direction of $\mathbf{v}_B$ is given, as that shown in Fig. 3.5. Since the directions of $\mathbf{v}_A$ and $\mathbf{v}_B$ are already given, the two lines $a_{(12)}$ and $b_{(12)}$ are determined. It is possible to rotate the vector $\mathbf{v}_A$ through 90° into the line of $a_{(12)}$ to get $\bar{\mathbf{v}}_A$. A line drawn from the arrowhead of $\bar{\mathbf{v}}_A$ parallel to AB, to intersect $b_{(12)}$ will yield $\bar{\mathbf{v}}_B$. Rotating $\bar{\mathbf{v}}_B$ back through 90° will yield $\mathbf{v}_B$. The intersection of $a_{(12)}$ and $b_{(12)}$ is P. The points A_0 and B_0 can be chosen

Dimensional synthesis of four-bar linkages --- body guidance problems

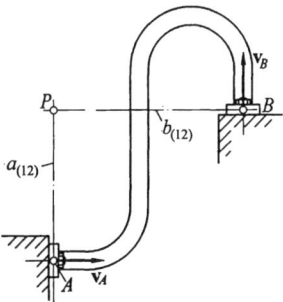

Fig. 3.4. A practical example of guiding a body through two infinitesimally separated positions.

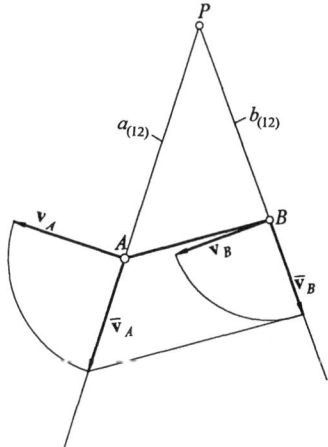

Fig. 3.5. A solution when v_A is given, and the direction of v_B is also given.

arbitrarily on $a_{(12)}$ and $b_{(12)}$ respectively. If only the magnitude of v_B is given, the procedure of the construction can be carried out as shown in Fig. 3.6. Draw a circle with B as centre and the magnitude v_B of v_B as radius. A line drawn from the arrowhead of $\bar{v}_A$ parallel to AB cuts the circle in two points. Take one of these two points as the arrowhead of $\bar{v}_B$. The vector normal to $\bar{v}_B$ is then v_B, and its direction is readily determined. Both $b_{(12)}$ and P are determined correspondingly. Choosing as before A_0 and B_0 on $a_{(12)}$ and $b_{(12)}$ respectively, we can complete the required four-bar linkage.

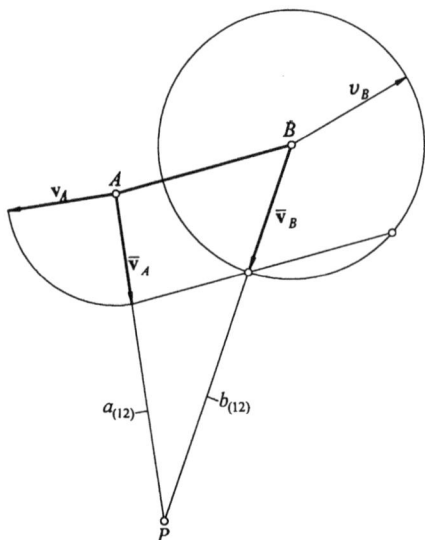

Fig. 3.6. A solution when $\mathbf{v}_A$ is given, and the magnitude v_B of $\mathbf{v}_B$ is also given.

3.2.2 Equations of fixed and moving polodes of a four-bar linkage

Equations of the polodes of a four-bar linkage as that shown in Fig. 1.1 were first given by (Müller, 1903). The derivation procedure which will be introduced here, is somewhat different from that of Müller. In Fig. 3.7, A_0ABB_0 is a given four-bar linkage. Take an xy-coordinate system, with A_0 as the origin, and A_0B_0 as the

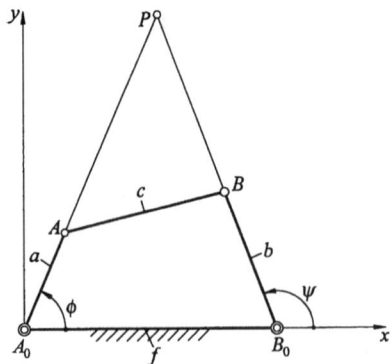

Fig. 3.7. Coordinate system for derivation of equation of fixed polode π_f.

Dimensional synthesis of four-bar linkages --- body guidance problems 59

x-axis. The equation of the straight line A_0AP is then

$$\cos\phi = \frac{x}{y}\sin\phi \tag{3.5}$$

and the equation of the straight line B_0BP is

$$\cos\psi = \frac{x-f}{y}\sin\psi \tag{3.6}$$

However, the relation between ψ and ϕ is coupled by equation (2.1), or

$$R_1\cos\phi - R_2\cos\psi = R_3 - \cos(\phi - \psi) \qquad [(2.1)]$$

Substituting equations (3.5) and (3,6) into equation (2.1) gives

$$R_1\frac{x}{y}S_\phi - R_2\frac{x-f}{y}S_\psi = R_3 - \left[\frac{x(x-f)}{y^2}+1\right]S_\phi S_\psi \tag{3.7}$$

In equation (3.7) S_ϕ stands for $\sin\phi$, and S_ψ for $\sin\psi$ for brevity. Equation (3.7) can be rewritten as

$$LS_\phi - MS_\psi = K' - NS_\phi S_\psi \tag{3.8}$$

where

$$L = R_1\frac{x}{y}$$

$$M = R_2\frac{x-f}{y}$$

$$K' = R_3$$

$$N = \frac{x(x-f)}{y^2}+1$$

Rearranging equation (3.8), and squaring both sides of the equation until only S_ϕ^2 and S_ψ^2 remain, yields

$$N^4S_\phi^4S_\psi^4 - 2L^2N^2S_\phi^4S_\psi^2 - 2M^2N^2S_\phi^2S_\psi^4 + L^4S_\phi^4 + M^4S_\psi^4$$
$$- 2(L^2M^2 + K'^2N^2 - 4LMK'N)S_\phi^2S_\psi^2 - 2L^2K'^2S_\phi^2 - 2K'^2M^2S_\psi^2 + K'^4 = 0 \tag{3.9}$$

From equations (3.5) and (3.6) we have

$$S_\phi^2 = \frac{y^2}{x^2+y^2} \tag{3.10}$$

$$S_\psi^2 = \frac{y^2}{(x-f)^2 + y^2} \qquad (3.11)$$

Substituting equations (3.10) and (3.11) into equation (3.9), we get the equation of the fixed polode π_f:

$$\pi_f: [U^2 - R_1^2 Vx^2 - R_2^2 W(x-f)^2 + R_3^2 VW]^2 \\ - VW[2R_3 U - 2R_1 R_2 x(x-f)]^2 = 0 \qquad (3.12)$$

where

$$U = x(x-f) + y^2 = (x^2 + y^2) - fx = W - fx \\ V = (x-f)^2 + y^2 = (x^2 + y^2) - 2fx + f^2 = W - f(2x - f) \\ W = x^2 + y^2$$

Since U, V, W are all quadratic terms, equation (3.12) is of the 8th degree. Its term of highest degree contains the factor $(x^2 + y^2)^2$. Hence π_f is a bicircular curve of 8th degree passing through the circular points $I(1,i,0)$ and $J(1,-i,0)$. (Please refer to Section A1.2). Rewriting equation (4.12) in a homogeneous form $F(x,y,z) = 0$, and substituting the coordinates of I, J into the expressions of $\partial F/\partial x, \partial F/\partial y, \partial F/\partial z$, we can see that π_f satisfies equation (A2.18) in Appendix 2. Therefore π_f possesses at each point I, J a double point. Further substitution of the coordinates of I, J into equation (A2.19) will yield the equations of the tangents at these circular points, which are the asymtotes of π_f at I and J:

$$[(R_1^2 - R_2^2)(x \pm iy) + f R_1^2]^2 = 0 \qquad (3.13)$$

In equation (3.13), the ($\pm$) signs correspond respectively to the asymtote at I and J. From equation (3.13) it is clear that the two tangents at I (or J) coincide, hence the double point at I (or J) becomes a cusp (imaginary). To find the linear equation of this tangent, substituting the definitions of R_1 and R_2 in equations (2.2) and (2.3) into equation (3.13) yields

$$x \pm iy = \frac{a^2 f}{a^2 - b^2}$$

This is the equation of the two asymptotes. These two lines intersect the x-axis in the same point, which is $(a^2 f/(a^2 - b^2), 0)$ (please refer to Fig. 3.7). From Section A2.5 we know this is the real focus of the fixed polode π_f.

To find the equation of the moving polode π_c, it is only necessary to exchange the rôles of f and c, while the equation (3.12) remains unchanged. In other words, in equation (3.12), apart from replacing f with c, we may also replace R_1, R_2, R_3

Dimensional synthesis of four-bar linkages --- body guidance problems

respectively by R_1', R_2', R_3', where $R_1' = c/b$, $R_2' = c/a$, $R_3' = (a^2 + b^2 - f^2 + c^2)/(2ab)$. Hence π_c is also a bicircular equation of 8th degree.

In special cases, equation (3.12) may be simplified, as will be shown by the two following examples.

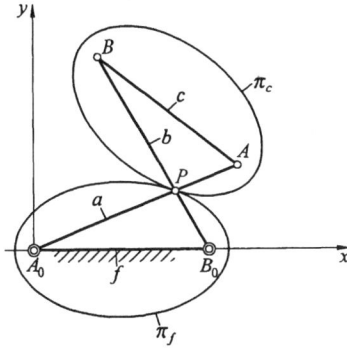

Fig. 3.8.   The two polodes of a non-parallel equal crank linkage.

Example: Fig. 3.8 shows a well-known non-parallel equal crank linkage. In this mechanism, $a = b$, $c = f$. We transform first equation (3.12) into a homogeneous equation in (x, y, z) (please refer to Section A2.4). It is only necessary to replace here f by fz. We have now $R_1 = R_2 = f/a = f/b$, $R_3 = 1$. Hence equation (3.12) is simplified to

$$\pi_f:\ z^2 y^4 [R_1^2(1-R_1^2)(2x-fz)^2 + 4R_1^2 y^2 - (1-R_1^2)f^2 z^2] = 0$$

In the above equation $z^2 = 0$ represents a two-fold line at infinity, being the location of P when the four-bar linkage keeps a parallelogram configuration as $A_0 A$ and $B_0 B$ are parallel. $y^4 = 0$ represents a four-fold line of $A_0 B_0$. (Please note that this line does not belong to the original fixed polode π_f.) The quadratic equation in the square bracket is the ellipse in Fig. 3.8, with A_0 and B_0 as its two foci, and the length of the link a as its major axis.

Example: Fig.3.9 shows a well-known double slider-crank mechanism. π_f is a circle with $c\ (=\overline{AB})$ as radius, and π_c is a circle with c as its diameter. This is the so-called *Cardan circle-pair*. Because the origin A_0 is now removed to infinity, equation (3.12) cannot express the circle of π_f, but the equation of π_c can be obtained according to the procedure mentioned above, by exchanging f and c in equation (3.12). We have now $R_1' = R_2' = R_3' = 0$. Hence equation (3.12) becomes

$$U^4 = 0$$

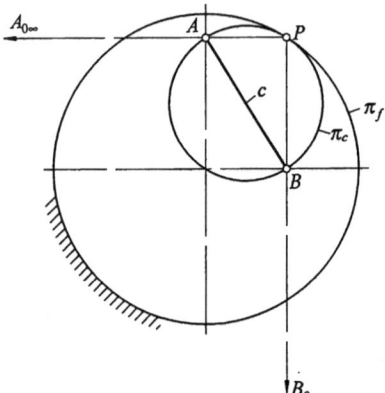

Fig. 3.9. The two polodes of a double-slider crank mechanism.

or

$$\pi_c: (x'^2 + y'^2 - cx')^4 = 0$$

This means that, in Fig. 3.9, π_c is a four-fold moving polode, being a circle with c as its diameter.

3.3 Guiding a body through three finitely separated positions --- geometrical method

3.3.1 General case

Suppose three positions c_1, c_2, c_3 of a body c are given. The problem is, how to synthesize a four-bar linkage to guide the body c through these three positions? This

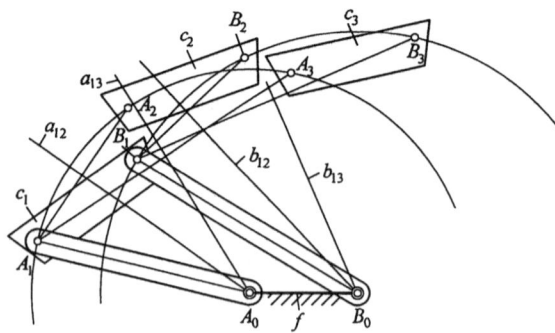

Fig. 3.10. Guiding a body through three finitely separated positions.

Dimensional synthesis of four-bar linkages --- body guidance problems 63

problem can be solved by elementary geometry. As shown in Fig. 3.10, let a point A_1 be chosen on c_1. The positions A_2 and A_3 are thus correspondingly determined. Such three points A_1, A_2, A_3 are called *homologous points*, or *correlated points*. Erect the perpendicular bisector a_{12} of $\overline{A_1 A_2}$ and the perpendicular bisector a_{13} of $\overline{A_1 A_3}$. The intersection of a_{12} and a_{13} is A_0. Similarly, choose arbitrarily another point B_1 on c_1. The points B_2 and B_3 are also determined. From the three homologous points B_1, B_2, B_3 the point B_0 can be found as the intersection of the perpendicular bisector b_{12} of $\overline{B_1 B_2}$ and the perpendicular bisector b_{13} of $\overline{B_1 B_3}$. The four-bar linkage $A_0 A_1 B_1 B_0$ is the required one. The points $A_0, B_0, \ldots$ will be called hereafter *centre-points*. These points are on the fixed body. The points $A_1, B_1, \ldots$ will be called *circle-points*. They are on the moving body which is in its position c_1.

Since any point on c_1 can be chosen as A_1, there are altogether ∞^2 ways of choosing A_1. Similarly, any point on c_1 can be chosen as B_1; there are also ∞^2 ways of choosing B_1. Therefore there are ∞^4 solutions to this problem.

With regard to this problem, there are some geometrical concepts which have to be introduced, because these are essential in solving specific problems. Fig. 3.11(a)

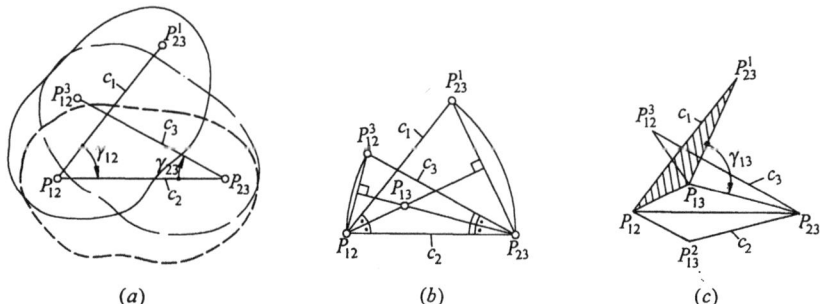

Fig. 3.11. The three poles P_{12}, P_{23}, P_{13} and poletriangle for three finitely separated positions c_1, c_2, c_3 of a body c.

shows three given positions c_1, c_2, c_3 of a certain body c. We can find as before the pole P_{12} between c_1, c_2, and the pole P_{23} between c_2, c_3. The angle γ_{12} is the angle of rotation from c_1 to c_2, and γ_{23} is the angle of rotation from c_2 to c_3. All counterclockwise angles are considered as positive, hence the angles γ_{12}, γ_{23} in Fig. 3.11 are negative. It is now required to find the pole P_{13} between c_1 and c_3. Join first the line $P_{12} P_{23}$, and this line represents c_2, because obviously both P_{12} and P_{23} are points on c_2. Rotate $P_{12} P_{23}$ back with P_{12} as centre through an angle

$-\gamma_{12}$ to the position c_1. P_{23} will then be moved to P_{23}^1. Next rotate $P_{23}P_{12}$ with P_{23} as centre through an angle γ_{23} to the position of c_3; P_{12} will then be moved to P_{12}^3. For clarity, the points in Fig. 3.11(a) are redrawn as in Fig. 3.11(b). In this figure, considering P_{23}^1 and P_{23} as positions of P_{23} on c_1 and c_3 respectively, erect the perpendicular bisector of $\overline{P_{23}^1 P_{23}}$. Next, consider P_{12}, P_{12}^3 as positions of P_{12} on c_1, c_3 respectively. Erect the perpendicular of $\overline{P_{12} P_{12}^3}$. The intersection of these two perpendicular bisectors is P_{13}. The triangle $\triangle P_{12}P_{23}P_{13}$ is called the *pole triangle*. A comparison between Figs. 3.11(a) and (b) shows that (please refer to Fig. 3.12):

The apex angle of $\triangle P_{12}P_{23}P_{13}$ at P_{12} is $\gamma_{12}/2$, represented by $\alpha = |\gamma_{12}/2|$;
the apex angle of $\triangle P_{12}P_{23}P_{13}$ at P_{23} is $\gamma_{23}/2$, represented by $\beta = |\gamma_{23}/2|$;
the supplementary angle of the apex angle of $\triangle P_{12}P_{23}P_{13}$ at P_{13} is $\gamma_{13}/2$, represented by $\eta = 180° - |\gamma_{13}/2|$.

It is obvious from Fig. 3.12 that

$$\gamma_{12}/2 + \gamma_{23}/2 = \gamma_{13}/2$$

Fig. 3.11(c) is redrawn by taking points from Fig. 3.11(b). It can be seen that, c_1 is represented by $\triangle P_{12}P_{13}P_{23}^1$, c_2 is represented by $\triangle P_{12}P_{13}^2 P_{23}$, and c_3 is represented by $\triangle P_{12}^3 P_{13}P_{23}$. The angle of rotation from c_1 to c_3 is, of course, γ_{13}. Furthermore, it can be seen from the figure the following relations:

$\triangle P_{12}P_{23}^1 P_{13}$ is the mirror image of $\triangle P_{12}P_{23}P_{13}$ with respect to side $P_{12}P_{13}$, belonging to c_1;

$\triangle P_{12}P_{23}P_{13}^2$ is the mirror image of $\triangle P_{12}P_{23}P_{13}$ with respect to side $P_{12}P_{23}$, belonging to c_2;

$\triangle P_{12}^3 P_{23}P_{13}$ is the mirror image of $\triangle P_{12}P_{23}P_{13}$ with respect to side $P_{13}P_{23}$, belonging to c_3.

In Fig. 3.12, let a poletriangle $\triangle P_{12}P_{23}P_{13}$ be given. The position A_1 of a circle point A on the moving body in its first position is also given. Let it be required to find the homologous points A_2, A_3 of A_1. Joining $P_{12}A_1$, and rotating this line about P_{12} through an angle γ_{12} we get the point A_2. Since $\angle P_{13}P_{12}P_{23} = \gamma_{12}/2$, and $\angle A_1 P_{12} A_2 = \gamma_{12}$, we do not have to join $P_{12}A_1$, but simply fold A_1 with respect to $P_{12}P_{13}$ to get its mirror image A_g, and further fold A_g with respect to $P_{12}P_{23}$ to get A_2. Similarly, we can get A_3. The point A_g is called the *cardinal point*. We have the following geometrical relations:

Folding A_g with respect to $P_{12}P_{13}$ to the opposite side, we will get A_1.

Dimensional synthesis of four-bar linkages --- body guidance problems

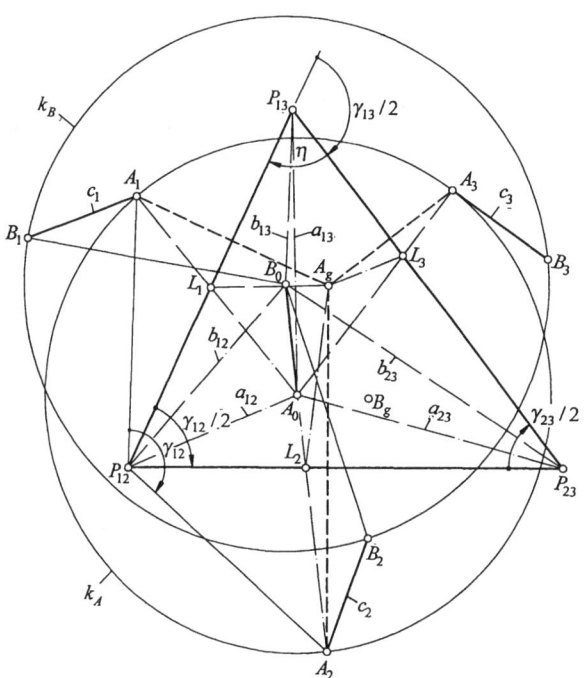

Fig. 3.12. Poletriangle, cardinal point, circle-point and centre-point.

Folding A_g with respect to $P_{12}P_{23}$ to the opposite side, we will get A_2.

Folding A_g with respect to $P_{13}P_{23}$ to the opposite side, we will get A_3.

It is to be noted that, the common digit belonging to two poles of a certain side (e.g. "2" of $P_{12}P_{23}$) signifies that specific homologous point (e.g. A_2).

The centre of the circle k_A passing through the three points A_1, A_2, A_3 is the centre point A_0 corresponding to A_1. Similarly for another point B_1, its homologous points B_2, B_3 can be found in the same way, and the centre B_0 of the circle k_B passing through the three points B_1, B_2, B_3 can also be found. We have

$$\angle A_1P_{12}A_0 = \angle B_1P_{12}B_0 = \gamma_{12}/2$$

Therefore
$$\angle A_0P_{12}B_0 = \angle A_1P_{12}B_1$$

just as stated in theorem 7. Similar relations apply to other poles. For instance,

$$\angle A_0P_{13}B_0 = \angle A_1P_{13}B_1 = \angle A_3P_{13}B_3$$

etc. Hence we have

$$\measuredangle A_1 P_{12} A_0 = \gamma_{12}/2 = \measuredangle B_1 P_{12} B_0$$

$$\measuredangle A_1 P_{12} B_1 = \measuredangle A_0 P_{12} B_0$$

This is the same as the statement of Theorem 7.

Example: This example is taken from (Beyer, 1953, Figs. 28,29). Fig. 3.13(a) shows a thread feeding mechanism of a sewing machine. $A_0 A$ is the crank, and the

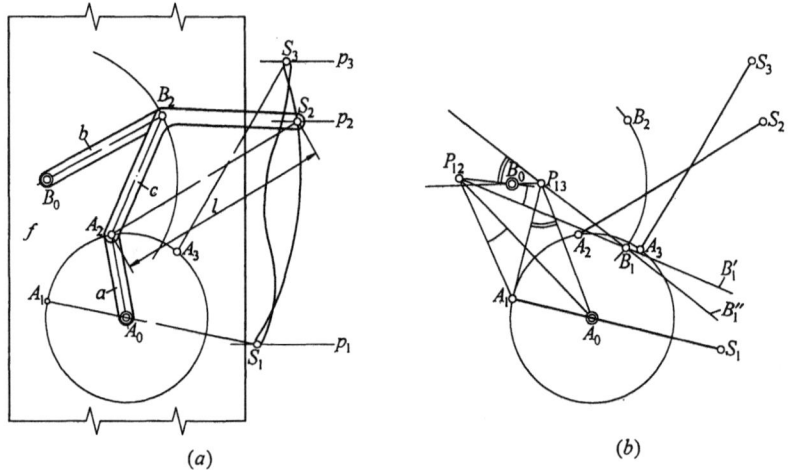

Fig. 3.13. (a) (b) Example: thread feeding mechanism of a sewing machine.

point A_0 and the three homologous points A_1, A_2, A_3 of point A are given. S is the thread feeding hole. It is expected that while A is at A_1, A_2, A_3, the height level of S should be at p_1, p_2, p_3 respectively. Choose first a proper length $\overline{AS} = l$, and draw arcs with A_1, A_2, A_3 as centres and l as the radius, cutting the lines p_1, p_2, p_3 in the points S_1, S_2, S_3 respectively. The question is: With the three positions $A_1 S_1, A_2 S_2, A_3 S_3$ of the body c ($= AS$) given, how can we find B_1 for a prescribed B_0.

Solution: As shown in Fig. 3.13(b), locate first the two poles P_{12} and P_{13} of AE. Make $\measuredangle B_0 P_{12} B_1' = \measuredangle A_0 P_{12} A_1$ according to theorem 7, then the point B_1 should lie on the line B_1'. Next make $\measuredangle B_0 P_{13} B_1'' = \measuredangle A_0 P_{12} A_1$, then B_1 should also lie on the line B_1''. Hence B_1 is the intersection of the two lines B_1' and B_1''. It can be seen from the figure that $\measuredangle B_0 P_{13} B_1 = 180° - \measuredangle A_0 P_{13} A_1$, because the sense of reading $A_0 B_0$ about P_{13} is opposite to that of reading $A_1 B_1$.

Please note that the synthesized four-bar linkage $A_0 A_1 B_1 B_0$ should be a

Dimensional synthesis of four-bar linkages --- body guidance problems   67

crank-rocker, or, that it must satisfy the Grashof inequality which is, in the present case, $a + f \leq b + c$.

The sequence of synthesis of this problem is: $1 \rightarrow A_1S_1, A_2S_2, A_3S_3 \rightarrow P_{12}, P_{13} \rightarrow B_1', B_1'' \rightarrow B_1$

Example: This example is also taken from (Beyer, 1953, Figs. 30, 31). In Fig. 3.14(a), a slider-crank mechanism A_0AC is given. The three positions C_1, C_2, C_3 of the point C corresponding respectively to A_1, A_2, A_3 of the point A are precisely on a straight line g. Hence the three positions A_1C_1, A_2C_2, A_3C_3 of the coupler c are known. It is required to find the corresponding coupler point B_1 for a prescribed B_0.

The solution is similar to that of the preceding example. Locate first the poles P_{12} and P_{13}. Make $\sphericalangle B_0 P_{12} B_1' = \sphericalangle A_0 P_{12} A_1$ and $\sphericalangle B_0 P_{13} B_1'' = \sphericalangle A_0 P_{13} A_1$. The intersection of the two lines B_1' and B_1'' is B_1. If the original slider-crank A_0AC is replaced by the four-bar linkage A_0ABB_0, the path of the point C will be the γ_C as shown in the figure. Of course the three positions C_1, C_2, C_3 of C lie precisely on a straight line.

In Fig. 3.14(b), the original slider-crank is retained, but use is made of the point B_1 just found by joining it to the point B_0 by a link e. As the slider-crank moves, the path of B is γ_B, as shown in the figure. The three positions B_1, B_2, B_3 of the point B lie of course precisely on the circle k_B drawn with B_0 as its centre and $\overline{B_0B_1}$ as its radius. Further connect B_0 to a fixed joint B_{00} by a link h, then as B moves through the positions of B_1, B_2, B_3, link h remains nearly standstill. This is a coupler dwell mechanism. But as B moves to the lower part of γ_B, link h will swing through an angle ε. In Fig. 3.14(b), the point P is the instantaneous velocity pole of the coupler c of the slider-crank mechanism in its position A_1C_1. It can be seen from the figure that B_0 does not lie on the line PB_1. However, the closer the point B_0 is to PB_1, the better is the osculation of the path γ_B to the circle k_B.

In Fig. 3.12, joining the radii A_0A_1, A_0A_2, A_0A_3 cuts the three sides of the poletriangle $\Delta P_{12}P_{23}P_{13}$ in the three points L_1, L_2, L_3 respectively. Further join the lines A_gL_1, A_gL_2, A_gL_3. Since $\overline{A_gL_1} = \overline{A_1L_1}$, we have

$$\overline{A_0L_1} + \overline{A_gL_1} = \overline{A_0L_2} + \overline{A_gL_2} = \overline{A_0L_3} + \overline{A_gL_3} = \overline{A_0A_1}$$

Hence the three points L_1, L_2, L_3 lie on an ellipse with A_0, A_g as its two foci. Furthermore, since

$$\sphericalangle A_1L_1P_{13} = \sphericalangle A_gL_1P_{13} = \sphericalangle A_0L_1P_{12}$$

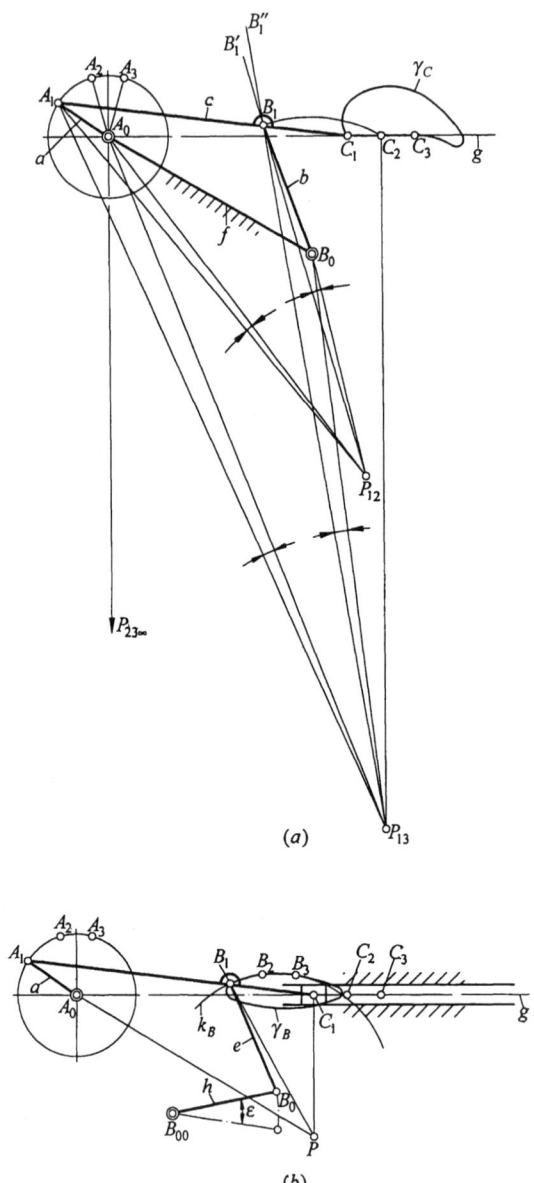

Fig. 3.14. Example: (a) Substitution of slider-crank A_0AC by a four-bar linkage A_0ABB_0. (b) Coupler dwell mechanism.

Dimensional synthesis of four-bar linkages --- body guidance problems

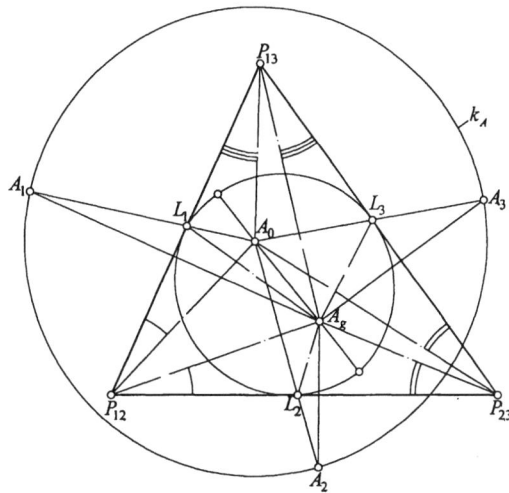

Fig. 3.15. Inscribed ellipse of a poletriangle.

$P_{12}P_{13}$ is the tangent of this ellipse at L_1, as shown in Fig. 3.15. Similarly, $P_{12}P_{23}, P_{13}P_{23}$ are respectively the tangents of this ellipse at L_2 and L_3. In other words, the poletriangle $\triangle P_{12}P_{23}P_{13}$ is the cirscribing triangle of the ellipse. It is clear from the characteristics of the ellipse that:

The rays originating from a pole (P_{12}, P_{23} or P_{13}) to A_0 and A_g build with the corresponding sides of the poletriangle at the apex (P_{12}, P_{23} or P_{13}) equal angles.

According to these relations, if the poletriangle is known, A_g can be found for a given A_0. We have thus proved again theorem 7. It can also be proved that (please refer to Fig. 3.12):

A_0B_0 and A_gB_g subtend equal angles at a pole (P_{12}, P_{23} or P_{13}), but in opposite senses.

3.3.2 Three homologous points on a straight line

We can easily find a four-bar linkage to guide a body through three given positions by employing the method mentioned in Section 3.3.1. If however, the problem is to find a point on the body so that its three homologous points lie on a straight line, then trial-and-error approach would no longer be adequate. A definite method becomes necessary. In Fig. 3.16, let $\overline{P_{12}P_{23}} = s$, $\overline{P_{12}A_0} = r_0$, $\sphericalangle P_{23}P_{12}A_0 = \theta$, and $v = \pi - \theta$, $\sphericalangle P_{12}P_{23}A_0 = \sigma$, $\alpha = \sphericalangle P_{23}P_{12}P_{13} = |\gamma_{12}/2|$, $\beta = \sphericalangle P_{12}P_{23}P_{13} = |\gamma_{23}/2|$. (Please note that, in Fig. 3.16, both γ_{12} and γ_{23} are negative.) Applying sine

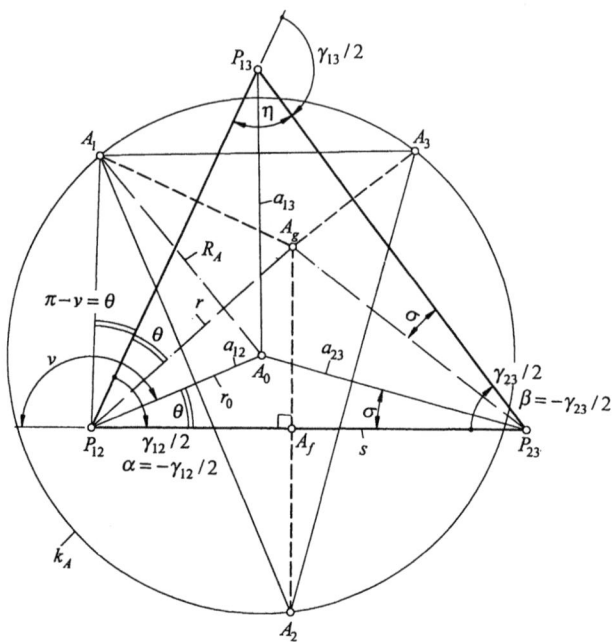

Fig. 3.16. Geometrical relations of A_0, A_g with respect to the poletriangle.

law to $\Delta P_{12}P_{23}A_0$ gives

$$\frac{s}{r_0} = \frac{\sin(v - \sigma)}{\sin \sigma} \tag{3.14}$$

Let further $\overline{P_{12}A_g} = r$, $\overline{A_gA_2} \times \overline{P_{12}P_{23}} = A_f$. Since $\sphericalangle A_gP_{12}P_{13} = \sphericalangle P_{23}P_{12}A_0 = \theta = \pi - v$, thus $\overline{A_fA_g} = r\sin[\alpha - (\pi - v)]$. From the right triangle $\Delta P_{23}A_fA_g$ we have

$$\tan(\beta - \sigma) = \frac{\overline{A_fA_g}}{\overline{P_{23}A_f}} = \frac{r\sin(\alpha - \pi + v)}{s - r\cos(\alpha - \pi + v)} \tag{3.15}$$

Eliminating σ from equations (3.14) and (3.15) yields

$$\frac{s}{r_0} = \frac{s\sin(v - \beta) - r\sin\eta}{s\sin\beta + r\sin(\eta - v)} \tag{3.16}$$

The angle $\eta = \pi + \gamma_{13}/2$ (where γ_{13} is negative) in equation (3.16) is the apex angle of the poletriangle at P_{13}.

In case the three points A_1, A_2, A_3 lie on a straight line, the circle k_A becomes a

Dimensional synthesis of four-bar linkages --- body guidance problems 71

straight line, and A_0 goes to infinity and r_0 approaches to infinity. So let r_0 in equation (3.16) become infinity, or

$$s \sin(v - \beta) - r \sin \eta = 0$$

or

$$r = \frac{s}{\sin \eta} \sin(v - \beta) = \frac{\overline{P_{23}P_{13}}}{\sin \beta} \sin(v - \beta) \qquad (3.17)$$

In Fig. 3.16, $\theta = \sphericalangle P_{13}P_{12}A_1 = \sphericalangle P_{13}P_{12}A_g = \sphericalangle P_{23}P_{12}A_0 = \pi - v$, therefore $v = \pi - \theta$. Equation (3.17) can be written as

$$r = \frac{\overline{P_{23}P_{13}}}{\sin \beta} \sin(\beta + \theta) \qquad (3.18)$$

Fig. 3.17 shows a poletriangle $\Delta P_{12}P_{23}P_{13}$ with its three poles P_{12}, P_{23}, P_{13} folded with respect to their respective opposite sides to get their mirror images $P_{12}^3, P_{23}^1, P_{13}^2$. Denote the circumscribed circle of $\Delta P_{12}P_{23}P_{13}$ by k, and those of $\Delta P_{12}P_{13}P_{23}^1$, $\Delta P_{12}P_{23}P_{13}^2$, $\Delta P_{13}P_{23}P_{12}^3$ by k_1, k_2, k_3 respectively, these being the mirror images of the circumscribed circle k of the poletriangle with respect to the three sides. As according to Fig. 3.16, $r = \overline{P_{12}A_g} = \overline{P_{12}A_1}$, therefore equation (3.18)

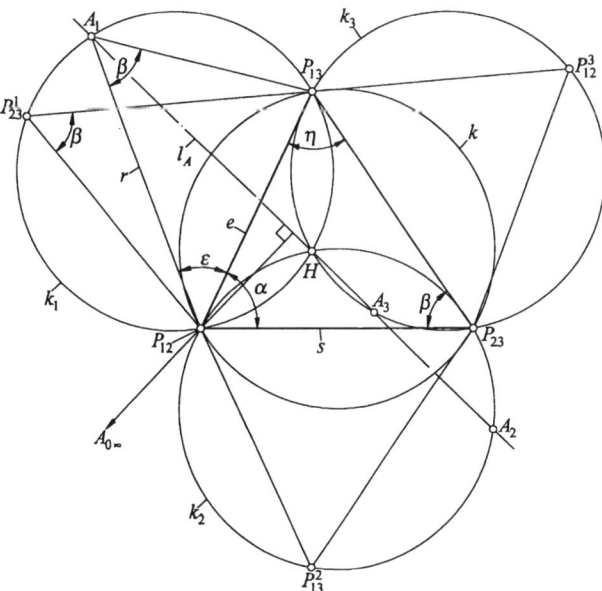

Fig. 3.17.   Three homologous points A_1, A_2, A_3 on a straight line l_A.

implies that in case A_1, A_2, A_3 are on a straight line, the circle k_1 is the locus of A_1. The common line of A_1, A_2, A_3 is denotd by l_A.

Similarly it can be proved that A_2 is on the circle k_2, and A_3 is on k_3. Hence we have:

Theorem 8: If three homologous point A_1, A_2, A_3 lie on a straight line, then they are respectively on the mirror images k_1, k_2, k_3 of the circumsribed circle k of the poletriangle, and the line l_A joining the three points A_1, A_2, A_3 passes through the orthocentre H of the poletriangle $\Delta P_{12}P_{23}P_{13}$.

It is not difficult to prove the latter part of theorem 8. We reserve it as Exercise 3.4.

Example: This example is taken from (Hirschhorn, 1962, Fig. 11-18). Suppose the three positions $c_1 = A_1B_1$, $c_2 = A_2B_2$, $c_3 = A_3B_3$ of the coupler c of a four-bar linkage A_0ABB_0 are given, as shown in Fig. 3.18. It is required to find a point C on c, whose three homologous points C_1, C_2, C_3 lie on a straight line.

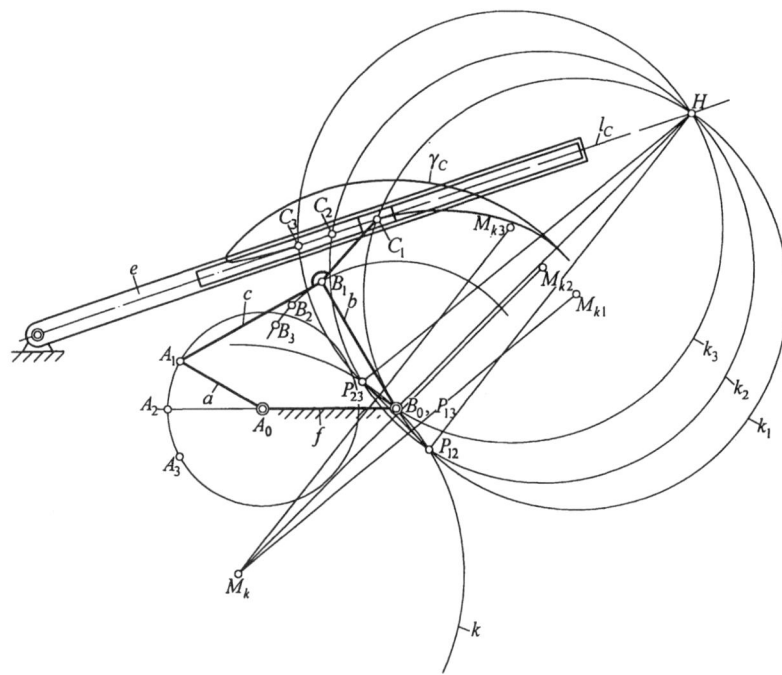

Fig. 3.18. Example: three homologous points C_1, C_2, C_3 on a line l_C.

Solution: The solution procedure is as follows. Since A_1B_1, A_2B_2, A_3B_3 are known, the poletriangle $\Delta P_{12}P_{23}P_{13}$ and its orthocentre H can be found. Draw the circumscribed circle k and its three mirror circles k_1, k_2, k_3 of the poletriangle $\Delta P_{12}P_{23}P_{13}$. Draw any line l_C passing through H intersecting each of k_1, k_2, k_3 in one

Dimensional synthesis of four-bar linkages --- body guidance problems 73

point. These are the required C_1, C_2, C_3.

The sequence of synthesis for this problem is: $\Delta P_{12}P_{23}P_{13} \to H \to k \to k_1, k_2, k_3 \to l_C \to C_1, C_2, C_3$.

During the motion of the four-bar linkage, as the coupler c passes through its three positions c_1, c_2, c_3, the path γ_C of C passes through C_1, C_2, C_3, being an approximate straight line (the three points C_1, C_2, C_3 lying precisely on a straight line). Hence a dwell mechanism can be constructed, in which the link e, during the motion of point A passing through A_1, A_2, A_3, remains approximately standstill.

3.3.3 Three homologous lines passing through a fixed point

For three given positions of a body, the question to be asked is: which line on the body the three homologous (or correlated) positions of which pass through a point on the fixed body? This question is the *dual* of the one in Section 3.3.2. We first present the answer and then demonstrate why it is so: A poletriangle $\Delta P_{12}P_{23}P_{13}$ and its circumscribed circle k are given as shown in Fig. 3.19. All three mirror images k_1, k_2, k_3 of k pass through the orthocentre H of $\Delta P_{12}P_{23}P_{13}$. (The proof of this statement is left as Exercise 3.5, but let us accept it for the time being.) Hence

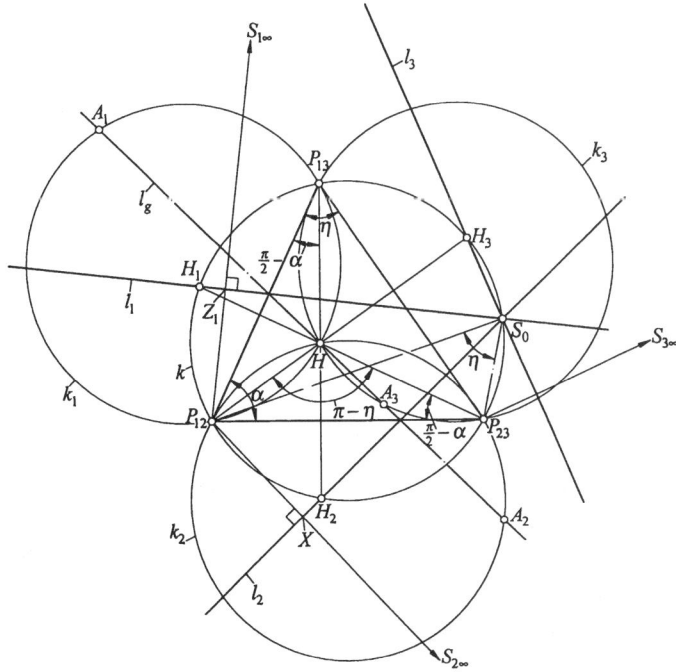

Fig. 3.19. Three homologous lines l_1, l_2, l_3 passing through a fixed point S_0.

the three mirror images H_1, H_2, H_3 of the *cardinal point H* lie on the circle k. In order to find H_1, H_2, H_3, just drop perpendiculars from the three apexes to the corresponding opposite sides, and their intersections with k are the required points. Choose any point S_0 on the circle k, and join S_0H_1, S_0H_2, S_0H_3. These are the three *homologous* (or *correlated*) lines.

To prove this, draw from P_{12}

$$P_{12}Z_1 \perp S_0H_1$$

$$P_{12}X \perp S_0H_2$$

Since the arc lengths $\overset{\frown}{P_{12}H_1} = \overset{\frown}{P_{12}H_2}$, we have $\sphericalangle H_1S_0P_{12} = \sphericalangle H_2S_0P_{12} = \pi/2 - \alpha$. Consequently

$$\overline{P_{12}Z_1} = \overline{P_{12}X}$$

thus $\qquad \sphericalangle Z_1P_{12}X = 2\alpha = |\gamma_{12}|$

This proves that X is the second homologous point of Z_1, or $X = Z_2$; in other words, S_0H_2 is the second *homologous* (or *correlated*) line of S_0H_1. Similarly it can be proved that S_0H_3 is the third homologous line of S_0H_1. So that the three lines S_0H_1, S_0H_2, S_0H_3 can be renamed as l_1, l_2, l_3. However, it should be noted that, as l_1 moves to l_2, the point on l_2 coinciding with S_0 is not the second homologous point of the point on l_1 coinciding with S_0, and as l_1 moves to l_3, the point on l_3 coinciding with S_0 is also not the third homologous point of the point on l_1 coinciding with S_0. We have the following theorem:

Theorem 9: The three lines l_1, l_2, l_3 passing through any point on the circumcircle k of the poletriangle and the points H_1, H_2, H_3 are homologous.

Example: This example is also taken from (Hirschhorn, 1962, Fig. 11-19). In Fig. 3.20, the three positions A_1S_0, A_2S_0, A_3S_0 of a line AS_0 on the coupler c of a given swinging block linkage pass through a common point S_0. The point S_0 is identical with B_0, being the centre of rotation of the block b. The poletriangle $\Delta P_{12}P_{23}P_{13}$ and its circumcircle k can be found as before. The points A_2, A_3 are symmetrically disposed with respect to A_0B_0, hence P_{23} is identical with B_0, and the centre M_k of the circle k lies on A_0B_0. Since the three lines A_1S_0, A_2S_0, A_3S_0 pass through a common point, the intersections of these three lines with the circle k are H_1, H_2, H_3. Choose another point S_0' on the circle k, and join $S_0'H_1, S_0'H_2, S_0'H_3$. This is another set of three homologous lines on the body c passing through the fixed point S_0'.

The sequence of synthesis of this problem is: $\Delta P_{12}P_{23}P_{13} \rightarrow k \rightarrow H_1, H_2, H_3 \rightarrow S_0'$

Dimensional synthesis of four-bar linkages --- body guidance problems 75

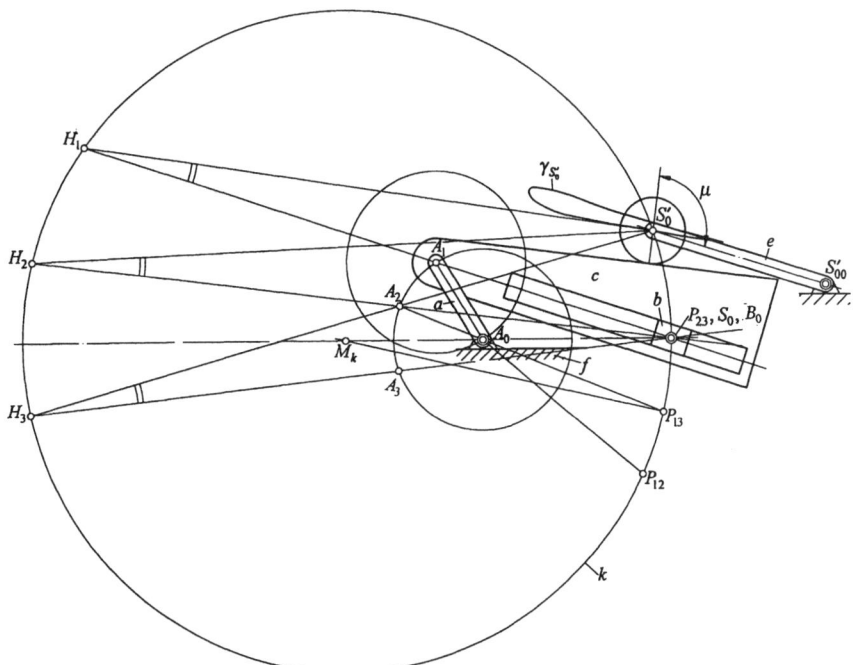

Fig. 3.20. Example: three homologous lines l_1, l_2, l_3 passing through a fixed point S_0.

→ $S'_0H_1, S'_0H_2, S'_0H_3$.

It can be seen from the figure that the circular arc on circle k subtending at the three points H_1, H_2, H_3 is $\widehat{S_0 S'_0}$, which means that the circumferential angles at H_1, H_2, H_3 are equal. If the two lines $S_0 H_1$ and $S'_0 H_1$ were rigid rods fixed on and moving together with body c, then as the point A moves to the positions A_1, A_2, A_3, these two lines would pass through the two points S_0, S'_0 precisely. A dwell mechanism can therefore be constructed as that shown in the figure. A roller is made with S'_0 as its centre, and a flat surface is made on the body c which is parallel to and at a normal distance equal to the radius of the roller from $H_1 S'_0$ in the position shown. Choose any point S'_{00} as the fixed centre of rotation of link e to connect S'_0 and S'_{00}. As A passes through A_1, A_2, A_3, link e remains approximately standstill.

Draw a line passing through S'_0 perpendicular to $H_1 S'_0$. The angle μ between this perpendicular and $S'_0 S'_{00}$ is the *transmission angle* (please refer to the later Section 7.5). It is preferable to have the angle μ as close to 90° as possible. If the resulting value of μ is not satisfactory, we can choose another point for S_0'.

In doing an inversion of this mechanism by fixing c, the points A_0, B_0, S_0' are all points on f moving together with f. Point A_0 makes a circular motion about A_1, and B_0 makes a rectilinear motion along H_1A_1. A portion of the path $\gamma_{S_0'}$ of the point S_0' nearly coincides with the straight line along H_1S_0'.

A substituting mechanism may be derived from the above example. In Fig. 3.21,

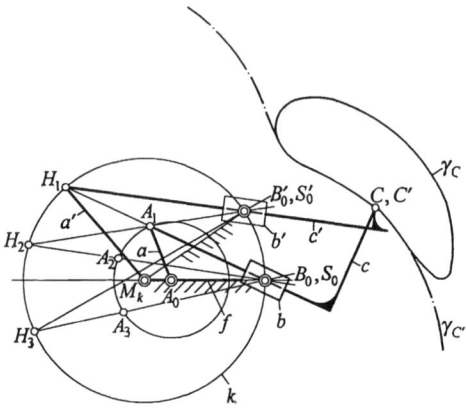

Fig. 3.21. Substitution of $acbf$ by the mechanism $a'c'b'f$.

a swinging block linkage $acbf$ is given. The path γ_C of a coupler point C is as shown. Suppose it is required for some design reason to change the position of the bearing from $B_0 (= S_0)$ to another place. We can choose any point M_k on A_0B_0, and draw a circle k with $\overline{M_kB_0}$ as radius. Choose any point S_0' on circle k as the position B_0' of a new bearing. Construct as shown a new swinging block linkage $a'c'b'f$ (note that the lengths $a' = f$). Now take a point C' which is coinciding with the original C as a coupler point of the new coupler c'. The path of C' is $\gamma_{C'}$. With this new mechanism, as the crank M_kH moves from the position M_kH_1 to M_kH_3, γ_C and $\gamma_{C'}$ osculate each other closely. This means that the old swinging block can be replaced by the new swinging block. However, please note that the new swinging block cannot make complete rotations, the link M_kH being able to swing only within a certain range.

With regard to osculating curves please refer to the later Sections 6.9.4 and 6.12.

From the geometrical relations in Fig. 3.19, we can derive the following interesting results: Folding l_1 with respect to $P_{12}P_{13}$ we get l_g. Obviously l_g

passes through H, but the point S_0 on l_1 is folded to A_1. Similarly, folding l_2, l_3 with respect to $P_{12}P_{23}$ and $P_{13}P_{23}$ respectively, we will also get l_g, but the point S_0 on l_2, l_3 are folded to A_2 and A_3 respectively. Therefore l_g may be considered as the *cardinal line* of l_1, l_2, l_3, and S_0 is the cardinal point of the three homologous points A_1, A_2, A_3.

A comparison between Figs. 3.19 and 3.17 shows that, if the points A_1, A_2, A_3 in both figures are identical, the line l_A in Fig. 3.17 is exactly the line l_g in Fig. 3.19. Consequently we have the following theorem:

Theorem 10: If l_A is the common line of three homologous points A_1, A_2, A_3, then l_A is also the cardinal line of three homologous lines l_1, l_2, l_3 that pass through a common point S_0, and S_0 lies on the circumcircle k of the poletriangle, being at the same time the cardinal point A_g of A_1, A_2, A_3.

A further geometrical relationship can be derived from Fig. 3.19, namely, the locus of the centre point, as the corresponding circle point goes to infinity. This can be found by setting in equation (3.16) $r \to \infty$. Then equation (3.16) becomes

$$\frac{s}{r_0} = -\frac{\sin \eta}{\sin(\eta - v)}$$

or, written in the form

$$r_0 = \frac{s \sin(v - \eta)}{\sin \eta} \quad (3.19)$$

What equation (3.19) expresses is that the locus of the centre point (represented by r_0) is the circle k. (Please do not confuse the symbols A_1, A_2, A_3 in both Figs. 3.16 and 3.19.) Another geometrical relationship that can be derived is the directions in which the homologous points $S_{1\infty}, S_{2\infty}, S_{3\infty}$ lie which correspond to the centre point S_0. This can be found by applying theorem 7, or by observing from Fig. 3.16 that $\sphericalangle A_0 P_{12} A_1 = \sphericalangle S_0 P_{12} S_1 = \alpha = |\gamma_{12}/2|$. Hence in Fig. 3.19, $S_{1\infty}$ lies at infinity in the direction of $P_{12} Z_1$, or

$S_{1\infty}$ lies at infinity in the direction normal to l_1.

Similarly we have

$S_{2\infty}$ lies at infinity in the direction normal to l_2;

$S_{3\infty}$ lies at infinity in the direction normal to l_3.

Consequently we have

Theorem 11: The three homologous circle-points $S_{1\infty}, S_{2\infty}, S_{3\infty}$ that correspond to the centre point S_0 on the circumcircle k of the poletriangle lie at infinity in the directions normal respectively to $S_0 H_1, S_0 H_2, S_0 H_3$.

3.3.4 The R_M- and R^1- curves

From the geometrical relationships mentioned in foregoing sections, we know that for a given poletriangle, if a point A_1 is prescribed, the points A_g, A_2, A_3 can be found, and the centre of the circle passing through the three points A_1, A_2, A_3 is A_0. Conversely, if A_0 is prescribed, A_g and hence A_1, A_2, A_3 can also be found (see Exercise 3.3). We now deal with the problem where both A_g and A_0 are not prescribed, and only the radius R_A of circle k_A is prescribed. We want to find the loci of A_0 and A_1. This problem appeared first in (Alt, 1921). We shall derive the equations of these two loci, but by using a procedure somewhat different from that of Alt. In Fig. 3.22, let $\sphericalangle P_{23}P_{12}A_0 = \theta$, $\sphericalangle P_{12}P_{23}A_0 = \sigma$, $p_0 = \overline{P_{12}A_0}$, $p = \overline{P_{12}A_2}$. Comparing with Fig. 3.16, we have

$$\sphericalangle A_2 P_{12} A_0 = |\gamma_{12}/2| = \alpha$$
$$\sphericalangle A_2 P_{23} A_0 = |\gamma_{23}/2| = \beta$$

(In Fig. 3.16, both γ_{12} and γ_{23} are clockwise, being negative.) Applying sine law to $\Delta P_{12}P_{23}A_0$ and $\Delta P_{12}P_{23}A_2$, we get

$$\frac{s}{\sin(\theta+\sigma)} = \frac{p_0}{\sin\sigma} \qquad (3.20)$$

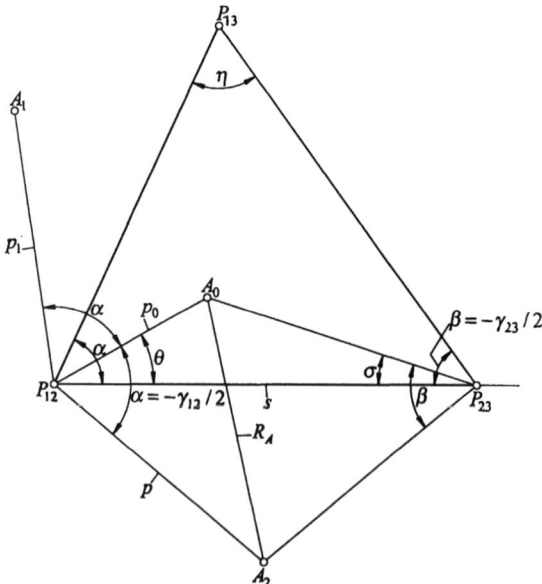

Fig. 3.22. Relation of coordinates of A_0, A_1.

$$\frac{s}{\sin(\alpha - \theta + \beta - \sigma)} = \frac{p}{\sin(\beta - \sigma)} \tag{3.21}$$

Eliminating σ from equations (3.20) and (3.21) and rearranging, we get

$$\frac{E}{p} + \frac{F}{p_0} + \frac{K}{pp_0} + G = 0 \tag{3.22}$$

where

$$\left.\begin{array}{l} E = \sin(\beta + \theta) \\ F = \sin(\alpha + \beta - \theta) \\ K = -s\sin\beta \\ G = -\sin(\alpha + \beta)/s \end{array}\right\} \tag{3.23}$$

E and F are functions of θ, while K, G are constants. Equation (3.22) can be solved for p or p_0, as

$$p = -\frac{K + Ep_0}{F + Gp_0} \tag{3.24}$$

or

$$p_0 = -\frac{K + Fp}{E + Gp} \tag{3.25}$$

Applying law of cosine to $\Delta P_{12} A_2 A_0$ gives

$$R_A^2 = p^2 + p_0^2 - 2pp_0 \cos\alpha \tag{3.26}$$

Substituting equation (3.24) into equation (3.26), we get

$$R_M : m_4 p_0^4 + m_3 p_0^3 + m_2 p_0^2 + m_1 p_0 + m_0 = 0 \tag{3.27}$$

where

$$\begin{array}{l} m_4 = G^2 \\ m_3 = 2G(F + E\cos\alpha) \\ m_2 = -R_A^2 G^2 + E^2 + F^2 + 2(KG + EF)\cos\alpha \\ m_1 = 2(EK - R_A^2 FG + KF\cos\alpha) \\ m_0 = K^2 - F^2 R_A^2 \end{array}$$

Equation (3.27) is the polar coordinate equation of the R_M-curve, or the locus of the point A_0 for a prescribed R_A. For a given value of θ, both E and F can be calculated from equations (3.23). Substituting these into equation (3.27), we can find p_0, thus yielding the polar coordinates $A_0(p_0, \theta)$ of the centre point, with P_{12} as the origin.

In Fig. 3.22, the rectangular coordinates of $A_0(x_0, y_0)$ in the coordinate system with P_{12} as origin and $P_{12}P_{23}$ as the x-axis are:

$$x_0 = p_0 \cos\theta \\ y_0 = p_0 \sin\theta$$

Equation (3.27) can be rewritten as

$$R_M : (x_0^2 + y_0^2)[(x_0 - s)^2 + y_0^2]\left[\left(x_0 - s\frac{S_\beta}{S_\eta}C_\alpha\right)^2 + \left(y_0 - s\frac{S_\beta}{S_\eta}S_\alpha\right)^2\right]$$
$$= R_A^2[(x_0^2 + y_0^2) - sx_0 - sy_0 \cot\eta]^2 \tag{3.28}$$

The angle $\eta = 180° - (\alpha + \beta)$ in equation (3.28) is the apex angle of $\Delta P_{12}P_{23}P_{13}$ at P_{13}, S_β stands for $\sin\beta$, and C_α for $\cos\alpha$, etc. Equation (3.28) is identical with the R_M - equation given by Alt. It can be seen from this equation that the R_M - curve is a *tricircular sextic* (please refer to Section A1.2). In other words, this curve passes three times through each of the two circular points I and J at infinity, and with P_{12}, P_{23}, P_{13} as its three double points. As mentioned before, equation (3.27) is the polar coordinate equation of R_M. For a given value of θ, there are four values of p_0, because any line passing through the origin P_{12} cuts R_M in two points at P_{12} (i.e. $p_0 = 0,0$), altogether cutting R_M in six points.

Since the coupler curve of a four-bar linkage is also a tricircular sextic (please refer to the later Section 6.7), R_M could be generated by a coupler point of a four-bar linkage. We shall discuss this aspect in Section 6.7.

Example: Fig. 3.23 shows an R_M - curve with $s = \overline{P_{12}P_{23}} = 30$, $\alpha = |\gamma_{12}/2| = 34.8°$, $\beta = |\gamma_{23}/2| = 100.2°$, and $R_A = 36.6$.

Besides, substituting equation (3.25) into equation (3.26) gives

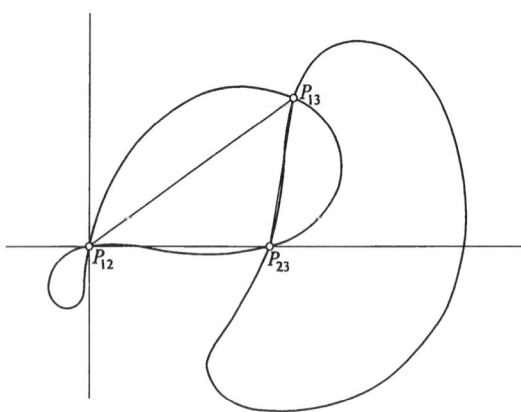

Fig. 3.23. Example: R_M-curve. $s = \overline{P_{12}P_{23}} = 30$, $\alpha = 34.8°$, $\beta = 100.2°$, $R_A = 36.6$.

Dimensional synthesis of four-bar linkages --- body guidance problems 81

$$R^1: b_4 p^4 + b_3 p^3 + b_2 p^2 + b_1 p + b_0 = 0 \qquad (3.29)$$

where

$$b_4 = G^2$$
$$b_3 = 2G(E + F\cos\alpha)$$
$$b_2 = -R_A^2 G^2 + E^2 + F^2 + 2(KG + EF)\cos\alpha$$
$$b_1 = 2(FK - R_A^2 EG + KE\cos\alpha)$$
$$b_0 = K^2 - E^2 R_A^2$$

Equation (3.29) is the polar coordinate equation of the R^1-curve, being the equation of the locus of A_1 for a prescribed R_A. For a given θ, the values of E and F can be found from equations (3.23), and by substituting these values into equation (3.29) we can obtain p, and hence get the polar coordinates $A_1(p, \theta + \alpha)$, or $(p, \theta - \gamma_{12}/2)$ with P_{12} as the origin, as can clearly be seen in Fig. 3.22. (Please note it is not (p, θ)). The reason that this curve is called the R^1- curve, is because the locus of such a point A_2 on c_2, is identical with the locus of such a point A_3 on c_3, and is identical with the locus of A_1 on c_1, or R^1. We just take R^1 as a representative for all three curves.

Similar to equation (3.27), equation (3.29) can also be transformed into an equation in rectangular coordinates (not shown here) like equation (3.28). Similar to the R_M- curve, R^1- curve is also a tricircular sextic, with P_{12}, P_{13}, P_{23}^1 as its three double points.

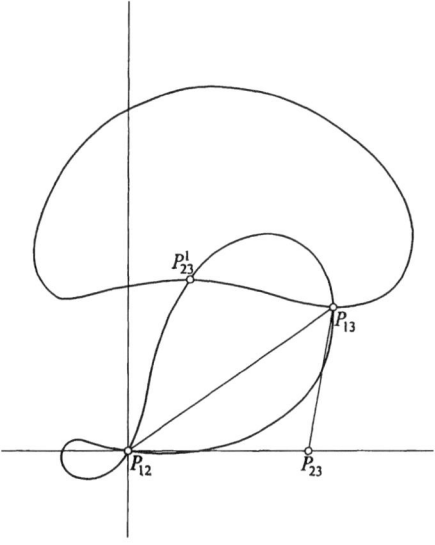

Fig. 3.24. Example: R^1-curve. $s = \overline{P_{12}P_{23}} = 30$, $\alpha = 34.8°$, $\beta = 100.2°$, $R_A = 36.6$.

Example: Fig. 3.24 shows an R^1- curve with $s = \overline{P_{12}P_{23}} = 30$, $\alpha = |\gamma_{12}/2| = 34.8°$, $\beta = |\gamma_{23}/2| = 100.2°$, $R_A = 36.6$.

It can be seen from Figs. 3.23 and 3.24 of the above two examples that R_M - and R^1 - curves are symmetrically disposed with respect to $P_{12}P_{13}$.

3.4 Guiding a body through three finitely separated positions --- algebraic method

The method described in the preceding section was a calculation method based on the geometrical relationships. In the present section, we shall explain an algebraic method without relating to a poletriangle. What we are going to introduce here, is a technique using matrix algebra. Mechanism synthesis using primarily matrix algebra appeared first in (Suh & Radcliffe, 1967). The method described here, however, is somewhat different from that method.

3.4.1 Angle of rotation of a body displacement

Fig. 3.25(a) shows a body c displacing from c_1 to c_j. The angle between two positions of any line on this body, is the angle of rotation of this body. Suppose the coordinates of two points E_1, F_1 on c_1 and those of the same two points E_j, F_j on c_j are known. The angle γ_{1j} between the two lines E_1F_1 and E_jF_j is the angle of rotation of the body c. Using scalar quantities exclusively, we have

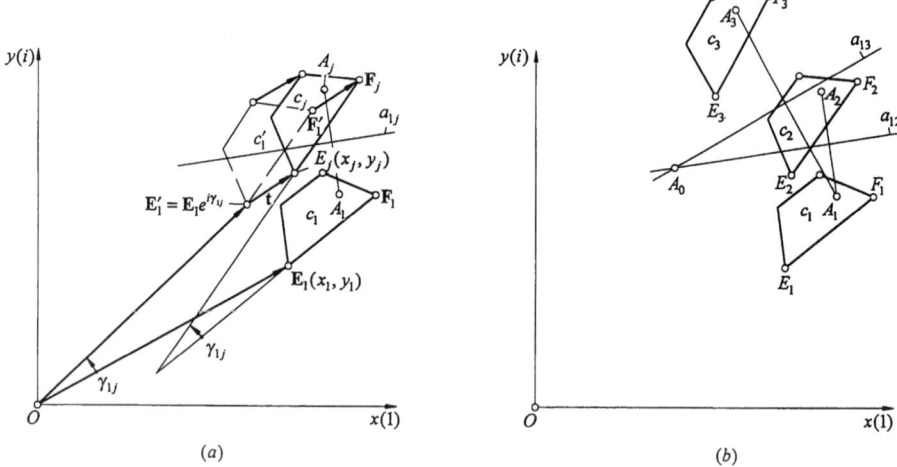

Fig. 3.25. (a) Displacement of a body from c_1 to c_j = rotation $(c_1 \to c_1')$ + translation $(c_1' \to c_j)$. (b) Intersection of a_{12}, a_{13} is A_0.

$$\gamma_{1j} = \tan^{-1}\frac{y_{Fj} - y_{Ej}}{x_{Fj} - x_{Ej}} - \tan^{-1}\frac{y_{F1} - y_{E1}}{x_{F1} - x_{E1}} \tag{3.30}$$

Let the position vector $\overrightarrow{OE_1}$ of E_1 with respect to a certain origin O be denoted by $\mathbf{E}_1$, and the x-axis of a certain coordinate system xy represent the real axis and the y-axis represent the imaginary axis ($i = \sqrt{-1}$). We have then

$$e^{i\gamma_{1j}} = \cos\gamma_{1j} + i\sin\gamma_{1j} = \frac{\mathbf{F}_j - \mathbf{E}_j}{\mathbf{F}_1 - \mathbf{E}_1} \tag{3.31}$$

In equation (3.31), $\mathbf{E}_1 = x_{E1} + iy_{E1}$, etc.

3.4.2 The displacement matrix

The displacement of a body c from c_1 to c_j, as that shown in Fig. 3.25(a), can be considered as a combination of a *rotation* and a *translation*. It does not matter to carry out rotation first and then translation, or translation first and then rotation. If rotation precedes translation, the centre of rotation can be chosen arbitrarily, but the direction and distance of translation vary, depending on the centre of rotation chosen.

In Fig.3.25(a), the origin is arbitrarily chosen, as mentioned before, and the coordinate system is also arbitrarily chosen. Only specifying the initial coordinates $E_1(x_1,y_1)$ and the final coordinates $E_j(x_j,y_j)$ of a point E on c, and the angle of rotation γ_{1j} of c, the displacement is completely defined. In other words, the displacement matrix is defined. Take as shown in the figure the point O as the centre of rotation and rotate c from c_1 through an angle γ_{1j} to a position c'_1 which is parallel to c_j. Counterclockwise sense of γ_{1j} is considered positive. Next translate c from c'_1 to c_j. The vector of translation is $\mathbf{t} = \overrightarrow{E'_1 E_j}$. The coordinates of $A_j(x_{Aj},y_{Aj})$ of a point $A_1(x_{A1},y_{A1})$ on c_1 after c has been displaced to c_j are:

$$\begin{bmatrix} x_{Aj} \\ y_{Aj} \\ 1 \end{bmatrix} = \begin{bmatrix} d_{11j} & d_{12j} & d_{13j} \\ d_{21j} & d_{22j} & d_{23j} \\ d_{31j} & d_{32j} & d_{33j} \end{bmatrix} \begin{bmatrix} x_{A1} \\ y_{A1} \\ 1 \end{bmatrix} \tag{3.32}$$

in which the displacement matrix is written as

$$\mathbb{D}_{1j} = \begin{bmatrix} d_{11j} & d_{12j} & d_{13j} \\ d_{21j} & d_{22j} & d_{23j} \\ d_{31j} & d_{32j} & d_{33j} \end{bmatrix} = \begin{bmatrix} \cos\gamma_{1j} & -\sin\gamma_{1j} & x_j - x_1\cos\gamma_{1j} + y_1\sin\gamma_{1j} \\ \sin\gamma_{1j} & \cos\gamma_{1j} & y_j - x_1\sin\gamma_{1j} - y_1\cos\gamma_{1j} \\ 0 & 0 & 1 \end{bmatrix} \tag{3.33}$$

The four elements $d_{11j}, d_{12j}, d_{21j}, d_{22j}$ in the matrix in equation (3.33) are due to rotation only, while d_{13j}, d_{23j} are due to translation. It can be seen from Fig. 3.25(a) that d_{13j}, d_{23j} are just the x and y components of the vector **t**, because the coordinates of $\mathbf{E}'_1$ are $(x_1 \cos\gamma_{1j} - y_1 \sin\gamma_{1j}, x_1 \sin\gamma_{1j} + y_1 \cos\gamma_{1j})$, while

$$d_{13j} = x_j - (x_1 \cos\gamma_{1j} - y_1 \sin\gamma_{1j}) \tag{3.34}$$

$$d_{23j} = y_j - (x_1 \sin\gamma_{1j} + y_1 \cos\gamma_{1j}) \tag{3.35}$$

3.4.3 The point A_0 corresponding to A_1, A_2, A_3

Having found $\mathbb{D}_{1j}$, we can choose a point $A_1(x_{A1}, y_{A1})$, and can then readily find $A_j(x_{Aj}, y_{Aj})$. The perpendicular bisector a_{1j} of $\overline{A_1 A_j}$, can be considered as the locus of $A_0(x_0, y_0)$, its equation being

$$a_{1j}: (x_{A1} - x_0)^2 + (y_{A1} - y_0)^2 = (x_{Aj} - x_0)^2 + (y_{Aj} - y_0)^2 \tag{3.36}$$

Substituting equations (3.32) and (3.33) into equation (3.36) yields

$$a_{1j}: G_j x_0 + S_j y_0 + T_j = 0 \tag{3.37}$$

Where

$$\left. \begin{aligned} G_j &= -\text{vers}\,\gamma_{1j} x_{A1} - \sin\gamma_{1j} y_{A1} + d_{13j} \\ S_j &= \sin\gamma_{1j} x_{A1} - \text{vers}\gamma_{1j} y_{A1} + d_{23j} \\ T_j &= e_{13j} x_{A1} + e_{23j} y_{A1} - (e_{13j}^2 + e_{23j}^2)/2 \\ e_{13j} &= x_1 - x_j \cos\gamma_{1j} - y_j \sin\gamma_{1j} \\ e_{23j} &= y_1 + x_j \sin\gamma_{1j} - y_j \cos\gamma_{1j} \end{aligned} \right\} \tag{3.38}$$

e_{13j}, e_{23j} are just the two elements of the matrix $\mathbb{D}_{j1}$ shown in equation (3.42a).

Setting in equation (3.36) $j = 2,3$, we obtain equations of the two perpendicular bisectors a_{12} and a_{13}. Put these two equations together as simultaneous equations, i.e.:

$$\left. \begin{aligned} a_{12}: G_2 x_0 + S_2 y_0 + T_2 &= 0 \\ a_{13}: G_3 x_0 + S_3 y_0 + T_3 &= 0 \end{aligned} \right\} \tag{3.39}$$

The solution of these equations is the intersection of a_{12} and a_{13}, or $A_0(x_0, y_0)$, as shown in Fig. 3.25(b):

$$x_0 = Z_1 / Z_3 \qquad y_0 = Z_2 / Z_3 \tag{3.40}$$

where

$$Z_1 = S_2 T_3 - S_3 T_2 \;,\; Z_2 = T_2 G_3 - T_3 G_2 \;,\; Z_3 = G_2 S_3 - G_3 S_2 \tag{3.41}$$

3.4.4 The point A_1 corresponding to A_0

Suppose in Fig. 3.26, the three positions c_1, c_2, c_3 of the body c are known. To find the point A_1 corresponding to a prescribed A_0, we may make use of the concept of kinematic inversion. Imagine first that A_0 were a point on c_2, and move c_2 back to

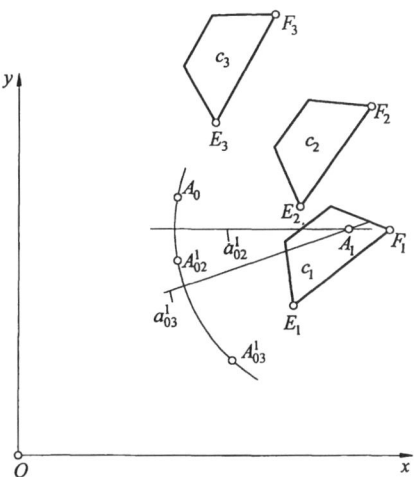

Fig. 3.26. Point A_1 corresponding to A_0.

c_1. A_0 would then be displaced together with c_1 to a position A_{02}^1. Next imagine that A_0 were a point on c_3, and displace c_3 back to c_1. Now A_0 would be displaced to a position A_{03}^1. The centre of the circle passing through A_0, A_{02}^1, A_{03}^1 is the required A_1 on c_1.

For an algebraic approach, we use a matrix $\mathbb{D}_{j1}$ to find the coordinates of A_{0j}^1. Applying $\mathbb{D}_{j1}$ to the coordinates of a point A_j on c_j, we will get the coordinates of A_1, i.e.

$$\mathbb{D}_{j1} \begin{bmatrix} x_{Aj} \\ y_{Aj} \\ 1 \end{bmatrix} = \begin{bmatrix} x_{A1} \\ y_{A1} \\ 1 \end{bmatrix} \qquad (3.42)$$

A comparison between equations (3.42) and (3.32) shows that

$$\mathbb{D}_{j1} = \mathbb{D}_{1j}^{-1} = \begin{bmatrix} \cos\gamma_{1j} & \sin\gamma_{1j} & x_1 - x_j\cos\gamma_{1j} - y_j\sin\gamma_{1j} \\ -\sin\gamma_{1j} & \cos\gamma_{1j} & y_1 + x_j\sin\gamma_{1j} - y_j\cos\gamma_{1j} \\ 0 & 0 & 1 \end{bmatrix} \qquad (3.42a)$$

Applying $\mathbb{D}_{j1}$ to A_0, we can obtain the coordinates of A_{0j}^1. The perpendicular bisector of $\overline{A_0 A_{0j}^1}$ can also be written in the form of equation (3.37) which is the later equation (3.43). Setting respectively $j = 2,3$, we get the equations of the perpendicular bisectors a_{02}^1 of $\overline{A_0 A_{02}^1}$ and a_{03}^1 of $\overline{A_0 A_{03}^1}$. Setting equations of a_{02}^1 and a_{03}^1 as two simultaneous equations, we can get their intersection A_1.

Despite the fact that the above mentioned procedure is correct, and that all computations can be left to a computer, since we are employing an algebraic approach, we can do away with the geometrical inversion as follows:

Equation (3.37) can readily be rewritten as

$$a_{0j}^1 : L_j x_{A1} + Q_j y_{A1} + H_j = 0 \tag{3.43}$$

where

$$\left. \begin{array}{l} L_j = -\text{vers}\gamma_{1j} x_0 + \sin\gamma_{1j} y_0 + e_{13j} \\ Q_j = -\sin\gamma_{1j} x_0 - \text{vers}\gamma_{1j} y_0 + e_{23j} \\ H_j = d_{13j} x_0 + d_{23j} y_0 - (d_{13j}^2 + d_{23j}^2)/2 \end{array} \right\} \quad (j = 2,3) \tag{3.44}$$

Setting in equation (3.43) $j = 2,3$ respectively, we get the equations of the two perpendicular bisectors a_{02}^1 and a_{03}^1. Solving these two equations as simultaneous equations, i.e.

$$\left. \begin{array}{l} a_{02}^1: L_2 x_{A1} + Q_2 y_{A1} + H_2 = 0 \\ a_{03}^1: L_3 x_{A1} + Q_3 y_{A1} + H_3 = 0 \end{array} \right\} \tag{3.45}$$

we get the intersection of a_{02}^1 and a_{03}^1, or $A_1(x_{A1}, y_{A1})$, as shown in Fig. 3.26:

$$x_{A1} = W_1 / W_3 \quad , \quad y_{A1} = W_2 / W_3 \tag{3.46}$$

where

$$W_1 = Q_2 H_3 - Q_3 H_2 \; , \; W_2 = H_2 L_3 - H_3 L_2 \; , \; W_3 = L_2 Q_3 - L_3 Q_2 \tag{3.47}$$

Please note that $d_{13j}^2 + d_{23j}^2 = e_{13j}^2 + e_{23j}^2$.

3.4.5 The R_M- and R^1- curves

The R_M- and R^1- curves mentioned in Section 3.3.4 can also be computed by matrix method. Let the radius of the circle k_A passing through A_1, A_2, A_3 be denoted by R_A. We have then

$$(x_{A1} - x_0)^2 + (y_{A1} - y_0)^2 = R_A^2 \tag{3.48}$$

Substituting equation (3.46) into equation (3.48) results in

Dimensional synthesis of four-bar linkages --- body guidance problems 87

$$R_M : (W_1 - W_3 x_0)^2 + (W_2 - W_3 y_0)^2 - R_A^2 W_3^2 = 0 \tag{3.49}$$

Equation (3.49) is a *tricircular sextic* in the variables x_0, y_0.
Similarly, substituting equation (3.40) into equation (3.48) results in

$$R^1 : (Z_3 x_{A1} - Z_1)^2 + (Z_3 y_{A1} - Z_2)^2 - R_A^2 Z_3^2 = 0 \tag{3.50}$$

Equation (3.50) is also a tricircular sextic, but in the variables x_{A1}, y_{A1}. The R_M- and R^1- curves have already been shown in the two preceding examples (Figs. 3.23 and 3.24).

3.4.6 The displacement matrix based on coordinates of the pole

The displacement matrix $\mathbb{D}_{12}$ for $j = 2$ in equation (3.33) was built on the basis of four coordinates $E_1(x_1, y_1)$ and $E_2(x_2, y_2)$ of two positions E_1, E_2 of a point E on the body c, and the angle of rotation γ_{12}. In case E_1 and E_2 coincide or, the application of $\mathbb{D}_{12}$ on E_1 results in E_1 itself, then the point $E_1 = E_2$ is the pole P_{12} mentioned in Section 3.1. That is

$$\mathbb{D}_{12} \begin{bmatrix} x_{P12} \\ y_{P12} \\ 1 \end{bmatrix} = \begin{bmatrix} \cos\gamma_{12} & -\sin\gamma_{12} & x_2 - x_1\cos\gamma_{12} + y_1\sin\gamma_{12} \\ \sin\gamma_{12} & \cos\gamma_{12} & y_2 - x_1\sin\gamma_{12} - y_1\cos\gamma_{12} \\ 0 & 0 & 1 \end{bmatrix} \begin{bmatrix} x_{P12} \\ y_{P12} \\ 1 \end{bmatrix} = \begin{bmatrix} x_{P12} \\ y_{P12} \\ 1 \end{bmatrix} \tag{3.51}$$

The values of (x_{P12}, y_{P12}) can be calculated from equation (3.51):

$$\left. \begin{aligned} x_{P12} &= \frac{(1-\cos\gamma_{12})(x_2 + x_1) - (y_2 - y_1)\sin\gamma_{12}}{2(1-\cos\gamma_{12})} \\ y_{P12} &= \frac{(1-\cos\gamma_{12})(y_2 + y_1) + (x_2 - x_1)\sin\gamma_{12}}{2(1-\cos\gamma_{12})} \end{aligned} \right\} \tag{3.52}$$

After this transformation, the displacement matrix $\mathbb{D}_{12}$ can be simplified on the basis of (x_{P12}, y_{P12}) and γ_{12}, into

$$\mathbb{D}_{12} = \begin{bmatrix} \cos\gamma_{12} & -\sin\gamma_{12} & x_{P12}(1-\cos\gamma_{12}) + y_{P12}\sin\gamma_{12} \\ \sin\gamma_{12} & \cos\gamma_{12} & -x_{P12}\sin\gamma_{12} + y_{P12}(1-\cos\gamma_{12}) \\ 0 & 0 & 1 \end{bmatrix} \tag{3.53}$$

$(x_{P12}, y_{P12}, 1)$ is just the *eigenvector* of the matrix $\mathbb{D}_{12}$ at its *eigenvalue* = +1. After this transformation, equation (3.38) becomes the later equation (3.95), and (3.44) becomes the later (3.101).

3.4.7 Selection of coordinate system

An adequately selected coordinate system can greatly simplify the equations. From the positions of E_1, E_2, E_3 in the original coordinate system, and from the angles γ_{12}, γ_{13}, we can find the positions of P_{12}, P_{13} by means of equation (3.52). We can also find the angle γ_{23} by using equation (3.31) as a formula and find the position of P_{23} according to equation (3.52). With the points P_{12} and P_{23} known, select a coordinate system such as that shown in Fig. 3.27, with P_{12} as origin, and $P_{12}P_{23}$ as the direction of the x-axis. Equation (3.53) can be further simplified in this way to

$$\mathbb{D}_{12} = \begin{bmatrix} \cos\gamma_{12} & -\sin\gamma_{12} & 0 \\ \sin\gamma_{12} & \cos\gamma_{12} & 0 \\ 0 & 0 & 1 \end{bmatrix} \quad (3.54)$$

Corresponding to equation (3.53) we can also write

$$\mathbb{D}_{13} = \begin{bmatrix} \cos\gamma_{13} & -\sin\gamma_{13} & x_{P_{13}}\text{vers}\gamma_{13} + y_{P_{13}}\sin\gamma_{13} \\ \sin\gamma_{13} & \cos\gamma_{13} & -x_{P_{13}}\sin\gamma_{13} + y_{P_{13}}\text{vers}\gamma_{13} \\ 0 & 0 & 1 \end{bmatrix} \quad (3.55)$$

or, based on P_{23} and γ_{23} in Fig. 3.27, we can write

$$\mathbb{D}_{13} = \mathbb{D}_{23}\mathbb{D}_{12} = \begin{bmatrix} \cos\gamma_{13} & -\sin\gamma_{13} & s\,\text{vers}\,\gamma_{13} \\ \sin\gamma_{13} & \cos\gamma_{13} & -s\sin\gamma_{13} \\ 0 & 0 & 1 \end{bmatrix} \quad (3.56)$$

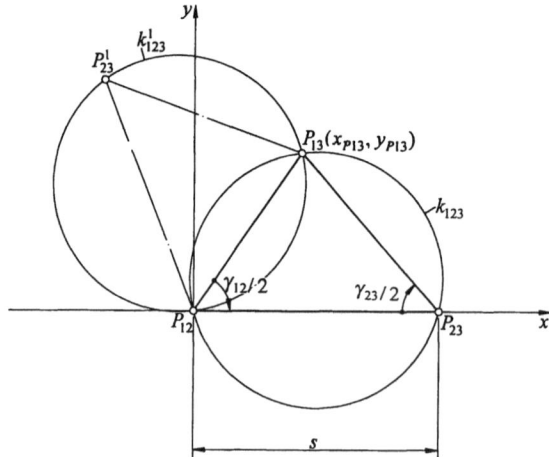

Fig. 3.27. Selected coordinate system.

Dimensional synthesis of four-bar linkages --- body guidance problems 89

where $s = \overline{P_{12}P_{23}}$. Please note that, in Fig. 3.27, both γ_{12} and γ_{23} are negative.

3.5 Guiding a body through three infinitesimally separated positions

What will happen to the geometrical relationships mentioned in Section 3.4, as the three positions through which a body is guided approach one another indefinitely?

3.5.1 Geometrical method

Fig. 3.28 shows a body c ($= AB$) taking consecutively the positions $c_1, c_2, c_3, c_4, \ldots$ As the angles of rotation $\gamma_{12}, \gamma_{23}, \ldots$ of the body c gradually diminish,

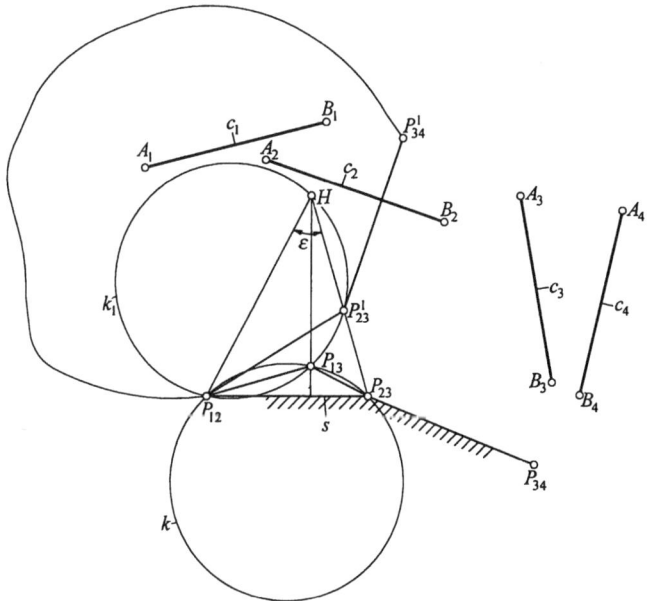

Fig. 3.28. As three finitely separated positions approach closer to one another.

the poletriangle $\Delta P_{12} P_{23} P_{13}$ becomes flatter and flatter, shrinking into a single point. This is because $\alpha = |\gamma_{12}/2|$, $\beta = |\gamma_{23}/2|$, and $\sphericalangle P_{12}P_{13}P_{23} = \eta = \pi - |\gamma_{13}/2|$. In Fig. 3.28, we show only the circles k and k_1 of Fig. 3.19. The orthocentre H of the poletriangle $\Delta P_{12} P_{23} P_{13}$ is on the outside of the triangle. For simplicity, let

$$\gamma_{12} = \Delta\gamma$$
$$\gamma_{23} = \Delta\gamma + d(\Delta\gamma)$$

As the three positions approach one another indefinitely, $\lim \overline{P_{12}P_{23}} = ds$ and

$$\gamma_{12} \approx d\gamma$$
$$\gamma_{23} \approx d\gamma + d^2\gamma$$

and also

$$\sphericalangle P_{23}P_{12}P_{13} = \alpha, \quad \lim_{ds \to 0} \alpha = -d\gamma/2$$
$$\sphericalangle P_{23}P_{12}P_{13} = \beta, \quad \lim_{ds \to 0} \beta = -(d\gamma + d^2\gamma)/2$$

Further let $\sphericalangle P_{12}HP_{23}=\varepsilon$. It can be seen from Fig. 3.19 that $\varepsilon = \sphericalangle P_{12}HP_{23} = \pi - \eta = \alpha + \beta$. Therefore

$$\lim_{ds \to 0} \varepsilon = -d\gamma - d^2\gamma/2$$

By the law of sine, $\overline{P_{12}H} = s\cos\alpha / \sin\varepsilon$. In approaching the limit,

$$\lim_{ds \to 0} \overline{P_{12}H} = \lim_{ds \to 0} \frac{ds\left[1 - \frac{1}{2} \cdot \frac{d\gamma^2}{4}\right]}{-d\gamma - \frac{d^2\gamma}{2}} = \frac{ds}{-d\gamma} = \frac{ds/dt}{-d\gamma/dt} = \frac{u}{-\omega} = \delta$$

In the above equation, t represents time, u is the pole changing velocity, ω is the angular velocity of the body, clockwise sense being considered negative, and δ is the diameter of the inflection circle (please refer to equation (1.10)). Furthermore,

$$\lim_{ds \to 0} \sphericalangle P_{23}P_{12}H = \lim_{ds \to 0} \left(\frac{\pi}{2} - \beta\right) = \pi/2$$

Therefore the limiting position of H is the inflection pole W as mentioned in Section 1.3, and that of the circle k_1 is the inflection circle k_W. In fact, the limiting position of all three image circles k_1, k_2, k_3 is the inflection circle, while the limiting position of the circle k is the return circle k_R.

Let us see how equation (3.22) varies. In equation (3.23), in approaching the limit, since $\alpha \to -d\gamma/2$, $\beta \to -(d\gamma + d^2\gamma)/2$, $s \to ds$, we have

$$E \approx \sin\theta$$
$$F \approx -\sin\theta$$
$$K \approx ds\left|\frac{d\gamma_{23}}{2}\right| \approx 0$$
$$G \approx \frac{d\gamma + d^2\gamma/2}{ds} \approx \frac{d\gamma}{ds} = -\frac{1}{\delta}$$

Equation (3.22) becomes

$$\left(\frac{1}{p} - \frac{1}{p_0}\right)\sin\theta = \frac{1}{\delta}$$

which is precisely the Euler-Savary equation (1.14).

Let us examine the limit of the case in which the three homologous points A_1, A_2, A_3 lie on a straight line as mentioned in Section 3.3.2. As stated in theorem 8, the three points lie respectively on the circles k_1, k_2, k_3, and their common line l_A passes through the orthocentre H of the poletriangle. Since the three circles k_1, k_2, k_3 combine into one and become the inflection circle k_W in the limit, and since the limit position of H is the return pole W, the tangent of the path of any point on k_W is the limit of l_A, and this tangent passes through W. Consequently, just as we have mentioned in Section 1.3, the tangent of the path of an inflection point touches the curve in three points. Hence we can take a short portion of this path to approximate a straight line motion.

Next consider the limiting condition of the case in which three homologous lines l_1, l_2, l_3 pass through a fixed point S_0 mentioned in Section 3.3.3. As stated in theorem 9, this point S_0 lies on the circle k, and the three lines l_1, l_2, l_3 ($= S_0H_1$, S_0H_2, S_0H_3) pass respectively through H_1, H_2, H_3. In the limiting condition, circle k becomes the return circle k_R, and the three points H_1, H_2, H_3 become a single point, the return pole R.

Fig. 3.29(a) shows two Cardan circles, the small circle π_c rolling without slip on the inside of the large circle π_f. The inflection circle k_W belonging to c and the return circle k_R on the opposite side are known. Choose any point S_0 on k_R, and join S_0 and R. The line S_0R is the limting position of the three homologous lines l_1, l_2, l_3. Denote this line by h, and consider h as a line belonging to and moving together with c. Next consider h as a generating line, its envelope h_b exhibiting a cusp at S_0. Please note that h_b is on the fixed body f. During the motion of the moving body c, the tangent point between h and h_b changes its position along h_b. It can be imagined that this *tangent point* moves towards the return circle until it reaches S_0 and then moves backwards. The line h coincides with S_0R as the *tangent point* moves to S_0. This is the reason that the circle k_R is called *return circle* or *cusp circle*.

According to Fig. 1.29 of Section 1.10, the centre of curvature of the envelope of a generating line on a moving body should lie on the return circle k_R. If this envelope cuts k_R in a real point, then the centre of curvature of the envelope at this intersection is itself. Hence this intersection point must be a cusp. In order to clarify this point, please refer to Fig. 3.29(a,b). Before π_c reaches its position shown in Fig. 3.29(a), the line h on it has not yet passed through the point R. As shown in Fig.

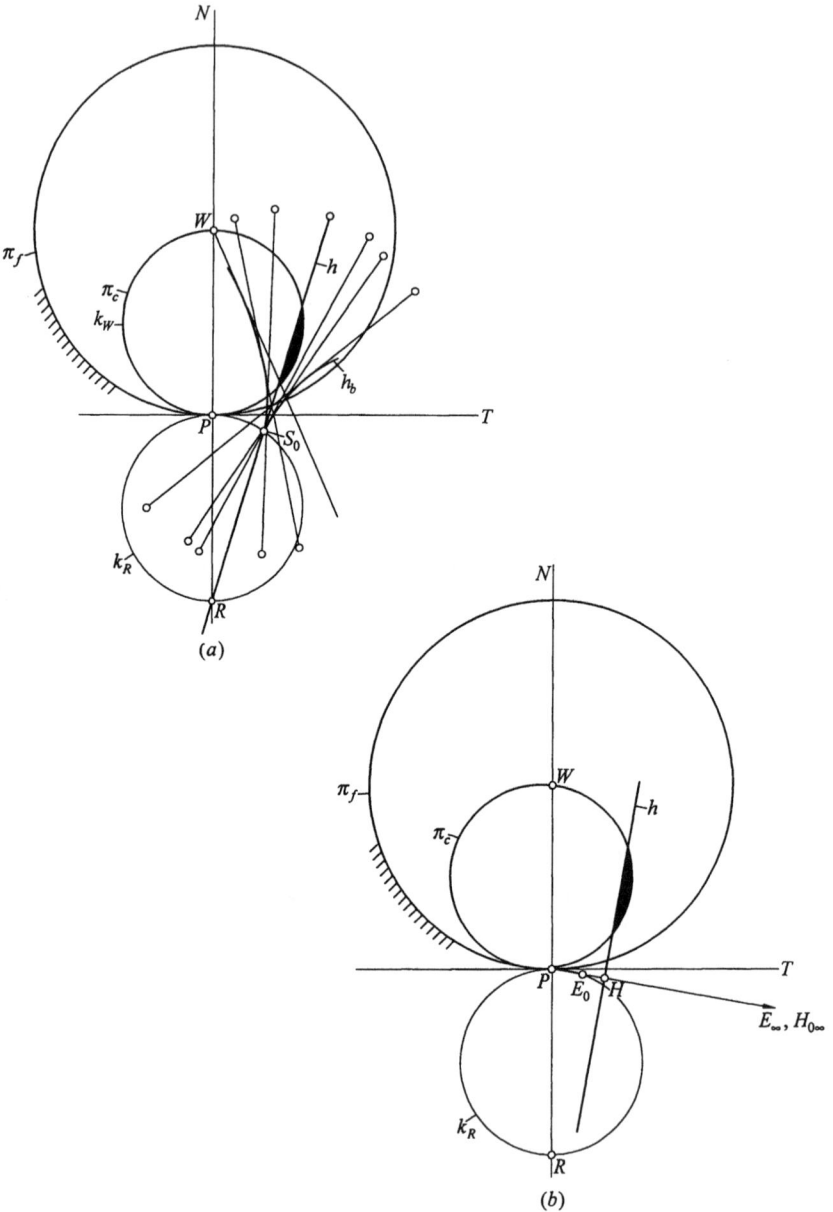

Fig. 3.29. (a) Envelope h_b of straight line h and return circle (or cusp circle) k_R.
(b) Radius of curvature $\overline{E_0 H}$ of envelope of h at H.

Dimensional synthesis of four-bar linkages --- body guidance problems 93

1.29, draw a perpendicular to h from P. This perpendicular intersects h in H, and k_R in C_0. The radius of curvature of the envelope h_b (not shown in Fig. 3.29(b)) is $\overline{C_0H}$. As π_c reaches the position shown in Fig. 3.29(a), $\overline{C_0H}$ diminishes to zero, hence the envelope h_b exhibits a cusp at S_0. For this reason, the circle k_R is also called a *cusp circle* (Hunt, 1978, p.128).

As h moves with c reaching the position S_0R, the viewer may have a visual feeling that it were undergoing a temporary dwell. In fact the whole body c rotates about the instantaneous velocity pole P. The motion of h is a rotation with a radius $\overline{PS_0}$. The point on h that coincides with S_0 has a finite velocity along the direction S_0R.

3.5.2 Algebraic method

In the present case the three points P_{12}, P_{23}, P_{13} approach to a single point. In Fig. 3.27, we may write $\gamma_{12} \approx d\gamma$, $\gamma_{23} \approx d\gamma + d^2\gamma$, $s \approx ds$. Since there are three rather than two positions approaching one another, the term $d^2\gamma$ in the matrix $\mathbb{D}_{13}$ has to be reserved instead of using the approximation $\cos d\gamma \approx 1$. The following approximations are used:

$$\sin d\gamma \approx d\gamma$$
$$\sin(d\gamma + d^2\gamma) \approx d\gamma + d^2\gamma$$
$$\cos d\gamma \approx \cos(d\gamma + d^2\gamma) \approx 1 - d\gamma^2/2$$
$$\text{vers}\, d\gamma \approx d\gamma^2/2$$

The coordinate system adopted in Section 3.4.7 is the one with P as the origin, and direction PT as the x-axis. In this book we shall keep using the direction of *pole changing velocity* $\mathbf{u}$ as the positive x-direction. Equation (3.54) becomes now

$$\mathbb{D}_{12} = \begin{bmatrix} 1 - d\gamma^2/2 & -d\gamma & 0 \\ d\gamma & 1 - d\gamma^2/2 & 0 \\ 0 & 0 & 1 \end{bmatrix} \qquad (3.57)$$

while equation (3.56) becomes

$$\mathbb{D}_{13} = \begin{bmatrix} 1 - 2d\gamma^2 & -2d\gamma - d^2\gamma & 0 \\ 2d\gamma + d^2\gamma & 1 - 2d\gamma^2 & -ds d\gamma \\ 0 & 0 & 1 \end{bmatrix} \qquad (3.58)$$

To find the locus of A_1 for three homologous points A_1, A_2, A_3 lying on a straight line, we can set in equation (3.50) $R_A \to \infty$, or $Z_3^2 = 0$, which is

$$G_2 S_3 - G_3 S_2 = 0$$

or

$$(-d\gamma\, x_{A1} - 2y_{A1})[(2d\gamma + d^2\gamma)x_{A1} - 2d\gamma^2 y_{A1} - ds d\gamma]$$
$$-[-2d\gamma^2 x_{A1} + (-2d\gamma + d^2\gamma) y_{A1}](2x_{A1} - d\gamma\, y_{A1}) = 0$$

Rearranging this equation and taking $ds/d\gamma = -\delta$ into consideration, we get

$$x_{A1}^2 + y_{A1}^2 - \delta y_{A1} = 0 \qquad (3.59)$$

This is just equation (1.5) of the inflection circle.

Similarly, to find the locus of A_0 for A_1 lying at infinity, we may set in equation (3.49) $R_A \to \infty$, or $W_3^2 = 0$, yielding

$$L_2 Q_3 - L_3 Q_2 = 0$$

or

$$(d\gamma\, x_0 - 2y_0)[(2d\gamma + d^2\gamma)x_0 + 2d\gamma^2 y_0 - (1 - 2d\gamma^2) ds d\gamma]$$
$$-[2d\gamma^2 x_0 - (2d\gamma + d^2\gamma) y_0 - (2d\gamma + d^2\gamma) ds d\gamma](d\gamma\, y_0 + 2x_0) = 0$$

Rearranging this equation, we obtain

$$x_0^2 + y_0^2 + \delta y_0 = 0 \qquad (3.60)$$

This is equation (1.29) of the return circle.

3.5.3 The curve of equal radius of curvature of the first kind --- the ρ- and ρ_M - curves

For a body performing a plane motion, if at an instant, the radius of curvature of the path of a point on the body is equal to a prescribed value ρ, the locus of this point is called the ρ- curve, and the locus of the corresponding centre of curvature is called the ρ_M -curve (Alt, 1932a, 1932b). From the Euler-Savary equation (1.14)

$$\left(\frac{1}{\overline{PA}} - \frac{1}{\overline{PA_0}}\right) \sin\theta_A = \frac{1}{\delta} \qquad [(1.14)]$$

we have

$$(\overline{PA_0} - \overline{PA}) \delta \sin\theta_A = \overline{PA} \cdot \overline{PA_0} \qquad (3.61)$$

where

$$\overline{PA_0} - \overline{PA} = \pm\rho \qquad (3.62)$$

Dimensional synthesis of four-bar linkages --- body guidance problems 95

In equation (3.62) ρ is the prescribed radius of curvature, being a scalar quantity. As stated in Section 1.6, both $\overline{PA}$ and $\overline{PA_0}$ are directed line segments. In the Euler-Savary equation, the range of θ_A is taken within $0 \le \theta_A \le 180°$. As can be seen from Fig. 1.12, for a θ_A within this range, if the point A is inside the inflection circle k_W, then $\overline{PA_0}$ is (+), both A and A_0 being on the same side of P, and $\overline{PA_0} - \overline{PA}$ is (+). Otherwise, if A is outside the inflection circle k_W, then $\overline{PA_0}$ is (−), A and A_0 being on opposite sides of P, and $\overline{PA_0} - \overline{PA}$ is (−). We can therefore ascertain that, the (+) sign in equation (3.62) indicates that A and A_0 are on the same side of P, and that the (−) sign indicates that A and A_0 are on the opposite sides of P. Eliminating $\overline{PA_0}$ from equations (3.62) and (3.61), we get

$$\overline{PA}^2 \pm \rho(\overline{PA} - \delta \sin\theta_A) = 0 \tag{3.63}$$

Solving this quadratic equation for the unknown $\overline{PA}$, we get

$$\left.\begin{aligned}\overline{PA_\mathrm{I}}, \overline{PA_\mathrm{II}} &= \frac{\rho}{2}\left(-1 \pm \sqrt{1 + \frac{4\delta}{\rho}\sin\theta_A}\right), (A_\mathrm{I}, A_\mathrm{I0}; A_\mathrm{II}, A_\mathrm{II0} \text{ on the same side of } P \text{ respectively}) \\ \overline{PA_\mathrm{III}}, \overline{PA_\mathrm{IV}} &= \frac{\rho}{2}\left(1 \pm \sqrt{1 - \frac{4\delta}{\rho}\sin\theta_A}\right), (A_\mathrm{III}, A_\mathrm{III0}; A_\mathrm{IV}, A_\mathrm{IV0} \text{ on the opposite sides of } P \text{ respectively})\end{aligned}\right\} \tag{3.64}$$

This means that there are four points of A (A_I, A_II, A_III, A_IV) on a line passing through

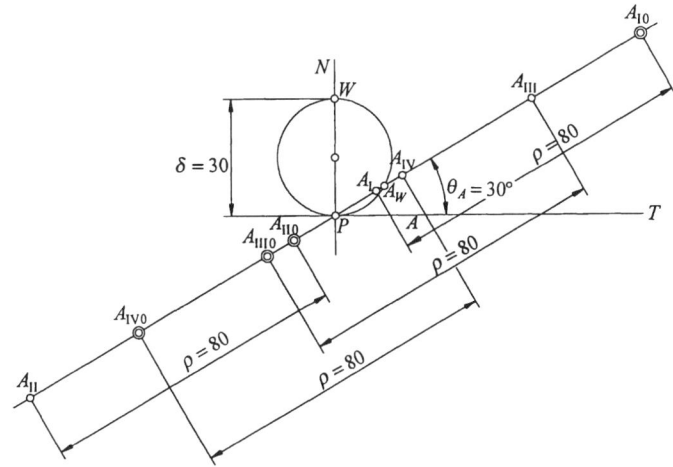

Fig. 3.30. The four point-pairs A, A_0 for $\delta = 30$ mm, $\rho = 80$ mm, $\theta_A = 30°$.

P, the radii of curvature of whose paths equal to the prescribed value ρ. It can be seen from equation (3.64) that, if $\rho < 4\delta \sin\theta_A$, the two points A_{III}, A_{IV} are imaginary. Fig. 3.30 shows four pairs of A, A_0 (eight) points, as calculated from equation (3.64) for $\delta = 30$ mm, $\rho = 80$ mm, $\theta_A = 30°$.

Fig. 3.31 shows the locus of A for a prescribed value of $\rho(<4\delta)$ within the range $0° \leq \theta_A \leq 180°$. This curve is called after (Alt, 1932a) the ρ-curve. In the present case the ρ-curve is of a two-part type. Fig. 3.32 shows a ρ-curve for a prescribed value of $\rho(>4\delta)$. In this case the ρ-curve is of a three-part type, including $(\rho_a), (\rho_W)$ and (ρ'_W).

Taking the poletangent PT as the x-axis and the polenormal PN as the y-axis, and using the following transformation equations

$$\left.\begin{array}{l}\overline{PA}\cos\theta_A = x\\ \overline{PA}\sin\theta_A = y\end{array}\right\}$$

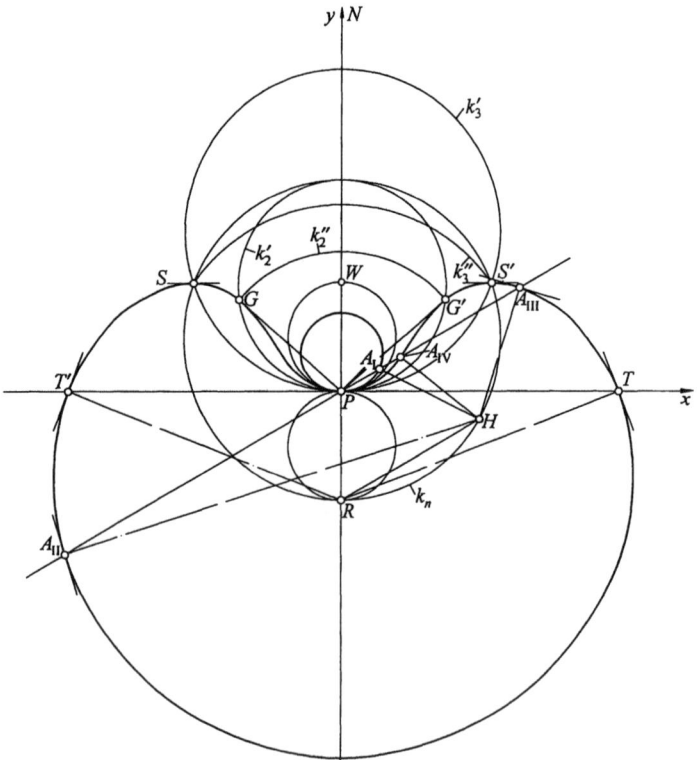

Fig. 3.31. The ρ-curve for prescribed ρ value in case $\rho < 4\delta$ ($\delta = 30$, $\rho = 80$).

Dimensional synthesis of four-bar linkages --- body guidance problems 97

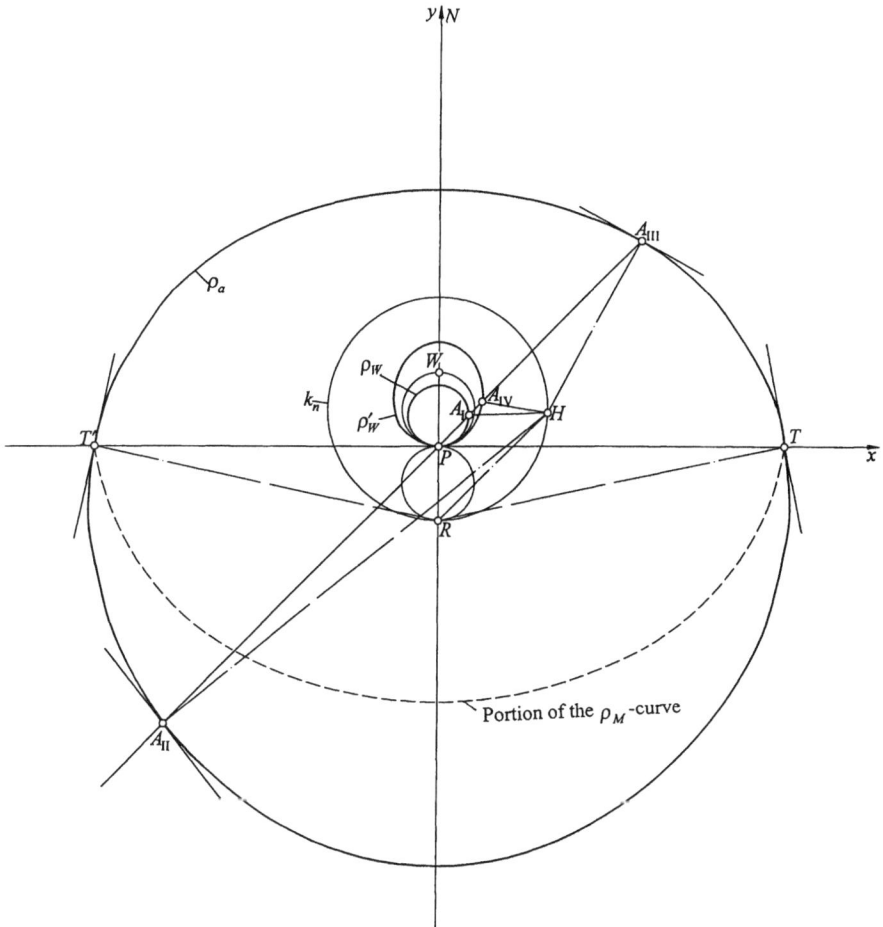

Fig. 3.32. The ρ-curve for prescribed ρ value in case $\rho > 4\delta$ ($\delta = 10$, $\rho = 48$), including three parts $(\rho_a), (\rho_W), (\rho'_W)$.

we can transform equation (3.63) into a rectangular coordinate equation:

$$\rho: (x^2 + y^2)^3 - \rho^2(x^2 + y^2 - \delta y)^2 = 0 \qquad (3.65)$$

Equation (3.65) indicates that the ρ-curve is a tricircular sextic (please refer to Section A1.2) with the inflection circle k_W

$$k_W: x^2 + y^2 - \delta y = 0$$

as its circle of curvature at the origin P.

Nowadays, the ρ-curve can be drawn pretty fast by a computer according to equation (3.64). However, in order to explain and examine the geometrical characteristics of the ρ-curve, we introduce in the following some construction methods to help inspection of the curve drawn.

(a) Resolve equation (3.65) into the following two parametric equations:

$$x^2 + y^2 = \lambda^2 \qquad (3.66a)$$

$$x^2 + \left(y - \frac{\delta}{2}\right)^2 = \frac{\delta^2}{4} \pm \frac{\lambda^3}{\rho} \qquad (3.66b)$$

In these equations, λ is a parameter. For a given value of λ, equation (3.66a) represents a circle with P as its centre and λ as its radius; while equation (3.66b) represents a circle with $(0, \delta/2)$ (centre of the circle k_W) as its centre, and $(\delta^2/4 \pm \lambda^3/\rho)^{1/2}$ as its radius. The intersections of these two circles are the points on the ρ-curve.

(b) Resolve equation (3.65) into the following two parametric equations:

$$k'_\nu : x^2 + y^2 = \nu \delta y \qquad (3.67a)$$

$$k''_\nu : x^2 + y^2 = \left[\rho\left(1 - \frac{1}{\nu}\right)\right]^2 \qquad (3.67b)$$

In the above two equations, ν is the parameter. For a given value of ν, equation (3.67a) represents a circle tangent to the x-axis and with $(\nu\delta)/2$ as its radius. Equation (3.67b) represents a circle with the origin P as its centre and $\rho(1-1/\nu)$ as its radius. The intersections of these two circles are the points on the ρ-curve. In particular, we would like to mention points for $\nu = 2$ and $\nu = 3$.

(bi) $\nu = 2$. In this case the circles k'_2 and k''_2 intersect in the two points G, G', as shown in Fig. 3.31. $\overline{PG} = \overline{PG'} = \rho/2$ is the radius of the circle k''_2. The radius of the circle k'_2 is the diameter δ of the inflection circle. The lines PG, PG' are just the tangents of the ρ-curve at G, G'.

(bii) $\nu = 3$. The circles k'_3 and k''_3 intersect in S, S' as shown in Fig. 3.31. The tangents of the ρ-curve at S, S' are parallel to the x-axis. $\overline{PS} = \overline{PS'} = (2\rho)/3$ is the radius of k''_3.

(c) Draw a circle k_n with the centre of k_W as its centre, and $(3\delta)/2$ as its radius. k_n passes through the return pole R. Draw a line from R parallel to PA (the common line of the four points A_I, A_II, A_III, A_IV), to intersect k_n in H. Join HA_I, HA_II, HA_III, HA_IV. These four lines are respectively the normals of the ρ-curve at A_I, A_II, A_III, A_IV. In the special case of PA being PT, the points H and R coincide, therefore RT, RT' are respectively the normals of the ρ-curve at T, T'.

Rewriting equation (3.62) in the form

Dimensional synthesis of four-bar linkages --- body guidance problems 99

$$\overline{PA} = \overline{PA_0} \mp \rho$$

and substituting it into equation (3.61) gives

$$\overline{PA_0}^2 \mp \rho \cdot \overline{PA_0} \mp \rho \delta \sin \theta_A = 0 \qquad (3.68)$$

Transforming equation (3.68) into rectangular coordinates, i.e. substituting the relations

$$\left. \begin{array}{l} \overline{PA_0} \cos \theta_A = x_0 \\ \overline{PA_0} \sin \theta_A = y_0 \end{array} \right\}$$

into equation (3.68), we get then the rectangular coordinate equation of the point

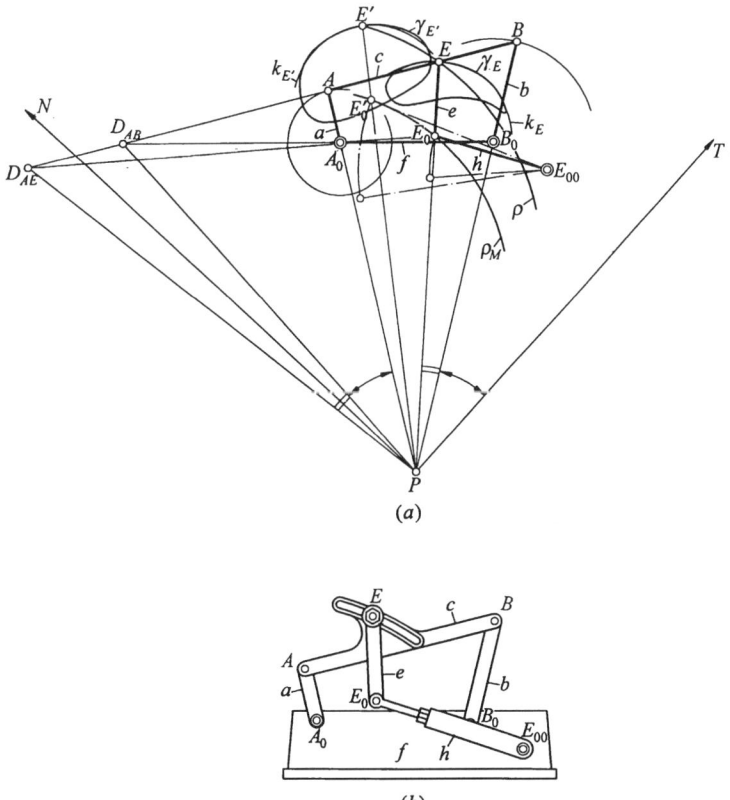

Fig. 3.33. (a) An application example of ρ-curve and ρ_M-curve.
(b) Adjustable dwell mechanism designed according to (a).

$A_0(x_0, y_0)$:

$$\rho_M : (x_0^2 + y_0^2)^3 - \rho^2 (x_0^2 + y_0^2 + \delta y_0)^2 = 0 \tag{3.69}$$

Equation (3.69) is called the ρ_M-curve, being the locus of the point A_0. It can be seen that ρ_M is also a tricircular sextic. ρ_M-curve is the mirror image of the ρ-curve with respect to the x-axis. In Fig. 3.32 only a part of the ρ_M-curve is shown, which is symmetrical to a part of the ρ-curve with respect to the x-axis.

Example: Fig. 3.33(a) shows a given four-bar linkage $A_0 A B B_0$. In its position shown in the figure, the ρ-curve can be drawn for a prescribed length of ρ. Obviously, before constructing the ρ-curve, we will have to locate the axes PT and PN by means of Bobillier construction. Choose a suitable point E on the ρ-curve, and join PE. Take on PE the length $\overline{EE_0} = \rho$, then E_0 must lie on the ρ_M-curve. Make a link e to join EE_0, and connect E_0 to another fixed point E_{00} by another link h. The path γ_E of E is also shown in the figure. It can be seen that the path γ_E is tangent to the circle of curvature k_E in three points. As the point E moves in the vicinity of the position as shown in the figure, the point E_0 exhibits an approximate standstill, hence enabling the link h to have a period of dwell.

In case the point E is chosen at E', its corresponding centre point is at E'_0. It can be seen that the angle of swing of h is larger than before. In this way it is possible to construct an adjustable dwell mechanism as shown in Fig. 3.33(b). The curved slot at E on the coupler c is made according to the shape of the ρ-curve. The length $\overline{E_0 E} = \rho$ is constant, but the length $\overline{E_0 E_{00}}$ is adjustable.

3.5.4 The curve of equal radius of curvature of the second kind --- the q_1- and q_M (q_{M1})- curves

For two finitely separated positions c_1, c_2 of the coupler c of a mechanism, the corresponding positions of a point E on c are E_1, E_2. If the radii of curvature of the paths of E_1, E_2 are equal, and both paths curl to the same side, the locus of such point E_1 is called the q_1-curve. This problem was first investigated by (Alt, 1932a,b). To find the q_1-curve, we can proceed as follows:

(a) Algebraic method

Fig. 3.34 shows a four-bar linkage $A_0 A B B_0$ moving from its first position to a second position. In its first position, the coupler $c_1 (= A_1 B_1)$ is shown together with its velocity pole P_1, poletangent $P_1 T_1$, inflection circle k_{W1} and the diameter δ_1 of k_{W1}. In its second position $c_2 (= A_2 B_2)$ of the coupler c, its velocity pole P_2, poletangent $P_2 T_2$, and inflection circle k_{W2} together with the diameter δ_2 of k_{W2} are also shown. Now move $P_2, P_2 T_2, k_{W2}$ together with $A_2 B_2$ back to the

Dimensional synthesis of four-bar linkages --- body guidance problems 101

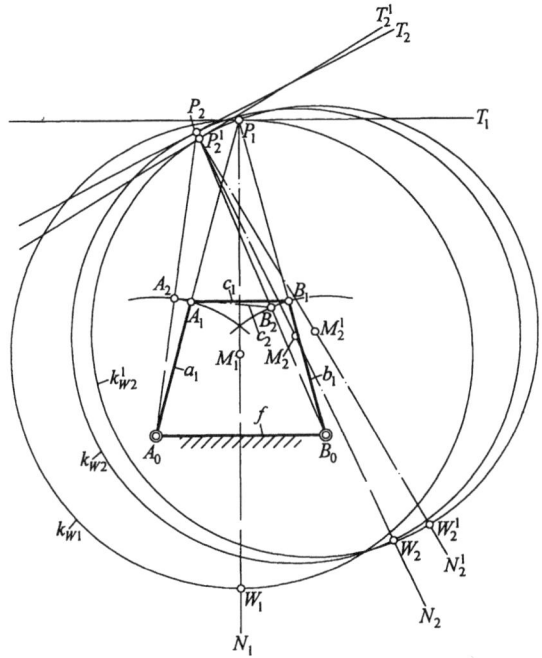

Fig. 3.34. Positions P_2, P_2T_2, k_{W2} when coupler c is in position c_2, and positions P_2^1, $P_2^1 T_2^1$, k_{W2}^1, when c is in c_1.

position $A_1 B_1$, reaching respectively to $P_2^1, P_2^1 T_2^1, k_{W2}^1$. In this inversion conception, one simply imagines that the moving coupler were kept stationary, while $P_1, P_1 T_1, k_{W1}$ were moved respectively to $P_2^1, P_2^1 T_2^1, k_{W2}^1$. For clarity, the essential part in Fig. 3.34 is redrawn in Fig. 3.35. In this figure, a certain point A_1 is in its first position, and the centre and radius of curvature of its path are A_{01} and $\rho_1 = \overline{A_1 A_{01}}$ respectively. In the second position, $A_2^1 = A_1$, and the centre of curvature of its path is A_{02}^1 and the radius of curvature of its path is $\rho_2 = \overline{A_1 A_{02}^1}$. We obtain from the Euler-Savary equation (1.14)

$$\left(\frac{1}{p} - \frac{1}{p_0}\right) \sin\theta = \frac{1}{\delta}$$

that

$$p_0 - p = \rho = \frac{pp_0}{\delta \sin\theta}$$

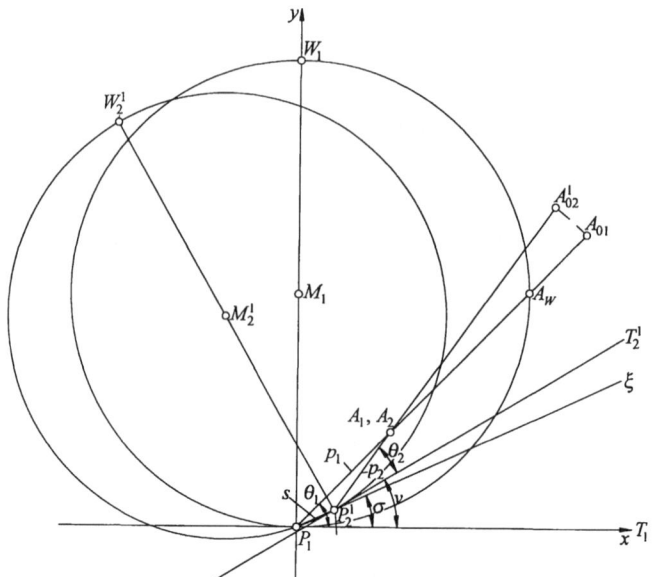

Fig. 3.35. Essential parts of Fig. 3.34.

or
$$\rho = \frac{p^2}{\delta \sin\theta - p}$$

Applying this equation to the point A_1 in Fig. 3.35, we get

$$\rho_1 = \frac{p_1^2}{\delta_1 \sin\theta_1 - p_1} \qquad (3.70)$$

where $p_1 = \overline{P_1 A_1}$, $\theta_1 = \sphericalangle T_1 P_1 A_1$.

Similarly, for the point $A_2^1 (= A_1)$, we have

$$\rho_2 = \frac{p_2^2}{\delta_2 \sin\theta_2 - p_2} \qquad (3.71)$$

where $p_2 = \overline{P_2^1 A_1}$, $\theta_2 = \sphericalangle T_2^1 P_2^1 A_1$. Since the condition assigned at present is $\rho_1 = \rho_2$, we have

$$\frac{p_1^2}{\delta_1 \sin\theta_1 - p_1} = \frac{p_2^2}{\delta_2 \sin\theta_2 - p_2} \qquad (3.72)$$

In order to find the locus of $A_1(p_1, \theta_1)$, we have to express the variables (p_2, θ_2) in equation (3.72) in terms of (p_1, θ_1). Let ν represent the angle between T_2^1 and T_1, and σ represent the angle between $P_1 P_2^1 (= \xi$-axis) and T_1. Then

Dimensional synthesis of four-bar linkages --- body guidance problems

$$p_2^2 = p_1^2 + s^2 - 2p_1 s \cos(\theta_1 - \sigma) \quad (3.73a)$$
$$p_2 \sin \theta_2 = p_1 \sin(\theta_1 - v) + s \sin(v - \sigma) \quad (3.73b)$$

In the above two equations s, v, σ are all constants. Rewriting equation (3.72) in the form

$$p_2^3 (\delta_1 \sin \theta_1 - p_1) = p_1^2 p_2 (\delta_2 \sin \theta_2 - p_2)$$

and substituting equation (3.73a,b) into it, we get

$$q_1: [p_1^2 + s^2 - 2p_1 s \cos(\theta_1 - \sigma)]^{3/2} (\delta_1 \sin \theta_1 - p_1) - p_1^2 \delta_2 [p_1 \sin(\theta_1 - v) + s \sin(v - \sigma)] + p_1^2 [p_1^2 + s^2 - 2p_1 s \cos(\theta_1 - \sigma)] = 0 \quad (3.74)$$

Equation (3.74) appeared in (Beyer, 1953, equation (153)), and was not included in the original Alt's paper. However, in practical application, equation (3.74) has to be developed first, as the following:

$$q_1: a_7 p_1^7 + a_6 p_1^6 + a_5 p_1^5 + a_4 p_1^4 + a_3 p_1^3 + a_2 p_1^2 + a_1 p_1 + a_0 = 0 \quad (3.75)$$

where

$a_7 = 2[\delta_2 \sin(\theta_1 - v) - s \cos(\theta_1 - \sigma) - \delta_1 \sin \theta_1]$

$a_6 = s^2 [1 + 8 \cos^2 (\theta_1 - \sigma)] + \delta_1 \sin \theta_1 [\delta_1 \sin \theta_1 + 12 s \cos(\theta_1 - \sigma)]$
$\quad - \delta_2 [\delta_2 \sin^2(\theta_1 - v) + 4s \sin(\theta_1 - v) \cos(\theta_1 - \sigma) - 2s \sin(v - \sigma)]$

$a_5 = -8s^3 \cos(\theta_1 - \sigma)[1 + \cos^2(\theta_1 - \sigma)]$
$\quad - 6s \delta_1 \sin \theta_1 [\delta_1 \sin \theta_1 \cos(\theta_1 - \sigma) + 4s \cos^2(\theta_1 - \sigma) + s]$
$\quad + 2\delta_2 s \{s[\sin(\theta_1 - v) - 2\cos(\theta_1 - \sigma) \sin(v - \sigma)] - \delta_2 \sin(\theta_1 - v) \sin(v - \sigma)\}$

$a_4 = 3s^2 \delta_1^2 \sin^2 \theta_1 [1 + 4\cos^2(\theta_1 - \sigma)] + 8s^3 \delta_1 \sin \theta_1 \cos(\theta_1 - \sigma)[3 + 2\cos^2(\theta_1 - \sigma)]$
$\quad + 2s^4 [1 + 6\cos^2(\theta_1 - \sigma)] + \delta_2 s^2 \sin(v - \sigma)[2s - \delta_2 \sin(v - \sigma)]$

$a_3 = -6s^3 \delta_1 \sin \theta_1 \{2 \cos(\theta_1 - \sigma)[\delta_1 \sin \theta_1 + 2s \cos(\theta_1 - \sigma)] + s\}$
$\quad - 2s^3 \cos(\theta_1 - \sigma)[4\delta_1^2 \sin^2 \theta_1 \cos^2(\theta_1 - \sigma) + 3s^2]$

$a_2 = s^4 \{3\delta_1 \sin \theta_1 [\delta_1 \sin \theta_1 (1 + 4\cos^2(\theta_1 - \sigma)) + 4s \cos(\theta_1 - \sigma)] + s^2\}$

$a_1 = -2s^5 \delta_1 \sin \theta_1 [3\delta_1 \sin \theta_1 \cos(\theta_1 - \sigma) + s]$

$a_0 = s^6 \delta_1^2 \sin^2 \theta_1$

The above eight coefficients $a_7, a_6, \ldots, a_0$ are all functions of θ_1. For a given value of $\theta_1 (0 \le \theta_1 \le 180°)$, 7 values of p_1 can be calculated, so that the q_1-curve can be drawn.

In rectangular coordinates, substituting the following transformation equations

$$p_1 \cos \theta_1 = x_1$$
$$p_1 \sin \theta_1 = y_1$$

into equation (3.74) results in:

q_1: $2[-sU + \delta_2 V - \delta_1 y_1](x_1^2 + y_1^2)^4$ + terms of 8th and lower degrees (3.76)

where

$$U = x_1 \cos \sigma + y_1 \sin \sigma$$
$$V = y_1 \cos v - x_1 \sin v$$

Because of the complicated form of the complete equation (3.76), it is given in Appendix 3 for the reference of the readers. However, it should be noted that the coordinate system is according to Fig. 3.31, equation (3.76) being referred to a system with P_1 as the origin, $P_1 T_1$ as the x-axis, and $P_1 N_1$ as the y-axis.

It can be seen from equation (3.76) that, since both U and V are linear, as the highest term of q_1 indicates, the q_1-curve is a quadruple circular curve of 9th degree (please refer to Section A1.2). This curve is self-tangent at P_1 and P_2^1, which means that each of P_1 and P_2^1 can be considered as a double point. That equation (3.75) is of the 7th degree, is because any line passing through the origin P_1 intersects q_1 twice in P_1, and these two points are not included in equation (3.75) (please refer to Fig. 3.40(a)).

(b) Geometrical method

The following description is according to (Alt, 1932a,b). Equation (3.63), or $\overline{PA}^2 \pm \rho(\overline{PA} - \delta \sin \theta_A) = 0$, which is identical with equation (1.18), can be rewritten as:

$$\overline{PA}^2 = \rho |\overline{AA_W}| \qquad [(1.18)]$$

or

$$\overline{PA}^3 = \rho |\overline{AA_W} \cdot \overline{PA}|$$

or

$$\overline{PA}^3 = \rho \pi^2 \qquad (3.77)$$

$\pi^2 = |\overline{AA_W} \cdot \overline{PA}|$ is called the *potency* of A with respect to the circle k_W. Please refer to Fig. 3.36. Rewrite now equation (3.77) in the form

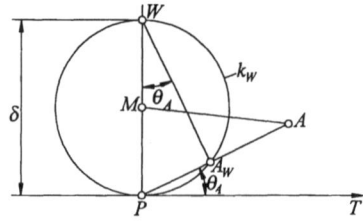

Fig. 3.36. $\pi^2 = |\overline{AA_W} \cdot \overline{PA}|$.

Dimensional synthesis of four-bar linkages --- body guidance problems   105

$$\overline{P_1 A_1}^3 = \rho \pi_1^2 \tag{3.78}$$

Similarly, let π_2^2 represent the potency of A_1 with respect to the circle k_{W2}^1, then we have

$$\overline{P_2^1 A_1}^3 = \rho \pi_2^2 \tag{3.79}$$

Since in the two positions 1 and 2, both ρ are equal, combining equations (3.78) and (3.79) gives

$$\frac{\overline{P_1 A_1}^3}{\overline{P_2^1 A_1}^3} = \frac{\pi_1^2}{\pi_2^2} \tag{3.80}$$

Introducing now a new parameter ζ, we can resolve equation (3.80) into the following two equations

$$\overline{P_1 A_1}^3 = \zeta^3 \overline{P_2^1 A_1}^3 \tag{3.81}$$

and

$$\pi_1^2 = \zeta^3 \pi_2^2 \tag{3.82}$$

For different values of ζ, equations (3.81) and (3.82) represent respectively a pencil of circles. For a particular value of ζ, the circle of equation (3.81) intersects the circle of equation (3.82). These intersection points are the points on the q_1-curve. We shall consider them separately.

(1) Pencil of circles of equation (3.81). These will be called hereafter the first pencil of circles. Transform equation (3.81) into

$$\overline{P_1 A_1}^2 = \zeta^2 \overline{P_2^1 A_1}^2 \tag{3.81a}$$

Now choose $P_1 P_2^1$ as the ξ-axis, and take the middle point O' of $\overline{P_1 P_2^1}$ as the origin, and the line passing through O' and perpendicular to the ξ-axis as the η-axis. The coordinates of point A_1 with respect to this $\xi\eta$-system are (ξ, η). Since $\overline{P^1 P_2^1} = s$, equation (3.81a) can be written as

$$\left(\xi + \frac{s}{2}\right)^2 + \eta^2 = \zeta^2 \left[\left(\xi - \frac{s}{2}\right)^2 + \eta^2\right] \tag{3.81b}$$

and further reduced to an equation of a single circle

$$\left(\xi - \frac{s}{2}\frac{\zeta^2 + 1}{\zeta^2 - 1}\right)^2 + \eta^2 = \frac{s^2}{4}\left[\left(\frac{\zeta^2 + 1}{\zeta^2 - 1}\right)^2 - 1\right] \tag{3.81c}$$

Draw a circle k_0 with O' as its centre and passing through the two points P_1, P_2^1. Equation (3.81c) indicates that the centres of the circles of the first pencil always lie on the ξ-axis, and at a distance $[(\zeta^2+1)/(\zeta^2-1)]\,s/2$ from O', and all circles are orthogonal to the circle k_0, with a radius equal to the tangent from the centre of the circle to k_0 (please refer to Fig. 3.37). The circles $\zeta = m$ and $\zeta = -m$ are

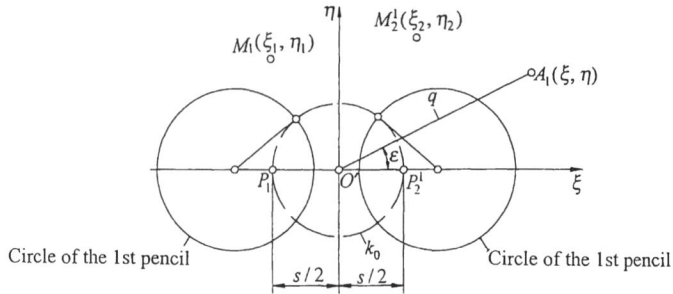

Fig. 3.37. Definition of $\xi\eta$-coordinate system.

identical. In case $m > 1$, then $1/m < 1$; the circle for $\zeta = 1/m$ is equal to the circle for $\zeta = m$, but on the negative side of the ξ-axis, both circles being at the same distance from O'. For $\zeta = 0$, the circle shrinks into a single point P_1, and for $\zeta \to \infty$, the circle shrinks into a single point P_2^1.

(2) Pencil of circles of equation (3.82). These will be called hereafter the second pencil of circles (please refer to Fig. 3.36). By the definition of π in equation (3.77), we have

$$\pi^2 = \left|\overline{PA} \cdot \overline{AA_W}\right| = \overline{PA}(\overline{PA} - \delta\sin\theta_A) = \overline{MA}^2 - \left(\frac{\delta}{2}\right)^2 \qquad (3.83)$$

In Fig. 3.38, let the centres of the circles k_{W1} and k_{W2}^1 be M_1 and M_2^1 respectively. Construct a (u, v) coordinate system, with $M_1 M_2^1$ as the u-axis, the middle point O'' of $\overline{M_1 M_2^1}$ as the origin, and the line passing through O'' and perpendicular to the u-axis as the v-axis. Further let $\overline{M_1 M_2^1} = d$. The coordinates of the point A_1 with respect to this coordinate system are (u, v). Substituting equation (3.83) into equation (3.82) gives

$$\left(u + \frac{d}{2}\right)^2 + v^2 - \left(\frac{\delta_1}{2}\right)^2 = \zeta^3 \left[\left(u - \frac{d}{2}\right)^2 + v^2 - \left(\frac{\delta_2}{2}\right)^2\right] \qquad (3.84)$$

Reducing again this equation to an equation of a single circle gives

Dimensional synthesis of four-bar linkages --- body guidance problems   107

$$\left(u - \frac{d}{2}\frac{\zeta^3+1}{\zeta^3-1}\right)^2 + v^2 = \left(\frac{d}{2}\right)^2\left[\left(\frac{\zeta^3+1}{\zeta^3-1}\right)^2 - 1\right] - \frac{1}{\zeta^3-1}\left[\left(\frac{\delta_1}{2}\right)^2 - \zeta^3\left(\frac{\delta_2}{2}\right)^2\right] \quad (3.84a)$$

Equation (3.84a) indicates that the centres of the circles of the second pencil always lie on the u-axis at a distance equal to $(d/2)\,[(\zeta^3+1)/(\zeta^3-1)]$. We consider the following two cases:

(i) In case the circles k_{W1} and k_{W2}^1 intersect in two real points, all circles of the second pencil pass through these two intersections.

(ii) In case the two circles k_{W1} and k_{W2}^1 do not intersect in real points, make a circle k_e, as shown in Fig. 3.38,

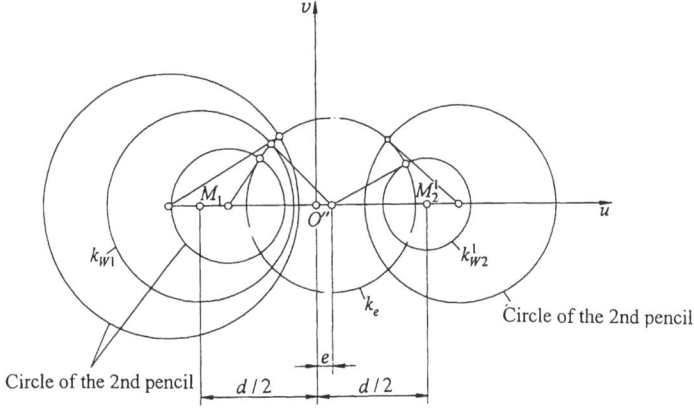

Fig. 3.38.   Definition of uv-coordinate system.

$$k_e: (u-e)^2 + v^2 = \left\{\frac{d}{2} + \frac{1}{2d}\left[\left(\frac{\delta_1}{2}\right)^2 - \left(\frac{\delta_2}{2}\right)^2\right]\right\}^2 - \left(\frac{\delta_1}{2}\right)^2$$

whose centre is at $(e, 0)$, where $e = [(\delta_1/2)^2-(\delta_2/2)^2]/(2d)$. The circle k_e is orthogonal to both k_W and k_{W2}^1, and all circles of the second pencil are orthogonal to k_e, the radius being equal to the length of the tangent from the centre of that circle to k_e. If the centre of the circle is inside k_e, that circle does not exist.

For example, in constructing the q_1-curve, suppose $\zeta = 1.5$, the centre of the circle of the first pencil is at 2.600 $s/2$ on the ξ-axis, and the centre of the circle of the second pencil is at 1.842 $d/2$ on the u-axis. Next assume $\zeta = -1.5$, then the centre of the circle of the first pencil is again at 2.600 $s/2$ on the ξ-axis and identical with the circle drawn before, but the centre of the circle of the second pencil is at

0.543 $d/2$ on the u- axis. Hence one circle of the first pencil corresponds to two circles of the second pencil. There are altogether four intersections. These intersections are points on the q_1-curve.

Although the present method is considered as geometrical, the points on the q_1-curve are not found by geometrical construction, but by calculation. To draw the curve by a computer, the simplest way is still by using the $\xi\eta$-coordinate system in Fig. 3.37, as that shown in Fig. 3.39. The equation of the circles of the first pencil is still the equation (3.81b)

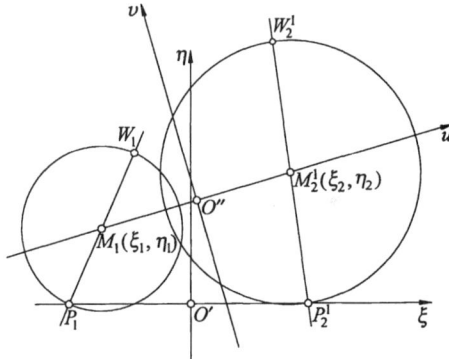

Fig. 3.39. Coordinates of M_1 and M_2^1 in $\xi\eta$-coordinate system.

$$\left(\xi+\frac{s}{2}\right)^2 + \eta^2 = \zeta^2 \left[\left(\xi-\frac{s}{2}\right)^2 + \eta^2\right] \qquad [(3.81b)]$$

The equation of the circles of the second pencil, with respect to the $\xi\eta$-coordinate system, is

$$(\xi-\xi_1)^2 + (\eta-\eta_1)^2 - \left(\frac{\delta_1}{2}\right)^2 = \zeta^3 \left[(\xi-\xi_2)^2 + (\eta-\eta_2)^2 - \left(\frac{\delta_2}{2}\right)^2\right] \qquad (3.85)$$

where $(\xi_1,\eta_1),(\xi_2,\eta_2)$ are respectively the coordinates of the centres M_1, M_2^1 of the circles k_{W1}, k_{W2}^1 with respect to the (ξ,η) coordinate system. It can be seen from Fig. 3.35 that

$$\begin{aligned} \xi_1 &= \left(\frac{\delta_1}{2}\right)\sin\sigma - \frac{s}{2} \\ \eta_1 &= \left(\frac{\delta_1}{2}\right)\cos\sigma \end{aligned} \qquad (3.86)$$

$$\left.\begin{array}{l}\xi_2 = \dfrac{s}{2} - \left(\dfrac{\delta_2}{2}\right)\sin(v-\sigma) \\ \eta_2 = \left(\dfrac{\delta_2}{2}\right)\cos(v-\sigma)\end{array}\right] \tag{3.87}$$

Eliminating ζ from equations (3.81b) and (3.85), we get

$$\left[\frac{(\xi-\xi_1)^2+(\eta-\eta_1)^2-(\delta_1/2)^2}{(\xi-\xi_2)^2+(\eta-\eta_2)^2-(\delta_2/2)^2}\right]^2 = \left[\frac{(\xi+s/2)^2+\eta^2}{(\xi-s/2)^2+\eta^2}\right]^3 \tag{3.88}$$

Developing equation (3.88) and taking the polar coordinates (q,ε) of A_1 with respect to the polar line $O'\xi$, i.e.

$$\left.\begin{array}{l}\xi = q\cos\varepsilon \\ \eta = q\sin\varepsilon\end{array}\right]$$

into consideration, we get another polar equation of the q_1-curve:

$$b_9 q^9 + b_8 q^8 + b_7 q^7 + b_6 q^6 + b_5 q^5 + b_4 q^4 + b_3 q^3 + b_2 q^2 + b_1 q + b_0 = 0 \tag{3.89}$$

The coefficients in equation (3.89) are given as follows:
Assume that

$$\begin{array}{l}m_1 = -2(\xi_1\cos\varepsilon + \eta_1\sin\varepsilon) \\ n_1 = \xi_1^2 + \eta_1^2 - (\delta_1/2)^2 \\ m_2 = -2(\xi_2\cos\varepsilon + \eta_2\sin\varepsilon) \\ n_2 = \xi_2^2 + \eta_2^2 - (\delta_2/2)^2 \\ m_3 = s\cos\varepsilon \\ n_3 = (s/2)^2\end{array}$$

In the above expressions m_1, m_2, m_3 are functions of ε, while n_1, n_2, n_3 are all constants.

$$b_9 = 2(m_1 - m_2) - 6m_3$$

$$b_8 = (m_1^2 - m_2^2) + 2(n_1 - n_2) - 6m_3(m_1 + m_2)$$

$$b_7 = 2(m_1 n_1 - m_2 n_2) - 3m_3(m_1^2 + m_2^2) - 6m_3(n_1 + n_2)$$
$$+ 6(n_3 + m_3^2)(m_1 - m_2) - 2m_3(6n_3 + m_3^2)$$

$$b_6 = (n_1^2 - n_2^2) - 6m_3(m_1 n_1 + m_2 n_2) + 3(n_3 + m_3^2)(m_1^2 - m_2^2)$$
$$+ 6(n_3 + m_3^2)(n_1 - n_2) - 2m_3(6n_3 + m_3^2)(m_1 + m_2)$$

$$b_5 = -3m_3(n_1^2 + n_2^2) + 6(n_3 + m_3^2)(m_1 n_1 - m_2 n_2) - m_3(6n_3 + m_3^2)(m_1^2 + m_2^2)$$
$$- 2m_3(6n_3 + m_3^2)(n_1 + n_2) + 6n_3(n_3 + m_3^2)(m_1 - m_2) - 6m_3 n_3^2$$

$$b_4 = 3(n_3 + m_3^2)(n_1^2 - n_2^2) - 2m_3(6n_3 + m_3^2)(m_1 n_1 + m_2 n_2)$$
$$+ 3n_3(n_3 + m_3^2)(m_1^2 - m_2^2) + 6n_3(n_3 + m_3^2)(n_1 - n_2) - 6m_3 n_3^2(m_1 + m_2)$$

$$b_3 = -m_3(6n_3 + m_3^2)(n_1^2 + n_2^2) + 6n_3(n_3 + m_3^2)(m_1 n_1 - m_2 n_2)$$
$$- 3m_3 n_3^2(m_1^2 + m_2^2) - 6m_3 n_3^2(n_1 + n_2) + 2n_3^3(m_1 - m_2)$$

$$b_2 = 3n_3(n_3 + m_3^2)(n_1^2 - n_2^2) - 6m_3 n_3^2(m_1 n_1 + m_2 n_2)$$
$$+ n_3^3(m_1^2 - m_2^2) + 2n_3^3(n_1 - n_2)$$

$$b_1 = -3m_3 n_3^2(n_1^2 + n_2^2) + 2n_3^3(m_1 n_1 - m_2 n_2)$$

$$b_0 = n_3^3(n_1^2 - n_2^2)$$

For a given value of ε, the 9th degree equation (3.89) can be solved for q, hence the q_1-curve is obtained.

Another method may be even more simple. Take the equations (3.81c) of the first pencil and (3.85) of the second pencil of circles. For an assumed value of ζ, two circles intersect in two real points. The line passing through these two points is

$$\left[s(\zeta^3 - 1)\frac{\zeta^2 + 1}{\zeta^2 - 1} - 2(\zeta^3 \xi_2 - \xi_1) \right]\xi - 2(\zeta^3 \eta_2 - \eta_1)\eta$$
$$= \left[\zeta^3 \left(\frac{\delta_2}{2}\right)^2 - \left(\frac{\delta_1}{2}\right)^2 \right] - (\zeta^3 \xi_2^2 - \xi_1^2) - (\zeta^3 \eta_2^2 - \eta_1^2) + \frac{s^2}{4}(\zeta^3 - 1) \qquad (3.90)$$

Eliminating η from equations (3.81c) and (3.90) results in a quadratic equation in the unknown ξ, the solution of which will yield two sets of values of (ξ, η), or the q_1-curve.

Having found the q_1-curve, a point E_1 can be chosen on q_1, and the corresponding E_2 can readily be found. The paths of these two points E_1, E_2 should curl to the same side, otherwise they are not usable. The following example will illustrate this.

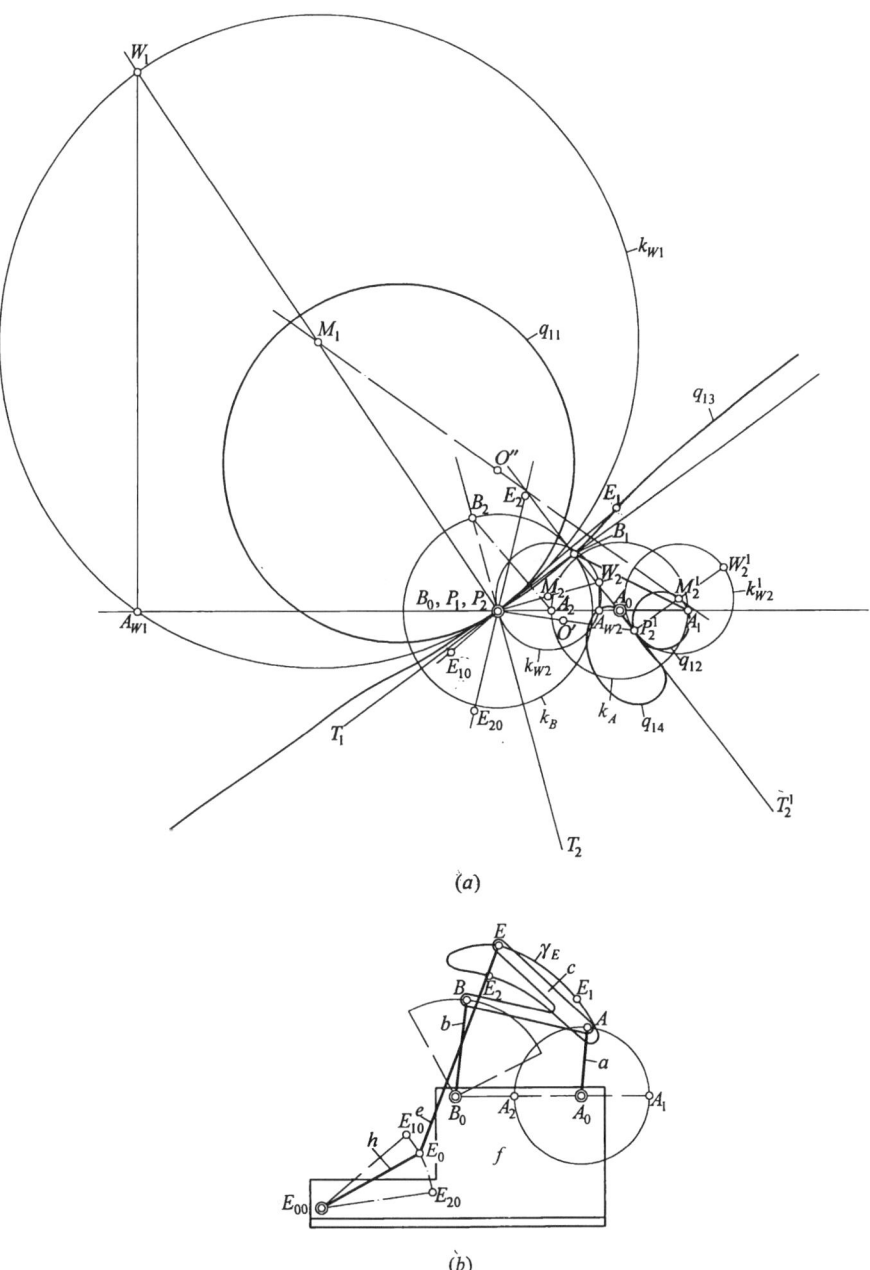

Fig. 3.40. Example: (a) q_1-curve contains four parts q_{11}, q_{12}, q_{13}, q_{14}; selection of E_1 on q_{13} or q_{14}. (b) Actual construction of the mechanism.

Example: A four-bar linkage $A_0 ABB_0$ is given as shown in Fig. 3.40(a). The lengths of the links are: $a = 14$, $b = 20$, $c = 26$, $f = 26$. It is obvious that this mechanism is a crank-rocker, a being able to make complete rotations. It is required to find a point E on the coupler c, so that as A is in positions A_1 and A_2, i.e. as a and f are collinear, the follower (the link $E_0 E_{00}$ to be synthesized) should have respectively one dwell.

The method of solution is to find the q_1-curve for the two positions $A_1 B_1$ and $A_2 B_2$ of the coupler, and then choose a point E_1 on q_1, and locate the corresponding E_2 on $A_2 B_2$. The radii of curvature of the paths at E_1 and E_2 are equal.

Now in the two positions $A_0 A_1 B_1 B_0$ and $A_0 A_2 B_2 B_0$ of the four-bar linkage, the poles P_1, P_2 of AB are at B_0, and $B_0 B_1, B_0 B_2$ are respectively the poletangents. By using equation (1.18) the points A_{W1} on k_{W1} and A_{W2} on k_{W2} can be found, so that the points W_1 and W_2 can also be found. The radii of the inflection circles are respectively $\delta_1/2 = 66.1534$ and $\delta_2/2 = 11.5897$. Consider P_2 as a point on $A_2 B_2$, and let it follow the movement of the latter back to $A_1 B_1$ to reach the position P_2^1. The position of this P_2^1 can be found by using the displacement matrix in equation (3.33). We can also calculate the angles $\sigma = -43.0441°$ and $\nu = -86.0882°$ in Fig. 3.40(a). The coordinates (ξ_1, η_1) and (ξ_2, η_2) can thus be found from equations (3.86), (3.87), and subsequently the q_1-curve can be obtained.

The q_1-curve shown in Fig. 3.40(a) includes four parts, namely, q_{11}, q_{12}, q_{13} and q_{14}. If a point E_1 is taken on q_{11} or q_{12}, the point E_2 corresponding to E_1 will have a path which curls to the different side from that of E_1, and is of course not usable. We have only to choose a point E_1 on either q_{13} or q_{14}, because the path of the corresponding point E_2 of which will then curl to the same side as that of E_1. As shown in the figure, a point E_1 has been chosen on q_{13}, and its corresponding second position is E_2. The centre of curvature of the path of E_1 is at E_{10}, and that of the path of E_2 is at E_{20}. $\overline{E_1 E_{10}} = \overline{E_2 E_{20}} = 45.8$. A link e can therefore be made to connect E_1 and E_{10}, as shown in Fig.3.40(b). Choose a point on the perpendicular bisector of $\overline{E_{10} E_{20}}$ as E_{00}. Taking E_{00} as a fixed joint, we can make a link h to connect $E_{10} E_{00}$, so that as the point A moves from A_1 to A_2, the link h exhibits respectively a dwell. Fig. 3.40(b) shows a practical construction of this mechanism.

The locus of all centres of curvature E_{10} corresponding to the point E_1 on the q_1-curve, is also a curve of 9th degree, which is called after Alt the q_M-curve. However, since this is the locus of E_{10}, not that of E_{20}, it is the locus of the centre of curvature of the curve of equal radius of curvature of the second kind for the moving body in its first position, hence it would be appropriate to call it q_{M1}.

If it is required to have the link h exhibiting three dwells within one cycle of motion of the four-bar linkage A_0ABB_0, we have then to find for three positions c_1, c_2, c_3 of the coupler c first the q_1-curve $q_{1(12)}$ between c_1, c_2, and then the q_1-curve $q_{1(13)}$ between c_1, c_3. The intersection between $q_{1(12)}$ and $q_{1(13)}$ is the required coupler point E_1. Points E_2, E_3 can then readily be found from E_1. The three points E_1, E_2, E_3 have each at its own position the respective centres of curvature E_{10}, E_{20}, E_{30} of their paths. The centre of the circle passing through these three centres is the point E_{00} which can be taken as the pivot centre of the link h. Although the above procedure is theoretically correct, a satisfactory solution may not easily be found in practice.

3.6 Guiding a body through four finitely separated positions --- the centre-point curve and circle-point curve

3.6.1 The centre-point curve

Suppose four positions of a body c are given. For an arbitrarily chosen point A_1 on c_1, it is by no means certain that the four homologous points A_1, A_2, A_3, A_4 should lie on a circle. To find out those points A_1 on c_1, whose four homologous points lie on a circle, it is appropriate to use a logical procedure rather than by trial-and-error. The locus of such points on the body satisfying the condition of four homologous points lying on a circle is called the *circle-point curve*. This locus is the same, no matter on which one of the four positions c_1, c_2, c_3, c_4 it lies. Hence we have only to consider the one that lies on c_1, and denote it by k_{1234}. For the same reason, we can also not arbitrarily choose on the fixed body a centre of the circle passing through four homologous points. There is only one *centre-point* corresponding to one *circle-point*. The locus of such centre points on the fixed plane is called the *centre-point curve*, and will be denoted by m_{1234}. Centre-point curve and *circle-point curve* are called together *Burmester curves* (Burmester, 1876, 1877a, b). The terminology centre-point curve or circle-point curve refers to four finitely separated positions. This is because, for three separated positions of a moving body, any point on the body can be chosen as the circle-point and any point on the fixed body can be chosen as the centre-point; and for five finitely separated positions, only finite number of points can be chosen as centre-points and circle-points, their loci constituting no curves.

Let us see how to find the centre-point curve. In Fig. 3.41, suppose four positions c_1, c_2, c_3, c_4 of a body c are given. A is a point on body c, whose four homologous points A_1, A_2, A_3, A_4 do lie on a circle k_A with centre A_0. We shall

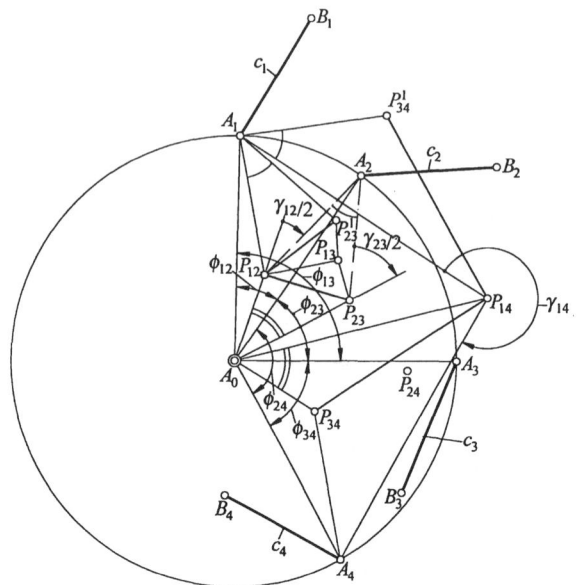

Fig. 3.41. For the derivation of centre-point curve and circle-point curve.

consider hereafter $A_0 A$ as a crank, the crank angle notations being $\phi_{12} = \sphericalangle A_1 A_0 A_2$, $\phi_{23} = \sphericalangle A_2 A_0 A_3$, $\phi_{34} = \sphericalangle A_3 A_0 A_4$. Please do not confuse the angle ϕ with γ, the angle of rotation of the body c itself. From the four positions B_1, B_2, B_3, B_4 of another point B on the body c, the six poles among c_1, c_2, c_3, c_4 can be found:

$$P_{12} \quad P_{23} \quad P_{34}$$
$$P_{13} \quad P_{24}$$
$$P_{14}$$

It can be seen from Fig. 3.41 that

P_{12}, P_{23}, P_{34} lie respectively on the bisectors of the angles $\phi_{12}, \phi_{23}, \phi_{34}$;

P_{13}, P_{24}, P_{14} lie respectively on the bisectors of the angles $\phi_{13}, \phi_{24}, \phi_{14}$.

and further from the figure that

$$\sphericalangle P_{12} A_0 P_{23} = \sphericalangle P_{12} A_0 A_2 + \sphericalangle A_2 A_0 P_{23}$$
$$= \phi_{12}/2 + \phi_{23}/2 = \phi_{13}/2$$

and

$$\sphericalangle P_{14} A_0 P_{34} = \sphericalangle P_{14} A_0 A_4 - \sphericalangle P_{34} A_0 A_4$$
$$= \phi_{14}/2 - \phi_{34}/2 = \phi_{13}/2$$

Dimensional synthesis of four-bar linkages --- body guidance problems

Therefore we have

$$\angle P_{12}A_0P_{23} = \angle P_{14}A_0P_{34} \tag{3.91}$$

Equation (3.91) implies that: the angles subtended by $\overline{P_{12}P_{23}}$ and $\overline{P_{14}P_{34}}$ at A_0 are equal. These four poles $P_{12}, P_{23}, P_{14}, P_{34}$ constitute a quadrilateral which possesses a special notation that none of the four subscripts of any two diagonal poles are identical. Take for instance, the poles P_{12}, P_{34} or P_{13}, P_{24} as one pair of diagonal poles. Next take from the rest four poles the two without repeating subscript as the

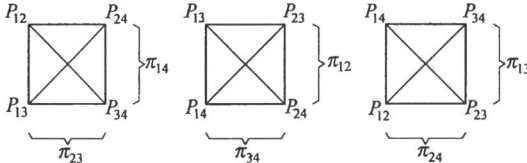

Fig. 3.42. Three oppisite-pole quadrilaterals.

second pair of diagonal poles, as shown in Fig. 3.42. Such quadrilaterals are called *opposite-pole quadrilaterals*. There are altogether three different opposite-pole quadrilaterals from the six poles. From equation (3.91) we can establish the following theorem:

Theorem 12: The opposite sides of an opposite-pole quadrilateral subtend (1) equal angles, or (2) supplementary angles (please refer to method 2 of Section 3.6.2) at a point on the centre-point curve, and the reverse is also true.

The latter part of theorem 12 means that, any point at which the opposite sides of an opposite-pole quadrilateral subtend equal (or supplementary) angles lies on the centre-point curve. In the foregoing we have only proved the former part of the theorem. The latter part is left as Exercise 3.17. Moreover, if the condition (3.91) is satisfied, the other pair of opposite sides $\overline{P_{12}P_{14}}$ and $\overline{P_{23}P_{34}}$ of the quadrilateral must also subtend equal angles at A_0. Hence it is not necessary to specify the particular pair of opposite sides. Although there are three opposite-pole quadrilaterals, any one can be used to construct the centre-point curve. The result is the same.

3.6.2 Construction methods of centre-point curve

(*a*) Method 1

The method shown in Fig. 3.43 is taken from a practical example of (Kraus, 1956, Fig. 79). In this example, the mechanism to be synthesized is a straight-line four-bar linkage. Given four collinear positions E_1, E_2, E_3, E_4 of a coupler point E, a centre point A_0 of a crank A_0A, and the crank angle ϕ_{14} corresponding to the coupler point displacement $E_1 \to E_4$. It is required to synthesize the four-bar

116 **Kinematics and Design of Planar Mechanisms**

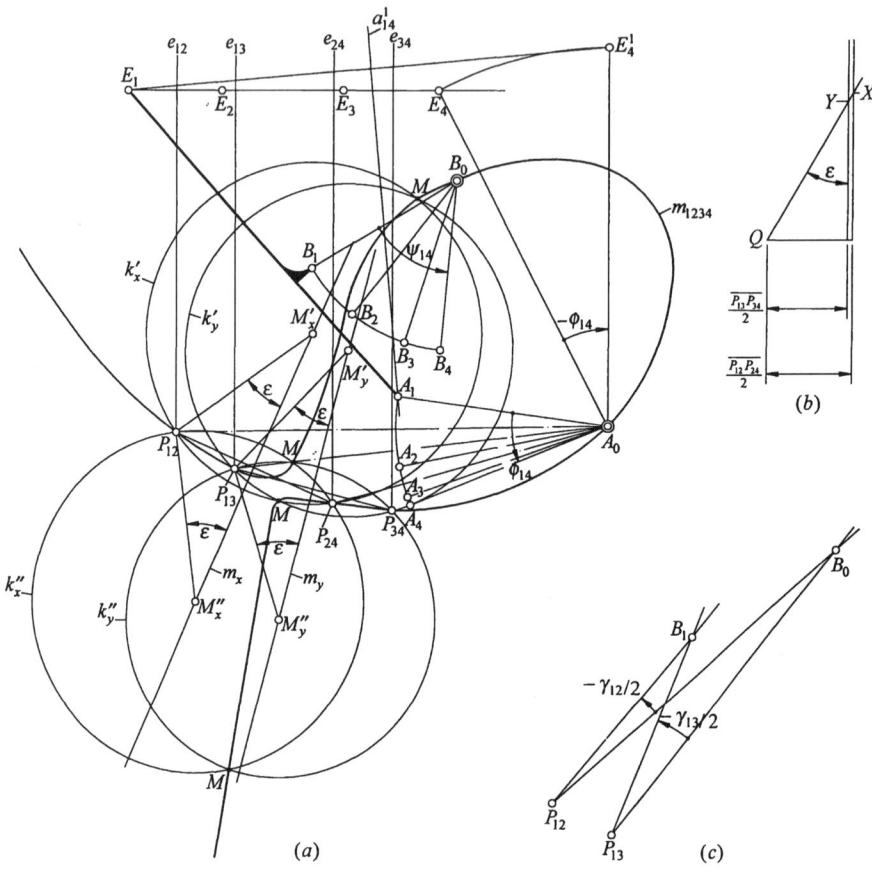

Fig. 3.43. Construction of centre-point curve m_{1234}--- method 1.

linkage. Please note that the points A_1, B_1, A_0 are not known yet.

The solution is as follows: In Fig. 3.43(a), first revolve the point E_4 through an angle $-\phi_{14}$ about A_0 to E_4^1. Make the perpendicular bisector a_{14}^1 of $\overline{E_1 E_4^1}$. Choose arbitrarily a point A_1 on a_{14}^1 (please refer to Section 6.1). We are able to construct the crank circle of A. From the length of $\overline{E_1 A_1}$, we can determine the points A_2, A_3, A_4. Knowing the four positions $A_1 E_1, A_2 E_2, A_3 E_3, A_4 E_4$ of the coupler, we can find the two opposite sides $\overline{P_{12} P_{24}}, \overline{P_{13} P_{34}}$ of the opposite-pole quadrilateral $P_{12} P_{24} P_{13} P_{34}$. Draw the perpendicular bisectors m_x of $\overline{P_{12} P_{24}}$ and

Dimensional synthesis of four-bar linkages --- body guidance problems 117

m_y of $\overline{P_{13}P_{34}}$. Construct another diagram 3.43(b), in which we draw two vertical sides of two right triangles with $\overline{P_{12}P_{24}}/2$ and $\overline{P_{13}P_{34}}/2$ as bases. Lay an inclined line from a point Q on the base line, intersecting the two vertical lines in two points X, Y. Draw arcs with P_{12} as centre and $\overline{QX}$ as radius, cuttng m_x in two centres M'_x, M''_x. With these two centres draw the two circles k'_x, k''_x passing through P_{12} and P_{24}. Next draw arcs with P_{13} as centre and $\overline{QY}$ as radius, cutting m_y in two centres M'_y, M''_y. Draw two circles k'_y, k''_y about these two centres passing through P_{13}, P_{34}. The circles k'_x and k'_y intersect in two M points, and k''_x and k''_y also intersect in two M points. These four M's are the points on the centre-point curve m_{1234}. The reason is: Let the angle between the inclined line and the vertical lines in Fig. 3.43(b) be denoted by ε. The central angles subtended by the arcs $\overset{\frown}{P_{12}P_{24}}$ and $\overset{\frown}{P_{13}P_{34}}$ are both 2ε, and the circumferential angles subtended by the arc $\overset{\frown}{P_{12}P_{24}}$ or $\overset{\frown}{P_{13}P_{34}}$ at any point on the circumferences of the circles $k'_x, k'_y; k''_x, k''_y$ are all equal to ε (or $180° - \varepsilon$). Therefore the M's are points on the centre-point curve. In this way we can draw another set of four circles by varying the slope of the inclined line passing through point Q in Fig. 3.43(b), and get another set of four M points. The whole centre-point curve m_{1234} can thus be completed.

In Fig. 3.43(a), the centre point A_0 is already given, therefore it must be on the curve m_{1234}. Now choose another centre-point B_0 on m_{1234}. Extending theorem 7 and applying it, we can find the corresponding circle-point B_1 in Fig. 3.43(c). Join first $P_{12}B_0$ and $P_{13}B_0$, and build respectively the angles $-\gamma_{12}/2$, $-\gamma_{13}/2$ (please note that because γ_{12}, γ_{13} are clockwise, $-\gamma_{12}, -\gamma_{13}$ are counterclockwise). We then readily obtain the point B_1. If, however, the angle of rotation ψ_{14} of B_0B is prescribed, we cannot arbitrarily choose the centre-point B_0 on m_{1234}. It can be seen from Fig. 3.41 that the angle subtended by $\overline{P_{12}P_{23}}$ at A_0 is $\phi_{13}/2$, and the angle subtended by $\overline{P_{14}P_{24}}$ at A_0 is $\phi_{14}/2$. Applying this principle to Fig. 3.43, and taking $\varepsilon = \psi_{14}/2$, we can get the intersection of the two circles k'_x, k'_y thus obtained, which is the required centre-point B_0.

(b) Method 2

Fig. 3.44 shows a construction method taken from (Beyer, 1953, Fig. 117). Two opposite sides $\overline{P_{13}P_{23}}, \overline{P_{14}P_{24}}$ of an opposite-pole quadrilateral are given. Draw as before the perpendicular bisectors m_x and m_y of $\overline{P_{13}P_{23}}$ and $\overline{P_{14}P_{24}}$ respectively. Let the intersection of $P_{13}P_{23}$ and $P_{14}P_{24}$ be denoted by π_{12}. Draw the circumcircle k_{12} of $\Delta P_{23}P_{24}\pi_{12}$ and choose arbitrarily a point K_{12} on k_{12}. Join

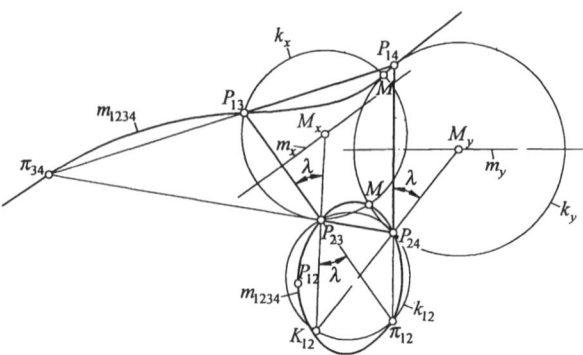

Fig. 3.44. Construction of centre-point curve m_{1234}--- method 2.

$K_{12} P_{23}$, $K_{12} P_{24}$, cutting respectively m_x, m_y in M_x, M_y. Draw circles k_x, k_y with M_x, M_y as centres passing respectively through P_{13}, P_{23} and P_{14}, P_{24}. The other two intersections of k_x, k_y are points M on the centre-point curve m_{1234}. The reason is the same as that for method 1. Denoting $\sphericalangle K_{12} P_{23} \pi_{12}$ by λ, we see that the central angles subtended by the arcs $\overset{\frown}{P_{13}P_{23}}$ and $\overset{\frown}{P_{14}P_{24}}$ are both $2(\pi/2 - \lambda)$, that the circumferential angles subtended by the chords $\overline{P_{13}P_{23}}$ and $\overline{P_{14}P_{24}}$ at a point on the larger arcs of the circles k_x, k_y are both $\pi/2 - \lambda$, and that those at a point on the smaller arcs of these two circles are both $\pi - (\pi/2 - \lambda) = \pi/2 + \lambda$. Therefore the angles subtended by $\overline{P_{13}P_{23}}$ at the two M's in Fig. 3.44 are $\pi/2 - \lambda$, while the angles subtended by $\overline{P_{14}P_{24}}$ at the M's are $\pi/2 + \lambda$. These two angles are supplementary. We can therefore conclude that, the angles subtended by the opposite sides of an opposite-pole quadrilateral at a centre-point from the same side are equal, otherwise the angles are supplementary. This is the origin of the above mentioned theorem 12. In fact, as long as the two corresponding circles intersect, their intersection points are on the centre-point curve, no matter what the angular relation is.

In Fig. 3.45, all six poles $P_{12}, P_{13}, P_{14}, P_{23}, P_{24}, P_{34}$ are present. For $\Delta P_{12} P_{23} P_{13}$, we have known from Fig. 3.12 that $\sphericalangle P_{23} P_{12} P_{13} = |\gamma_{12}/2|$. Similarly, for $\Delta P_{12} P_{24} P_{14}$, we have also $\sphericalangle P_{24} P_{12} P_{14} = |\gamma_{12}/2|$. Hence $\overline{P_{13}P_{23}}$ and $\overline{P_{14}P_{24}}$ subtend equal angles at P_{12}. Therefore P_{12} is a point on the centre-point curve. Similarly, it can be proved that the other five poles are also points on the centre-point curve. Moreover, the angles subtended by $\overline{P_{13}P_{23}}$ and $\overline{P_{14}P_{24}}$ at the intersection point π_{34} of $P_{13}P_{14}$ and $P_{23}P_{24}$ are also equal. Similarly, the other five

Dimensional synthesis of four-bar linkages --- body guidance problems 119

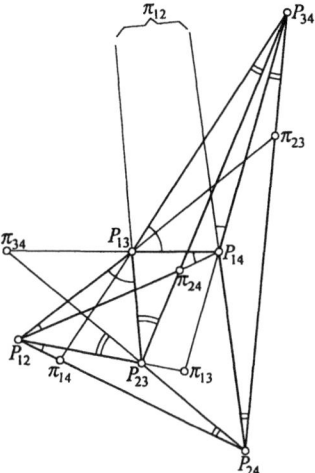

Fig. 3.45. Angular relationship of the six poles $P_{12}, P_{13}, P_{14}, P_{23}, P_{24}, P_{34}$.

intersection points $\pi_{12}, \pi_{13}, \pi_{14}, \pi_{23}, \pi_{24}$ of the opposite sides of the opposite-pole quadrilateral are all points on the centre-point curve. This means that there have already been 12 points of the centre-point curve before the curve is drawn. (Keller, 1965) shows how to sketch a centre-point curve from a given opposite-pole quadrilateral in a simple way.

From the construction method of a centre-point curve it is well-known in geometry that it is a circular cubic. Therefore we arrive at the following theorem:

Theorem 13 The centre-point curve is a circular cubic passing through the six poles and the six intersections of the opposite sides of the three opposite-pole quadrilaterals.

For the definition of a circular curve, please refer to Section A1.2. For the explanation of a circular curve becoming a circular cubic, please refer to the later Section 3.6.6.

3.6.3 The circle-point curve

Please refer again to Fig. 3.41. It can be seen that

$$\sphericalangle P_{12}A_2P_{23} = \sphericalangle P_{12}A_2A_0 + \sphericalangle A_0A_2P_{23}$$
$$= (|\gamma_{12}/2|-\phi_{12}/2) + (|\gamma_{23}/2|-\phi_{23})$$
$$= (|\gamma_{13}|-\phi_{13})/2$$

and

$$\sphericalangle P_{14}A_4P_{34} = \sphericalangle P_{14}A_4A_0 - \sphericalangle A_0A_4P_{34}$$
$$= (|\gamma_{14}/2|-\phi_{14}/2) - (|\gamma_{34}/2|-\phi_{34}/2)$$
$$= (|\gamma_{13}|-\phi_{13})/2$$

Therefore

$$\sphericalangle P_{12}A_2P_{23} = \sphericalangle P_{14}A_4P_{34} \tag{3.92}$$

A comparison with Fig.3.17 shows that the point P_{23}^1 is the mirror image of P_{23} with respect to $P_{12}P_{13}$. Rotating $\Delta A_2P_{12}P_{23}$ about P_{12} through an angle of $-\gamma_{12}$ will reach the position $\Delta A_1P_{12}P_{23}^1$. Similarly, P_{34}^1 is the mirror image of P_{34} with respect to $P_{13}P_{14}$. Rotating $\Delta A_4P_{14}P_{34}$ about P_{14} through an angle $-\gamma_{14}$ will reach the position $\Delta A_1P_{14}P_{34}^1$. Hence equation (3.92) can be rewritten as

$$\sphericalangle P_{12}A_1P_{23}^1 = \sphericalangle P_{14}A_1P_{34}^1 \tag{3.93}$$

Equation (3.93) implies that, the angles subtended by the two opposite sides $\overline{P_{12}P_{23}^1}$ and $\overline{P_{14}P_{34}^1}$ of the opposite-pole quadrilateral $P_{12}P_{23}^1P_{34}^1P_{14}$ at A_1 are equal. There are also three opposite-pole quadrilaterals of this kind, as shown in the following theorem 14. Similar to theorem 12, we have for the circle-point curve (the locus of the circle-point A_1, denoted by k_{1234} §):

Theorem 14: The angles subtended by the opposite sides of any one of the opposite-pole quadrilaterals shown in Fig. 3.46 at a point on the circle-point curve are (1) equal, or (2) supplementary.

Corresponding to theorem 13, we have also

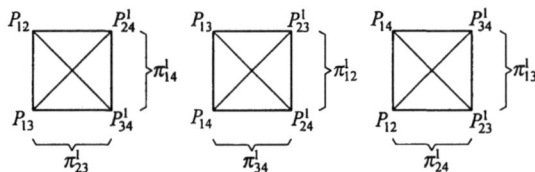

Fig. 3.46. Three opposite-pole quadrilaterals containing $P_{12}, P_{13}, P_{14}, P_{23}^1, P_{24}^1, P_{34}^1$.

Theorem 15: The circle-point curve is a circular cubic passing through the six poles $P_{12}, P_{13}, P_{14}, P_{23}^1, P_{24}^1, P_{34}^1$, and the six intersections $\pi_{12}^1, \pi_{13}^1, \pi_{14}^1, \pi_{23}^1, \pi_{24}^1, \pi_{34}^1$, of the opposite sides of the three opposite-pole quadrilaterals in Fig. 3.46.

In general, to guide a body through four finitely separated positions, having established the circle-point curve k_{1234}, we can choose arbitrarily a point A_1 on the

---

§ The precise notation should be say, $k_{1234(1)}$, to specify that the curve is on body c_1, not on c_2, c_3, or c_4. However, for simplicity reason, we write it as k_{1234}.

curve. There are ∞^1 choices. Next choose another point B_1 on the same curve. Again there are ∞^1 choices. A_0 is determined by A_1, and B_0 is determined by B_1. There are altogether ∞^2 solutions.

Please note that, the synthesized four-bar linkage should be checked to see (1) if the linkage can move through the prescribed positions without obstructions (i.e. without dismantling and reassembling), and (2) if the body to be guided can be moved in the prescribed sequence. This sort of problems are called *branch problems* and *order problems*. Please refer to (Filmon, 1972; Modler, 1972; Waldron, 1976; Waldron & Strong, 1978).

3.6.4 The break-ups of centre-point curve and of circle-point curve

In special cases, the centre-point curve or circle-point curve may degenerate. The curve, being a circular cubic, may break up into a circle and a straight line, or into three straight lines. In the latter case these are often two straight lines and the line at infinity.

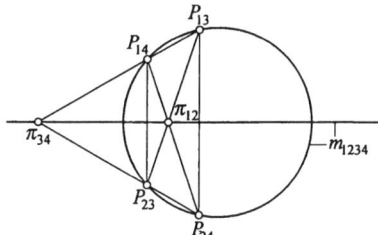

Fig. 3.47. Break up of centre-point curve m_{1234} into a circle and a line.

Fig. 3.47 shows a disposition of the four poles $P_{13}, P_{24}, P_{23}, P_{14}$ in a right trapezoid configuration. The points π_{12}, π_{34} lie on the centre line of the trapezoid. The centre-point curve m_{1234} breaks up into the circumcircle of the trapezoid and its

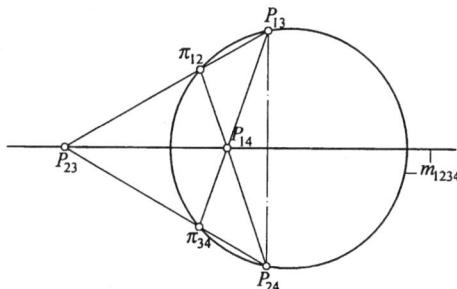

Fig. 3.48. Break up of centre-point curve m_{1234} into a circle and a line.

centre line. In Fig. 3.48 the disposition of the four poles is that $P_{13}P_{24}$ forms the base of the right trapezoid, and P_{14}, P_{23} lie on its centre line; the centre-point curve breaks up again into the circumcircle of the trapezoid and its centre line.

Fig. 3.49 shows the six poles disposed on two orthogonal straight lines. In this case the centre-point curve breaks up into these two lines and the line at infinity.

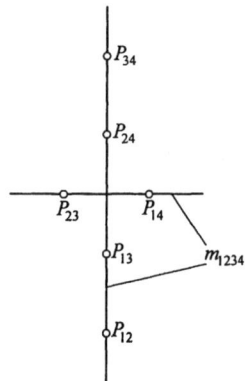

Fig. 3.49. Break up of centre-point curve m_{1234} into two orthogonal lines and the line at infinity.

Example: Adjustable dwell mechanism

This example is taken from (Beyer, 1953, Figs. 131,132). The synthesis problem shown in Fig. 3.50(a) is to make use of a slider-crank mechanism A_0AB to produce a dwell motion. As the crank pin travels through the arc $\widehat{A_1A_4}$, the slider pin B moves from B_1 towards the left and then backwards to the right to the initial position, or $B_4 = B_1$. The coupler point E passes through the four points E_1, E_2, E_3, E_4. If E_1 belongs to a point on the circle-point curve, the centre E_0 of the circle k_E on which the four E's lie should be a centre-point. Make a link e joining $E_1 E_0$, and a link h joining E_0 to another fixed joint E_{00}, to enable h swinging through a prescribed angle β, so that during the motion of E from E_1 to E_4, the link h may remain to be almost standstill. A_0, ϕ_{14}, β and E_{00} are given.

This problem can be solved as follows: Draw a vertical line from E_{00} perpendicular to a horizontal line passing through A_0, its foot being denoted by E_c. Dispose the angle β equally on both sides of $E_{00}E_c$ to obtain the two points E_0, E_0'. The crank length is then $a = \overline{A_0A} = \overline{E_0E_0'}/2$. Draw the crank circle k_A with A_0 as centre and a as radius. Dispose the angle ϕ_{14} equally on the upper and lower sides of the horizontal line to get the two points A_1, A_4. We then arbitrarily

Dimensional synthesis of four-bar linkages --- body guidance problems 123

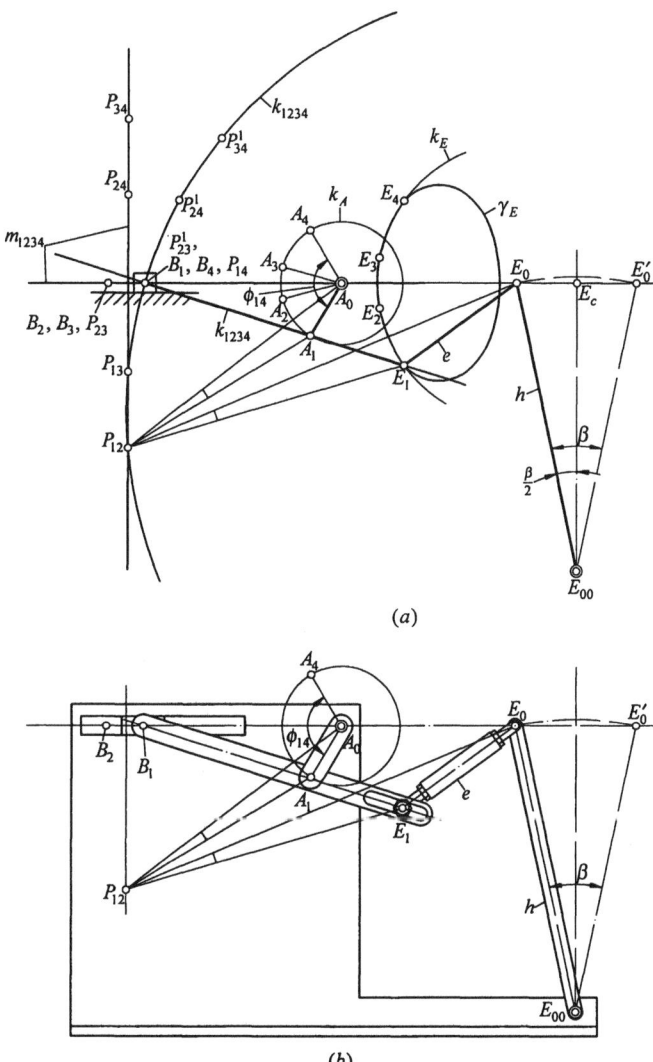

Fig. 3.50. Example: adjustable dwell mechanism. (*a*) Method of construction. (*b*) Practical construction.

choose the two points A_2, A_3 such that they are symmetrical to each other with respect to the horizontal line. Choosing any coupler length $c = \overline{AB}$, we get the two points $B_1 = B_4, B_2 = B_3$. Finally draw the perpendicular bisector of $\overline{B_1 B_2}$. Now, for the four positions of the coupler $c = AB$, the disposition of the six poles

$P_{12}, P_{13}, P_{14}, P_{23}, P_{24}, P_{34}$ are just as that shown in Fig. 3.49. P_{14} coincides with B_1, B_4, and P_{23} coincides with B_2, B_3. As mentioned before, the centre-point curve m_{1234} breaks up into the two orthogonal lines and the line at infinity.

For the points on the circle-point curve, besides the three given poles P_{12}, P_{13}, P_{14}, the point P_{23}^1 is the mirror image of P_{23} with respect to $P_{12}P_{13}$, coinciding with B_1, B_4, and P_{24}^1 is the mirror image of P_{24} with respect to $P_{12}P_{14}$, and P_{34}^1 is the mirror image of P_{34} with respect to $P_{13}P_{14}$. The disposition of the four poles $P_{12}, P_{13}, P_{24}^1, P_{34}^1$ is symmetrical with respect to A_1B_1, just like the case shown in Fig. 3.47. The circle-point curve k_{1234} therefore breaks up into a circle passing through these four poles and its diameter A_1B_1. (Please note that in general, the circle does not necessarily pass through B_1.)

With the known centre-point E_0, to find the corresponding circle-point E_1, we may apply again theorem 7 by building $\sphericalangle A_1P_{12}E_1 = \sphericalangle A_0P_{12}E_0$. The intersection of $P_{12}E_1$ and A_1B_1 is E_1. During the motion of E, the four homologous points E_1, E_2, E_3, E_4 lie indeed on a circle k_E. However, although the actual path γ_E of E and the circle k_E intersect in these four points, the path is not a circle. As the crank A_0A rotates through the angle ϕ_{14}, the link h remains almost standstill.

The sequence of the above synthesis is: $A_0, E_{00} \to E_c \to E_0, E_0' \to a \to k_A \to A_1, A_4 \to A_2, A_3 \to c \to A_1B_1, A_2B_2, A_3B_3, A_4B_4 \to m_{1234}', k_{1234} \to E_1$.

Fig. 3.50(b) shows a practical construction of this mechanism. If we vary the angle ϕ_{14}, but keep the length $\overline{AB}$ unchanged, all positions of B_1, B_2, P_{12} will vary, and the position of E_1 on A_1B_1 will also vary, but the position of E_0 will remain unchanged. The length $\overline{E_0E_1}$ will be changed. The practical way is to make a slot at E_1 on the link AB, and the link e is made of an adjustable screw buckle with identification marks. The linkage thus constructed is an adjustable dwell mechanism.

3.6.5 Four homologous lines passing through a fixed point and four homologous points on a line

Theorems 8 and 9 can now be extended. As mentioned before, for four finitely separated positions there are six poles. For these six poles there are four poletriangles and therefore four circumcircles. To facilitate recognition, the circumcircle of $\Delta P_{12}P_{23}P_{13}$ and its orthocenter will be denoted by k_{123} and H_{123} respectively, etc. In Fig. 3.51 are shown four circumcircles $k_{123}, k_{124}, k_{134}, k_{234}$ intersecting in a point S_0. (The fact that the four circles intersect in a single point is left as Exercise 3.23.) Corresponding to Fig. 3.19, the three mirror images k_1, k_2, k_3 of the circle k_{123} are now designated as $k_{123}^1, k_{123}^2, k_{123}^3$ respectively; and the three mirror images H_1,

Dimensional synthesis of four-bar linkages --- body guidance problems 125

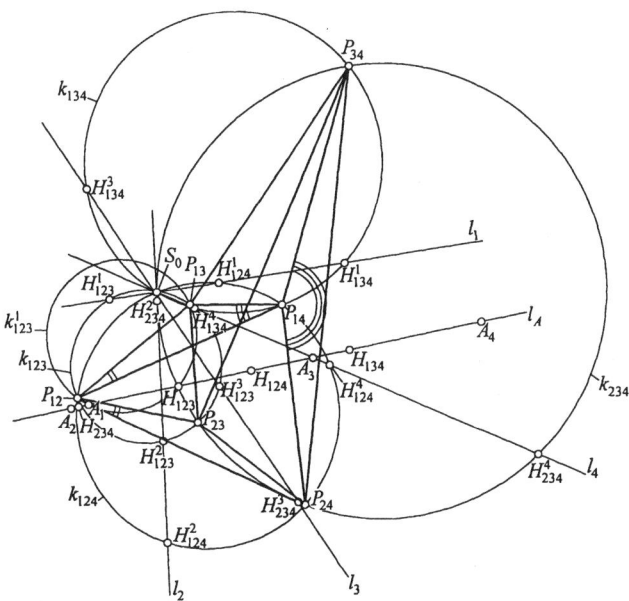

Fig. 3.51. Four homologous lines l_1, l_2, l_3, l_4 passing through a fixed point S_0, and four homologous points A_1, A_2, A_3, A_4 lying on a straight line l_A.

H_2, H_3 of H_{123} are now designated as $H^1_{123}, H^2_{123}, H^3_{123}$ respectively. For clarity, the collinear points are listed in Table 3.1, and the concurrent lines are listed in Table 3.2.

Table 3.1 Four homologous lines passing through a point

	S_o	S_o	S_o	S_o	
H_{123}	H^1_{123}	H^2_{123}	H^3_{123}		
H_{124}	H^1_{124}	H^2_{124}		H^4_{124}	
H_{134}	H^1_{134}		H^3_{134}	H^4_{134}	
H_{234}		H^2_{234}	H^3_{234}	H^4_{234}	
Line	l_1	l_2	l_3	l_4	Common point S_0

Let us start from four lines intersecting in a point. In Fig. 3.19, any point on the circle k can be chosen as the concurrent point S_0 of three homologous lines.

Table 3.2 Four homologous points lying on a line

Circle	Image circle				Common line
k_{123}	k_{123}^1	k_{123}^2	k_{123}^3		
k_{124}	k_{124}^1	k_{124}^2		k_{124}^4	
k_{134}	k_{134}^1		k_{134}^3	k_{134}^4	
k_{234}		k_{234}^2	k_{234}^3	k_{234}^4	
Intersection point	A_1	A_2	A_3	A_4	l_A

However, in Fig. 3.51, there is only one concurrent point S_0 for four homologous lines; and, as mentioned above, the point is the intersection point of the four circumcircles of the poletriangles. Next, we deal with four collinear points. According to Fig. 3.19, where any point on the mirror image k_1 of k can be chosen as the first point A_1 of three collinear homologous points, the only point in Fig. 3.51 which can be chosen as the first point A_1 of four collinear homologous points is the intersection point of the three image circles $k_{123}^1, k_{124}^1, k_{134}^1$. (In order to avoid confusion, only the circle k_{123}^1 is shown in Fig. 3.51, and the common line of the four points A_1, A_2, A_3, A_4 is denoted by l_A.) We have therefore the three following theorems, the proof of which is left as Exercises 3.24, 3.25 and 3.26.

Theorem 16 The common point S_0 of four concurrent homologous lines l_1, l_2, l_3, l_4 is the intersection point of the four circumcircles of the four poletriangles. S_0 belongs to a point on the centre-point curve.

Theorem 17 The first line l_1 of four concurrent homologous lines l_1, l_2, l_3, l_4 passes through the three points $H_{123}^1, H_{124}^1, H_{134}^1$.

Theorem 18 The first point A_1 of four collinear homologous points A_1, A_2, A_3, A_4 is the intersection point of the three mirror images $k_{123}^1, k_{124}^1, k_{134}^1$ of the three circumcircles of the poletriangles. The common line l_A passes through the four orthocentres $H_{123}, H_{124}, H_{134}, H_{234}$ of the four poletriangles. A_1 belongs to a point on the circle-point curve.

3.6.6 Circle-point curve and centre-point curve by algebraic method

Based on the algebraic method depicted in Section 3.4, the equations of circle-point curve and of centre-point curve can easily be formulated. Now because of the fourth position, we can write down, besides the two equations in (3.39), another one for the perpendicular bisector a_{14}. All three equations are written as follows:

Dimensional synthesis of four-bar linkages --- body guidance problems

$$\left.\begin{aligned} G_2 x_0 + S_2 y_0 + T_2 &= 0 \\ G_3 x_0 + S_3 y_0 + T_3 &= 0 \\ G_4 x_0 + S_4 y_0 + T_4 &= 0 \end{aligned}\right\} \quad (3.94)$$

Since we have selected the displacement matrix in equation (3.53), therefore

$$\left.\begin{aligned} G_j &= -\text{vers}\gamma_{1j} x_{A1} - \sin\gamma_{1j} y_{A1} + (x_{P1j}\text{vers}\gamma_{1j} + y_{P1j}\sin\gamma_{1j}) \\ S_j &= \sin\gamma_{1j} x_{A1} - \text{vers}\gamma_{1j} y_{A1} + (-x_{P1j}\sin\gamma_{1j} + y_{P1j}\text{vers}\gamma_{1j}) \\ T_j &= (x_{P1j}\text{vers}\gamma_{1j} - y_{P1j}\sin\gamma_{1j}) x_{A1} + (x_{P1j}\sin\gamma_{1j} + y_{P1j}\text{vers}\gamma_{1j}) y_{A1} \\ &\quad - (x_{P1j}^2 + y_{P1j}^2)\text{vers}\gamma_{1j} \end{aligned}\right\} \quad (j=2,3,4)$$

(3.95)

According to the theory of linear equations, the three equations in (3.94) include two unknowns x_0, y_0. The determinant of the coefficients must vanish, if non-trivial solution exists, or

$$\begin{vmatrix} G_2 & S_2 & T_2 \\ G_3 & S_3 & T_3 \\ G_4 & S_4 & T_4 \end{vmatrix} = 0 \quad (3.96)$$

Each of the elements G_j, S_j, T_j is linear in the two variables x_{A1}, y_{A1}, therefore equation (3.96) is a cubic equation in x_{A1}, y_{A1}, i.e. the equation of the circle-point curve in the variables x_{A1}, y_{A1}.

Moreover, since we have selected the coordinate system of Fig. 3.27, except for G_4, S_4, T_4, the expressions $G_2, S_2, T_2; G_3, S_3, T_3$ are simplified. The expressions of the nine elements in the determinant in equation (3.96) are given in Appendix 4.

Expanding equation (3.96) gives

$$k_{1234}: \begin{aligned}(d_1 y_{A1} + d_2 x_{A1})(x_{A1}^2 + y_{A1}^2) \\ + d_7 y_{A1}^2 + d_8 x_{A1} y_{A1} + d_9 x_{A1}^2 + d_5 y_{A1} + d_6 x_{A1} = 0\end{aligned} \quad (3.97)$$

The coefficients $d_1, d_2, \ldots d_9$ in equation (3.97) are due to the coefficients G_j, S_j, T_j. For convenience, all these coefficients are listed in Appendix 5. It can be seen from equation (3.97) that, since we have chosen P_{12} as the origin, and the circle-point curve k_{1234} is passing through the origin, therefore $x_{A1} = 0, y_{A1} = 0$ satisfy equation (3.97). This equation clearly shows that the circle-point curve is a circular cubic, which explains theorem 15. (Please refer to Appendix 1.)

In practical curve plotting, it is more convenient to use polar coordinates than rectangular coordinates. Since k_{1234} passes through the origin, its polar equation can be reduced to a quadratic equation. This is because any line passing through the origin will meet the k_{1234} curve besides the one point at the origin only in two more points. To transform the rectangular coordinates into polar coordinates, we can use the equations

$$x_{A1} = p_{A1}\cos\theta_{A1} \brace y_{A1} = p_{A1}\sin\theta_{A1}} \quad (3.98)$$

according to Fig. 3.52, and obtain

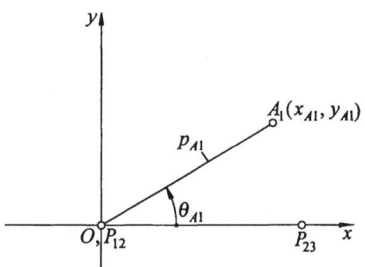

Fig. 3.52. Transformation of coordinates.

$$k_{1234}: D_2 p_{A1}^2 + D_1 p_{A1} + D_0 = 0 \quad (3.99)$$

The expressions of the coefficients D_2, D_1, D_0 in equation (3.99) are also listed in Appendix 5; they are all functions of θ_{A1}. For an assumed value of θ_{A1}, substituting this value into equation (3.99) will yield two A_1 points. The whole k_{1234} curve can be plotted out in the same way.

Similar to the two equations (3.45), we can write another equation for the displacement from position 1 to position 4, i.e. by setting $j = 2, 3, 4$ in equation (3.43). All three equations are now listed below:

$$\left. \begin{array}{l} L_2 x_{A1} + Q_2 y_{A1} + H_2 = 0 \\ L_3 x_{A1} + Q_3 y_{A1} + H_3 = 0 \\ L_4 x_{A1} + Q_4 y_{A1} + H_4 = 0 \end{array} \right\} \quad (3.100)$$

Since we have adopted the displacement matrix in equation (3.53), therefore

$$\left. \begin{array}{l} L_j = -\text{vers}\gamma_{1j} x_0 + \sin\gamma_{1j} y_0 + (x_{P1j}\text{vers}\gamma_{1j} - y_{P1j}\sin\gamma_{1j}) \\ Q_j = -\sin\gamma_{1j} x_0 - \text{vers}\gamma_{1j} y_0 + (x_{P1j}\sin\gamma_{1j} + y_{P1j}\text{vers}\gamma_{1j}) \\ H_j = (x_{P1j}\text{vers}\gamma_{1j} + y_{P1j}\sin\gamma_{1j}) x_0 + (-x_{P1j}\sin\gamma_{1j} + y_{P1j}\text{vers}\gamma_{1j}) y_0 \\ \qquad - (x_{P1j}^2 + y_{P1j}^2)\text{vers}\gamma_{1j} \end{array} \right\} \quad (3.101)$$

The centre-point curve is obtained by setting the determinant of the coefficients of equations (3.101) to zero, or

$$\begin{vmatrix} L_2 & Q_2 & H_2 \\ L_3 & Q_3 & H_3 \\ L_4 & Q_4 & H_4 \end{vmatrix} = 0 \quad (3.102)$$

Dimensional synthesis of four-bar linkages --- body guidance problems 129

The elements L_j, Q_j, H_j are all linear in the two variables x_0, y_0, hence equation (3.102) is a cubic in x_0, y_0. After applying the coordinate system in Fig. 3.27, the expressions of the nine elements in equation (3.102) are listed in Appendix 6.

Expanding equation (3.102) gives

$$m'_{1234} : (m_1 y_0 + m_2 x_0)(x_0^2 + y_0^2) \\ + m_7 y_0^2 + m_8 x_0 y_0 + m_9 x_0^2 + m_5 y_0 + m_6 x_0 = 0 \quad (3.103)$$

The coefficients in equation (3.103) are due to L_j, Q_j, H_j. They are listed in Appendix 7. It can be seen from equation (3.103) that, since P_{12} is taken as the origin, and the centre-point curve m_{1234} passes through the origin, therefore $x_0 = 0, y_0 = 0$ satisfy equation (3.103). Furthermore, since m_{1234} passes through P_{23}, the point $x_0 = s$, $y_0 = 0$ also satisfies equation (3.103). This equation also shows clearly that the centre-point curve is a circular cubic, which verifies theorem 13.

To find the polar coordinate equation of (3.103), substituting the following transformation equations according to equation (3.98)

$$\left. \begin{array}{l} x_0 = p_0 \cos\theta_0 \\ y_0 = p_0 \sin\theta_0 \end{array} \right\}$$

yields

$$M_2 p_0^2 + M_1 p_0 + M_0 = 0 \quad (3.104)$$

The expressions of the coefficients M_2, M_1, M_0 are also listed in Appendix 7; they are all functions of θ_0. For an assumed value of θ_0, substituting this value into equation (3.104) will give two A_0 points. The whole centre-point curve m_{1234} can then be plotted out in this way.

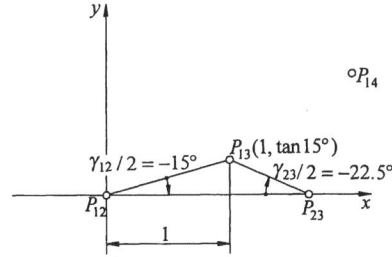

Fig. 3.53. Example.

Example: As shown in Fig. 3.53, the three displacements of a body are given: P_{12} (0,0), P_{13} (1, tan 15°), P_{14} (2,1) ; and the rotation angles $\gamma_{12} = -30°$, $\gamma_{13} = -75°$, $\gamma_{14} = -90°$. It is required to find the circle-point curve.

Solution: Substituting the given data into equations in Appendices 4 and 5, we get the values of the coefficients in equation (3.97):

$$d_1 = -0.039680, \quad d_2 = -0.357517, \quad d_5 = -1.936411$$
$$d_6 = -0.347910, \quad d_7 = 1.522042, \quad d_8 = 0.296777$$
$$d_9 = 1.072552$$

Substituting these coefficients into equations (A5.1) results in the circle-point curve calculated according to equation (3.99), as shown in Fig. 3.54.

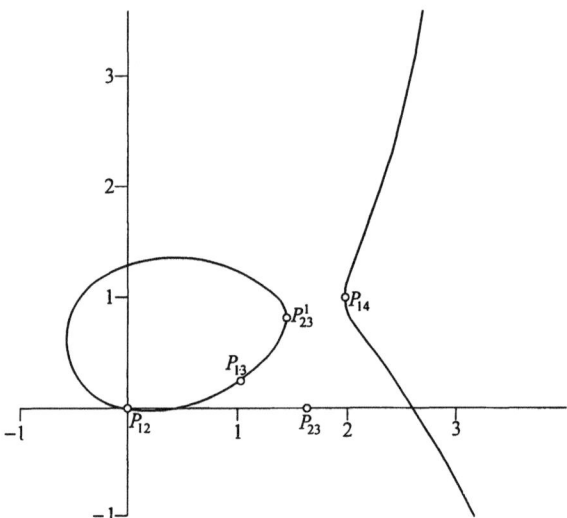

Fig. 3.54. Example: circle-point curve k_{1234}.

3.6.7 Four homologous points on a line by algebraic method

The geometrical concept of four homologous points on a straight line mentioned in Section 3.6.5 can be expressed algebraically for the ease of computer processing. Suppose there are four homologous points lying on a circle. The first point A_1 must lie on the circle-point curve k_{1234}. Furthermore, if the three points A_1, A_2, A_3 are collinear, then in the equation of the R^1-curve, (3.50), it should be $R_A \to \infty$, which means that $Z_3 = 0$, or by equation (3.41),

$$Z_3 = G_2 S_3 - G_3 S_2 = 0$$

By using the coordinate system in Fig. 3.27, the expressions of G_2, G_3, S_2, S_3 can be found from Appendix 4 and substituted into the above equation, to become

Dimensional synthesis of four-bar linkages --- body guidance problems   131

$$(\sin\gamma_{12} - \sin\gamma_{13} + \sin\gamma_{23})(x_{A1}^2 + y_{A1}^2) + s[-\sin\gamma_{12} + \sin\gamma_{23} + \sin(\gamma_{12} - \gamma_{23})]x_{A1}$$
$$+ s[1 - \cos\gamma_{12} - \cos\gamma_{23} + \cos(\gamma_{12} - \gamma_{23})]y_{A1} = 0 \qquad (3.105)$$

This is just the equation of the circle k_1 in Fig. 3.17 (k_{123}^1 in Fig. 3.51). (Please note that the signs of γ_{12}, γ_{23} in Fig. 3.19 are the same as those in Fig. 3.16, being both negative.) Substituting equations (3.98) into (3.105) will transform the latter into a polar coordinate equation, as shown in Fig. 3.52.

$$k_{123}^1 : p_1 = (-C_1 \cos\theta_1 - C_0 \sin\theta_1)/C_2 \qquad (3.106)$$

where

$$\left.\begin{array}{l} C_2 = \sin\gamma_{12} - \sin\gamma_{13} + \sin\gamma_{23} \\ C_1 = s[-\sin\gamma_{12} + \sin\gamma_{23} + \sin(\gamma_{12} - \gamma_{23})] \\ C_0 = s[(1 - \cos\gamma_{12} - \cos\gamma_{23} + \cos(\gamma_{12} - \gamma_{23})] \end{array}\right\} \qquad (3.107)$$

Substituting equation (3.106) into equation (3.99) will yield a cubic equation in the unknown $\tan\theta_{A1}$. Let us count the number of intersections of the two curves. k_{123}^1 is a circle, being a quadratic curve, while k_{1234} is a circular cubic. These two curves should have six intersections, among which the known ones are: the origin P_{12}, the poles P_{13}, P_{23}^1 (please refer to Fig. 3.17), a pair of imaginary roots (the circular points I, J (see Appendix 1)), and the required, unknown A_1. Having found the cubic equation mentioned above, the known roots $\tan(-\gamma_{12}/2)$ corresponding to P_{13} and $\tan(-\gamma_{12})$ corresponding to P_{23}^1 have to be eliminated. The remaining linear equation will supply the required root $\tan\theta_{A1}$.

Obviously the above procedure is quite cumbersome. Since it is shown in Table 3.2 that the point A_1 is the intersection of $k_{123}^1, k_{124}^1, k_{134}^1$, all we have to do is to write down the equation of k_{124}^1 or k_{134}^1, and set it together with equation (3.106) as simultaneous equations to solve for the point A_1. Apparently k_{124}^1 can conveniently be used as it is a circle passing through P_{12}, P_{14}, P_{24}^1. Selecting the coordinate system in Fig. 3.27, with P_{12} as the origin, we can find the points P_{12}, P_{24}. The point P_{24}^1 can be reached by revolving P_{24} about P_{12} through an angle $-\gamma_{12}$. The simultaneous equations including k_{124}^1 and (3.106) can be solved for the value $\tan\theta_{A1}$ of the orientation angle θ_{A1}, and hence the coordinates of the point A_1 can be found. In this procedure, there are no extraneous roots involved in the simultaneous equations.

Example: Find the first point A_1 of the four collinear homologous points in last example.

Solution: (a) By the intersection of k_{123}^1 and k_{1234}

We calculate first from the given data $\gamma_{23} = -45°$, $s = 1.646887$, and then

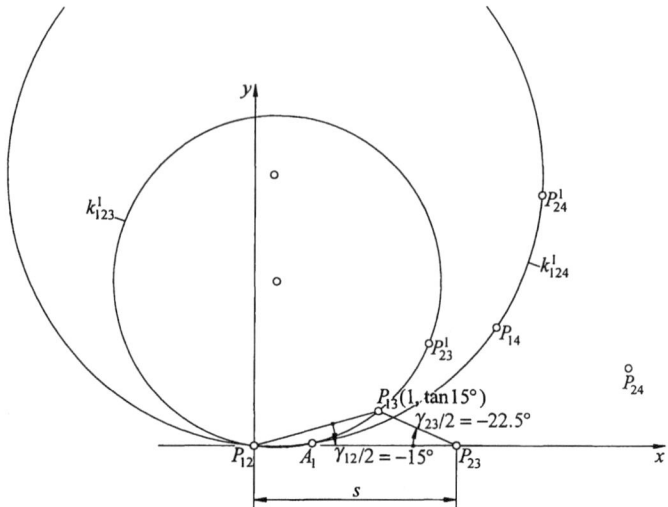

Fig. 3.55. Example: determination of A_1 for four homologous points lying on a line in Fig. 3.53.

P_{23}^1 (1.426246, 0.823443), as shown in Fig. 3.55. The equation of k_{123}^1 thus found is

$$k_{123}^1 : p_{A1} = 0.353113\cos\theta_{A1} + 2.682163\sin\theta_{A1}$$

Setting together this equation of k_{123}^1 simultaneously with that of k_{1234} found in last example, and eliminating p_1, we get a cubic equation in the unknown $t = \tan\theta_{A1}$

$$1.860494t^3 - 1.661593t^2 + 0.362982t - 0.013756 = 0$$

Eliminating from this equation the known roots $t = \tan 15°$ and $t = \tan 30°$ corresponding respectively to P_{13} and P_{23}^1, we get finally $1.860494\ t - 0.088919 = 0$, or $t = \tan\theta_{A1} = 0.047793$, or A_1 (0.48075, 2.736°).

(b) By the intersection of k_{123}^1 and k_{124}^1

To find the equation of k_{124}^1, we have to find the poles P_{24} and P_{24}^1 first. We can find P_{24} as follows. Find $\mathbb{D}_{24}$ from $\mathbb{D}_{14} = \mathbb{D}_{24} \mathbb{D}_{12}$, and then get $x_{P24} = 3.098076$, $y_{P24} = 0.633975$ from $\gamma_{24} = \gamma_{14} - \gamma_{12} = -60°$. Next get $x_{P24}^1 = 2.366025$, $y_{P24}^1 = 2.098076$. Finally we obtain

$$k_{124}^1 : p_{A1} = 0.267949\cos\theta_{A1} + 4.464102\sin\theta_{A1}$$

Solving this equation simultaneously with the equation of k_{123}^1 found previously, we get $\tan\theta_{A1} = 0.047793$, which is identical with the result found in (a).

3.7 Guiding a body through four infinitesimally separated positions --- the circling-point curve and centering-point curve

3.7.1 The circling-point curve

In Section 3.5.1 we have proved the Euler-Savary equation

$$\left(\frac{1}{p} - \frac{1}{p_0}\right) \sin \theta_A = \frac{1}{\delta}$$

from which we can derive the expression of the radius of curvature of the path of a moving point A as

$$p_0 - p = \rho = \frac{p p_0}{\delta \sin \theta_A}$$

and rewrite it as

$$\rho = \frac{p^2}{\delta \sin \theta_A - p} \qquad (3.108)$$

This equation is originally derived on the basis of guiding a body through three infinitesimally separated positions. In other words, as the three homologous points A_1, A_2, A_3 approach one another indefinitely, ρ becomes the radius of curvature of the path of the moving point A. Considering the case of four infinitesimally separated positions of a body, suppose the four homologous points A_1, A_2, A_3, A_4 are on a circle, then we may say that the radius ρ of the circle on which the three points A_1, A_2, A_3 lie, is also the same radius ρ of the circle on which the three points A_2, A_3, A_4 lie. In other words, differentiating ρ with respect to a certain independent variable, and putting the differential coefficient equal to zero should give the locus of such a circle-point A. Let the independent variable be the length s along the polode. (Please note that s no longer represents the distance $\overline{P_{12}P_{23}}$ in Fig. 3.28, which should now be written as ds.) Setting dρ/ds = 0 will lead to the locus of A. This locus is called the *circling-point curve*, it is therefore also called the *cubic of stationary curvature*. Strictly speaking, this last term may cause some confusion, because the object curve which exhibits a stationary curvature is the path of the moving point, not the cubic itself.

Fig. 3.56(a) shows a body rotating about the pole P_{12}, and Fig. 3.56(b) shows the same body having rotated through an angle dγ to a position where P_{23}^1 and P_{23} coincide. Differentiating equation (3.108) with respect to s and setting it equal to zero yields

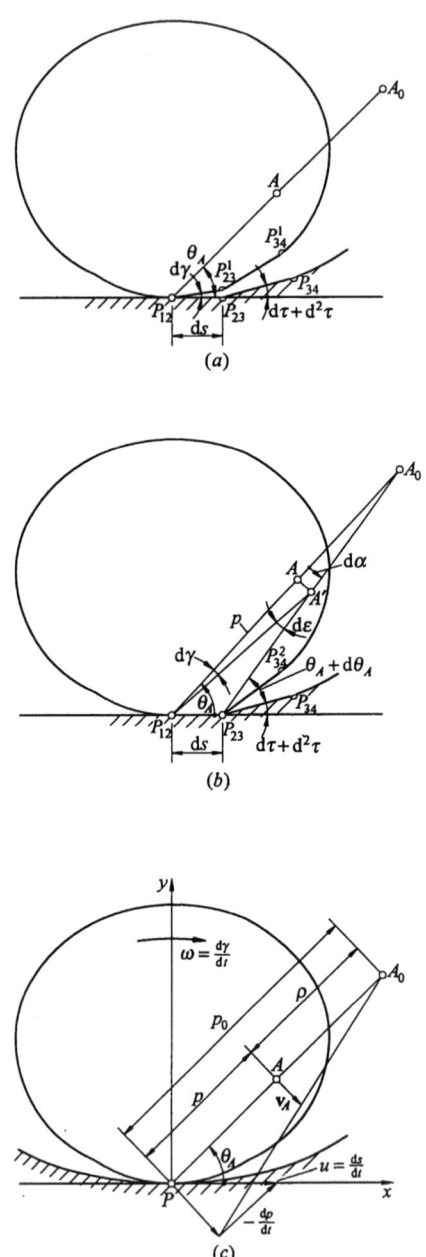

Fig. 3.56. Geometrical relationship in the derivation of circling-point curve. (*a*) Initial position. (*b*) After the body rotated through an angle dγ. (*c*) Limiting case.

Dimensional synthesis of four-bar linkages --- body guidance problems

$$\frac{dp}{ds} = \frac{1}{(\delta \sin\theta_A - p)^2}[2(\delta\sin\theta_A - p)pp' - p^2(\delta\cos\theta_A \cdot \theta'_A + \delta'\sin\theta_A - p')]$$
$$= 0 \qquad (3.109)$$

where the symbol (') represents a differential coefficient with respect to s. In this equation there are p', θ_A' and δ' which have to be dealt with separately.

(a) p'

In approaching limit, all differential geometrical quantities may be replaced by velocities, as shown in Fig. 3.56(c). Let time be represented by t. We have then

$$\frac{dp}{dt} = -\frac{ds}{dt}\cos\theta_A$$

therefore

$$p' = \frac{dp}{ds} = -\cos\theta_A \qquad (3.110)$$

It can be seen from Fig. 3.56(c) that the length of p decreases due to an increase ds of s, therefore p' is negative.

(b) $p\theta'_A$

We define first the term *contingence angle*. This is the angle between two small line segments in front and behind a point on the polode. Suppose the contingence angle at P_{12} is $d\tau$, then the contingence angle at P_{23} can be written as $d\tau + d^2\tau$. As shown in Fig. 3.56(b) (please note that $d\gamma$ is negative.) we have

$$\theta_A + d\theta_A + (d\tau + d^2\tau) = (\theta_A + d\gamma) + d\varepsilon$$

Eliminating higher order terms gives

$$d\theta_A \approx d\varepsilon + d\gamma - d\tau$$
$$\approx \frac{ds\sin\theta_A}{p} + d\gamma - d\tau$$

therefore
$$pd\theta_A = ds\sin\theta_A + pd\gamma - pd\tau$$

or
$$p\theta'_A = \sin\theta_A + p(\gamma' - \tau') \qquad (3.111)$$

(c) δ'

From equation (1.10), we have $\delta = -u/\omega = -ds/d\gamma = -1/\gamma'$, hence

$$\delta' = \frac{\gamma''}{\gamma'^2} \qquad (3.112)$$

Substituting the three equations (3.110), (3.111), (3.112) into equation (3.109), we get

$$p[\gamma'(2\gamma' - \tau')\cos\theta_A - \gamma''\sin\theta_A] + 3\gamma'\sin\theta_A\cos\theta_A = 0 \quad (3.113)$$

Setting

$$\left.\begin{array}{r}\dfrac{3\gamma'}{\gamma''} = m \\[6pt] -\dfrac{3}{2\gamma' - \tau'} = l\end{array}\right\} \quad \begin{array}{c}(3.114)\\[6pt](3.115)\end{array}$$

we can transform equation (3.113) into

$$k_u : \frac{1}{p} = \frac{1}{m\cos\theta_A} + \frac{1}{l\sin\theta_A} \quad (3.116)$$

Equation (3.116) is the polar coordinate equation of the circling-point curve, called Müller's equation and is denoted by k_u, where m, l are constants. Substituting according to equations (3.98) $x_A = p\cos\theta_A$, $y_A = p\sin\theta_A$ into equation (3.116) gives

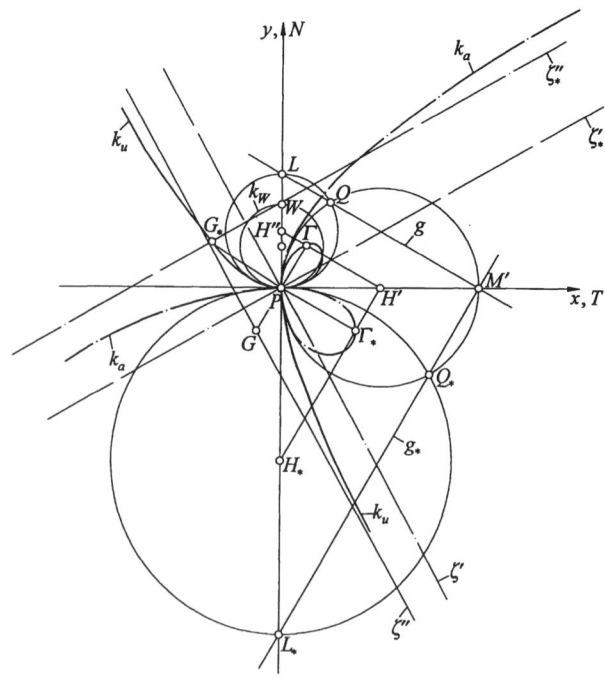

Fig. 3.57. Circling-point curve k_u and centering-point curve k_a.

Dimensional synthesis of four-bar linkages --- body guidance problems 137

$$k_u : (x_A^2 + y_A^2)(mx_A + ly_A) - mlx_A y_A = 0 \tag{3.117}$$

Equation (3.117) is the rectangular coordinate equation of the circling-point curve. Fig. 3.57 shows a circling-point curve k_u with $m = 42$ and $l = 24$.

3.7.2 The centering-point curve

Substituting $\delta = -1/\gamma'$ into the Euler-Savary equation in Section 3.7.1, we can rewrite the latter as

$$p = \frac{p_0 \sin\theta_A}{-p_0\gamma' + \sin\theta_A} \tag{3.118}$$

Substituting equation (3.118) into equation (3.109) results in

$$p_0[\gamma'(\tau' + \gamma')\cos\theta_A + \gamma''\sin\theta_A] - 3\gamma'\sin\theta_A \cos\theta_A = 0 \tag{3.119}$$

Let

$$\frac{3}{\gamma' + \tau'} = l_* \tag{3.120}$$

then equation (3.119) can be written as

$$k_a : \frac{1}{p_0} = \frac{1}{m\cos\theta_A} + \frac{1}{l_* \sin\theta_A} \tag{3.121}$$

Equation (3.121) is the polar coordinate equation of the centering-point curve, denoted by k_a. Substituting $x_0 = p_0 \cos\theta_A$, $y_0 = p_0 \sin\theta_A$ into equation (3.121) gives

$$k_a : (x_0^2 + y_0^2)(mx_0 + l_* y_0) - ml_* x_0 y_0 = 0 \tag{3.122}$$

In Fig. 3.57 is also shown a centering-point curve with $m = 42$, $l_* = -70$.

Please note that the four equations (3.116), (3.117), (3.121), (3.122) apply only to the coordinate system with the poletangent as the x-axis (x_0-axis), and the polenormal as the y-axis (y_0-axis). From the characteristics of these equations, we have the following theorems:

Theorem 19 The circling-point curve and the centering-point curve are circular cubics, each having a double point at the velocity pole P, and with the poletangent and polenormal as two tangents.

From the definitions of l and l_*, equations (3.115) and (3.120), we have

$$\frac{1}{l} - \frac{1}{l_*} = \frac{1}{\delta} \tag{3.123}$$

Please note that (Fig. 1.13)

$$\tau' = \frac{d\tau}{ds} = \frac{1}{\overline{PM_0}} \quad (3.124)$$

is the curvature of the fixed polode π_f, and

$$\frac{d\tau}{ds} - \frac{d\gamma}{ds} = \frac{1}{\overline{PM}} \quad (3.125)$$

is the curvature of the moving polode π_c. Furthermore, from the definitions of l and l_* in equations (3.115) and (3.120), and the relations (1.15, 1.15a) we have

$$\left. \begin{array}{c} \dfrac{1}{l} + \dfrac{2}{l_*} = \dfrac{1}{\overline{PM_0}} = \dfrac{1}{\rho_f} \\[2mm] \dfrac{2}{l} + \dfrac{1}{l_*} = \dfrac{1}{\overline{PM}} = \dfrac{1}{\rho_c} \end{array} \right\} \quad (3.126)$$

and from these and equation (1.15a) we have

$$\left. \begin{array}{c} m = -\dfrac{3\rho_c \rho_f}{(\rho_f - \rho_c)\delta} \\[2mm] l = \dfrac{3\rho_c \rho_f}{2\rho_f - \rho_c} \\[2mm] l_* = \dfrac{3\rho_c \rho_f}{2\rho_c - \rho_f} \end{array} \right\} \quad (3.127)$$

There are a few geometrical properties of the circling-point curve and centering-point curve which have to be mentioned here. In Fig. 3.57, take a distance $\overline{PM'} = m$ on the x-axis, and a distance $\overline{PL} = l$ on the y-axis. Denote the line $M'L$ by g. Take $m/2$ to get H', and $l/2$ to get H''. Join $H'H''$. Draw a perpendicular from P to $H'H''$, intersecting it in Γ. Γ is called the *singular focus* of k_u (please refer to Exercise 3.33). k_u passes through Γ. Extend ΓP until point G, where $|\overline{PG}| = \overline{P\Gamma}$, then the asymptote of k_u is a line ζ'' passing through G and of a slope $-m/l$. The line ζ' passing through P and parallel to ζ'' is called the *focal axis* of k_u. The two circles of curvature of k_u at P are two circles with diameters m and l respectively. (Please refer to Exercise 3.31(a) (b).)

Similarly, take the distance $\overline{PL_*} = l_*$ on the y-axis and denote the line $M'L_*$ by g_*. Take $l_*/2$ to get H_*. Draw a perpendicular from P to $H'H_*$ intersecting it in Γ_*. Γ_* is the singular focus of k_a. Extend $\Gamma_* P$ to G_*, until $\overline{PG_*} = \overline{\Gamma_* P}$. The asymptote of k_a is a line ζ_*'' passing through G_* with a slope $-m/l_*$. The line

Dimensional synthesis of four-bar linkages --- body guidance problems

ζ'_* passing through P and parallel to ζ''_* is called the focal axis of k_a. The circles of curvature of k_a at P are two circles with diameters m and l_* respectively.

3.7.3 Construction of circling-point curve and of centering-point curve

The equations of circling-point curve and centering-point curve are of the same type. Hence, for construction, we only have to consider any one of them. In the following, we shall describe the construction methods of the circling-point curve.

With modern technology of computer graphics, the circling-point curve of equation (3.116) or (3.117) can easily be traced out as soon as the two constants m and l are given. The following methods require finding the poletangent and polenormal first, before applying equations (3.116) and (3.117).

(*a*) Method 1

Fig. 3.58 shows a partial view of Fig. 3.57. Choose arbitrarily a point M_ε on the line ζ'. Draw a circle k_ε with M_ε as centre and $\overline{M_\varepsilon P}$ as radius. The circle k_ε meets the line $M_\varepsilon \Gamma$ (denoted by g_ε) in two points A, A' which are points on the circling-point curve k_u. This is because the coordinates of Γ are

$$x_\Gamma = \frac{ml^2}{2(m^2+l^2)}, \quad y_\Gamma = \frac{m^2 l}{2(m^2+l^2)} \qquad (3.128)$$

and the equation of g_ε is

$$g_\varepsilon : y - y_\Gamma = \varepsilon(x - x_\Gamma)$$

where ε is a parameter, the slope of the line g_ε. The coordinates of the intersection point M_ε of g_ε and ζ' are therefore

$$x_{M\varepsilon} = \frac{l(\varepsilon x_\Gamma - y_\Gamma)}{m + \varepsilon l}, \quad y_{M\varepsilon} = -\frac{m(\varepsilon x_\Gamma - y_\Gamma)}{m + \varepsilon l} \qquad (3.129)$$

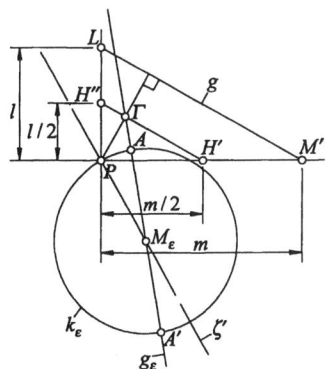

Fig. 3.58. Construction of circling-poing curve k_u --- method 1.

so that the equation of k_ε is

$$k_\varepsilon : (x - x_{M\varepsilon})^2 + (y - y_{M\varepsilon})^2 = x_{M\varepsilon}^2 + y_{M\varepsilon}^2$$

Setting equations of k_ε and g_ε simultaneously and eliminating ε, we get equation (3.117) (x_A, y_A are replaced here by x, y respectively).

(b) Method 2

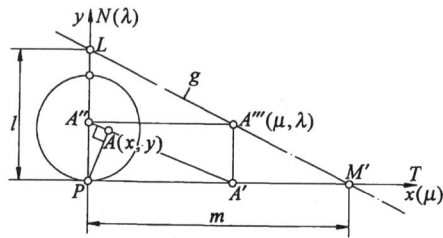

Fig. 3.59. Construction of circling-poing curve k_u --- method 2..

As shown in Fig. 3.59, the line $M'L$ is denoted by g as in Fig. 3.57. Choose any point A''' (μ, λ) on g. In other words, considering (μ, λ) as variables, we can write the linear equation of g as

$$g : m\lambda + l\mu - ml = 0$$

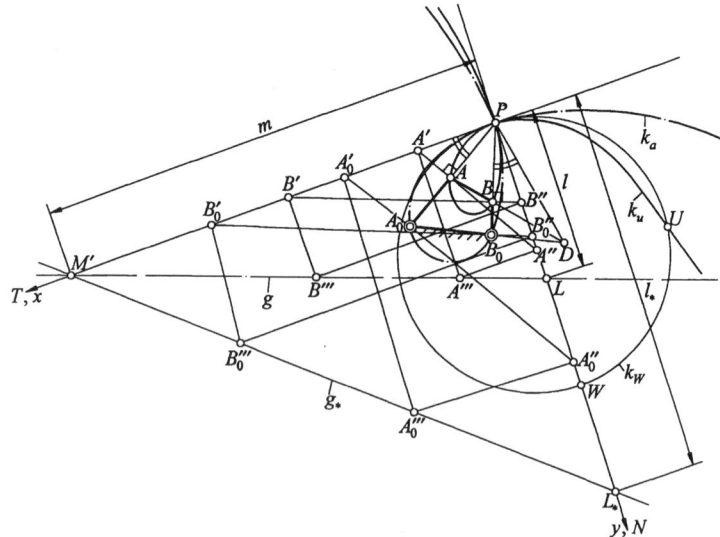

Fig. 3.60. Determination of k_u-curve of the coupler of a four-bar linkage by method 2.

Project A''' onto x-axis to get A', and project A''' onto y-axis to get A''. Join $A'A''$. Draw a perpendicular from P to $A'A''$, the foot point A of this perpendicular being a point on the circling-point curve k_u. It can easily be proved that the coordinates of A are equation (3.117).

By means of this method, we can find the circling-point curve of the coupler AB of a given four-bar linkage such as that shown in Fig. 3.60. We find first by Bobillier construction the poletangent PT as the x-axis, and the polenormal PN as the y-axis. Since the points A,B are known points on the circling-point curve k_u, we can find the points A''', B''' by reversing the steps shown in Fig. 3.59. The line joining A''' and B''' is g, which cuts the x- and y- axes to yield m and l.

Applying the same method to the points A_0, B_0, we can find the points A_0''', B_0'''. The line joining A_0''', B_0''' is g_*, which should pass through the point M'. The intersection of g_* and y-axis is L_*, and $\overline{PL_*} = l_*$. We can find the whole centering-point curve k_a of the coupler AB.

(c) Method 3

This method appears in (Kraus, 1954). In the following, we shall illustrate a method of constructing the centering-point curve k_a. Before describing the method, let us consider first a special case, in which the four positions of a body c are neither completely finitely, nor infinitesimally separated, but its two positions c_1, c_2 are infinitesimally separated; the two positions c_2, c_3 are finitely separated; and the two

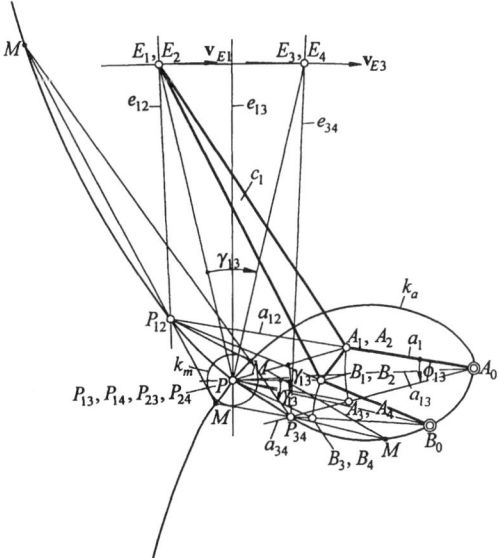

Fig. 3.61. For the derivation of method 3 of constructing centering-point curve k_a.

positions c_3, c_4 are again infinitesimally separated. In Fig. 3.61, the first two positions E_1, E_2 of a moving point on the body coincide, and the two positions E_3, E_4 coincide too, but the direction of the velocity $\mathbf{v}_{E1}$ of E_1 and that of the velocity $\mathbf{v}_{E3}$ of E_3 are given. The first position a_1 of the crank and the position of the crank centre A_0 are also given. As the point E moves from E_1 to E_3, the crank a should rotate correspondingly through a prescribed angle ϕ_{13}. Hence we can freely choose the point A_1 on a_1, and consequently determine A_3. We shall now construct the centre-point curve k_a and the four-bar linkage A_0ABB_0.

Draw first the perpendicular bisectors e_{12} and e_{34}, orthogonal to $\mathbf{v}_{E1}$ and $\mathbf{v}_{E3}$ respectively. a_{12} and a_{34} are already known. The intersection point of e_{12} and a_{12} is P_{12}, and that of e_{34} and a_{34} is P_{34}. In the present case the four poles $P_{13}, P_{14}, P_{23}, P_{24}$ shrink into one point, denoted by P as shown In Fig. 3.61. P is also the intersection of the perpendicular bisector e_{13} of $\overline{E_1E_3}$ and the perpendicular bisector a_{13} of $\overline{A_1A_3}$. So all six poles are found. Use still the two opposite sides $\overline{P_{12}P_{13}}$, $\overline{P_{24}P_{34}}$ of the opposite-pole quadrilateral $P_{12}P_{34}P_{13}P_{24}$ for construction as follows: Draw a circle k_m about P with any radius. Lay two tangents from each pole P_{12}, P_{34} to the circle k_m. These four tangents intersect in four M points. All four M's are points on the centre-point curve k_a. This fact can easily be shown, because if we join any one of the M's with P, the angles subtended by $\overline{P_{12}P}$ and $\overline{PP_{34}}$ at M are equal. In this way, we can plot out the entire centre-point curve k_a. It can be seen from this construction method that the two tangents of k_a at P are mutually orthogonal, and $\overline{P_{12}P}$ and $\overline{PP_{34}}$ are symmetrically disposed with respect to these two tangents. The proof of this statement is left as Exercise 3.36.

Having found k_a, we can choose any point on it as B_0. In order to find the corresponding B_1, it is not necessary to draw the circle-point curve k_u. According again to Theorem 7, with $P_{13} = P$ as the pole of rotation, we see that $\sphericalangle E_1PE_3$ = $\sphericalangle A_1PA_3 = \sphericalangle B_1PB_3 = \gamma_{13}$. Join the two lines PB_0 and $P_{12}B_0$, and build with P as apex two angles $\gamma_{13}/2$ on both sides of PB_0. As shown in the figure, the intersection of the upper line and $P_{12}B_0$ is the required $B_1(=B_2)$.

We come now to the method 3. This is an extension of the construction mentioned above. In Fig. 3.62, suppose a centering-point curve k_a is given, whose equation is (3.121). It can be shown that:

Theorem 20 For two points E_0 and F_0 on a centering-point curve k_a whose position angles are symmetrical with respect to the poletangent PT (or polenormal PN), (i.e. position angle of E_0 is $\sphericalangle TPE_0 = \theta_E$, and position angle of F_0 is $\sphericalangle TPF_0 = 180° - \theta_E$.) the angle ε subtended by $\overline{PE_0}$ and the angle η

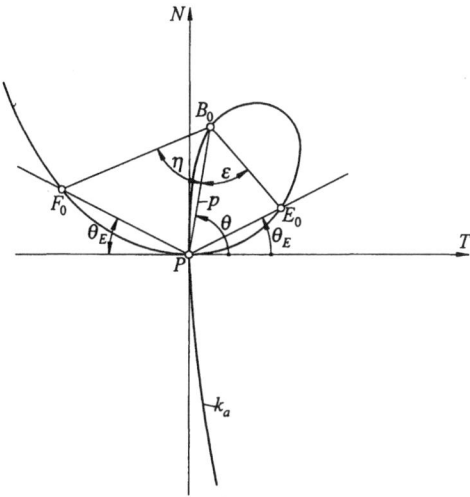

Fig. 3.62. Principle of method 3 of constructing centering-point curve k_a.

subtended by $\overline{PF_0}$ at any point B_0 on k_a are equal.

The proof of this theorem is left as Exercise 3.38. Applying this theorem to the case where *PT*, *PN*, and E_0, B_0 are given while k_a is unknown yet, we can construct the point F_0 as follows:

(1) Fold PE_0 with respect to *PN*;
(2) Fold B_0E_0 with respect to PB_0, the intersection with the line drawn in (1) being the required F_0.

This is the basis of method 3. In Fig. 3.63, a four-bar linkage A_0ABB_0 is given. Find first the poletangent *PT* and polenormal *PN* by Bobillier construction. The known points on the centering-point curve k_a of the coupler are A_0 and B_0. Considering A_0 as the point E_0 in the above mentioned procedure, we can find F_0. According to Theorem 20, the set of points P_{12}, P_{34} in Fig. 3.61 correspond exactly the set of points E_0, F_0 in Fig. 3.62. Therefore as the six poles in Fig. 3.63 shrink into one point *P*, and although P_{12}, P_{34} coincide with point *P*, their rôles can be replaced by E_0, F_0. The construction is carried out by drawing a circle k_m about *P* with any radius, and then by lying two tangents on k_m from each of $A_0(=E_0)$ and F_0. These four tangents intersect in four *M*'s, and these are points on k_a. In this way, we can plot out the entire k_a.

Having found the centering-point curve k_a as that shown in Fig. 3.63, we can choose an arbitrary point C_0 on it as a centre-point. The corresponding circle-point

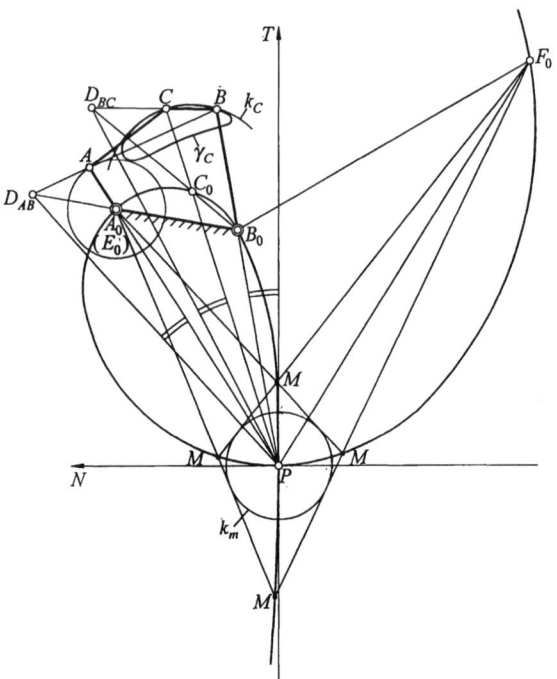

Fig. 3.63. Construction of centering-poing curve k_a --- method 3.

C can be found by Bobillier construction. The path γ_C of point C and its circle of curvature are in a four-point contact in the position shown in the figure.

3.7.4 The velocity pole as a moving point

It has been mentioned in Section 1.5 that the path of the velocity pole is along the direction of the polenormal. In general, the path of P forms a cusp. However, this statement needs qualification. Whereas the point P is the velocity pole, a point is not a velocity pole both before reaching the point P and after passing the point P. The point is indeed a moving point on the body. It only instantaneously becomes the velocity pole during its motion and is denoted by P just for convenience.

In principle, the path of a moving point should be left to Chapter 6 where path generation will be discussed. However, here we discuss a few cases of the motion of a velocity pole as they are related to the breaking up of centre-point curves and of circle-point curves.

Fig. 3.64 shows the path of P as successive circle arcs $\widehat{P'P''}$, $\widehat{P''P'''}$, $\cdots$ when the moving body makes infinitesimal rotations about the points $P_{12}, P_{23}, P_{34}, \ldots$ on

Dimensional synthesis of four-bar linkages --- body guidance problems 145

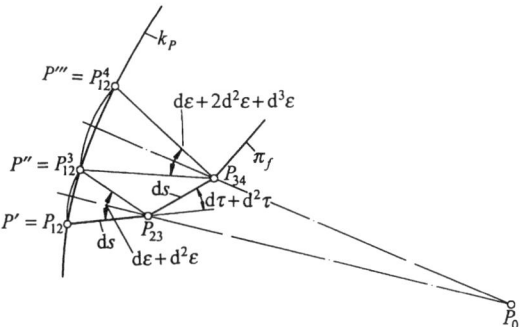

Fig. 3.64. Velocity pole P as a moving point.

the fixed polode π_f. The contingence angles of the fixed polode π_f at P_{12} and P_{23} are successively $d\tau$, $d\tau + d^2\tau$. In fact, P'' is P_{12}^3, and P''' is P_{12}^4. For clarity, consider $\Delta P_{23}P''P_{34}$ as the moving body. In the position as shown in the figure, this triangle has just rotated about the centre P_{23}, and it is going to rotate about the centre P_{34}.

The angles of rotation of the moving body are successively $d\varepsilon$, $d\varepsilon + d^2\varepsilon$, $d\varepsilon + 2d^2\varepsilon + d^3\varepsilon$, ... This can be understood by examining Fig. 3.65. Here we assume a

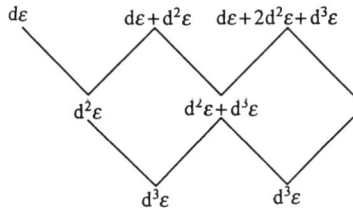

Fig. 3.65. Consecutive infinitesimal rotations of a body.

second order arithmetical progression with a second order difference $d^3\varepsilon$. Without loss of generality we may assume that the infinitesimal line segments $\overline{P_{12}P_{23}}$, $\overline{P_{23}P_{34}}$ are of equal lengths denoted by ds. Next let the centre of the circle k_p passing through the three points P', P'', P''' be denoted by P_0. In the isosceles triangle $\Delta P''P_{23}P_{34}$, we have

$$\angle P_{23}P_{34}P'' = \frac{1}{2}(d\varepsilon + d^2\varepsilon + d\tau + d^2\tau)$$

and

$$\sphericalangle P_{34}P_{23}P_0 = \frac{d\varepsilon + d^2\varepsilon}{2} + d\tau + d^2\tau$$

hence

$$\sphericalangle P_{23}P_0P_{34} = \sphericalangle P_{23}P_{34}P'' + \frac{1}{2}\sphericalangle P''P_{34}P''' - \sphericalangle P_{34}P_{23}P_0$$

$$= \frac{1}{2}(d\varepsilon + d^2\varepsilon + d\tau + d^2\tau) + \frac{1}{2}(d\varepsilon + 2d^2\varepsilon + d^3\varepsilon) - \frac{d\varepsilon + d^2\varepsilon}{2} - d\tau - d^2\tau$$

$$= \frac{1}{2}d\varepsilon + d^2\varepsilon - \frac{d\tau + d^2\tau}{2}$$

Applying the sine law to $\triangle P_{23}P_0P_{34}$, we get

$$\overline{P_0P_{34}} = ds\frac{\sin \sphericalangle P_{34}P_{23}P_0}{\sin \sphericalangle P_{23}P_0P_{34}} \approx ds\frac{d\varepsilon + d^2\varepsilon + 2d\tau + 2d^2\tau}{d\varepsilon + 2d^2\varepsilon - d\tau - d^2\tau} \qquad (3.130)$$

So far we have been using the symbol $\Delta\gamma$ to represent the angle of rotation of a body, and taking a counterclockwise sense as positive, therefore $d\varepsilon = -d\gamma$, $d^2\varepsilon = -d^2\gamma$. In approaching the limit, $\lim\limits_{ds\to 0} \overline{P_0P_{34}} = \rho_P$ is the radius of curvature of the path of P. Equation (3.130) becomes

$$\rho_P = \lim_{ds\to 0} ds\frac{d\gamma - 2d\tau + d^2\gamma - 2d^2\tau}{d\gamma + d\tau + 2d^2\gamma + d^2\tau} \qquad (3.131)$$

We shall consider the following different cases.

(a) General case: $d\gamma \neq -d\tau$

In this case, equation (3.131) becomes

$$\rho_P = \lim_{ds\to 0} \overline{P_0P_{34}} = 0$$

The path of P, just before and after becoming a velocity pole, exhibits a cusp, as shown in Fig. 1.6.

(b) Special case: $d\gamma = -d\tau$, but $2d^2\gamma \neq -d^2\tau$

In this case, equation (3.131) becomes

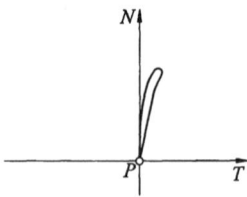

Fig. 3.66. Path of P exhibiting a beak cusp.

Dimensional synthesis of four-bar linkages --- body guidance problems

$$\rho_P = \lim_{ds \to 0} \frac{3 d\gamma \, ds}{2d^2 \gamma + d^2 \tau} = \frac{3\gamma'}{2\gamma'' + \tau''} \quad (3.132)$$

Now ρ_P does not vanish. According to equations (3.120), (3.123) and (3.126) we have $l_* \to \infty$, while $l = \delta$, hence (please refer to Fig. 1.13)

$$\overline{PM_0} = 2\overline{PM} \quad (3.133)$$

The path of P appears like a beak cusp, as shown in Fig. 3.66. We have therefore

Theorem 21 At a velocity pole, if the radius of curvature of the fixed polode is twice the radius of curvature of the moving polode, then the path of the velocity pole as a moving point exhibits a finite radius of curvature, with its centre of curvature lying on the poletangent, and the path appears not on the two sides but on only one side of the polenormal, exhibiting a beak shape.

Whenever the coupler of a mechanism is in such a position, it is called by Alt the *Cardan position* (Alt, 1944)[§]. This is because, as is well-known, the moving polode π_c of the coupler AB (= c) of a double-slider mechanism is a circle with $\overline{AB}$ as the diameter, as shown in Fig. 3.67, and the fixed polode π_f is a circle with

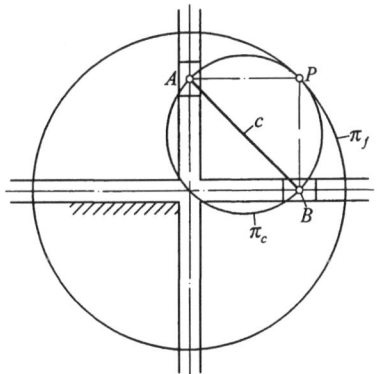

Fig. 3.67. Cardan circle-pair.

[§]The term *Cardan position* was first introduced by Rauh (Rauh, et al., 1938; Rauh, 1951). He claimed that, during the instantaneous motion of a body, the Cardan position occurs whenever the diameter of its inflection circle reaches an extreme (maximum or minimum). However, (Alt, 1944) disproved this statement by defining the *Cardan position* as that in which the condition $\overline{PM_0} = 2\overline{PM}$ is satisfied (equation (3.133)). These two different assertions were then investigated by (Freudenstein, 1960) who showed that, as far as path generation is concerned, only three infinitesimally separated positions (i.e. three-point contact, or third order synthesis as will be mentioned later in Section 6.12) are not sufficient for simulating a Cardan motion, and that in order to simulate a fourth order Cardan motion, both Rauh's and Alt's definitions have to be satisfied. However, at this moment we may stay with Alt's definition and call a body position which satisfies equation (3.133) the Cardan position, as (Meyer zur Capellen, 1949) did.

$\overline{AB}$ as radius. π_c equals just one half of π_f, and rolls on the inside of π_f. All this sort of mechanisms consisting of a pair of circles rolling relative to each other can be called *Cardan mechanisms*.

The Cardan position occurs in the inner dead-centre position, or outer dead-centre position of a four-bar linkage, as will be shown in the later Fig. 3.70(a)(b). Cardan position also occurs when the collineation axis is perpendicular to the line of centres, as that shown in Fig. 3.68. In this position, as mentioned in Theorem 6, the ratio ω_{cb}/ω_{ca} becomes a maximum or minimum. Since it is clear from Fig. 3.68 that

$$\frac{\overline{PB_0}}{\overline{PA_0}} = \frac{\cos\theta_B}{\cos\theta_A}$$

and that both A_0, B_0 satisfy the equation of the centering-point curve (3.121), therefore $1/l_* = 0$. In Fig. 3.68, according to equation (3.132) the radius of curvature of the path γ_P of P becomes (Müller, 1897, p.270; Beyer, 1953, p.134)

$$\rho_P = \frac{\overline{PD}}{\cos(\theta_A - \theta_B) - \sin(\theta_A + \theta_B)\tan\lambda} \quad (3.134)$$

(c) Special case: $d\gamma = -d\tau, 2d^2\gamma = -d^2\tau$

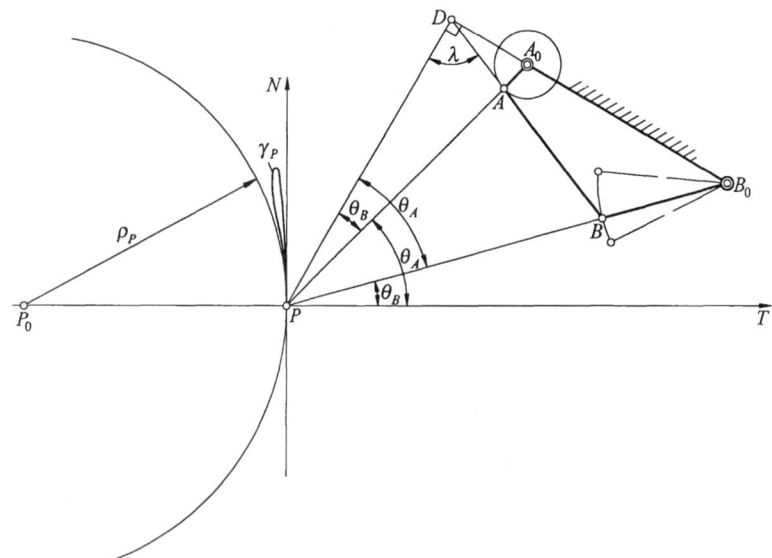

Fig. 3.68. As collineation axis PD is perpendicular to A_0B_0, coupler AB is in its Cardan position, and path γ_P of P exhibits a beak cusp.

In this case $\rho_P \to \infty$. The well-known Cardan-circles belong to this case. Due to $d^2\gamma = 0$, $d^2\tau = 0$, the path of the moving point P becomes a straight line, or $\rho_P \to \infty$.

3.7.5 Circling-point curve and centering-point curve of a given four-bar linkage by algebraic method

It can be seen from equations (3.116) and (3.121) that, the circling-point curve is determined only by the two constants m, l, and the centering-point curve is also determined only by two constants m, l_*. Hence for the coupler of a given four-bar linkage, in order to find the coordinates of the points A_0, B_0, A, B, it is only necessary to find first by Bobillier construction the directions of PT and PN. Since A, B belong to points on the circling-point curve, substituting the coordinates of these two points into equation (3.116) will result in two linear equations in the two unknowns $1/m$ and $1/l$, and values of m and l are determined. Similarly, A_0, B_0 belong to points on the centering-point curve. Substituting the coordinates of A_0, B_0 into equation (3.121) will result in two linear equations in the two unknowns $1/m$ and $1/l_*$, and values of m and l_* are determined.

3.7.6 The break-ups of circling-point curve and of centering-point curve

As the centering-point curve and circling-point curve are both circular cubics, in some cases, each of them can break up into a circle and a straight line, so that the construction can greatly be simplified. In short, it can be seen from equation (3.116) of k_u that, if $l \to \infty$, then k_u breaks up into a circle with diameter m and a straight line PT (poletangent). From equation (3.121) of k_a, if $l_* \to \infty$, then k_a also breaks up into a circle with diameter m and a straight line PT. If $m \to \infty$, then both k_u and k_a break up simultaneously. k_u becomes a circle with diameter l and a straight line PN (polenormal), while k_a breaks up into a circle with diameter l_* and a straight line PN. We explain now these cases separately:

(a) $\gamma' = -\tau'$

From equation (3.120) we get $l_* \to \infty$, therefore equation (3.121) becomes

$$\sin\theta_A (p_0 - m\cos\theta_A) = 0$$

k_a now breaks up into

$$\begin{aligned} p_0 - m\cos\theta_A &= 0 \\ \sin\theta_A &= 0 \end{aligned} \begin{aligned} &\text{(circle } k_a') \\ &\text{(line } k_a'' = PT) \end{aligned}$$

In this case the circling-point curve k_u does not break up. This is the case when a four-bar linkage is in its inner dead-centre position, as that shown in Fig. 3.69. The

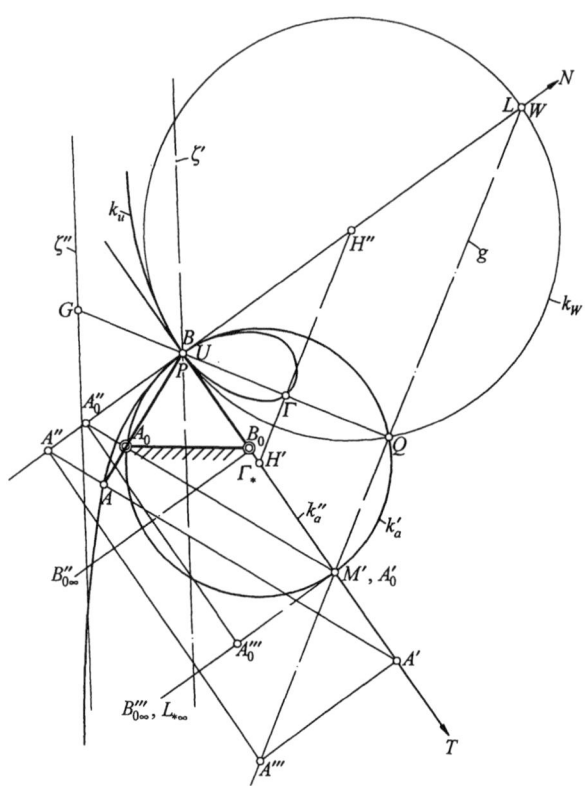

Fig. 3.69. As four-bar linkage A_0ABB_0 is in inner dead-centre position, k_a of the coupler breaks up into a circle k'_a and line k''_a, but k_u does not break up.

point B is then identical with P. According to Bobillier theorem, the poletangent coincides with B_0B. Following the construction method of Fig. 3.60, we see that the points B''_0, B'''_0 and L_* all lie at infinity in the direction of PN, and $A'_0 = M'$. From equation (3.115) we have

$$l = -\frac{1}{\gamma'} = \delta$$

and from equation (3.126) we have

$$\overline{PM_0} = 2\overline{PM}$$

Therefore the circle with $\overline{PL}$ as diameter is the inflection circle k_W, and L is the inflection pole W. The circling-point curve k_u does not break up, and the inflection

Dimensional synthesis of four-bar linkages --- body guidance problems 151

circle k_W is a circle of curvature of k_u. Γ_* is at the centre of the circle k'_a.

From the condition $\overline{PM_0} = 2\overline{PM}$ it is obvious that a Cardan circle pair also belongs to this case. Similarly, the coupler of a four-bar linkage in its outer dead-centre position also belongs to this case. This is put together with the inner dead-centre position of a four-bar linkage in Fig. 3.70(a)(b).

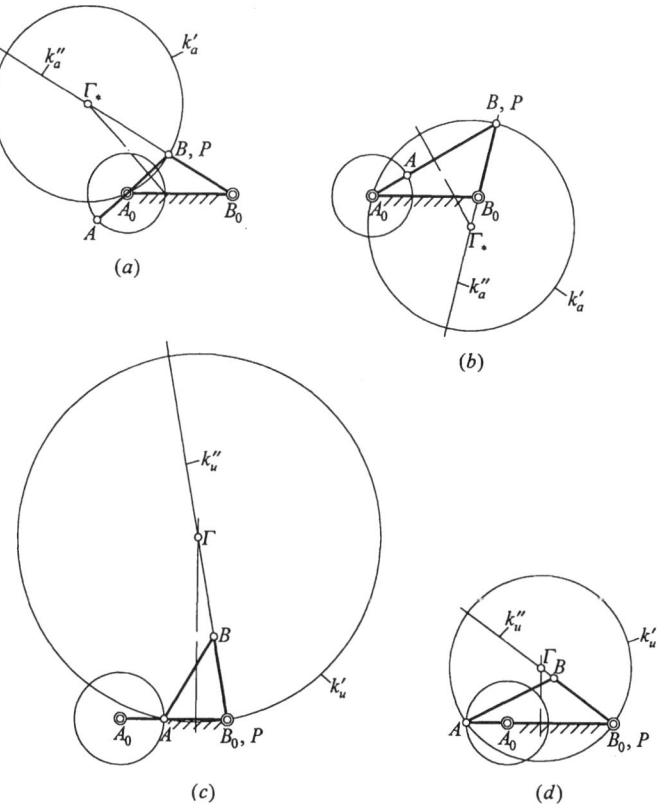

Fig. 3.70. (a) (b) k_a breaks up into k'_a and k''_a.
(c) (d) k_u breaks up into k'_u and k''_u.

Moreover, the position of a four-bar linkage in which its collineation axis is perpendicular to the line of centres, also belongs to this case, as explained in Section 3.7.4(b) and shown in Fig. 3.68.

(b) $2\gamma' = \tau'$

It can be seen from equation (3.115) that, in this case, $l \to \infty$. Equation (3.116) becomes

$$\sin\theta_A(p - m\cos\theta_A) = 0$$

hence k_u breaks up into

$$\left. \begin{array}{l} p - m\cos\theta_A = 0 \\ \sin\theta_A = 0 \end{array} \right\} \begin{array}{l} (\text{circle } k'_u) \\ (\text{line } k''_u = PT) \end{array}$$

The centering-point curve k_a does not break up.

The coupler of a four-bar linkage in its inner frame-position (Fig. 3.70(c)) and outer frame-position (Fig. 3.70(d)) belongs to this case. The point B_0 is P. According to Bobillier theorem we know that the poletangent coincides with $B_0 B$. Following the construction in Fig. 3.60 we can see that B'', B''' and L all lie at infinity in the direction of PN, and $A' = M'$. Furthermore, we have from equation (3.120)

$$l_* = \frac{1}{\gamma'} = -\delta$$

and from equation (3.126)

$$\overline{PM} = 2\overline{PM_0}$$

Therefore the circle with $\overline{PL_*}$ as diameter is the return circle k_R. L_* is the return pole R. The centering-point curve k_a does not break up, and the return circle k_R is one of the circles of curvature of k_a. Γ is at the centre of the circle k'_u.

For a Cardan circle-pair, if the smaller circle is kept fixed and the larger circle is rolling on the outside of the smaller circle, the motion of the larger circle belongs to this case.

For clarity, the four cases shown in Fig. 3.70 are tabulated in Table 3.3.

Table 3.3 Break-ups of circling-point curve k_u and centering-point curve k_a

Position of crank-rocker	Inner dead-centre	Outer dead-centre	Inner frame	Outer frame
Figure	3.70(a)	3.70(b)	3.70(c)	3.70(d)
k_u	Does not break up		Breaks up into a circle k'_u and a line k''_u	
k_a	Breaks up into a circle k'_a and a line k''_a		Does not break up	

Fig. 3.71 shows an adjustable coupler dwell mechanism designed on the basis of the broken-up centering-point curve shown in Fig. 3.70(b). Choose a point C_0 on the circle k'_a as the centre point. The circle-point C corresponding to C_0 can be located by means of Bobillier construction. Or alternatively, according to Fig. 3.60, the point A''' can be found through A', A''. Then $A''' M'$ is the line g. Find a small

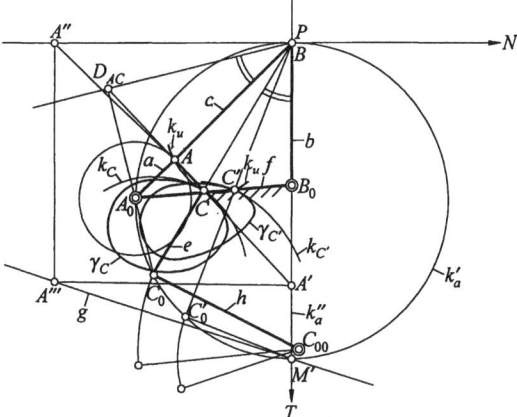

Fig. 3.71. An adjustable coupler dwell mechanism designed according to Fig. 3.70(b).

portion of the circling-point curve k_u according to the method shown in Fig. 3.59. The intersection of PC_0 and k_u is C. Make a link e to connect CC_0, and another link $h = C_0C_{00}$ to a fixed joint C_{00}. In this way, during the motion, the path γ_C of C and its circle of curvature k_C will be in a four-point contact, and the joint C_0 and link h will have quite a long period of approximate dwell. If we wish to vary the length of dwell of h and its swinging angle, we can choose different centre-points $C_0', \ldots$ on the circle k_a', so as to obtain different corresponding circle-points $C', \ldots$ However, we should vary the lengths of e and that of h correspondingly. Fig. 3.72 shows a practical construction of this machine. The centre line of the curved slot on the coupler c is a part of the circling-point curve k_u. The link e is made in a threaded sleeve shape, and an elongated slot is made on link h for adjusting the length. However, ear marks have to be made on the three adjusting places to ensure correct corresponding adjustments.

(c) $\gamma'' = 0$

From equation (3.114), it is now $m \to \infty$. The circling-point curve k_u and centering-point curve k_a break up simultaneously. The position of the four-bar linkage A_0ABB_0 shown in Fig. 3.73 is such that the collineation axis PD is perpendicular to B_0B. According to Bobillier Theorem, the poletangent PT is perpendicular to A_0A. Hence A_0A is the polenormal PN. According to Fig. 3.60, the lines passing through A_0, A and perpenducular to PA_0, PA respectively intersect PT at infinity, hence A_0', A' lie at infinity in the direction of PT, therefore $m \to \infty$. Equation (3.116) becomes

$$\cos\theta_A (p - l\sin\theta_A) = 0$$

The circling-point curve k_u breaks up into

$$\left.\begin{array}{r}p - l\sin\theta_A = 0 \\ \cos\theta_A = 0\end{array}\right\} \begin{array}{l}(\text{circle } k_u') \\ (\text{line } k_u'' = PN)\end{array}$$

Equation (3.121) becomes

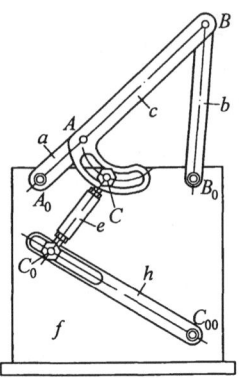

Fig. 3.72. Actual construction of the mechanism shown in Fig. 3.71.

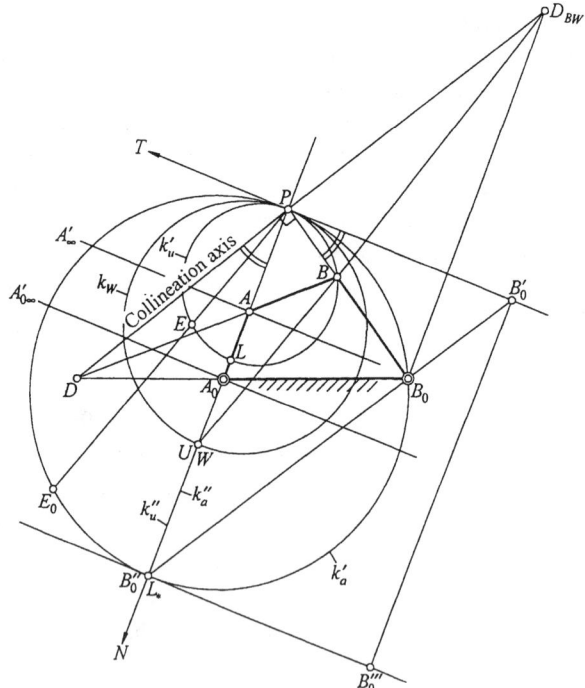

Fig. 3.73. k_u breaks up into circle k_u' and line k_u'', and k_a breaks up into circle k_a' and line k_a''.

Dimensional synthesis of four-bar linkages --- body guidance problems

$$\cos\theta_A (p_0 - l_* \sin\theta_A) = 0$$

Therefore the centering-point curve k_a breaks up into

$$\begin{matrix} p_0 - l_* \sin\theta_A = 0 \\ \cos\theta_A = 0 \end{matrix} \begin{matrix} (\text{circle } k'_a) \\ (\text{line } k''_a = PN) \end{matrix}$$

If a point E is taken now on k'_u as the circle-point, then joining PE to meet the circle k'_a will give the centre-point E_0. On the contrary, if the centre-point E_0 is given first, the circle-point E can also be found without difficulty. The path of E and its circle of curvature (the circle with E_0 as centre and $\overline{E_0 E}$ as radius) are in a four-point contact.

From equation (3.114), the definition of $m = 3\gamma'/\gamma''$, since δ is non zero, and $\gamma' = -1/\delta$ does not approach to infinity, γ'' must vanish. This means that in this position, γ' is stationary. In other words, the inflection circle is stationary (Please refer to the footnote on page 147).

(d) $\gamma'' = 0$, $\gamma' = -\tau'$

This is a combination of the above two cases (a) and (c). It can be seen from case (c) that the condition $\gamma'' = 0$ alone is sufficient to cause a simultaneous breaking up of k_u and k_a. Now in addition to that, the second condition $\gamma' = -\tau'$, i.e. $l_* \to \infty$, causes for k_a a further break of k'_a or $p_0 - l_* \sin\theta_A = 0$ into the poletangent PT and the line at infinity. Furthermore, because of $l = \delta$, the equation of the circle k'_u, or $p - l \sin\theta_A = 0$ is that of the inflection circle k_W.

The neutral position of a central slider-crank mechanism shown in Fig. 3.74 belongs to this case. The diameter of the inflection circle can be found by means of Euler-Savary equation, as $\delta = |c(a+c)/a|$.

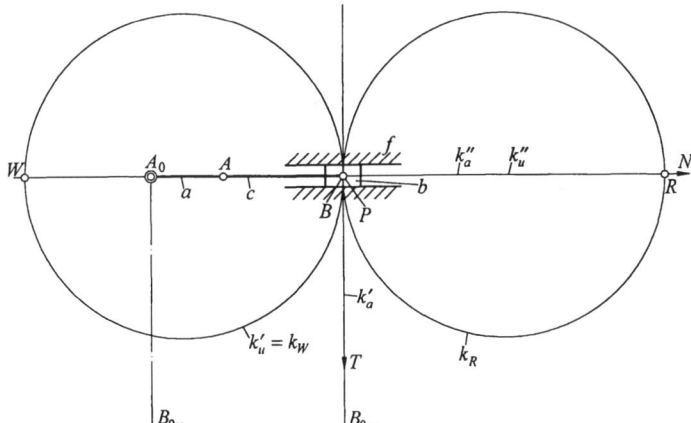

Fig. 3.74. k_u breaks up into circle k'_u and line k''_u, and k_a breaks up into lines k'_a, k''_a and the line at infinity.

(e) $\gamma'' = 0, 2\gamma' = \tau'$

This is a combination of the above two cases (b) and (c). Similarly in addition to $\gamma'' = 0$, the condition $2\gamma' = \tau'$, i.e. $l \to \infty$, causes for k_u a further break up of the circle k'_u or $p - l\sin\theta_A = 0$, into the poletangent PT and the line at infinity. Furthermore, because of $l_* = -\delta$, the equation of the circle k'_a, or $p_0 - l_* \sin\theta_A = 0$ is that of the return circle k_R.

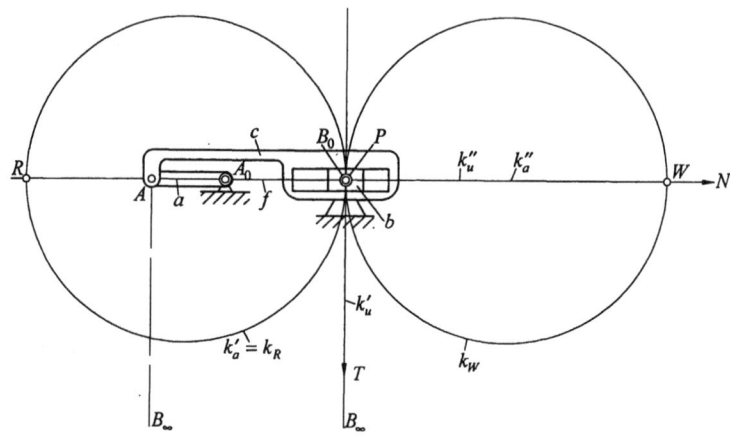

Fig. 3.75. k_u breaks up into lines k'_u, k''_u and the line at infinity; k_a breaks up into circle k'_a and line k''_a.

Fig. 3.75 shows an inversion of the mechanism in Fig. 3.74, which is the neutral position of a central swinging block linkage, belonging to this case. The diameter of the return circle can be found by means of Euler-Savary equation, as $\delta = |f(a+f)/a|$.

3.7.7 The Ball point (four infinitesimally separated homologous points on a line) and four infinitesimally separated homologous lines passing through a fixed point

On the circling-point curve k_u, there must be a certain point the radius of curvature of whose path is infinity. This should be the intersection between k_u and the inflection circle k_W. Rewrite equation (1.5) in terms of present symbols,

$$p = \delta \sin\theta_A$$
$$= -\frac{1}{\gamma'}\sin\theta_A \qquad [(1.5)]$$

Substituting this equation into equation (3.116) of k_u, and taking equations (3.114), (3.115) and (3.120) into consideration, we get

Dimensional synthesis of four-bar linkages --- body guidance problems   157

$$\tan\theta_A = \tan\theta_U = -\frac{\gamma'(\gamma'+\tau')}{\gamma''} = -\frac{m}{l_*} \qquad (3.135)$$

In Fig. 3.76 is shown a four-bar linkage A_0ABB_0, the line g_* of which is determined according to the method of Fig. 3.60. This line intersects the poletangent PT in the point M', yielding $\overline{PM'} = m$, and intersects the polenormal PN in L_*, yielding $\overline{PL_*} = l_*$. The slope of the line g_* is $-l_*/m$. Folding the normal of g_* with respect to PN, we get the line ζ'_*. The slope of ζ'_* is $-m/l_*$. Therefore the intersection between ζ'_* and the inflection circle k_W is the required point, denoted by U. This is why the symbol θ_A in equation (3.135) is written as θ_U. The point U is called *Ball point*, after the English mathematician Sir R. S. Ball (Ball, 1849), and also called *undulation point*. Since the path at a Ball point is in a four-point contact with its tangent, it exhibits a longer approximate straight portion than any other inflection point. It can be seen in Fig. 3.76 that quite a long portion of

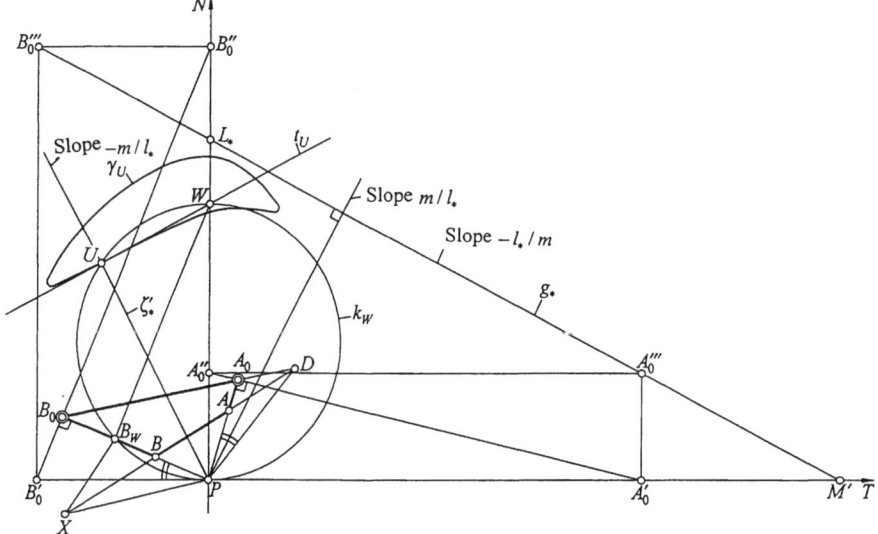

Fig. 3.76.   Position of Ball point U.

the path γ_U of U is approximately straight.

In Fig. 3.73, since the circling-point curve k_u is broken up into a circle k'_u and a straight line k''_u, the intersection of k''_u and the inflection circle k_W is the Ball point U, which is also the inflection pole W. Also in Fig. 1.20, since k_u breaks up in the same way, the inflection pole W is the Ball point U.

Fig. 3.77 is an illustration partly redrawn from Fig.3.57. As mentioned before, here $\overline{PM'} = m$, $PL = l$, $\overline{PL_*} = l_*$. The two circles with $\overline{PM'}$ and $\overline{PL}$ as diameters respectively intersect in Q, and the two circles with $\overline{PM'}$ and $\overline{PL_*}$ as diameters respectively intersect in Q_*. The middle point of $\overline{PQ}$ is Γ, and the middle point of $\overline{PQ_*}$ is Γ_*. The line obtained by folding PQ_* with respect to PN is ζ'_*. The intersection between ζ'_* and k_W is U. The coordinates of Ball point can be derived from equations (3.135) and (3.116):

$$x_U = -\frac{ml_*\delta}{m^2+l_*^2}, \quad y_U = \frac{m^2\delta}{m^2+l_*^2} \qquad (3.136)$$

Since the circling-point curve k_u and centering-point curve k_a are respectively the limiting cases of circle-point curve k_{1234} and centre-point curve m_{1234}, we should be able to observe the gradual change from the case of four collinear homologous points and four concurrent homologous lines as shown in Fig. 3.51 to the limiting case as shown in Fig. 3.77.

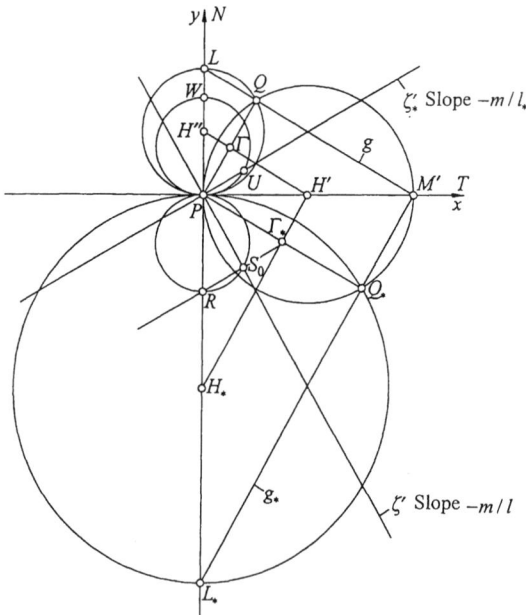

Fig. 3.77. U and S_0 are respectively limiting cases of A_1, A_2, A_3, A_4 and S_0 in Fig. 3.51.

Consider first the case of four points on a line. The Ball point can be considered as the limiting case of four homologous points A_1, A_2, A_3, A_4 on a line l_A. We may therefore also call the four collinear points in Fig. 3.51 the Ball-point in a broader sense. The line on which the four points lie is, according to Theorem 18, the line l_A

Dimensional synthesis of four-bar linkages --- body guidance problems

passing through the four orthocentres $H_{123}, H_{124}, H_{134}, H_{234}$. In approaching the limit, the four points A_1, A_2, A_3, A_4 there shrink into one point U. Because the four orthocentres combine into the inflection pole W, therefore l_A becomes the line UW in Fig. 3.77 (not shown in the figure).

We come now to the case of four lines passing through a point. In Fig. 3.51, the common point S_0 of the four homologous lines l_1, l_2, l_3, l_4 is the intersection of the four circumcircles $k_{123}, k_{124}, k_{134}, k_{234}$ of the four poletriangles. Now in the limiting case these four circles combine into the return circle k_R, and it would be impossible to find their intersection. However, according to Theorem 16, the point S_0 should lie on the centering-point curve k_a, so it should be the intersection of k_R and k_a. Rewriting equation (1.29) of the return circle in terms of present symbols, we have

$$p_0 = -\delta \sin\theta_A$$
$$= \frac{1}{\gamma'}\sin\theta_A$$

Substituting these expressions into equation (3.121) gives

$$\tan\theta_{AS} = -\frac{m}{l} \qquad (3.137)$$

The slope of the line ζ' in Fig. 3.77 is $-m/l$. Hence S_0 is the intersection of the line ζ' and the return circle k_R. The coordinates of S_0 are

$$x_{S_0} = \frac{ml\delta}{m^2 + l^2}, \quad y_{S_0} = -\frac{m^2\delta}{m^2 + l^2} \qquad (3.138)$$

Consider again Fig. 3.69. The point P coincides with B, being just the intersection between the inflection circle k_W and the circling-point curve k_u, i.e. the Ball point U. However, since the path of B is already a circular arc with B_0 as its centre, how could B generate a straight path? This paradox is due to the component, the straight line part k_a'' of the centering-point curve k_a. On the line k_a'' there are infinite points which can be chosen as the centre-point, including the point at infinity. But there is only one circle-point corresponding to these infinite centre-points, which is the point P itself. This Ball point is of no practical use.

3.7.8 The Ball curve

Fig. 3.78 shows a four-bar linkage A_0ABB_0. As this four-bar linkage moves to different positions, the inflection circles of its coupler c (= AB) are respectively $k_{W1}, k_{W2}, ...$, and the Ball points are respectively $U_1, U_2, ...$ Move each of these inflection circles and Ball points together with each coupler back to its first position

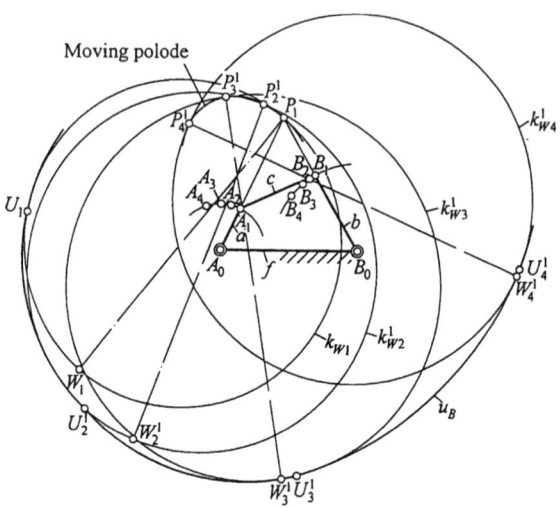

Fig. 3.78. The Ball curve.

c_1, to get $k_{W1}, k_{W2}^1, k_{W3}^1, ...$, and $U_1, U_2^1, U_3^1, ...$ The curve passing through $U_1, U_2^1, U_3^1,...$ is called *Ball curve*. This curve is just the envelope of the family of circles $k_{W1}, k_{W2}^1, k_{W3}^1 ...$ On the other side, the different poles of the coupler c are moved to the positions $P_1, P_2^1, P_3^1, ...$ The curve passing through $P_1, P_2^1, P_3^1, ...$ is the moving polode, as has been shown in Fig. 1.1. This polode is just the other side envelope of $k_{W1}, k_{W2}^1, k_{W3}^1, ...$

3.8 Guiding a body through five finitely separated positions --- the Burmester points

Suppose five positions of a body c are given. To find the position A_1 of a point A on c such that its five homologous points A_1, A_2, A_3, A_4, A_5 are on a circle, we find first the circle-point curve k_{1234} for the four positions c_1, c_2, c_3, c_4, and then the circle-point curve k_{1235} for the four positions c_1, c_2, c_3, c_5. The intersection of the two circle-point curves k_{1234} and k_{1235} is the point A_1 on c_1 which lies with its other four homologous points on a circle. Such points are called *Burmester points*. Let us count the number of intersections of these two curves. These two curves, each being a circular cubic, intersect in 9 points. Besides the three known points P_{12}, P_{13}, P_{23}^1 and a pair of imaginary points I, J (the circular points at infinity, please refer to Section A1.2), there can at most be four real Burmester points.

Alternatively we can find the intersections of two centre-point curves m_{1234}

Dimensional synthesis of four-bar linkages --- body guidance problems 161

and m_{1235}. These two circular cubics also intersect in 9 points. Besides the three known points P_{12}, P_{23}, P_{13} and a pair of imaginary points I, J, there can at most be four real intersections, called *Burmester centres*. This method is more convenient than the above method of finding the Burmester circle-points, because in the following equation, the root $\tan\theta_{P_{23}} = 0$ corresponding to P_{23} is already known (please refer to Fig. 3.27). Fig. 3.79 shows a centre-point curve m_{1235} intersecting

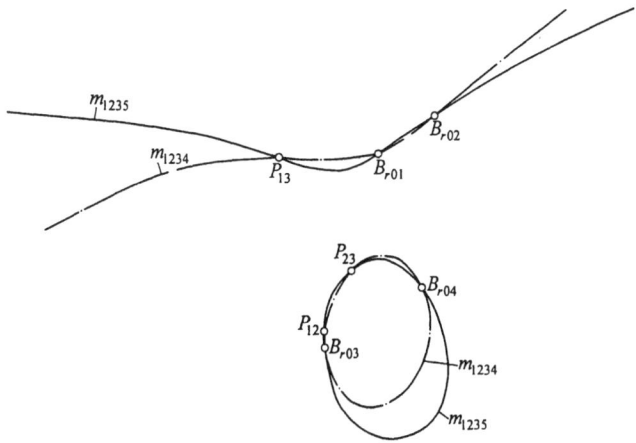

Fig. 3.79. Burmester centre-points $B_{r01}, B_{r02}, B_{r03}, B_{r04}$ are intersections of m_{1234} and m_{1235}, which intersect also in P_{12}, P_{23}, P_{13}.

the centre-point curve m_{1234} of Fig. 3.44. The intersections include, apart from the three known poles P_{12}, P_{23}, P_{13}, the four real points, i.e. the four Burmester centre-points $B_{r01}, B_{r02}, B_{r03}, B_{r04}$. Fig. 3.80 shows another centre-point curve m_{1235} intersecting the m_{1234} of Fig. 3.44. The intersections include no real points except the three poles P_{12}, P_{23}, P_{13}, which means there are no Burmester centres.

To find the Burmester centres by algebraic method, we can write according to equation (3.104) the following two equations of centre-point curves:

$$\left. \begin{array}{l} m_{1234}: M_2 p_0^2 + M_1 p_0 + M_0 = 0 \\ m_{1235}: E_2 p_0^2 + E_1 p_0 + E_0 = 0 \end{array} \right\} \quad (3.139)$$

where the coefficients E_2, E_1, E_0 of the equation of m_{1235} can be written with reference to M_2, M_1, M_0 in Appendix 7. Eliminating p_0 from equations (3.139) gives

$$\begin{vmatrix} M_2 & M_0 \\ E_2 & E_0 \end{vmatrix}^2 + \begin{vmatrix} M_2 & M_1 \\ E_2 & E_1 \end{vmatrix} \begin{vmatrix} M_0 & M_1 \\ E_0 & E_1 \end{vmatrix} = 0 \quad (3.140)$$

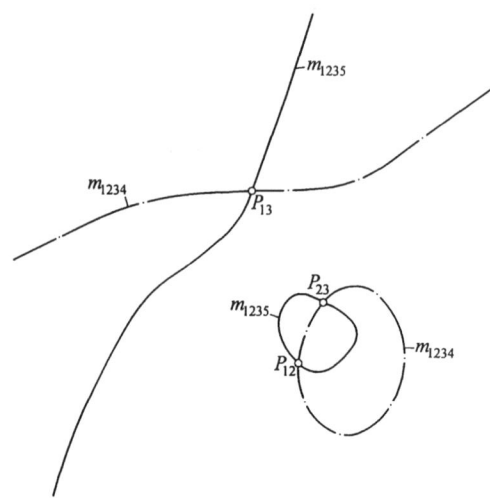

Fig. 3.80. Intersections of m_{1234} and m_{1235}, containing besides P_{12}, P_{23}, P_{13} no Burmester centre-points.

Equation (3.140) can be rewritten as

$$\frac{1}{\cos^2\theta_0}\begin{vmatrix} M_2/\cos\theta_0 & M_0/\cos\theta_0 \\ E_2/\cos\theta_0 & E_0/\cos\theta_0 \end{vmatrix}^2$$
$$+\begin{vmatrix} M_2/\cos\theta_0 & M_1/\cos^2\theta_0 \\ E_2/\cos\theta_0 & E_1/\cos^2\theta_0 \end{vmatrix}\begin{vmatrix} M_0/\cos\theta_0 & M_1/\cos^2\theta_0 \\ E_0/\cos\theta_0 & E_1/\cos^2\theta_0 \end{vmatrix}=0 \quad (3.141)$$

Expanding equation (3.141) results in a sextic equation in the unknown $\tau = \tan\theta_0$:

$$g_6\tau^6 + g_5\tau^5 + g_4\tau^4 + g_3\tau^3 + g_2\tau^2 + g_1\tau = 0 \quad (3.142)$$

The roots of equation (3.142) do not include the root corresponding to the origin P_{12}, but do include the roots corresponding to P_{23} ($\tau = 0$) and P_{13} ($\tau = \tan(-\gamma_{12}/2)$). Removing these two roots, we rewrite equation (3.142) as

$$g_6\tau^4 + g_{03}\tau^3 + g_{02}\tau^2 + g_{01}\tau + g_{00} = 0 \quad (3.143)$$

The coefficients in equation (3.143) are listed in Appendix 8, the number of its real roots being 4, 2, or 0. Having found the four roots by solving equation (3.143), we have the value of p_0 corresponding to each real root τ_0 as

$$p_0 = -\begin{vmatrix} M_2 & M_0 \\ E_2 & E_0 \end{vmatrix} \bigg/ \begin{vmatrix} M_2 & M_1 \\ E_2 & E_1 \end{vmatrix} \quad (3.144)$$

Dimensional synthesis of four-bar linkages --- body guidance problems

The expanded form of equation (3.144) is also listed in Appendix 8.

Having found the Burmester centre-point, to find the corresponding circle-point, we simply substitute the coordinates of the centre-point into equation (3.46). A pair of real Burmester centre-point and circle-point is called a *Burmester point-pair*. The number of possible four-bar linkages for guiding a body through five finitely separated positions is 6, 2 or 0, depending on whether the number of Burmester point-pairs is 4, 2, or 0.

3.9 Guiding a body through five infinitesimally separated positions

3.9.1 The general case

This is the limiting case of guiding a body through five finitely separated positions. The purpose is to find out, on a body, those points whose five infinitesimally separated homologous points are on a circle. We shall use the same reasoning as we did in Section 3.7.1, i.e. assuming that A_1, A_2, A_3, A_4 are on a circle, and A_2, A_3, A_4, A_5 are also on a circle. In other words, the point A that we are looking for, shall not only satisfy equation (3.109) of $\rho' = 0$, but also satisfy the condition $\rho'' = 0$. The number of points that satisfy simultaneously both these two equations is finite. These points were originally called by (Müller, 1892) as *Burmester points*, and the definition of *Burmester points* that appears in Section 3.8 is in fact a later-date extended version.

To find the equation $\rho'' = 0$, we have to differentiate equation (3.109) with respect to s, and to equate it to zero. However, since the condition $\rho' = 0$ is to be satisfied simultaneously, we only have to differentiate either its numerator or equation (3.116) with respect to s. Rewrite now equation (3.116), or $\rho' = 0$, as follows:

$$k_u : \frac{1}{p} = \frac{1}{m\cos\theta_A} + \frac{1}{l\sin\theta_A} \qquad [(3.116)]$$

Differentiating this equation with respect to s, and equating it to zero, we get the following equation of the curve, denoted by k_v.

$$k_v : \frac{p'}{p^2} + \frac{\theta'_A \sin\theta_A}{m\cos^2\theta_A} - \frac{m'}{m^2 \cos\theta_A} - \frac{\theta'_A \cos\theta_A}{l\sin^2\theta_A} - \frac{l'}{l^2 \sin\theta_A} = 0 \qquad (3.145)$$

where
$$m' = \frac{3(\gamma''^2 - \gamma'\gamma''')}{\gamma''^2}, \quad l' = \frac{3(2\gamma'' - \tau'')}{(2\gamma' - \tau')^2} \qquad (3.145a,b)$$

Substituting equations (3.110) and (3.111) into equation (3.145) yields

$$k_v : -\frac{C_\theta}{p^2} - \frac{m'}{m^2 C_\theta} - \frac{l'}{l^2 S_\theta} + \left[\left(\frac{1}{mS_\theta} - \frac{1}{lC_\theta}\right) + \frac{1}{p}\left(\frac{S_\theta}{C_\theta} - \frac{C_\theta}{S_\theta}\right)\right]\left(\frac{S_\theta}{p} - \frac{2}{l} - \frac{1}{l_*}\right) = 0$$

(3.146)

In equation (3.146), $S_\theta = \sin\theta_A$, $C_\theta = \cos\theta_A$. The intersections of k_u and k_v are the required Burmester points. Eliminating p from equations (3.116) and (3.146) gives the following quartic equation in the unknown $\tan\theta_A = \tan\theta_V$, where θ_V is the argument of the polar coordinates of that Burmester point.

$$v_4 \tan^4\theta_V + v_3 \tan^3\theta_V + v_2 \tan^2\theta_V + v_1 \tan\theta_V + v_0 = 0 \qquad (3.147)$$

where

$$v_4 = \gamma''^2$$
$$v_3 = \gamma'\gamma''(\gamma' - 2\tau')$$
$$v_2 = 3\gamma'\gamma''' - 4\gamma''^2$$
$$v_1 = -3\gamma'(\gamma''\tau' - \gamma'\tau'')$$
$$v_0 = -\gamma'^2(2\gamma' - \tau')(\gamma' + \tau')$$

Similar to the case of guiding a body through five finitely separated positions, in the present case there are also only 4, 2 or 0 real Burmester points. For each real Burmester point, the corresponding Burmester centre-point can be found by means of Bobillier theorem or Euler-Savary theorem. Each Burmester point together with its centre-point is still called a Burmester point-pair. Hence the number of possible four-bar linkages to guide a body through five infinitesimally separated positions is 6, 2 or 0.

A Ball point which is at the same time a Burmester point is said to have an *excess* 1, because it possesses a five-point contacting tangent line. Similarly a Ball point with a six-point contacting tangent is said to have an *excess* 2, etc. Please refer to (Dijksman, 1976, p.101).

3.9.2 Special cases

The following special cases will now be considered.
(a) $\gamma' = -\tau'$

This is the case discussed in Sections 3.7.4(b) and 3.7.6(a). The term v_0 in equation (3.147) vanishes, hence one of its roots is $\tan\theta_V = 0$. Because the circling-point curve k_u does not break up, the only moving point corresponding to $\theta_V = 0$ is the velocity pole P (please refer to Fig. 3.69). The point P is a Burmester point.

(b) $\gamma'' = 0$

This is the case discussed in Section 3.7.6(c). It was concluded there that the circling-point curve k_u breaks up in this case into a circle k'_u and the polenormal k''_u (please refer to Fig. 3.73.). Therefore all Burmester points must lie on the circle

Dimensional synthesis of four-bar linkages --- body guidance problems 165

k'_u and the line k''_u. Now due to $\gamma'' = 0$, both coefficients v_4, v_3 in equation (3.147) vanish. Hence two roots are $1/\tan\theta_V = 0$, $1/\tan\theta_V = 0$. In other words, there are two Burmester points on the polenormal, and the rest two roots are on the circle k'_u.

(c) $\gamma'' = 0, 2\gamma' = \tau'$

This is the case discussed in Section 3.7.6(e). The circling-point curve k_u breaks up into the lines PT, PN and the line at infinity, while the centering-point curve k_a breaks up into the return circle k_R and the line PN, as shown in Fig. 3.75. The coefficients v_4, v_3 and v_0 in equation (3.147) vanish. Therefore one of the roots is $\tan\theta_V = 0$, and two other roots are $1/\tan\theta_V = 0$, $1/\tan\theta_V = 0$. The value $\theta_V = 90°$ of the two latter roots means that two Burmester points are on the polenormal. Now one of them is the known point A, but the other one has to be determined alternatively. Since all Burmester points are on the curve k_v, and now $1/m = 0$, $1/l = 0$, but $m'/m^2 = -\gamma'''/3\gamma'$, $l'/l^2 = -\tau''/3$, and $l_* = -\delta$, for $\theta_A = 90°$ equation (3.146) of the k_v curve becomes the following quadratic equation

$$\frac{\gamma'''\delta}{3}p^2 - \frac{1}{\delta}p - 1 = 0 \qquad (3.148)$$

Substituting the polar coordinates of the point A into equation (3.148), i.e. setting $p_A = -(a+f)$, we can find γ'''. Consequently the coordinate of the other Burmester point is

$$p = \frac{\delta(a+f)}{\delta - (a+f)}$$

Substituting the expression $\delta = (a+f)f/a$ of the diameter of the inflection circle into the above equation, we get the p value of the other Burmester point

$$p = \frac{f(a+f)}{f-a}$$

3.9.3 To find the Burmester points of a given four-bar linkage

As mentioned in Section 3.7.5, for a given four-bar linkage, we can calculate the values of m, l, l_*. Similarly, with the known Burmester points A, B, we can find the values of m', l' in equation (3.146). Rewriting equation (3.146), and noting that $\gamma' - \tau' = -2/l - 1/l_*$, we have

$$\frac{m'}{m^2 C_\theta} + \frac{l'}{l^2 S_\theta} = \left(\frac{S_\theta}{mC_\theta^2} - \frac{C_\theta}{lS_\theta^2}\right)\left(\frac{S_\theta}{p} - \frac{2}{l} - \frac{1}{l_*}\right) - \frac{C_\theta}{p^2} \qquad (3.149)$$

Substituting respectively the coordinates of the points A and B into equation (3.149), we obtain the values of m' and l'.

Equation (3.147) can be rewritten in the following form, with leading coefficient 1,

$$\tan^4\theta_V + b_3\tan^3\theta_V + b_2\tan^2\theta_V + b_1\tan\theta_V + b_0 = 0 \qquad (3.150)$$

where

$$b_3 = -m\left(\frac{1}{l} + \frac{1}{l_*}\right)$$

$$b_2 = -(1+m')$$

$$b_1 = -\frac{m}{l}\left(3 + m\frac{l'}{l}\right)$$

$$b_0 = \frac{m^2}{ll_*}$$

In the quartic equation (3.150), the points A, B are known Burmester points, therefore removing the corresponding roots $\tan\theta_A$ and $\tan\theta_B$ from the equation, and solving the remaining quadratic equation, we obtain the other two Burmester points.

3.10 Intermediate cases

The body guidance problems depicted in the above sections are divided substantially in two groups, namely, finitely separated and infinitesimally separated positions. Please note that Theorem 7 and the formulae in Section 3.3 are applicable only to cases of finitely separated positions, and that the Euler-Savary equation and Bobillier construction are applicable only to cases of infinitesimally separated positions. If we apply the wrong theorem, we shall not be able to obtain the correct results. Between these two cases, there are some intermediate cases including combinations of finitely and infinitesimally separated positions. We adopt the symbol "P" suggested by (Tesar, 1967) to signify a single position of a body. Two finitely separated positions are symbolized by P–P. Two infinitesimally separated positions are symbolized by PP. Three finitely separated positions are symbolized by P–P–P, and three infinitesimally separated positions are symbolized by PPP, etc. For further clarity, we use, in addition, subscript for each "P" to indicate the sequence of that position. For instance, five finitely separated positions are symbolized by $P_1 - P_2 - P_3 - P_4 - P_5$, and two circle-point curves each involving different sets of four positions are denoted by k_{1234} and k_{1235} respectively.

All cases of body guidance are listed as follows:

Two-position: $P_1 - P_2$ P_1P_2

Three-position: $P_1 - P_2 - P_3$, $P_1P_2 - P_3$, $P_1 - P_2P_3$, $P_1P_2P_3$

Four-position: $P_1 - P_2 - P_3 - P_4$, $P_1P_2 - P_3 - P_4$, $P_1P_2 - P_3P_4$,
$P_1P_2P_3 - P_4$, $P_1P_2P_3P_4$

Five-position: $P_1 - P_2 - P_3 - P_4 - P_5$, $P_1P_2 - P_3 - P_4 - P_5$,
$P_1P_2 - P_3P_4 - P_5$, $P_1P_2P_3 - P_4 - P_5$, $P_1P_2P_3 - P_4P_5$,
$P_1P_2P_3P_4 - P_5$, $P_1P_2P_3P_4P_5$

We shall discuss each case individually.

3.10.1 Cases $P_1P_2-P_3$ and $P_1-P_2P_3$

In both of these cases the poletriangle $\Delta P_{12}P_{23}P_{13}$ becomes a straight line. In the case $P_1P_2-P_3$, the form of the poletriangle is like the one shown in Fig. 3.81(a), while in the case $P_1-P_2P_3$, it is like the one shown in Fig. 3.81(b). From the

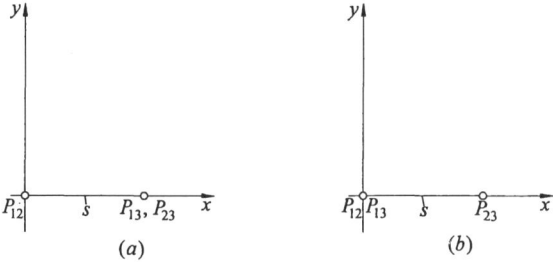

Fig. 3.81. (a) Situation of poles in the case $P_1P_2-P_3$. (b) Situation of poles in the case $P_1-P_2P_3$.

geometrical point of view, these two cases are not fundamentally different. However, as we have selected the coordinate system according to Section 3.4.7, the algebraic expressions in both cases are not completely identical. We shall consider them separately.

(a) $P_1P_2-P_3$

In this case, since γ_{12} becomes an infinitesimal, $\gamma_{12}=d\gamma$, hence $\sin\gamma_{12}\approx d\gamma$, $\cos\gamma_{12}\approx 1$. However, γ_{23} and $s=\overline{P_{12}P_{23}}$ remain finite. Therefore equation (3.54) becomes

$$\mathbb{D}_{12} = \begin{bmatrix} 1 & -d\gamma & 0 \\ d\gamma & 1 & 0 \\ 0 & 0 & 1 \end{bmatrix} \quad (3.151)$$

The form of $\mathbb{D}_{13}$ takes that of equation (3.55). The equation (3.49) of R_M becomes

$$R_M : (x_0^2+y_0^2)(H_3+W_3^*)^2 - R_A^2 W_3^{*2} = 0 \quad (3.152)$$

where $W_3^* = W_3/d\gamma = Q_3 y_0 + L_3 x_0$. Similarly, the equation (3.50) of R^1 becomes

$$R^1 : (x_{A1}^2+y_{A1}^2)(T_3+Z_3^*)^2 - R_A^2 Z_3^{*2} = 0 \quad (3.153)$$

where $Z_3^* = Z_3/d\gamma = S_3 y_{A1} + G_3 x_{A1}$.

(b) $P_1-P_2P_3$

The $\mathbb{D}_{12}$ becomes equation (3.54). γ_{23} becomes infinitesimal, $\gamma_{23}=d\gamma$, hence $\sin\gamma_{23}\approx d\gamma$, $\cos\gamma_{23}\approx 1$. However, γ_{12} and $s=\overline{P_{12}P_{23}}$ remain finite. For simplicity, let: $S_{12} = \sin\gamma_{12}$, $C_{12} = \cos\gamma_{12}$, $V_{12} = \text{vers}\,\gamma_{12}$, then the displacement matrix $\mathbb{D}_{13}$ in equation (3.56) becomes

$$\mathbb{D}_{13} = \begin{bmatrix} C_{12} - S_{12}d\gamma & -S_{12} - C_{12}d\gamma & 0 \\ S_{12} + C_{12}d\gamma & C_{12} - S_{12}d\gamma & -sd\gamma \\ 0 & 0 & 1 \end{bmatrix} \quad (3.154)$$

and

$$\mathbb{D}_{31} = \begin{bmatrix} C_{12} - S_{12}d\gamma & S_{12} + C_{12}d\gamma & sS_{12}d\gamma \\ -S_{12} - C_{12}d\gamma & C_{12} - S_{12}d\gamma & sC_{12}d\gamma \\ 0 & 0 & 1 \end{bmatrix} \quad (3.155)$$

In the present case,

$$\left.\begin{aligned} G_2 &= -V_{12}x_{A1} - S_{12}y_{A1} \\ S_2 &= S_{12}x_{A1} - V_{12}y_{A1} \\ T_2 &= 0 \end{aligned}\right\} \quad (3.156)$$

$$\left.\begin{aligned} G_3 &= (-V_{12} - S_{12}d\gamma)x_{A1} + (-S_{12} - C_{12}d\gamma)y_{A1} \\ S_3 &= (S_{12} + C_{12}d\gamma)x_{A1} + (-V_{12} - S_{12}d\gamma)y_{A1} - sd\gamma \\ T_3 &= sS_{12}d\gamma\, x_{A1} + sC_{12}d\gamma\, y_{A1} \end{aligned}\right\} \quad (3.157)$$

$$\left.\begin{aligned} L_2 &= -V_{12}x_0 + S_{12}y_0 \\ Q_2 &= -S_{12}x_0 - V_{12}y_0 \\ H_2 &= 0 \end{aligned}\right\} \quad (3.158)$$

$$\left.\begin{aligned} L_3 &= (-V_{12} - S_{12}d\gamma)x_0 + (S_{12} + C_{12}d\gamma)y_0 + sS_{12}d\gamma \\ Q_3 &= (-S_{12} - C_{12}d\gamma)x_0 + (-V_{12} - S_{12}d\gamma)y_0 + sC_{12}d\gamma \\ H_3 &= -sd\gamma\, y_0 \end{aligned}\right\} \quad (3.159)$$

In the present case the equation (3.49) of R_M becomes

$$R_M : (x_0^2 + y_0^2)[s^2 V_{12}^2 y_0^2 + (sS_{12}y_0 - W_3^{**})^2] - R_A^2 W_3^{**2} = 0 \quad (3.160)$$

where

$$W_3^{**} = W_3/d\gamma = -\text{vers}\,\gamma_{12}(x_0^2 + y_0^2) + s(\text{vers}\,\gamma_{12}x_0 + \sin\gamma_{12}y_0) \quad (3.161)$$

and the equation (3.50) of R^1 becomes

$$R^1 : (x_{A1}^2 + y_{A1}^2)\{s^2 V_{12}^2(S_{12}x_{A1} + C_{12}y_{A1})^2 + [sS_{12}(S_{12}x_{A1} + C_{12}y_{A1}) - Z_3^{**}]^2\} \\ - R_A^2 Z_3^{**2} = 0 \quad (3.162)$$

where

$$Z_3^{**} = Z_3/d\gamma = \text{vers}\,\gamma_{12}(x_{A1}^2 + y_{A1}^2) + s(\text{vers}\,\gamma_{12}x_{A1} + \sin\gamma_{12}y_{A1})$$

3.10.2 Intermediate cases of four positions

(a) $P_1 P_2 - P_3 - P_4$

Dimensional synthesis of four-bar linkages --- body guidance problems 169

Because of P_1P_2, equation (3.54) takes again the form of equation (3.151), and $\mathbb{D}_{13}$ and $\mathbb{D}_{14}$ remain those in the case of four finitely separated positions. In the equation (3.96) of the circle-point curve, all elements G_2, S_2, T_2 contain the common factor $d\gamma$, which means that $d\gamma$ should no longer appear in the development of equation (3.97). Hence the coefficients of G_2, S_2, T_2 in Appendix 4 should now take the following forms

$$\left.\begin{array}{lll} G_{2x} = 0 & G_{2y} = -1 & G_{2z} = 0 \\ S_{2x} = 1 & S_{2y} = 0 & S_{2z} = 0 \\ T_{2x} = 0 & T_{2y} = 0 & T_{2z} = 0 \end{array}\right\} \quad (3.163)$$

Similarly, in equation (3.102) of the centre-point curve the coefficients of L_2, Q_2, H_2 should be

$$\left.\begin{array}{lll} L_{2x} = 0 & L_{2y} = 1 & L_{2z} = 0 \\ Q_{2x} = -1 & Q_{2y} = 0 & Q_{2z} = 0 \\ H_{2x} = 0 & H_{2y} = 0 & H_{2z} = 0 \end{array}\right\} \quad (3.164)$$

To find the point A_1 of four homologous points on a line, we just set $R_A \to \infty$ in equation (3.153) and get

$$Z_3^* = S_3 y_{A1} + G_3 x_{A1} = 0 \quad (3.165)$$

Solving this equation together with the equation of k_{1234} simultaneously, i.e. finding the intersection of the curve k_{1234} and $Z_3^* = 0$, we get the point A_1.

(b) $P_1P_2 - P_3P_4$

This case may be considered as a combination of $P_1P_2 - P_3$ and $P_1 - P_3P_4$, as depicted in Section 3.10.1(a)(b). Due to P_1P_2, equation (3.54) takes again the form of equation (3.151). In general, the disposition of the poles is like that shown in Fig. 3.82, which is comparable with Fig. 3.81(a)(b). In $P_1P_2 - P_3$, the relative positions

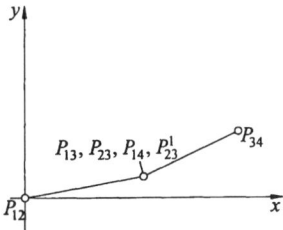

Fig. 3.82. Situation of poles in the case $P_1P_2 - P_3P_4$.

of P_{12}, P_{23}, P_{13} are somewhat like those shown in Fig. 3.81(a), P_{23} and P_{13} being combined together, but not necessarily on the x-axis. For positions P_1P_2, the

170   *Kinematics and Design of Planar Mechanisms*

coefficients of G_2, S_2, T_2 and L_2, Q_2, H_2 are still those in equations (3.163) and (3.164). The displacement matrix $\mathbb{D}_{13}$ remains that in the case of four finitely separated positions, hence the coefficients of G_3, S_3, T_3 in Appendix 4 and those of L_3, Q_3, H_3 in Appendix 6 are still valid, but those terms including s and γ_{23} are not valid. Moreover, since P_1 and P_2 coincide, the angles γ_{13} and γ_{23} are equal.

$P_1 - P_3 P_4$ indicates that P_{13} and P_{14} coincide. The angle γ_{34} becomes infinitesimal, but γ_{13} and $\overline{P_{13}P_{34}}$ remain finite. Our present purpose is to establish the displacement matrix $\mathbb{D}_{14}$. We may write

$$\mathbb{D}_{14} = \mathbb{D}_{34}\, \mathbb{D}_{13} \tag{3.166}$$

This indicates that a body is first displaced from P_1 through a finite displacement to position P_3, and successively from P_3 through an infinitesimally displacement to P_4. As mentioned before, $\mathbb{D}_{13}$ can be taken from equation (3.55). However, if the coordinate system in Fig. 3.27 is adopted, i.e. P_{23} is on the x-axis, then it can be taken from equation (3.56). As for $\mathbb{D}_{34}$, we can make use of equation (3.53). In that equation, we replace γ_{12} by γ_{34}, and P_{12} by P_{34}, and assume that $\gamma_{34} \approx \mathrm{d}\gamma$ without loss of generality. Hence $\sin\gamma_{34} \approx \mathrm{d}\gamma$, $\cos\gamma_{34} \approx 1$. Therefore

$$\mathbb{D}_{34} = \begin{bmatrix} 1 & -\mathrm{d}\gamma & y_{P_{34}}\mathrm{d}\gamma \\ \mathrm{d}\gamma & 1 & -x_{P_{34}}\mathrm{d}\gamma \\ 0 & 0 & 1 \end{bmatrix} \tag{3.167}$$

Hence we get

$$\mathbb{D}_{14} = \begin{bmatrix} C_{13} - S_{13}\mathrm{d}\gamma & -S_{13} - C_{12}\mathrm{d}\gamma & x_{P_{13}}V_{13} + y_{P_{13}}S_{13} + x_{P_{13}}S_{13}\mathrm{d}\gamma - y_{P_{13}}V_{13}\mathrm{d}\gamma + y_{P_{34}}\mathrm{d}\gamma \\ C_{13}\mathrm{d}\gamma + S_{13} & -S_{13}\mathrm{d}\gamma + C_{13} & x_{P_{13}}V_{13}\mathrm{d}\gamma + y_{P_{13}}S_{13}\mathrm{d}\gamma - x_{P_{13}}S_{13} + y_{P_{13}}V_{13} - x_{P_{34}}\mathrm{d}\gamma \\ 0 & 0 & 1 \end{bmatrix}$$
$$\tag{3.168}$$

where $S_{13} = \sin\gamma_{13}$, $C_{13} = \cos\gamma_{13}$, $V_{13} = \mathrm{vers}\,\gamma_{13}$. In the equation (3.96) of the present circle-point curve, after subtracting the second row from the third row, we see that each element contains a factor $\mathrm{d}\gamma$. Therefore we can replace the elements G_4, S_4, T_4 respectively by

$$G_4^* = (G_4 - G_3)/\mathrm{d}\gamma,\quad S_4^* = (S_4 - S_3)/\mathrm{d}\gamma,\quad T_4^* = (T_4 - T_3)/\mathrm{d}\gamma$$

or

Dimensional synthesis of four-bar linkages --- body guidance problems

$$G^*_{4x} = -\sin\gamma_{13}$$
$$G^*_{4y} = -\cos\gamma_{13}$$
$$G^*_{4z} = x_{P_{13}}\sin\gamma_{13} - y_{P_{13}}\text{vers}\gamma_{13} + y_{P_{34}}$$
$$S^*_{4x} = \cos\gamma_{13}$$
$$S^*_{4y} = -\sin\gamma_{13}$$
$$S^*_{4z} = x_{P_{13}}\text{vers}\gamma_{13} + y_{P_{13}}\sin\gamma_{13} - x_{P_{34}}$$
$$T^*_{4x} = -y_{P_{34}}\cos\gamma_{13} + x_{P_{34}}\sin\gamma_{13}$$
$$T^*_{4y} = y_{P_{34}}\sin\gamma_{13} + x_{P_{34}}\cos\gamma_{13}$$
$$T^*_{4z} = -(x_{P_{13}}\text{vers}\gamma_{13} + y_{P_{13}}\sin\gamma_{13})y_{P_{34}} + (-x_{P_{13}}\sin\gamma_{13} + y_{P_{13}}\text{vers}\gamma_{13})x_{P_{34}}$$

Similarly, in equation (3.102) of the centre-point curve, after subtracting the second row from the third row, we see that each element contains a factor $d\gamma$. Therefore we can replace the elements L_4, Q_4, H_4 respectively by

$$L^*_4 = (L_4 - L_3)/d\gamma, \quad Q^*_4 = (Q_4 - Q_3)/d\gamma, \quad H^*_4 = (H_4 - H_3)/d\gamma$$

or

$$L^*_{4x} = -\sin\gamma_{13}$$
$$L^*_{4y} = -\cos\gamma_{13}$$
$$L^*_{4z} = y_{P_{34}}\cos\gamma_{13} - x_{P_{34}}\sin\gamma_{13}$$
$$Q^*_{4x} = \cos\gamma_{13}$$
$$Q^*_{4y} = \sin\gamma_{13}$$
$$Q^*_{4z} = -y_{P_{34}}\sin\gamma_{13} - x_{P_{34}}\cos\gamma_{13}$$
$$H^*_{4x} = -x_{P_{13}}\sin\gamma_{13} + y_{P_{13}}\text{vers}\gamma_{13} + y_{P_{34}}$$
$$H^*_{4y} = -x_{P_{13}}\text{vers}\gamma_{13} - y_{P_{13}}\sin\gamma_{13} + x_{P_{34}}$$
$$H^*_{4z} = (x_{P_{13}}\text{vers}\gamma_{13} + y_{P_{13}}\sin\gamma_{13})y_{P_{34}} + (x_{P_{13}}\sin\gamma_{13} - y_{P_{13}}\text{vers}\gamma_{13})x_{P_{34}}$$

In the above expressions, the position of P_{13} ($= P_{23}$) is as that shown in Fig. 3.82. In case the position of P_{23} is moved to that shown in Fig. 3.81(a), then $x_{P_{13}} = s$, $y_{P_{13}} = 0$. The above expressions can further be simplified.

To find the point A_1 of four homologous points on a line, or the Ball point in a broader sense, we can still apply equation (3.165) to solve it simultaneously with the equation of the present circle-point curve to get the solution.

(c) $P_1P_2P_3 - P_4$

In the present case, the displacement matrices $\mathbb{D}_{12}$, $\mathbb{D}_{13}$ take respectively the forms of equations (3.57) and (3.58), while $\mathbb{D}_{14}$ takes the general form of equation (3.55) if we replace $x_{P13}, y_{P13}, \gamma_{13}$ respectively by $x_{P14}, y_{P14}, \gamma_{14}$, or

$$\mathbb{D}_{14} = \begin{bmatrix} \cos\gamma_{14} & -\sin\gamma_{14} & x_{P_{14}}\text{vers}\gamma_{14} + y_{P_{14}}\sin\gamma_{14} \\ \sin\gamma_{14} & \cos\gamma_{14} & -x_{P_{14}}\sin\gamma_{14} + y_{P_{14}}\text{vers}\gamma_{14} \\ 0 & 0 & 1 \end{bmatrix}$$

3.10.3 Intermediate cases of five positions

There are altogether five intermediate cases. Each one can be considered and solved as a combination of two four-position cases. Thus, for instance, for the case $P_1P_2P_3 - P_4P_5$, the intersections of circle-point curves of the two cases $P_1P_2P_3 - P_4$ and $P_1P_2 - P_3P_4$ yield the Burmester points. It is only a matter of applying the equations in Section 3.8. The readers will have no difficulty in listing the required equations.

3.11 Closing address

There are a number of methods for synthesizing a four-bar linkage to guide a body. There are methods that this book does not cover, e.g. the method of *loop equations* (Freudenstein & Sandor, 1961), the vector method (Erdman & Sandor, 1991), (Chen & Chiang, 1987), etc. In this book we are trying to introduce the major results and progress made by previous investigators. For instance, the method of opposite-pole quadrilateral was indeed the main thrust in the past. However, a further consideration reveals that, as we have done in this book, since four positions of a body are completely determined by three basic poles P_{12}, P_{13}, P_{14} and three rotation angles $\gamma_{12}, \gamma_{13}, \gamma_{14}$, the circle-point curve and the centre-point curve should be able to be determined by just these six data, and the handling of equations should be more systematic, and more adaptable for computer processing.

In the next chapter we shall introduce two other methods of guiding an instantaneous motion of a body.

Exercises

3.1 Show that in the (1:1) correspondence between the point set A_1, B_1, ... (called E_1) and the point set A_0, B_0, ... (called E_0) in guiding a body through three finitely separated positions (as in Fig. 3.12), there are the following exceptions:
(a) If the point in E_1 is taken at P_{12}, P_{13} or P_{23}^1, then the corresponding point in E_0 is not uniquely determined.
(b) If the point in E_0 is taken at P_{12}, P_{23} or P_{13}, then the corresponding point in E_1 is not uniquely determined.

3.2 Show that in guiding a body through three finitely separated positions (please refer to Fig. 3.12), if A_0 lies at infinity in a certain direction, then A_g lies

Dimensional synthesis of four-bar linkages --- body guidance problems 173

on the circumcircle k of the poletriangle $\Delta P_{12} P_{23} P_{13}$.

3.3 Show that in guiding a body through three finitely separated positions the point A_1 is uniquely determind for a prescribed A_0.
(a) Find A_1 from A_0 by means of the poletriangle.
(b) Find A_1 from A_0, by means of the concept of inversion on the basis of the three positions c_1, c_2, c_3 of the moving body c relative to the fixed body f.

3.4 Show that the four points A_1, A_2, A_3, H in Fig. 3.17 are collinear.

3.5 Show that the circumcircle k of the poletriangle $\Delta P_{12} P_{23} P_{13}$ in Fig. 3.19 passes through the three image points H_1, H_2, H_3 of the orthocenter H of the poletriangle.

3.6 The figure shows a guiding mechanism of an overhead garage door. A rolling wheel is mounted at the joint E to enable it to slide on a horizontal rail. It is required to replace the horizontal rail by a four-bar linkage $A_0 ABB_0$. Take three positions $A_1 E_1$, $A_2 E_2$, $A_3 E_3$ of AE. The location B_0 is known. Find B_2 by method of inversion.

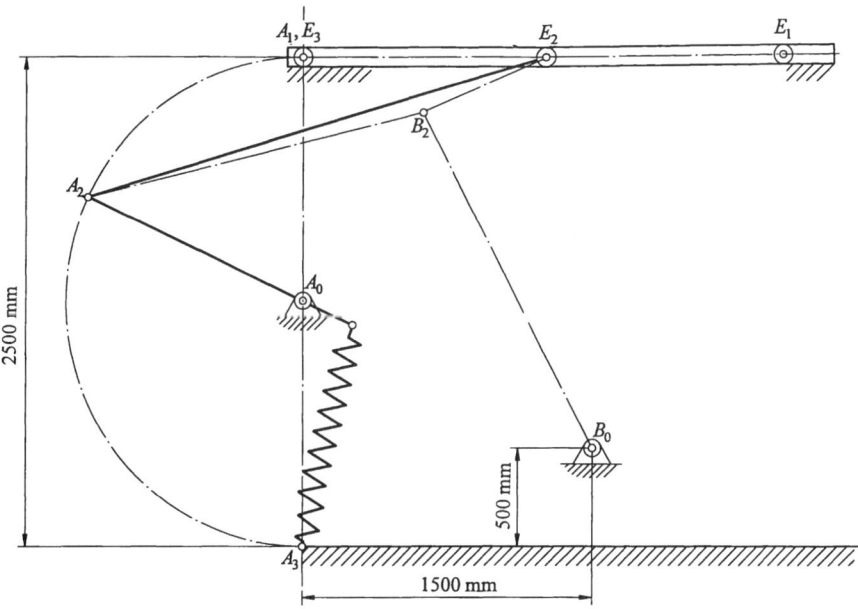

3.7 The figure shows a Roberts straight line mechanism, in which $\overline{A_0 B_0} = 2\overline{AB} = 100$, $\overline{A_0 A} = \overline{B_0 B} = \overline{AE} = \overline{BE} = 80$. Find the poletriangle and its three mirror images of the coupler AB from its three positions $A_1 B_1$, $A_2 B_2$, $A_3 B_3$. Assume it is proved that the four points P_{12}, P_{23}, P_{13}^2, E_2

lie on a circle. Show that the three points E_1, E_2, E_3 lie exactly on a straight line by means of the relation between the orthocentre and the three mirror images of the circumcircle of the poletriangle.

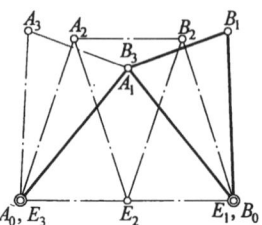

3.8 In the four-bar linkage shown, $\overline{A_0A} = \overline{B_0B}$ and the proportions of the link lengths are: $\overline{A_0B_0} : \overline{A_0A} : \overline{AB} = 4:5:2$. The three positions A_1B_1, A_2B_2, A_3B_3 of AB are as shown. Point E is the middle point of $\overline{AB}$. Explain that the three points E_1, E_2, E_3 are exactly on a straight line by means of the theorem of the poletriangle and its circumcircle.

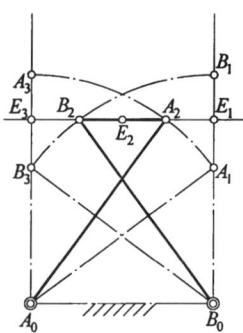

3.9 Find the four intersection points of the two circles

$$\left.\begin{array}{r}x^2 + y^2 = r^2 \\ (x-k)^2 + (y-h)^2 = R^2\end{array}\right\}$$

Hint: Please refer to Section A1.2 of Appendix 1.

3.10 In Fig. 3.16, if the point A_0 is taken as a cardinal point, and the three homologous points A_{01}, A_{02}, A_{03} are found correspondingly, the circle passing through the latter three points is called k_0. Show that
(a) the centre of k_0 is A_g;
(b) the radii of the two circles k_0 and k_A are equal, hence for a prescribed

Dimensional synthesis of four-bar linkages --- body guidance problems 175

value of R_A, both A_0 and A_g are points on the curve R_M.

3.11 Given the coordinates of two points E, F in the three positions c_1, c_2, c_3 of the body c:
$E_1(2, 1.5)$ $E_2(3, 1)$ $E_3(3.6464, 1.8536)$
$F_1(3, 1)$ $F_2(4, 0.5)$ $F_3(4.7071, 2.2071)$
Assume the coordinates of A_0 are $A_0(1, 0)$, and those of B_0 are $B_0(6, 0)$. Find A_1, B_1, and draw this four-bar linkage to guide c.

3.12 With respect to the three positions of the body c in Exercise 3.11, if A_0 is not prescribed, but $R_A = 2.201508$ is prescribed,
(a) find the equations of R_M- and R^1-curves;
(b) transform these two equations if the origin is shifted to $(6, 0)$.

3.13 Given the coordinates of two points E, F in the three positions c_1, c_2, c_3 of the body c:
$E_1(1, 1)$ $E_2(3, 1.5)$ $E_3(2, 2)$
$F_1(2, 1)$ $F_2(3.7071, 2.7071)$ $F_3(2, 3)$
Assume the coordinates of A_0 are $A_0(1.205, 2.616)$, and those of B_0 are $B_0(1.984, 0.533)$. Find A_1, B_1, and draw this four-bar linkage to guide c.

3.14 If in Exercise 3.13 the locations of A_0 and B_0 are not prescribed, but $R_A = 1.6424$ is prescribed,
(a) find the equations of R_M- and R^1-curves;
(b) shift the origin in these equations to P_{12}, and draw these two curves.
Hint: After shifting the origin, the equations in rectangular coordinates can be transformed into polar coordinate equations. The curve can be plotted by rotating the radial ray and solving quartic equations.

3.15 The displacement matrix of displacing a body from its 1st position to the 2nd position is:

$$\mathbb{D}_{12} = \begin{bmatrix} 0.9903 & 0.1392 & 0.1310 \\ -0.1392 & 0.9903 & 0.5539 \\ 0 & 0 & 1 \end{bmatrix}$$

Find the coordinates of the pole P_{12} of this displacement.

Hint: Applying $\mathbb{D}_{12}$ to the coordinates $\begin{bmatrix} x_{A1} \\ y_{A1} \\ 1 \end{bmatrix}$ of a certain point A_1 gives the

coordinates $\begin{bmatrix} x_{A2} \\ y_{A2} \\ 1 \end{bmatrix}$ of A_2 after the displacement. The definition of P_{12}

is that its position is unchanged after the displacement.

3.16 In the example shown in Fig. 3.40, $s = \overline{P_1 P_2^1} = 29.2331$, $d = \overline{M_1 M_2^1} = 91.4496$,

$\delta_1/2 = 66.1534$, $\delta_2/2 = 11.5897$. Substituting a certain value of ζ into equation (3.81c) gives a circle of the first pencil of circles in Fig. 3.40(a), and substituting the same value of ζ into equation (3.84a) gives a circle of the second pencil of circles. Inspect if the intersections of these two circles are on the q_1-curve in the figure.

3.17 Theorem 12 and Fig. 3.41 is to prove that, if A_0 is a point on the centre-point curve, then $\overline{P_{12}P_{23}}$ and $\overline{P_{14}P_{34}}$ subtend equal angles at A_0. Now suppose $\overline{P_{12}P_{23}}$ and $\overline{P_{14}P_{34}}$ subtend equal angles at a certain point B_0. Show that B_0 is a point on the centre-point curve.

3.18 The figure shows the method described in Section 3.6.2(a), or the method shown in Fig. 3.43 to find the centre-point curve. Show that P_{13} is a point on the centre-point curve.

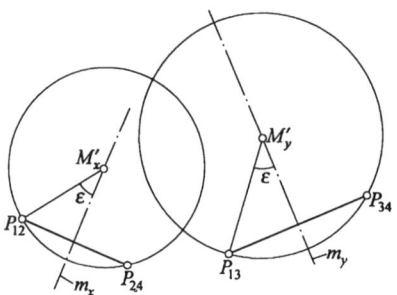

3.19 In Fig. 3.43(a), the points P_{24}^1 and P_{34}^1 (P_{24}^1 is the mirror image of P_{24} with respect to $P_{12}P_{14}$, and P_{34}^1 is the mirror image of P_{34} with respect to $P_{13}P_{14}$, hence P_{14} has to be found first.) are found first to get the two line segments $\overline{P_{12}P_{24}^1}$ and $\overline{P_{13}P_{34}^1}$. On the basis of these two line segments, draw the circle-point curve k_{1234} according to the method of Fig. 3.43 (a, b). This curve should pass through A_1, B_1 and E_1.

3.20 Suppose in Fig. 3.13(a), a fourth position A_4S_4 of the coupler AS is added as shown in Figure (a). Find B_0 and B_1, and hence the whole four-bar linkage A_0ABB_0.

Hint: As shown in Figure (b), the four poles P_{13}, P_{23}, P_{14}, P_{24} can be found from A_1S_1, A_2S_2, A_3S_3, A_4S_4. On the basis of the two opposite sides $\overline{P_{13}P_{23}}$, $\overline{P_{14}P_{24}}$ of this quadrilateral the centre-point curve m_{1234} can be drawn. Select a point B_0 on m_{1234} between P_{12} and P_{13}. The remaining procedure of construction is the same as that shown in Fig. 3.13.

Dimensional synthesis of four-bar linkages --- body guidance problems 177

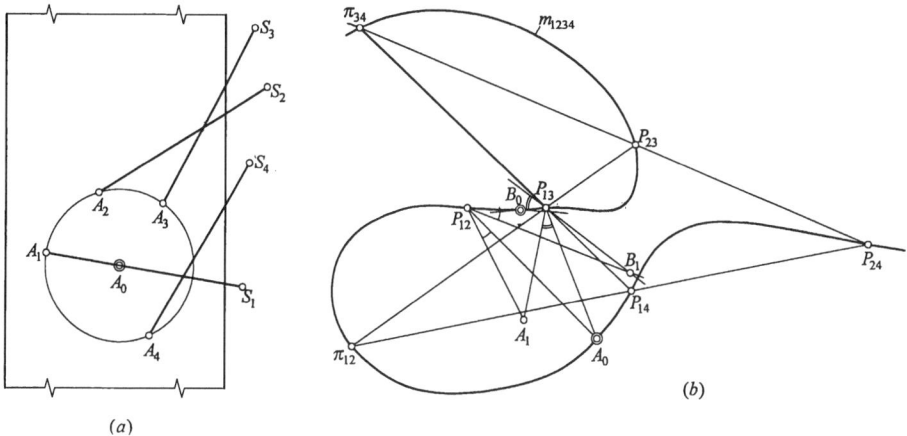

(a)   (b)

3.21 In the case of four finitely separated positions of a body, what is the number of such points (A_1, B_1, ...) that its four homologous points (e.g. A_1, A_2, A_3, A_4) lie on a circle of prescribed radius?
(Ans. : 6)

3.22 In the case of four finitely separated positions of a body, what is the number of such points (A_1, B_1, ...) that its four homologous points (e.g. A_1, A_2, A_3, A_4) lie on a straight line? Hence it can also be shown that the three circles k_{123}^1, k_{124}^1, k_{134}^1 intersect in one point (Theorem 18).

3.23 Show that the four circumcircles k_{123}, k_{124}, k_{134}, k_{234} in Fig. 3.51 intersect in one point S_0.

3.24 Prove Theorem 16.

3.25 Prove Theorem 17.

3.26 Prove Theorem 18.

3.27 In the construction of a centre-point curve as shown in the figure, the opposite sides $\overline{P_{13}P_{12}}$ and $\overline{P_{34}P_{24}}$ subtend an equal angle σ at a point M on the curve. Applying the notation for vectors and complex numbers, we use $\mathbf{M}$, $\mathbf{P}_{12}$, $\mathbf{P}_{13}$ .... to denote the position vectors of the respective points irrespective of the location of the origin to formulate

$$\frac{\mathbf{M}-\mathbf{P}_{12}}{\mathbf{M}-\mathbf{P}_{13}} \cdot \frac{\overline{\mathbf{M}}-\overline{\mathbf{P}}_{13}}{\overline{\mathbf{M}}-\overline{\mathbf{P}}_{12}} = e^{2i\sigma} \qquad \frac{\mathbf{M}-\mathbf{P}_{24}}{\mathbf{M}-\mathbf{P}_{34}} \cdot \frac{\overline{\mathbf{M}}-\overline{\mathbf{P}}_{34}}{\overline{\mathbf{M}}-\overline{\mathbf{P}}_{24}} = e^{2i\sigma}$$

This equation can be rewritten as $(\mathbf{M} - \mathbf{P}_{12})(\mathbf{M} - \mathbf{P}_{34}) \times (\mathbf{M} - \mathbf{P}_{24})(\mathbf{M} - \mathbf{P}_{13}) = 0$. Let $\mathbf{P}_{12} + \mathbf{P}_{34} = a$, $\mathbf{P}_{13} + \mathbf{P}_{24} = b$; $\overrightarrow{P_{12}P_{34}} = \mathbf{A}$, $\overrightarrow{P_{13}P_{24}} = \mathbf{B}$, and $\mathbf{A} - \mathbf{B} = c + id$, $a\mathbf{B} - \mathbf{A}b = e + if$ and $\mathbf{A} \times \mathbf{B} = -g$. Show that the locus of M is

$$(b-a)(x^2+y^2)y + 2cxy - d(x^2-y^2) + ey - fx - g = 0$$

Further show that this equation also includes such points M inside the quadrilateral as that shown in Fig. 3.44.

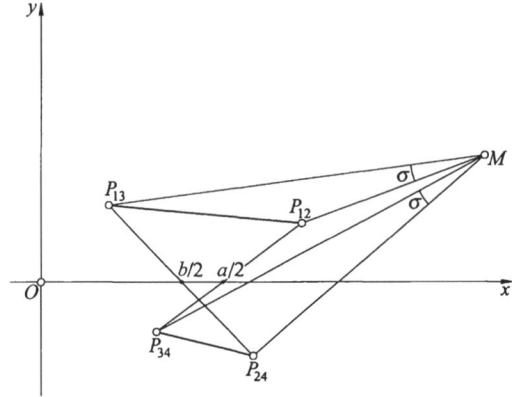

3.28 The four positions of a body are expressed in terms of the coordinates of a point E on the body and the angles of rotation γ_{1j} from its 1st position to the jth position as follows:

$$E_1 \quad (0, -0.5)$$
$$E_2 \quad (1, -1) \quad, \quad \gamma_{12} = 0°$$
$$E_3 \quad (1.6464, -0.1464) \quad, \quad \gamma_{13} = 45°$$
$$E_4 \quad (0.5, 0) \quad, \quad \gamma_{14} = 90°$$

(a) Find the equation of the centre-point curve (in the form of a determinant without expansion).
(b) Transfer the origin to P_{12}, and transform the above equation into a polar coordinate equation, and draw this centre-point curve.
(c) Choose two points A_0 and B_0 on this centre-point curve, and find the corresponding points A_1 and B_1. Construct a four-bar linkage to guide the body through the above four positions.

3.29 The four positions of a body are expressed in terms of the coordinates of a point F on the body and the angles of rotation γ_{1j} from its 1st position to the jth position as follows:

$$F_1 \quad (2, 1)$$
$$F_2 \quad (3, 0.5) \quad, \quad \gamma_{12} = 0°$$
$$F_3 \quad (3.7071, 2.2071) \quad, \quad \gamma_{13} = 45°$$
$$F_4 \quad (2, 3) \quad, \quad \gamma_{14} = 90°$$

(a) Find the equation of the centre-point curve (in the form of a determinant without expansion).

Dimensional synthesis of four-bar linkages --- body guidance problems 179

(b) Transfer the origin to P_{13}, and transform the above equation into a polar coordinate equation, and draw this centre-point curve.

(c) Choose two points A_0 and B_0 on this centre-point curve, and find the corresponding points A_1 and B_1. Construct a four-bar linkage to guide the body through the above four positions.

3.30 A given four-bar linkage A_0ABB_0 is as shown in the figure. The polar coordinates of the four joints are: A_0 (7, 30°), A (3, 30°), B (3.267, 60°), B_0 (6, 60°). Apply equations (3.116) and (3.121) to find the circling-point curve k_u and centering-point curve k_a of the coupler AB. Choose a point on k_u, and find the path of this point and the circle of curvature of this path.

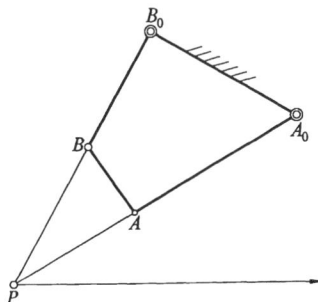

Hint: Find first the poletangent, and use the poletangent as a new coordinate axis. Substituting the above coordinates of the four point A_0, A, B, B_0, we can find the values of l, l_* and m.

3.31 (a) Equation (3.116) is the polar coordinate equation of a circling-point curve

$$k_u : \frac{1}{p} = \frac{1}{m\cos\theta} + \frac{1}{l\sin\theta}$$

where m, l are constants, and (p, θ) are variables.

The equation of a circle whose centre lies on the line PN and tangent to PT as shown in the figure is

$$p = d\sin\theta$$

where (p, θ) are still variables and d is the diameter of the circle. Find the intersection point of the circle with k_u, and consider d as a parameter. Show that the radius of curvature of k_u at P is $l/2$. Similarly show that another radius of curvature of k_u is $m/2$.

(b) Equation (3.117) can be written as

$$(x_{A1}^2 + y_{A1}^2 - mx_{A1})(mx_{A1} + ly_{A1}) + m^2 x_{A1}^2 = 0$$

or $$(x_{A1}^2 + y_{A1}^2 - ly_{A1})(mx_{A1} + ly_{A1}) + l^2 y_{A1}^2 = 0$$

From these two equations, show that the two radii of curvature of k_u at P are respectively $m/2$ and $l/2$.

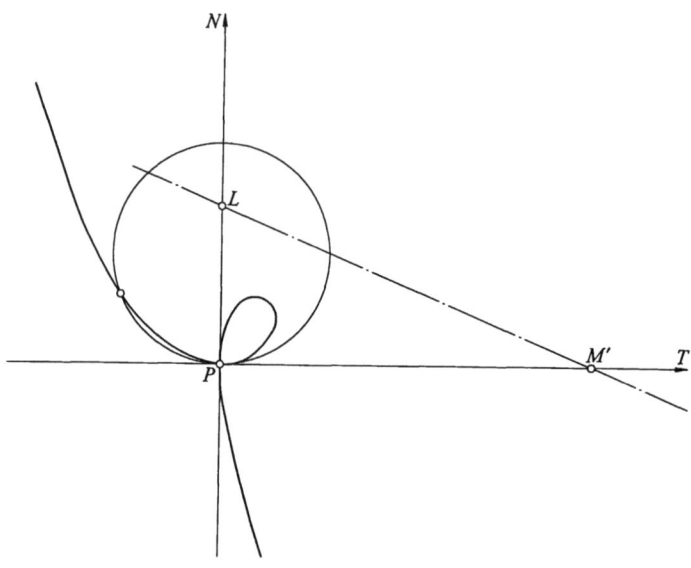

3.32 Show that the slope of the asymptote of the circling-point curve k_u (equation (3.117)) is $-m/l$, and that this asymptote passes through the point G in Fig. 3.57.

$$G\left[-\frac{ml^2}{2(m^2+l^2)}, -\frac{m^2l}{2(m^2+l^2)}\right]$$

3.33 Show that the point Γ in Fig. 3.57 whose coordinates are

$$x_\Gamma = \frac{ml^2}{2(m^2+l^2)}, \quad y_\Gamma = \frac{m^2l}{2(m^2+l^2)}$$

lies on the two imaginary asymptotes of the circling-point curve

$$(x^2+y^2-mx)(mx+ly)+m^2x^2 = 0$$

In other words, the point Γ is the singular focus of k_u.

Hint: This curve has three asymptotes, among which two are imaginary, and one is real. The slopes of the three asymptotes are respectively i, $-i$, $-m/l$.

3.34 In Fig. 3.59, the line g is determined by the two constants m, l. Choose any point $A'''(\mu, \lambda)$ on g. Find the equation of the locus of A.

3.35 Find, in Fig. 3.63, the circling-point curve k_u of the coupler AB by the same

Dimensional synthesis of four-bar linkages --- body guidance problems 181

method.

3.36 Show that the two tangents of the curve k_a in Fig. 3.61 are orthogonal, and that the two lines $P_{12}P$ and PP_{34} are symmetrically disposed with respect to these tangents.

Hint: Prove by means of the characteristics of constructing the intersections of the two tangents of the circle k_m.

3.37 The figure shows a method of constructing the centering-point curve similar to that shown in Fig. 3.61. As the circle k_m arround P becomes a maximum, the two tangents drawn to this circle from P_{12}, P_{34} become parallel to each other, or they intersect at infinity. Let $g = \overline{P_{12}P_{34}}$, $h = \overline{PP_{34}}$, $\beta = \sphericalangle P_{12}P_{34}P$. show that

$$\theta = \tan^{-1} \frac{g \sin \beta}{2h - g \cos \beta}$$
$$r = h \sin \theta$$

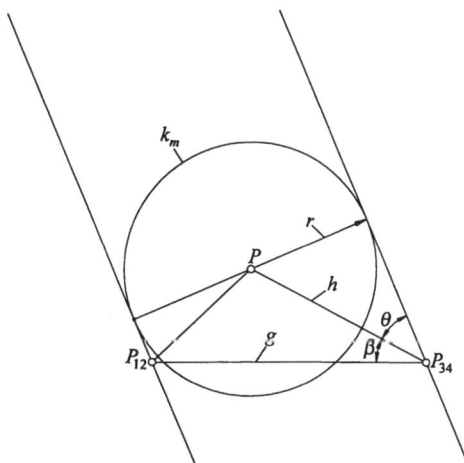

3.38 Show that, in Fig. 3.62, $\overline{PE_0}$ and $\overline{PF_0}$ subtend equal angles at any point B_0 on the curve k_a, or $\eta = \varepsilon$.

3.39 Comparing Figs. 3.68 and 3.69, show that the point L_* of the coupler AB in Fig. 3.68 also lies at infinity.

3.40 Explain, by means of equations (3.123) and (3.126), that Theorem 21 is valid for the positions of a four-bar linkage shown in Figs. 3.70 (a)(b) and 3.68.

3.41 A wheel of radius r rolls without slip on a horizontal straight track. Find the circling-point curve and centering-point curve of the wheel. Show how to locate the point C in Fig. 1.4(b).

3.42 Choose a point E_0 on the circle k'_a in Fig. 3.69 as a centre-point. Find the corresponding circle-point E.

3.43 Show that, in the inner frame position of a four-bar linkage as that shown in Fig. 3.70(c), the circling-point curve of the coupler breaks up into a circle k'_u and a straight line k''_u.

3.44 Prove the coordinates of S_0 given in equation (3.138).

3.45 In the four-bar linkage A_0ABB_0 shown in the figure, $\overline{AB} = \overline{B_0B} = \overline{CB}$. In the position shown, A_0, A, B_0 are collinear, $\triangle CAB = \triangle AB_0B$, and B_0C and AB intersect in E. Let $\overline{BF} = \overline{BE}$, and the length of the crank a be $\overline{A_0A} = \overline{EF}$. Show that the point C is the Ball point U. Hence a dwell mechanism can be so designed to allow the link h to exhibit a long period of standstill while the crank A_0A rotates.

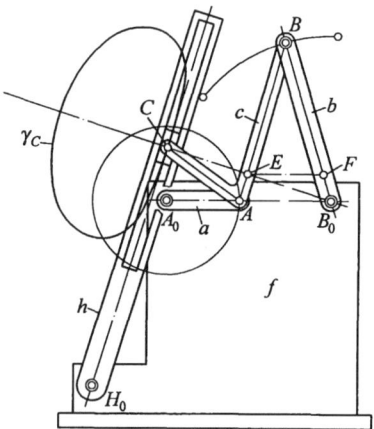

3.46 In the example shown in Fig. 3.53, a fifth position of the body is added: P_{15} (2.5, 1.5), $\gamma_{15} = -115°$. Find the Burmester circle-point and the Burmester centre-point.

$$\text{Ans: } \begin{pmatrix} B_{r1}(2.5298, 2.6704), B_{r01}(-0.3855, -0.2367); \\ B_{r2}(1.1359, 0.3757), B_{r02}(6.0425, 0.3486). \end{pmatrix}$$

3.47 Suppose in Fig. 3.75, $a = 10$, $f = 15$. Find the fourth Burmester point.

(Ans: (37.5, 0).)

3.48 Following the description of Section 3.9.3, find the other two Burmester points in Exercise 3.30.

(Ans: (1.0247, 2.4402), (2.6014, 1.9664).)

3.49 In the case of three positions P_1P_2–P_3 of a body AB, if the directions of the two velocities v_A, v_B of two points A, B on the body in $P_1(=P_2)$ are known, find the direction of the side $P_{13}P_{23}$ of the poletriangle $\Delta P_{12}P_{23}P_{13}$.

Hint: Knowing the directions of v_A, v_B, we can find P_{12}. The pole P_{13} can be found as usual. The poletriangle becomes in this case a straight line, as that shown in Fig. 3.81(a), but the angle $\gamma_{23}/2$ between $P_{23}P_{13}$ and $P_{12}P_{23}$, as that shown in Fig. 3.12, remains determinate.

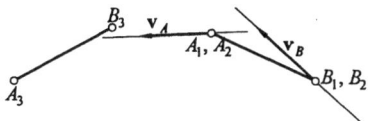

3.50 Let the four positions of a body be separated like P_1P_2–P_3P_4 (please refer to Fig. 3.82). Given: P_{12} (0, 0), $P_{13} = P_{23}$ (1, 0.18), $\gamma_{13} = 120°$, P_{34}(1.8, 0.50). Find the centre-point curve m_{1234} and the circle-point curve k_{1234}.

4
Two other methods of guiding an instantaneous motion of a body

In Chapter 3 we have depicted the methods of synthesis for body guidance. Specifically, to guide an instantaneous motion of a body, i.e., to guide a body through several infinitesimally separated positions, we can tackle the problems from two other approaches, namely, the *principle of instantaneous invariants* and *method of matching polode curvatures*. Roughly speaking, an instantaneous motion of a body, can be considered as P_1P_2, $P_1P_2P_3$, $P_1P_2P_3P_4$, or even $P_1P_2P_3P_4P_5$. It is evident that simulating a motion as a $P_1P_2P_3P_4P_5$ is much more accurate than as a P_1P_2.

4.1 The principle of instantaneous invariants

This concept originates in (Bottema, 1961) and (Veldkamp, 1963). Basically, it takes the rotation part out of the displacement matrix $\mathbb{D}_{12}$ of equation (3.33), and formulates the motion of a body c relative to a fixed body f according to Fig. 4.1 in the form of the following equation (4.1). Fig. 4.1(a) shows a coordinate system Σ_c bound to a moving body c. The coordinates of a point E on the body c with respect to

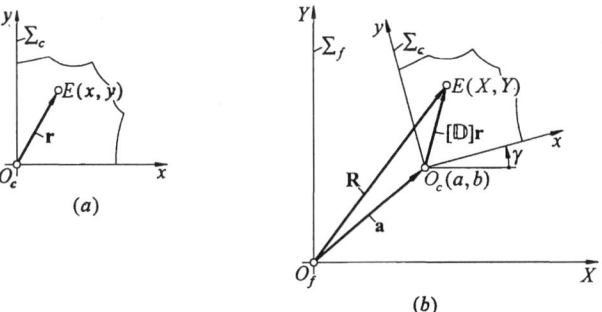

Fig. 4.1. Relative motion between two bodies c, f. (a) Moving body coordinate system Σ_c. (b) Relation between Σ_c and fixed body coordinate system Σ_f.

the origin O_c are (x, y), its position vector being represented by **r**. Fig. 4.1(b) shows the position of this moving coordinate system Σ_c relative to a fixed coordinate system Σ_f. The origin of the fixed coordinate system is denoted by O_f, and its axes are X, Y. At a certain moment the coordinates of O_c relative to O_f are $(a, b)^\S$, its position vector being **a**. The inclination angle of Σ_c relative to Σ_f is γ. The coordinates of E relative to Σ_f are (X, Y), its position vector being represented by **R**. We have then

$$\begin{bmatrix} X \\ Y \end{bmatrix} = \begin{bmatrix} \cos\gamma & -\sin\gamma \\ \sin\gamma & \cos\gamma \end{bmatrix} \begin{bmatrix} x \\ y \end{bmatrix} + \begin{bmatrix} a \\ b \end{bmatrix} \tag{4.1}$$

In equation (4.1), x, y are constants. Only the three quantities γ, a, b are variables, representing three degrees of freedom of the body performing plane motion. If the motion is a constrained one, there must exist a certain constraint relation among these three variables. We may therefore take γ as the independent variable, and a, b as functions of γ, i.e. $a = a(\gamma)$, $b = b(\gamma)$. Rewrite equation (4.1) in vector form,

$$[\mathbf{R}] = \mathbb{D} \; [\mathbf{r}] + [\mathbf{a}] \tag{4.2}$$

where

$$\mathbb{D} = \begin{bmatrix} \cos\gamma & -\sin\gamma \\ \sin\gamma & \cos\gamma \end{bmatrix}$$

$$[\mathbf{r}] = \begin{bmatrix} x \\ y \end{bmatrix}$$

$$[\mathbf{a}] = \begin{bmatrix} a(\gamma) \\ b(\gamma) \end{bmatrix}$$

$\mathbb{D}$ is the rotation part taken out from the displacement matrix of equation (3.33). Differentiating equation (4.2) n times with respect to γ gives

$$\left(\frac{d^n[\mathbf{R}]}{d\gamma^n}\right)_{\gamma=\gamma_0} = \left(\frac{d^n[\mathbb{D}]}{d\gamma^n}\right)_{\gamma=\gamma_0} [\mathbf{r}] + \left(\frac{d^n[\mathbf{a}]}{d\gamma^n}\right)_{\gamma=\gamma_0} \tag{4.3}$$

The last term in equation (4.3) is just the coefficient of the term for n taking successively 1, 2, ..., in the Taylor expansion of the function $[\mathbf{a}]$ in the vicinity of $\gamma = \gamma_0$,

$$[\mathbf{a}] = [\mathbf{a}]_{\gamma=\gamma_0} + \left(\frac{d[\mathbf{a}]}{d\gamma}\right)_{\gamma=\gamma_0} (\gamma - \gamma_0) + \left(\frac{d^2[\mathbf{a}]}{d\gamma^2}\right)_{\gamma=\gamma_0} \frac{(\gamma - \gamma_0)^2}{2!} + \cdots \tag{4.4}$$

where γ_0 is the value of γ in a specific position. If the vector $[\mathbf{a}]$ in equation (4.4) is resolved into two Taylor expansions of (a, b) individually, and for brevity, $(d^n a/d\gamma^n)_{\gamma=\gamma_0}$ is written as a_n, and $(d^n b/d\gamma^n)_{\gamma=\gamma_0}$ as b_n, then

---

[§] Please do not confuse these symbols (a,b) with the cranks a, b of a four-bar linkage.

$$a = a_0 + a_1(\gamma - \gamma_0) + a_2 \frac{(\gamma - \gamma_0)^2}{2!} + a_3 \frac{(\gamma - \gamma_0)^3}{3!} + \cdots$$
$$b = b_0 + b_1(\gamma - \gamma_0) + b_2 \frac{(\gamma - \gamma_0)^2}{2!} + b_3 \frac{(\gamma - \gamma_0)^3}{3!} + \cdots \quad (4.5)$$

The initial position of Σ_c relative to Σ_f, can yet be chosen arbitrarily. For simplicity, let O_c coincide in the initial position with O_f, and set this point at the velocity pole P. Furthermore, let the x-axis coincide with X-axis in the initial position, which means that $\gamma_0 = 0$. Rewrite equation (4.3) in two components separately, and use the symbols ('), ("), ('''), ... to indicate the respective differential coefficients of a certain function with respect to γ. We have

$$\begin{aligned}
X &= x\cos\gamma - y\sin\gamma + a, & Y &= x\sin\gamma + y\cos\gamma + b \\
X' &= -x\sin\gamma - y\cos\gamma + a', & Y' &= x\cos\gamma - y\sin\gamma + b' \\
X'' &= -x\cos\gamma + y\sin\gamma + a'', & Y'' &= -x\sin\gamma - y\cos\gamma + b'' \\
X''' &= x\sin\gamma + y\cos\gamma + a''', & Y''' &= -x\cos\gamma + y\sin\gamma + b''' \\
X^{IV} &= x\cos\gamma - y\sin\gamma + a^{IV}, & Y^{IV} &= x\sin\gamma + y\cos\gamma + b^{IV}
\end{aligned} \quad (4.6)$$

Please note the difference between a' and a_1. a' is a variable, while a_1 is a constant; $a_1 = (a')_{\gamma=\gamma_0}$. So are the differences between other symbols, for instance, those between a and a_0, a'' and a_2, b' and b_1. Among such constants as $a_0, a_1, \ldots$, $b_0, b_1, \ldots$, there are several of them which remain the same for any kind of instantaneous motion. These will be depicted separately as follows:

(1) Since in the initial position, O_c and O_f coincide, for the point $E = O_c$, $x = 0$, $y = 0$, and at $\gamma = \gamma_0 = 0$, $X = 0$, $Y = 0$, we have $a_0 = 0$, $b_0 = 0$.

(2) Because the velocity pole P is taken as the origin O_c in the initial position, and the velocity of P as a moving point is zero, or $X' = 0$, $Y' = 0$, we have $a_1 = 0$, $b_1 = 0$.

(3) Because in the initial position, x-axis coincides with X-axis, and the velocity pole P as a moving point has no acceleration component in the X direction, or $X'' = 0$, we have $a_2 = 0$.

Consequently, in the initial position or at the instant $\gamma = \gamma_0 = 0$, equations (4.6) take the following form

$$\begin{aligned}
X &= x, & Y &= y \\
X' &= -y, & Y' &= x \\
X'' &= -x, & Y'' &= -y + b_2 \\
X''' &= y + a_3, & Y''' &= -x + b_3 \\
X^{IV} &= x + a_4, & Y^{IV} &= y + b_4 \\
&\cdots\cdots & & \\
&\cdots\cdots & &
\end{aligned} \quad (4.7)$$

The quantities $(b_2; a_3,b_3; ...; a_n,b_n)$ are called *instantaneous invariants*. This terminology may sometimes cause confusion. Because for a certain motion, these values may be different, depending on the different coordinate systems assumed. Here, we offer the following explanation: Suppose for a given planar motion, the $2n - 3$ values of $b_2; a_3, b_3; ..., a_n, b_n$ are all known. If it is required to synthesize another motion to simulate this given motion up to the nth order, then this second motion should have, besides the same pole P and the same poletangent, also the same $2n - 3$ instantaneous invariants..

Please note that all instantaneous invariants are the values at the initial position, or at $\gamma = \gamma_0$.

It is clear from the two equations of X', Y' in (4.6) that the locus of points on Σ_c satisfying the conditions $X' = 0$, $Y' = 0$ is

$$\pi_c : \begin{matrix} x = a'\sin\gamma - b'\cos\gamma \\ y = a'\cos\gamma + b'\sin\gamma \end{matrix} \Big\} \qquad (4.8)$$

The variable γ in equations (4.8) is a parameter. These are the parametric equations of the moving polode π_c. Substituting equations (4.8) into the expressions of X, Y in equations (4.6), we get

$$\pi_f : \begin{matrix} X = a - b' \\ Y = a' + b \end{matrix} \Big\} \qquad (4.9)$$

Equations (4.9) are the parametric equations of the fixed polode π_f with parameter γ.

At the instant $\gamma = \gamma_0 = 0$, the locus of a point whose path generated on the fixed plane Σ_f possessing a zero curvature, should satisfy the condition $d^2Y/dX^2 = 0$, i.e. [please compare with the following equation (4.11)]

$$X'Y'' - X''Y' = 0 \qquad (4.10)$$

Substituting the expressions of X', Y'; X'', Y'' in equations (4.7) into equations (4.10), we get

$$x^2 + y^2 - b_2 y = 0 \qquad (4.11)$$

A comparison between equations (4.11) and (3.59) shows this is the equation of the inflection circle on Σ_c, with b_2 as the diameter of the inflection circle, which is the δ in equation (3.59). On the other hand, if we differentiate equations (4.9) with respect to γ, we obtain

$$\begin{matrix} X' = a' - b'' \\ Y = a'' + b' \end{matrix} \Big\}$$

In the initial position, $(X, Y) = (0, 0)$, and as already known, $a' = a_1 = 0$, $b' = b_1 = 0$, $a'' = a_2 = 0$, we have $(X')_{\gamma=\gamma_0} = -b_2$. This means that the magnitude of the *pole changing velocity* (although the differentiation is with respect to γ, not the time t) is

Two other methods of guiding an instantaneous motion of a body

$-b_2$. If the positive direction of the X-axis is taken in the direction of **u**, and the scale is so taken that $b_2 = -1$ (or $+1$), then the coordinate system is completely determined, and is the so-called *canonical system*. The inflection circle k_W in equation (4.11) should look like the one shown in Fig. 4.2. This circle lies on the negative side of the

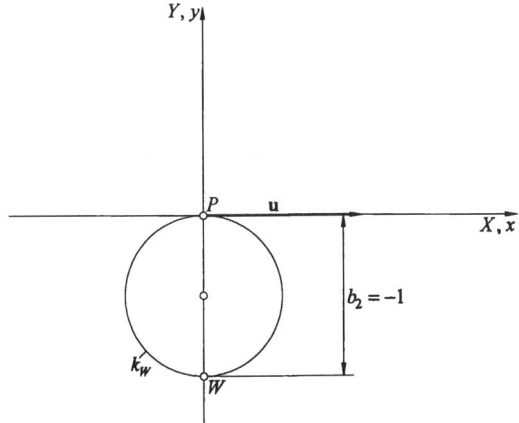

Fig. 4.2. The canonical system.

Y-axis. Please note that, this canonical system is not quite the same as the system used in Chapter 1, where the direction of pole changing velocity **u** is taken as the positive sense of x-axis, but the inflection circle is lying on the positive sense of y (in the present case, the sense of Y). It should be noted that, for a certain plane motion, the inflection circle **is** already determined, being always the same regardless of the direction of motion. However, the direction of all velocities (including the direction of **u**) can always be reversed. Since in this canonical system the inflection circle should lie on the negative side of the Y-axis, in case the sense of **u** is in a sense of $-X$, we can always reverse the directions of all velocities to convert the direction of **u** to a $+X$ sense. Furthermore, in the case where the scale is determined by other conditions, it is then no longer permissible to assume $b_2 = -1$, rather, the magnitude of b_2 has to be determined according to the scale assumed.

All concepts in planar instantaneous kinematics introduced previously can be interpreted in terms of instantaneous invariants. Consider first the curvature of the fixed polode π_f at the pole P. As is well-known, the curvature at a certain point of a curve expressed parametrically is

$$\kappa = \frac{X'Y'' - X''Y'}{(X'^2 + Y'^2)^{3/2}} \quad (4.12)$$

Now from equations (4.9) we have

$$X' = a' - b'' \ , \quad X'' = a'' - b'''$$
$$Y' = a'' + b' \ , \quad Y'' = a''' + b''$$

In the initial position, $(X, Y) = (0, 0)$, it is then

$$X' = a_1 - b_2 = 1 \ , \quad X'' = a_2 - b_3 = -b_3$$
$$Y' = a_2 + b_1 = 0 \ , \quad Y'' = a_3 + b_2 = a_3 - 1$$

Hence the curvature of π_f at the origin P is

$$\kappa_f = a_3 - 1 \tag{4.13}$$

Similarly, from the parametric equations (4.8) of the moving polode π_c, we can find at the origin $(x, y) = (0, 0)$

$$x' = 1, \quad y' = 0$$
$$x'' = -b_3, \quad y'' = -2 + a_3$$

Therefore the curvature of π_c at the origin P is

$$\kappa_c = a_3 - 2 \tag{4.14}$$

From equations (4.13) and (4.14) we have

$$\kappa_c - \kappa_f = -1$$

This is just equation (1.15), in which $\delta = |b_2|$.

Differentiating equation (4.12) with respect to γ, and equating it to zero yields

$$(X'^2 + Y'^2)(X'Y''' - X'''Y') - 3(X'Y'' - X''Y')(X'X'' + Y'Y'') = 0 \tag{4.15}$$

Substituting equations (4.7) into equation (4.15) gives

$$(x^2 + y^2)[(a_3 + 3b_2)x + b_3 y] - 3b_2^2 xy = 0 \tag{4.16}$$

This is the equation of the circling-point curve k_u, which corresponds to equation (3.117). Substituting equations (1.26a,b) into equation (4.16), and replacing ξ, η, δ respectively by X, Y, b_2, we get

$$(X^2 + Y^2)(a_3 X + b_3 Y) - 3b_2^2 XY = 0 \tag{4.17}$$

This is the equation of the centering-point curve k_a, corresponding to equation (3.122).

The intersection between the inflection circle (4.11) and the circling-point curve (4.16) is the Ball point whose coordinates are

$$\left(\frac{a_3 b_3}{a_3^2 + b_3^2}, -\frac{a_3^2}{a_3^2 + b_3^2} \right) \tag{4.18}$$

It is sometimes desirable to have the time t as the independent variable. In such case it is then necessary to know the values of the instantaneous angular velocity $\omega = d\gamma/dt$, and $\dot{\omega} = \alpha$, $\ddot{\omega} = \dot{\alpha}$, ... Suppose Φ is a function of γ, then

Two other methods of guiding an instantaneous motion of a body   191

$$\Phi' = d\Phi/d\gamma = (d\Phi/dt)/(dt/d\gamma) = \dot{\Phi}\omega^{-1}$$

$$\Phi'' = \ddot{\Phi}\omega^{-2} - \dot{\Phi}\alpha\omega^{-3}$$

$$\Phi''' = \dddot{\Phi}\omega^{-3} - 3\ddot{\Phi}\alpha\omega^{-4} + 3\dot{\Phi}\alpha^2\omega^{-5} - \dot{\Phi}\dot{\alpha}\omega^{-4}$$

$$\Phi'''' = \ddddot{\Phi}\omega^{-4} - 6\dddot{\Phi}\alpha\omega^{-5} - 4\ddot{\Phi}\dot{\alpha}\omega^{-5} + 15\ddot{\Phi}\alpha^2\omega^{-6} + 10\dot{\Phi}\alpha\dot{\alpha}\omega^{-6}$$
$$- 15\dot{\Phi}\alpha^3\omega^{-7} - \dot{\Phi}\ddot{\alpha}\omega^{-5}$$

Let $\Phi = X$, then according to equation (4.7), in the initial position, or at $\gamma = 0$, we have

$$\left.\begin{aligned}
X &= x \\
\dot{X} &= -y\omega \\
\ddot{X} &= -\omega^2 x - \alpha y \\
\dddot{X} &= -3\omega\alpha x + (\omega^3 - \dot{\alpha})y + \omega^3 a_3 \\
\ddddot{X} &= (\omega^4 - 4\omega\dot{\alpha} - 3\alpha^2)x + (6\omega^2\alpha - \ddot{\alpha})y + 6\omega^2\alpha a_3 + \omega^4 a_4
\end{aligned}\right\} \quad (4.19)$$

Next let $\Phi = Y$. Again according to equation (4.7), we have in the initial position

$$\left.\begin{aligned}
Y &= y \\
\dot{Y} &= \omega x \\
\ddot{Y} &= \alpha x - \omega^2 y + \omega^2 b_2 \\
\dddot{Y} &= (-\omega^3 + \dot{\alpha})x - 3\omega\alpha y + 3\omega\alpha b_2 + \omega^3 b_3 \\
\ddddot{Y} &= (-6\omega^2\alpha + \ddot{\alpha})x + (\omega^4 - 4\omega\dot{\alpha} - 3\alpha^2)y + (4\omega\dot{\alpha} + 3\alpha^2)b_2 \\
&\quad + 6\omega^2\alpha b_3 + \omega^4 b_4
\end{aligned}\right\} \quad (4.20)$$

Please note that equations (4.19), (4.20) are valid only for X, Y, x, y being coordinates in the *canonical system*.

In order to illustrate the application of instantaneous invariants in the synthesis of mechanisms, consider the following numerical example taken from (Roth & Yang, 1977).

Example: It is required to synthesize a four-bar linkage A_0ABB_0, with the following requirements:

(1) The coupler point E is a Ball point. In other words, E moves approximately on a horizontal line, with a linear velocity $v_E = 3$ cm/s, a linear acceleration $a_E = -3$ cm/s^2, and second acceleration $\dot{a}_E = 5$ cm/s^3.

(2) The angular velocity of the coupler is $\omega = 5$ rad/s (counterclockwise considered as positive). Its angular acceleration is $\alpha = 10$ rad/s^2, second angular acceleration is $\dot{\alpha} = -5$ rad/s^3.

(3) The crank A_0A rotates with a constant angular velocity.

Solution: Please refer to Fig. 4.3. We find first from v_E and ω that $\overline{EP} = 3/5 = 0.6$ cm, and locate the point P at 0.6 cm above E. With the point P as origin, we can

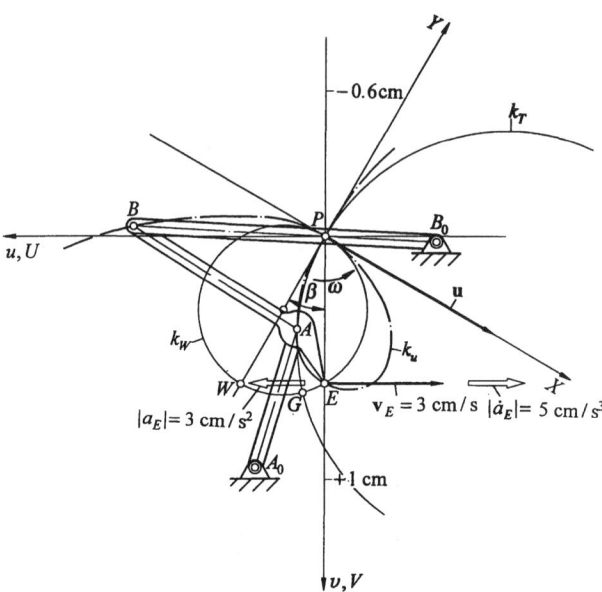

Fig. 4.3. Example: application of instantaneous invariants in dimensional synthesis.

construct the X, Y and x, y coordinate systems as shown in Fig. 4.1. In order to avoid confusion, we denote them here respectively as the coordinate systems U, V and u, v. In the initial position, the origins of these two coordinate systems coincide at P, and the U-axis and u-axis also coincide, as shown in the figure. However, as the present coordinate system is not a canonical one, we should not use equations (4.19) and (4.20); rather, we have to use the time derivatives corresponding to the third equations in (4.6), or

$$\left. \begin{array}{l} \ddot{U} = (-u\cos\gamma + v\sin\gamma)\omega^2 + (-u\sin\gamma - v\cos\gamma)\alpha + (a_P)_u \\ \ddot{V} = (-u\sin\gamma - v\cos\gamma)\omega^2 + (u\cos\gamma - v\sin\gamma)\alpha + (a_P)_v \end{array} \right\}$$

where the terms $(a_P)_u$ and $(a_P)_v$ correspond respectively to a'' and b'' in the third equations of (4.6), or the components of the linear acceleration of the moving point P along the U-axis and V-axis. Moreover, u, v in the above equations are the coordinates of the point E in the (u, v) coordinate system. Substituting the known values of the point E into the above equations gives

$$\left. \begin{array}{l} 3 = (-0.1 + 0.6 \cdot 0) \cdot 5^2 + (-0 \cdot 0 - 0.6 \cdot 1) \cdot 10 + (a_P)_u \\ 0 = (-0 \cdot 0 - 0.6 \cdot 1) \cdot 5^2 + (0.1 - 0.6 \cdot 0) \cdot 10 + (a_P)_v \end{array} \right\}$$

Hence $(a_P)_u = 9$ cm/s^2, $(a_P)_v = 15$ cm/s^2. From these we know the acceleration of P is $a_P = (9^2 + 15^2)^{1/2} = 17.4929$ cm/s^2. As $\mathbf{a}_P$ must be along the direction of the

Two other methods of guiding an instantaneous motion of a body

polenormal, the angle between this polenormal and the V-axis is $\beta = \tan^{-1}(9/15) = 30.96°$. Now as the orientation of the canonical system is determined, and ω is counterclockwise, we may take the sense of the polechanging velocity $\mathbf{u}$ as that of $+X$. The XY system is thus determined as shown in the figure.

With the canonical system thus determined, we can use the third equation of (4.20)

$$\ddot{Y} = -17.4929 = 10 \cdot 0 - 5^2 \cdot 0 + 5^2 b_2$$

and obtain $\qquad b_2 = -0.6997$

This is the inflection circle diameter $|\delta| = 0.6997$ cm. Hence the inflection circle k_W can be determined. The point E should lie on this inflection circle. The coordinates of E with respect to the xy-coordinate system are $E\,(0.3088, -0.5154)$.

Since it is known that $\ddot{U}_E = -5$ cm/s^3, $\ddot{V} = 0$, we have

$$\left.\begin{array}{l}\ddot{X}_E = 5\cos 30.96° = 4.2875 \text{ cm/s}^3 \\ \ddot{Y}_E = 5\sin 30.96° = 2.5725 \text{ cm/s}^3\end{array}\right\}$$

Substituting these known values into the fourth equations of (4.19) and (4.20), we get

$$\left.\begin{array}{l}4.2875 = -3\cdot 5\cdot 10\cdot 0.3087 + (5^3 + 5)\cdot(-0.5145) + 5^3 a_3 \\ 2.5725 = (-5^3 - 5)\cdot 0.3087 - 3\cdot 5\cdot 10\cdot(-0.5145) + 3\cdot 5\cdot 10\cdot(-0.6997) + 5^3 b_3\end{array}\right\}$$

hence
$$a_3 = 0.9398, \qquad b_3 = 0.5639$$

Substituting the values of b_2, a_3, b_3 into equation (4.16) gives the equation of the circling-point curve k_u.

$$k_u : (x^2 + y^2)(-1.1593x + 0.5639y) - 1.4687xy = 0$$

This curve is drawn as shown in the figure. Both points A, B should lie on the curve k_u.

Since the point A is moving with a constant velocity, it should also lie on the tangential circle k_T as explained in Section 1.4. As shown in equation (1.8), the diameter of k_T is $a_P/\alpha = 17.4929/10 = 1.7493$ cm. The centre of the circle k_T should lie on the poletangent, or the X-axis. However, whether it is on the side of $+X$ or of $-X$ depends on the location of the intersection point G of k_T and k_W (please refer to Fig. 1.8). In other words, the location of G should be such that the α about G should produce an acceleration a_P of P as that shown in Fig. 1.10. As the α is now of a counterslockwise sense, the k_T should be the one shown in the figure, or lying on the $+X$-axis. The intersection between k_T and k_u is A, whose coordinates can be found to be $(0.0936, -0.3937)$. Choose another point $B\,(-0.7084, -0.3767)$ on k_u. Finally using Euler-Savary equation (1.14) we can find the centre-points $A_0\,(0.2308, -0.9708)$, $B_0\,(0.4912, 0.2612)$.

After the synthesis, we can see that the point E is the intersection point of the circling-point curve k_u and inflection circle k_W, and hence is the Ball point.

4.2 The polode method

We have mentioned in Section 1.1 that the relative motion between two bodies can be replaced by the pure rolling motion of two polodes. In Section 3.2.2 we derived the equation of the fixed polode π_f as well as that of the moving polode π_c. It follows that, theoretically, given a fixed polode and a moving polode, if we could synthesize a four-bar linkage such that its fixed polode and moving polode not only possess respectively the same radii of curvature, but also the same respective rates of change of polode curvatures up to a certain order at the point P as those of the given polodes, we can replace the original relative motion with the synthesized four-bar linkage. The higher the order of simulation, the closer is the degree of approximation.

4.2.1 The Grübler-Hall equation

We have to derive in the first place the equations of the radii of curvature of the fixed polode and of the moving polode of a given four-bar linkage at the velocity pole P. These equations appeared first in (Grübler, 1884, 1889, 1892). (Hall, 1958) derived them thereafter in the following form (please refer to Fig. 4.4):

$$\rho_f = \frac{2\delta}{\left(\dfrac{3\overline{PA}}{AA_0}+1\right)(1-K) + \left(\dfrac{3\overline{PB}}{BB_0}+1\right)(1+K)} \qquad (4.21a)$$

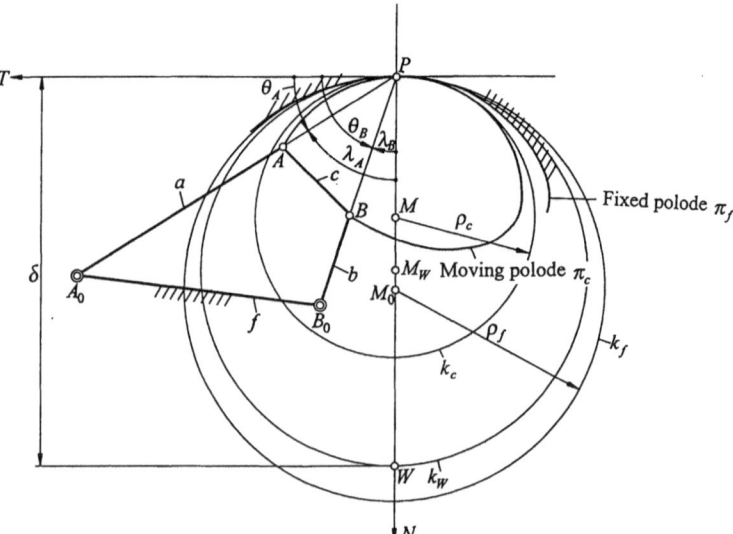

Fig. 4.4. For derivation of radii of curvature ρ_f, ρ_c of the polodes π_f, π_c.

Two other methods of guiding an instantaneous motion of a body

where $\overline{PA}$, $\overline{PB}$, $\overline{AA_0}$, $\overline{BB_0}$ are all directed distances, i.e. $\overline{PA} = -\overline{AP}$, and the direction of $\overline{PA}$ is considered as positive. δ is the diameter of the inflection circle, being always taken as positive, and

$$K = \frac{\sin(\lambda_B + \lambda_A)}{\sin(\lambda_B - \lambda_A)} \quad (4.21b)$$

$$\left.\begin{array}{l}\lambda_A = \theta_A - 90°\\ \lambda_B = \theta_B - 90°\end{array}\right\} \quad (4.22)$$

λ_A, λ_B are respectively the orientation angles of PA, PB with respect to PN, where a counterclockwise sense is considered as positive (please note that the λ_A, λ_B in Fig. 4.4 are all negative), and the definition of θ_A, θ_B is still the same as that defined in equation (1.14). For the derivation of equation (4.21a) the reader is referred to (Hirschhorn, 1962, equation (10-8)). Please note that if A, A_0 and B, B_0 are exchanged respectively, the result remains unchanged. Having found ρ_f, if we wish to find the radius of curvature ρ_c of the moving polode, we can make use of equation (1.15a). Furthermore, if ρ_f in equation (4.21a) is positive, then ρ_f and δ lie on the same side of the poletangent PT; if ρ_f is negative, then ρ_f and δ lie on opposite sides of PT.

The equations derived by (Grübler, 1889) were modified by (Grübler, 1892) and are now in the following forms

$$\frac{1}{\rho_f} = \frac{3}{\tan\lambda_A - \tan\lambda_B}\left(\frac{\sin\lambda_A}{\overline{PA_0}} - \frac{\sin\lambda_B}{\overline{PB_0}}\right) + \frac{1}{\delta} \quad (4.23a)$$

$$\frac{1}{\rho_c} = \frac{3}{\tan\lambda_A - \tan\lambda_B}\left(\frac{\sin\lambda_A}{\overline{PA}} - \frac{\sin\lambda_B}{\overline{PB}}\right) - \frac{1}{\delta} \quad (4.23b)$$

Equation (4.23a) is in fact identical with equation (4.21a).

Example: (taken from Hall, 1958) A given four-bar linkage A_0ABB_0 is shown in Fig. 4.4. $\overline{PA_0} = 7.80$, $\overline{PA} = 2.75$, $\overline{PB_0} = 4.97$, $\overline{PB} = 3.01$, $\angle APB = 40.1°$. Substituting these data into the Euler-Savary equation (1.14), we get $\theta_A = 31.9702°$, $\theta_B = 72.0702°$, $\delta = 8.0221$. Hence $\lambda_A = -58.0298°$, $\lambda_B = -17.9298°$. Substituting further these values into equations (4.23a,b) we obtain $\rho_f = 4.2644$, $\rho_c = 2.7843$.

Conversely, if the values of ρ_f, ρ_c are given, it is possible to synthesize a four-bar linkage so that the radii of curvature of its fixed and moving polodes match those of the given ρ_f, ρ_c. Please refer to Exercise 4.3.

4.2.2 Relative polodes between input link and output link of a four-bar linkage

The material of this section should in principle belong to the methods of synthesizing function generators in Chapter 7. However, in order to apply directly the equations of the radii of curvature of the polodes, we deal with it here. In the following we shall introduce the method of (Hall, 1958).

(a)Radii of curvature of relative polodes by lengths of links

In a four-bar linkage A_0ABB_0 as that shown in Fig. 2.1, there is a certain functional relationship between the relative motion of the input angle ϕ and output angle ψ, namely, $\psi = \psi(\phi)$, as that expressed in the Freudenstein equation (2.1). Alternatively, the functional relationship can be replaced by the relative rolling

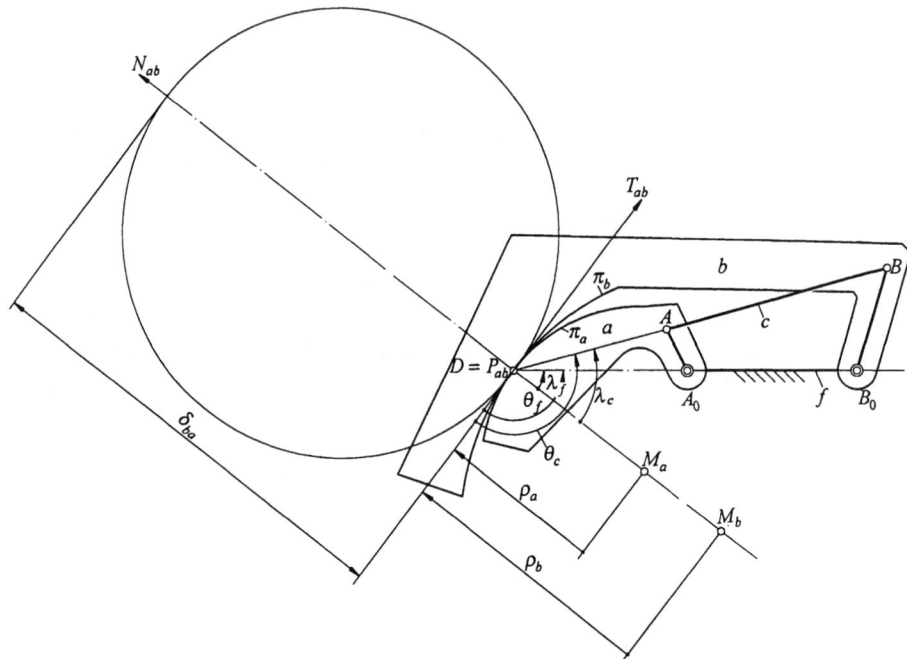

Fig. 4.5. Relative polodes π_a, π_b of a four-bar linkage.

contact between a pair of relative polodes π_a and π_b, as shown in Fig. 4.5. To find the radii of curvature of π_a and π_b at the relative pole P_{ab} (denoted previously as D), we write according to equation (4.21a) or (4.23a,b) the following equation:

$$\rho_a = \frac{2\delta_{ba}}{\left(\dfrac{3\overline{DB_0}}{\overline{B_0A_0}}+1\right)(1-K_{ab})+\left(\dfrac{3\overline{DB}}{\overline{BA}}+1\right)(1+K_{ab})} \quad (4.24a)$$

where

$$K_{ab} = \frac{\sin(\lambda_c + \lambda_f)}{\sin(\lambda_c - \lambda_f)} \quad (4.24b)$$

Two other methods of guiding an instantaneous motion of a body

The sign convention for the line segments in the denominator of equation (4.24a) is the same as that for equation (4.21a). The sign of $\overline{B_0 A_0}$ is negative if its sense of reading is contrary to that of $\overline{DB_0}$. The sign of $\overline{BA}$ is also negative if its sense of reading is contrary to that of $\overline{DB}$. The λ_c and λ_f in equation (4.24b) correspond respectively to the λ_B and λ_A in equation (4.21b), being the respective angles between the link c, f and the the relative polenormal DN_{ab}. Counterclockwise angles are considered positive. Furthermore, if the ρ_a in equation (4.24a) is positive, then ρ_a and δ_{ba} are on the same side of the relative poletangent $P_{ab}T_{ab}$; if ρ_a is negative, then ρ_a and δ_{ba} are on the opposite sides of $P_{ab}T_{ba}$. δ_{ab} is always considered positive. Please note that δ_{ba} is the inflection circle of the motion of b relative to a, not of the motion of a relative to b. Having found ρ_a, we can then find ρ_b according to equation (1.15a) with the following equation:

$$\frac{1}{\rho_b} - \frac{1}{\rho_a} = \frac{1}{\delta_{ba}} \qquad (4.25)$$

(b) Radii of curvature of the relative polodes by differential coefficients of ψ with respect to ϕ

Let $i_{ba} = \omega_b/\omega_a$, or $i_{ba} = \overline{DA_0}/\overline{DB_0} = \dot{\psi}/\dot{\phi} = d\psi/d\phi$. In the following we use the symbols ('), ("), etc.[§] to represent the differential coefficients of a certain function with respect to ϕ, i.e. $di_{ba}/d\phi = \psi''$, $d^2 i_{ba}/d\phi^2 = \psi'''$, ..., etc. The three applicable equations are:

$$\overline{DB_0} = \frac{1}{1-\psi'} f \quad \text{(positive if pointing rightwards)} \qquad (4.26)$$

$$\cot \lambda = -\frac{\psi''}{\psi'(1-\psi')} \qquad (4.27)$$

$$\rho_a = \overline{DM_a} = \frac{[\psi'^2(1-\psi')^2 + \psi''^2]^{3/2}}{[\psi'^2(1-\psi') + 2\psi''^2 - \psi'\psi'''](1-\psi')^3} f \qquad (4.28)$$

Equation (4.27) is identical with equation (2.16). Please refer to Figs. 1.17 and 2.7(a) for the angle λ. It can be seen according to Bobillier theorem that the angle $\lambda_f = \pm 90° - \lambda$ in Fig. 4.5. Please refer to (Hirschhorn, 1962, equation (10-15)) for the derivation of equation (4.28). In this equation, positive ρ_a indicates that the centre of curvature M_a of the polode π_a on the body a at the point D lies on the same side of the (relative) poletangent DT_{ab} as the point A_0; negative ρ_a indicates that they are on the opposite sides of the relative poletangent DT_{ab}.

[§]Please note that, the symbol (') used in Section 4.1 represents differential coefficients with respect to γ, and the symbol (') used in Section 4.2.2 represents differential coefficients with respect to the input angle ϕ of the four-bar linkage, while the symbol (') in Sections 4.2.3, 4.2.4, 4.2.5 and Table 4.1 represents those with respect to the curvilinear length s along the polode.

Example: (taken from Hall, 1958) Given $\psi' = -1$, $\psi'' = -0.57296/\text{rad}$, $\psi''' = 0$, we wish to synthesize a four-bar linkage that should exhibit these three differential coefficients.

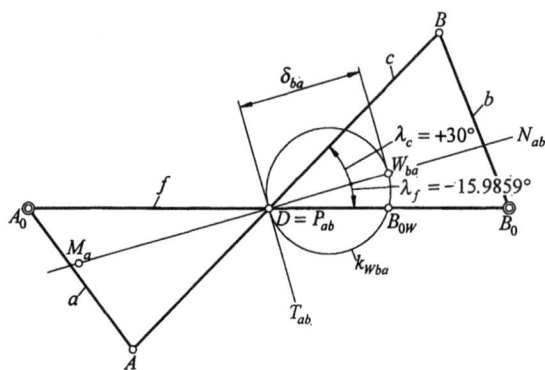

Fig. 4.6. Example: synthesis of four-bar linkage A_0ABB_0 for prescribed $\psi' = -1$, $\psi'' = -0.57296/\text{rad}$, $\psi''' = 0$.

Solution: The answer is shown in Fig. 4.6. Take first any length $f = \overline{A_0B_0}$. For example, let $f = 100$ mm. Since $\psi' = -1$, from equation (4.26) we know the point D ($= P_{ab}$) lies on the midpoint of f ($= \overline{A_0B_0}$). Next from equation (4.27) we get $\lambda_f = -15.9859°$. This determines the orientation of the polenormal $P_{ab}N_{ab}$ and that of poletangent $P_{ab}T_{ab}$, as shown in the figure. By means of the Euler-Savary equation (1.18) we can find

$$\overline{B_0B_{0W}} = \frac{\overline{P_{ab}B_0}^2}{\overline{B_0A_0}} = \frac{(0.5f)^2}{-f} = -25 \text{ mm}$$

Hence $\overline{P_{ab}B_{0W}} = 50 - 25 = 25$ mm. The diameter of the inflection circle is $\delta_{ab} = 25/\cos 15.9859° = 26.006$ mm. Consequently we can find the inflection circle of the motion of b relative to a.

We can then calculate from equation (4.28) $\rho_a = 0.4237 f = 42.37$ mm. Hence we can determine the centre of curvature M_a of the polode π_a bound on a at the point D. And since ρ_a is positive, the point M_a lies on the same side of DT_{ab} as A_0. Now we can freely assume the angle λ_c to be say, $\lambda_c = +30°$. We can find from equation (4.24b) $K_{ab} = 0.3367$. Substituting it into equation (4.24a), and setting ρ_a as negative (for ρ_a and δ_{ba} are on opposite sides of DT_{ab}), we get

$$-42.37 = \frac{2 \times 26.006}{\left(\dfrac{3 \times 50}{-100} + 1\right)(1 - 0.3367) + \left(\dfrac{3\overline{DB}}{-\overline{AB}} + 1\right)(1 + 0.3367)}$$

Two other methods of guiding an instantaneous motion of a body

Besides, the two points A, B should satisfy the (relative) Euler-Savary equation, or

$$\frac{1}{\overline{DB}} + \frac{1}{\overline{AB} - \overline{DB}} = \frac{1}{26.006\sin(90°+30°)}$$

Solving the above two equations simultaneously, we get $\overline{DB} = 50.81$ mm, $\overline{AB} = 91.26$ mm.

Since there are ∞^1 selections in choosing λ_c, there are ∞^1 solutions.

In the later Section 7.6.2, we shall introduce another method for solving this problem which can be compared with the present method.

4.2.3 The Sieker-Beyer equation

(Sieker-Beyer, 1943) derives the radii of curvature of the two polodes by means of *bipolar coordinates,* as adopted by (Frost, 1880-81). The method is as follows. In Fig. 4.7, let

$$\overline{PA_0} = r_1, \quad \overline{PB_0} = r_2, \quad \sphericalangle BPA = \varepsilon$$

then
$$\overline{PA} = r_1 - a, \quad \overline{PB} = r_2 - b$$

Please note that r_1, r_2, a,b are all directed line segments, i.e. $a = \overline{AA_0}$, $b = \overline{BB_0}$. If the sense of reading of a certain line segment is negative according to the convention

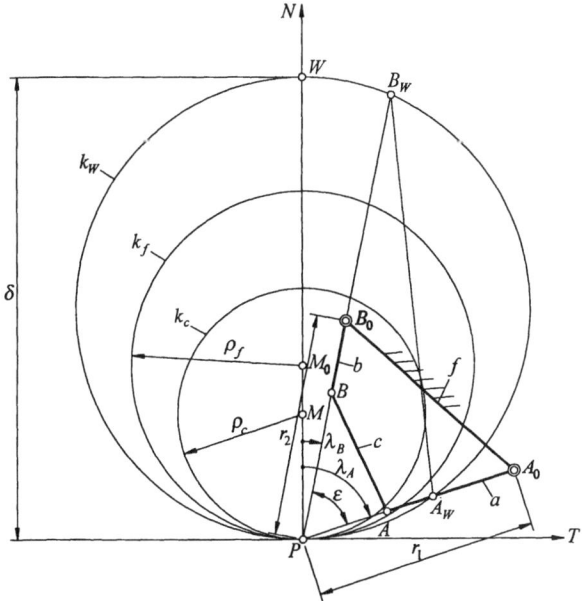

Fig. 4.7. Bipolar coordinates for the determination of radii of curvature ρ_f and ρ_c of fixed polode π_f and moving polode π_c of four-bar linkage A_0ABB_0.

of polar coordinates, then the line segment is negative (please refer to Exercise 4.6). According to the law of cosine of a triangle we have

$$f^2 = r_1^2 + r_2^2 - 2r_1 r_2 \cos\varepsilon \qquad (4.29)$$
$$c^2 = (r_1 - a)^2 + (r_2 - b)^2 - 2(r_1 - a)(r_2 - b)\cos\varepsilon \qquad (4.30)$$

Eliminating $\cos\varepsilon$ from equations (4.29) and (4.30), we obtain

$$\begin{aligned}F(r_1,r_2) &= br_1^3 + ar_2^3 - ar_1^2 r_2 - br_1 r_2^2 - ab(r_1^2 + r_2^2)\\ &\quad + (a^2 + b^2 - c^2 + f^2)r_1 r_2 - f^2(br_1 + ar_2) + abf^2 \\ &= 0\end{aligned} \qquad (4.31)$$

We are now in a position to derive the radius of curvature ρ_f of the fixed polode π_f at the point P by applying equation (A9.10) in Appendix 9.

$$\frac{1}{\rho_f} = \left[E^2\left(\frac{F_1}{r_1} + \frac{F_2}{r_2}\right) - \left(\frac{F_2}{r_1} + \frac{F_1}{r_2}\right) F_1 F_2 \sin^2\varepsilon \right.$$
$$\left. + (F_2^2 F_{11} - 2F_1 F_2 F_{12} + F_1^2 F_{22})\sin^2\varepsilon \right] \Big/ E^3 \qquad (4.32)$$

where

$$E^2 = F_1^2 + F_2^2 + 2F_1 F_2 \cos\varepsilon$$

From equation (4.31) the partial differential coefficients of F can be calculated:

$$\left.\begin{aligned}F_1 &= 3br_1^2 - 2ar_1 r_2 - br_2^2 - 2abr_1 + (a^2 + b^2 - c^2 + f^2)r_2 - f^2 b\\ F_2 &= 3ar_2^2 - 2br_1 r_2 - ar_1^2 - 2abr_2 + (a^2 + b^2 - c^2 + f^2)r_1 - f^2 a\\ F_{11} &= 6br_1 - 2ar_2 - 2ab\\ F_{12} &= -2ar_1 - 2br_2 + (a^2 + b^2 - c^2 + f^2)\\ F_{22} &= 6ar_2 - 2br_1 - 2ab\end{aligned}\right\} \qquad (4.33)$$

From equation (A9.4) we have

$$\cos\lambda_A = \frac{F_1 + F_2 \cos\varepsilon}{E}, \quad \sin\lambda_A = -\frac{F_2 \sin\varepsilon}{E} \qquad (4.34a,b)$$

$$\cos\lambda_B = \frac{F_1 \cos\varepsilon + F_2}{E}, \quad \sin\lambda_B = \frac{F_1 \sin\varepsilon}{E} \qquad (4.35a,b)$$

The inflection circle k_W intersects $A_0 A$, $B_0 B$ respectively in A_W, B_W. Let $\overline{PA_W} = q_A$, $\overline{PB_W} = q_B$, then

$$q_A = \delta\cos\lambda_A, \quad q_B = \delta\cos\lambda_B \qquad (4.36a,b)$$

We shall from now on always regard δ as positive; in other words, we shall regard the half plane on which the inflection circle lies as the upper half plane of the polar coordinate system. From Euler-Savary equation (1.14) we get

Two other methods of guiding an instantaneous motion of a body

$$q_A = \frac{r_1(r_1 - a)}{a}, \quad q_B = \frac{r_2(r_2 - b)}{b} \qquad (4.37a,b)$$

Next let $w = \overline{A_W B_W}$, then

$$w^2 = \overline{A_W B_W}^2 = q_A^2 + q_B^2 - 2q_A q_B \cos\varepsilon \qquad (4.38a)$$

hence

$$w = \delta \sin\varepsilon \qquad (4.38b)$$

Therefore

$$F_1 = 2ab(q_A - q_B \cos\varepsilon) \qquad (4.39a)$$

Similarly

$$F_2 = 2ab(q_B - q_A \cos\varepsilon) \qquad (4.39b)$$

Substituting the relation $\lambda_B = \varepsilon + \lambda_A$ into equation (4.35b) (ε is always taken as positive, and in Fig. 4.7, λ_A, λ_B are all negative), we get

$$F_1 = \frac{E(q_A - q_B \cos\varepsilon)}{\delta \sin^2\varepsilon} \qquad (4.40a)$$

$$F_2 = \frac{E(q_B - q_A \cos\varepsilon)}{\delta \sin^2\varepsilon} \qquad (4.40b)$$

A comparison of equations (4.40),(4.39) and (4.38b) shows that

$$E = 2ab\delta \sin^2\varepsilon = 2abw\sin\varepsilon \qquad (4.41)$$

Furthermore, the terms F_{11}, F_{12}, F_{22} in equation (4.33) can be written as

$$F_{11} = 2ab\left(\frac{3q_A}{r_1} - \frac{q_B}{r_2} + 1\right) \qquad (4.42a)$$

$$F_{12} = -2ab\cos\varepsilon\left(\frac{q_A}{r_1} + \frac{q_B}{r_2} + 1\right) \qquad (4.42b)$$

$$F_{22} = 2ab\left(\frac{3q_B}{r_2} - \frac{q_A}{r_1} + 1\right) \qquad (4.42c)$$

Substituting equations (4.40a,b) and (4.42a,b,c) into equation (4.32) gives

$$\frac{1}{\rho_f} = \frac{1}{\delta}(1+Q) \qquad (4.43)$$

where

$$Q = \frac{3q_A q_B}{w^2}\left[\frac{1}{r_1}(q_B - q_A \cos\varepsilon) + \frac{1}{r_2}(q_A - q_B \cos\varepsilon)\right] \qquad (4.44)$$

On the other hand, from the definition of $\rho_f = ds/d\tau = 1/\tau'$ (please refer to Fig. A9.1) and equations (4.43), (3.120) we have

$$Q = \tau'\delta - 1 = -\frac{\tau'}{\gamma'} - 1 = -\frac{\tau' + \gamma'}{\gamma'} = \frac{3\delta}{l_*}$$

which is

$$l_* = \frac{3\delta}{Q} \tag{4.45a}$$

and from the definitions of l, m in equations (3.115), (3.114) we have

$$l = \frac{3\delta}{Q+3} \tag{4.45b}$$

$$m = -\frac{3\delta}{\delta'} \tag{4.45c}$$

Moreover, from equation (1.15a) we have

$$\frac{1}{\rho_c} = \frac{1}{\delta}(2+Q) \tag{4.46}$$

For the convenience of calculation, let

$$V = \frac{\rho_c}{\rho_f}$$

we have then

$$Q = \frac{2V-1}{1-V} = \frac{2\rho_c - \rho_f}{\rho_f - \rho_c} \tag{4.47a}$$

$$l = \frac{3V}{2-V}\rho_f = \frac{3}{2-V}\rho_c \tag{4.47b}$$

$$l_* = \frac{3V}{2V-1}\rho_f = \frac{3}{2V-1}\rho_c \tag{4.47c}$$

4.2.4 Maximum and minimum values of the radii of curvature of the polodes

The content of this Section is taken from (Sieker & Beyer, 1943). When the curvature of the polode reaches an extreme value, the curvature is temporarily stationary, and the synthesized mechanism also exhibits a better approximation. Differentiate equations (4.43) and (4.46) respectively with respect to s (curvilinear length along the polode), or

$$\frac{d\rho_f}{ds} = \frac{\rho_f}{\delta}\left(\frac{d\delta}{ds} - \rho_f \frac{dQ}{ds}\right) \tag{4.48a}$$

$$\frac{d\rho_c}{ds} = \frac{\rho_c}{\delta}\left(\frac{d\delta}{ds} - \rho_c \frac{dQ}{ds}\right) \tag{4.48b}$$

The maximum or minimum values of ρ_f, ρ_c occur when $d\rho_f/ds = 0$, $d\rho_c/ds = 0$

Two other methods of guiding an instantaneous motion of a body

respectively. Assume that $\rho_f \neq 0$, $\rho_c \neq 0$, $1/\delta \neq 0$. From equations (4.48a,b) we get the following conditions

$$\left.\begin{array}{r}\dfrac{d\delta}{ds} - \rho_f \dfrac{dQ}{ds} = 0\end{array}\right\} \quad (4.49a)$$

$$\left.\begin{array}{r}\dfrac{d\delta}{ds} - \rho_c \dfrac{dQ}{ds} = 0\end{array}\right\} \quad (4.49b)$$

The two equations (4.49a,b) are two homogeneous linear equations in the two unknowns $d\delta/ds$, dQ/ds. Under the further assumption $\rho_f \neq \rho_c$, the solution is

$$\left.\begin{array}{r}\dfrac{d\delta}{ds} = 0\end{array}\right\} \quad (4.50a)$$

$$\left.\begin{array}{r}\dfrac{dQ}{ds} = 0\end{array}\right\} \quad (4.50b)$$

The two differential coefficients in equations (4.50a,b) can be calculated as follows (note that $\delta = -1/\gamma'$, and m can be calculated from Fig. 3.60):

$$\frac{d\delta}{ds} = \frac{\gamma''}{\gamma'^2} = -\frac{3\delta}{m} = \frac{3\delta}{\cot \lambda_A - \cot \lambda_B}\left(\frac{\cos \lambda_A}{r_1} - \frac{\cos \lambda_B}{r_2}\right)$$

$$= \frac{3(ar_2 - br_1)(q_A - q_B \cos \varepsilon)(q_B - q_A \cos \varepsilon)}{ab\delta^2 \sin^3 \varepsilon} \quad (4.51)$$

$$\frac{dQ}{ds} = \tau''\delta + \tau'\delta'$$

$$= \frac{3}{ab\delta^3 \sin^3 \varepsilon}[aq_A(q_A - q_B\cos\varepsilon)^2 - bq_B(q_B - q_A\cos\varepsilon)^2]$$

$$+ \frac{3}{a^2b^2\delta^5 \sin^3 \varepsilon}(br_1 - ar_2)(2q_Aq_B - \delta^2\cos\varepsilon) \quad (4.52)$$

$$[bq_B(2r_1 - a)(q_B - q_A\cos\varepsilon) + aq_A(2r_2 - b)(q_A - q_B\cos\varepsilon)]$$

$$+ \frac{3}{a^2b^2\delta^5 \sin^3 \varepsilon}q_Aq_B(q_B^2 - q_A^2)(br_1 - ar_2)^2$$

The condition in the two equations (4.50a,b) is the same as equating the right hand sides of equations (4.51), (4.52) to zero (assume for the time being $1/\delta \neq 0$). Let us start from equation (4.51). There can be three cases, namely, $ar_2 - br_1 = 0$, $q_A - q_B\cos\varepsilon = 0$ and $q_B - q_A\cos\varepsilon = 0$. We shall consider them separately.

(a) $ar_2 - br_1 = 0$

This condition is

$$\frac{r_1}{r_2} = \frac{a}{b}$$

Hence $(r_1 - a)/(r_2 - b) = a/b$, while

$$\frac{q_A}{q_B} = \frac{a}{b} \qquad (4.53)$$

Substituting condition (4.53) into equation (4.52), and then into equation (4.50b), we get

$$aq_A(q_A - q_B \cos\varepsilon)^2 - bq_B(q_B - q_A \cos\varepsilon)^2 = 0$$

Therefore

$$a = b, \quad r_1 = r_2, \quad q_A = q_B \qquad (4.54)$$

The four-bar linkage can then be made in the form of a right trapezoid, as shown in Fig. 4.8. In this case $\lambda_A = -\varepsilon/2$, $\lambda_B = +\varepsilon/2$.

$$q_A = q_B = \delta\cos(\varepsilon/2) \qquad (4.55)$$

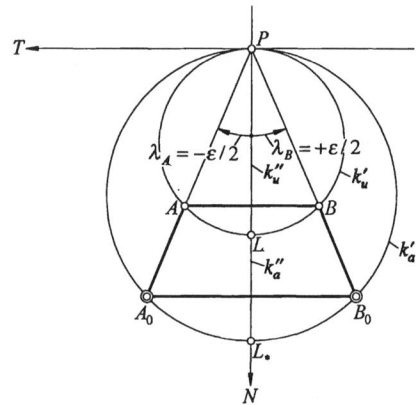

Fig. 4.8. Configuration of four-bar linkage becomes a right trapezoid when coupler is in a position of maximum or minimum radii of curvature of the polodes.

From equation (4.44) we get

$$Q = \frac{3\delta}{r_1}\cos\frac{\varepsilon}{2} \qquad (4.56)$$

Substituting equation (4.56) respectively into equations (4.45a,b), and taking equations (4.37a) and (4.36a) into consideration yields

$$l_* = \frac{r_1}{\cos\dfrac{\varepsilon}{2}}, \quad l = \frac{r_1 - a}{\cos\dfrac{\varepsilon}{2}} \qquad (4.57a,b)$$

Equations (4.57a,b) show that the locus of A_0, B_0 is the circle k'_a with l_* as the diameter, and that the locus of A and B is the circle k'_u with l as the diameter. The centres of these two circles k'_a and k'_u lie on the polenormal PN, and both circles

Two other methods of guiding an instantaneous motion of a body   205

are tangent to the poletangent *PT*. In Section 3.7.6(c) we mentioned that when $\gamma'' = 0$, but $1/\delta \neq 0$ ($\gamma' \neq 0$), the centering-point curve k_a and circling-point curve k_u will break up respectively into the polenormal *PN* and a circle, as shown in Fig. 3.73. This is what happens here.

Given ρ_c, ρ_f, we can easily calculate $V = \rho_c/\rho_f$. From equations (4.47 b,c) we can find l and l_*, and then obtain the two circles k'_u, k'_a in Fig. 4.8. Assuming a value of $\varepsilon/2$ we can determine the four points A_0, A, B, B_0. There are altogether ∞^1 solutions.

Please note that, the present sign convention is the one shown in Fig. 4.7. In other words, any one of the line segments ρ_c, ρ_f, l, l_*, on the positive side of *PN* is regarded as positive; otherwise, negative.

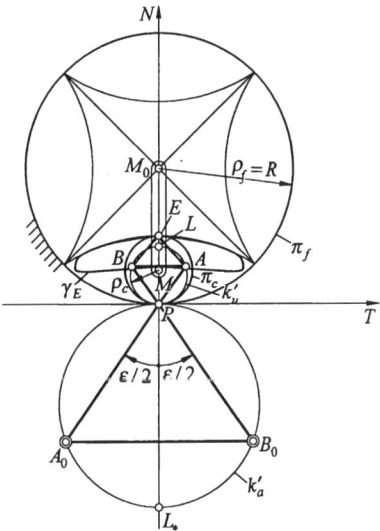

Fig. 4.9. Example: generating a coupler curve γ_E to approximate a hypocycloid.

Example: Fig. 4.9 shows a smaller circle rolling without slip on the inside of a fixed larger circle. The smaller circle is the moving polode π_c, and the larger circle is the fixed polode π_f. It is required to synthesize a four-bar linkage, so that the path of its coupler point *E* is quite close to the hypocycloid generated by a point on the smaller circle.

Solution: Assume the radius ρ_c of the smaller circle is 1/4 of the radius ρ_f of the larger circle, and let $\rho_f = R$. Then $V = \rho_c/\rho_f = 1/4$. From equations (4.47a,b,c) we have

206 Kinematics and Design of Planar Mechanisms

$$Q = -\frac{2}{3}, \quad l = \frac{3}{7}R, \quad l_* = -\frac{3}{2}R$$

We get from equation (4.43) $\delta = R/3$, and draw the circles k'_u, k'_a in the figure. k'_a is the locus of A_0, B_0, and k'_u is the locus of A, B. Choose any angle ε and thus determine the symmetrical four-bar linkage A_0ABB_0. It can be seen that the path γ_E of the coupler point E is quite close to the original hypocycloid.

(b) $q_A - q_B \cos \varepsilon = 0$

In this case, because of equation (4.40a), $F_1 = 0$, and because of equation (4.35b), $\lambda_B = 0$. Therefore B_0, B are all on the polenormal PN. Moreover, we have

$$\left. \begin{array}{l} q_B = \delta \\ q_A = \delta \cos \varepsilon \\ Q = \dfrac{3\delta}{r_1} \cos \varepsilon \\ l = \dfrac{r_1 - a}{\cos \varepsilon} \\ l_* = \dfrac{r_1}{\cos \varepsilon} \end{array} \right\} \qquad (4.58a,b,c,d,e)$$

In this case the disposition of the points is as shown in Fig. 4.10. The points A_0, A lie still respectively on the circles k'_a and k'_u. This is because we still have $\gamma'' = 0$, and k_a, k_u break up respectively into k'_a, k''_a and k'_u, k''_u. According to the

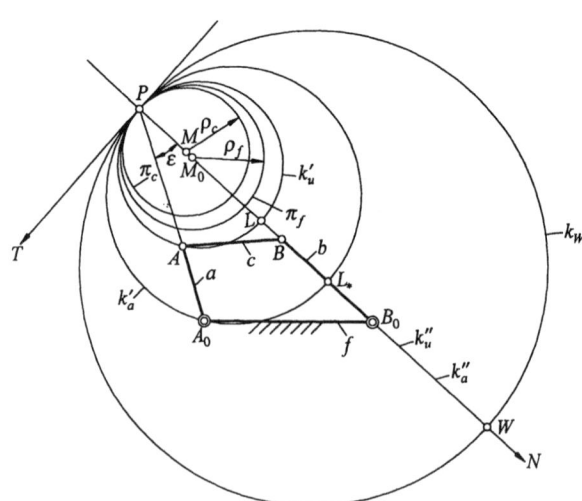

Fig. 4.10. $q_A - q_B \cos \varepsilon = 0$: k_u breaks up into k'_u, k''_u; k_a breaks up into k'_a, k''_a. B, B_0 are on the polenormal.

Two other methods of guiding an instantaneous motion of a body

requirement in equation (4.50b), substituting $q_A - q_B \cos \varepsilon = 0$ into equation (4.52) and setting it equal to zero, and taking equations (4.58) into consideration, we get

$$ab^2 \delta \sin^2 \varepsilon - (br_1 - ar_2)(3br_1 - ar_2 - ab) \cos \varepsilon = 0 \qquad (4.59)$$

On the other hand, in the present case, we have

$$a = \frac{3r_1}{Q+3}, \quad b = \frac{r_2^2}{r_2 + \delta}, \quad \cos \varepsilon = \frac{Qr_1}{3\delta} \qquad (4.60a,b,c)$$

Substituting these relations together with the above equation (4.58e) into equation (4.59), we get

$$r_1 = 3\delta r_2 \left(\frac{Q+3}{Q(2Qr_2 - 3\delta)[(2Q+3)r_2 - 3\delta]} \right)^{1/2} \qquad (4.61)$$

The angle ε should satisfy the following relation

$$\tan^2 \varepsilon = \frac{3}{Q(Q+3)r_2^2} [Q(Q+1)r_2^2 - \delta(4Q+3)r_2 + 3\delta^2] \qquad (4.62)$$

In solving a problem, given ρ_c, ρ_f, we can find δ from the Euler-Savary equation (1.15a). After finding V, we can calculate the values Q, l, l_* from equation (4.47a,b,c). These are all constants. For an assumed value of ε, we can use equations (4.58d,e) to determine the values of r_1, a. And then by solving a quadratic equation in the unknown r_2 from equation (4.61) or (4.62), we can determine r_2. Finally, we can determine b by equation (4.58a).

Another solving procedure is by assuming an r_2, and determining r_1 by equation (4.61). We determine the angle ε by equation (4.62), and finally obtain a and b by equations (4.58a, b).

(c) $q_B - q_A \cos \varepsilon = 0$

This is the case symmetrical to the last case (b) with respect to PN. All we have to do now is to interchange in all equations in (b) A with B, a with b, and r_1 with r_2. For instance it is now $q_B = \delta \cos \varepsilon$, $q_A = \delta$, $Q = (3\delta / r_2) \cos \varepsilon$, $l = (r_2 - b) / \cos \varepsilon$, $l_* = r_2 / \cos \varepsilon$, etc. It is unnecessary for us to discuss this case specifically.

4.2.5 Synthesis to match given radii of curvature of polodes and their rates of change

It was mentioned at the beginning of Section 4.2, that it is possible to synthesize a four-bar linkage to simulate an original rolling motion of two polodes by matching the predetermined radii of curvature of the polodes as well as the rates of change of these radii. In the following we shall explain the method in detail.

(a) The radii of curvature ρ_f of the fixed polode and ρ_c of the moving polode are prescribed

This problem is already discussed in Section 4.2.1 and will also be explained in the hint of Exercise 4.3. In other words, in case ρ_f and ρ_c are prescribed, and if it is required to synthesize a four-bar linkage to exhibit the same ρ_f and ρ_c, there still remain three selections, or there are ∞^3 solutions.

(b) ρ_f, ρ_c and δ' ($= d\delta/ds$) are prescribed

Substituting equations (4.45b,c) into equation (3.116), we can write the equation of the circling-point curve as

$$k_u : p\left[(Q+3)\cos\theta_A - \frac{d\delta}{ds}\sin\theta_A\right] - 3\delta\sin\theta_A \cos\theta_A = 0 \quad (4.63)$$

Similarly, substituting equations (4.45a, c) into equation (3.121), we can write the equation of the centering-point curve as

$$k_a : p_0\left[Q\cos\theta_A - \frac{d\delta}{ds}\sin\theta_A\right] - 3\delta\sin\theta_A \cos\theta_A = 0 \quad (4.64)$$

As defined before, the polar coordinates of a circle-point are (p, θ_A), and those of a centre-point are (p_0, θ_A). From equations (4.63) and (4.64) it is clear that prescribing ρ_c, ρ_f and $d\delta/ds$ is equivalent to guide a body through four infinitesimally separated positions, and there are ∞^2 solutions. The selection is to choose first a point A, and then another point B on k_u.

(c) ρ_f, ρ_c, $d\rho_f/ds$, $d\rho_c/ds$ and $d^2\delta/ds^2$ are prescribed

Differentiating equation (4.63) or (4.64) with respect to s, and setting it equal to zero, is equivalent to differentiating equation (3.116) in Section 3.7.1 with respect to s. However, as the term $d\theta_A/ds$ is involved here, it has to be found separately, despite of the known expression of the term $p\theta'_A$ in equation (3.111). Differentiating the Euler-Savary equation (1.14)

$$\left(\frac{1}{p} - \frac{1}{p_0}\right)\sin\theta_A = \frac{1}{\delta} \qquad [(1.14)]$$

with respect to s, and taking the relation $p' = -\cos\theta_A$ in equation (3.110) and $p'_0 = p'$ into consideration, and substituting the two equations (4.63) and (4.64) into it, we get

$$\frac{d\theta_A}{ds} = -\frac{2Q+3}{3\delta} - \frac{\tan\theta_A}{3\delta}\frac{d\delta}{ds} \quad (4.65)$$

Now differentiating equation (4.64) with respect to s, and substituting equations (3.110), (4.64), (4.65) into it and rearranging, we obtain

$$\left(\frac{d\delta}{ds}\right)^2 t^4 + \frac{d\delta}{ds}(2Q+3)t^3 + \left[2\left(\frac{d\delta}{ds}\right)^2 - 3\delta\frac{d^2\delta}{ds^2}\right]t^2 + 3\delta\frac{dQ}{ds}t + Q(Q+3) = 0 \quad (4.66)$$

Two other methods of guiding an instantaneous motion of a body

In equation (4.66), $t = \tan \theta_A$ is the unknown. This equation corresponds to equation (3.147), its four roots corresponding to four Burmester points. For each real root $\tan\theta_A$, the corresponding circle-point and centre-point can be found. In other words, to match prescribed ρ_f, ρ_c, $d\rho_f/ds$, $d\rho_c/ds$ and $d^2\delta/ds^2$ is equivalent to guide a body through five infinitesimally separated positions.

Example: (taken from Sieker, 1948) The instantaneous motion of a body is determined by the following quantities: the radius of curvature of the fixed polode ρ_f = 50 mm, that of the moving polode ρ_c = 40 mm, rate of change of these radii $d\rho_f/ds$ = 0.05, $d\rho_c/ds$ = 0.1, and the second derivative of the diameter of inflection circle with respect to s is $d^2\delta/ds^2$ = 0.015 mm^{-1}. It is required to synthesize a four-bar linkage so that the instantaneous motion of its coupler matches the above conditions.

Solution: Please refer to Fig. 4.11. The procedure of solution is as follows:

(1) From the given ρ_f, ρ_c, we can obtain by Euler-Savary equation (1.15a)

$$\delta = 200 \text{ mm}$$

(2) From the given $d\rho_f/ds = 0.05$ and $d\rho_c/ds = 0.1$, we can find from equations

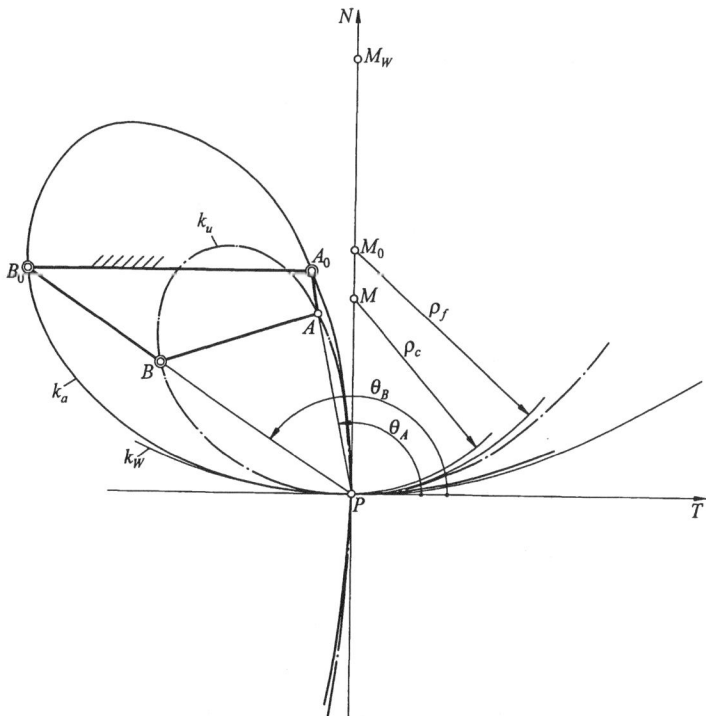

Fig. 4.11. Example: synthesis of four-bar linkage A_0ABB_0 for prescribed ρ_f=50mm, ρ_c = 40mm, $\rho'_f = 0.05$, $\rho'_c = 0.1$, $\delta'' = 0.015$mm^{-1}.

(4.48a,b)
$$d\delta/ds = 1.7, \quad dQ/ds = 0.03 \text{ mm}^{-1}$$

(3) By equation (4.43) or (4.46) we get
$$Q = 3$$

(4) Substituting the known values of Q, $d\delta/ds$ and δ into equation (4.63), we get
$$k_u : \frac{1}{p} = \frac{1}{100\sin\theta_A} - \frac{1.7}{600\cos\theta_A}$$

and into equation (4.64), we get
$$k_a = \frac{1}{p_0} = \frac{1}{200\sin\theta_A} - \frac{1.7}{600\cos\theta_A}$$

(5) Substituting the known values of δ, Q, $d\delta/ds$, dQ/ds and $d^2\delta/ds^2$ into equation (4.66) results in a quartic equation in the unknown $t = \tan\theta_A$
$$2.89t^4 + 15.3t^3 - 3.22t^2 + 18t + 18 = 0$$

Solving this equation gives two real roots
$$t_1 = \tan\theta_{A1} = -5.651746, \quad t_2 = \tan\theta_{A2} = -0.681954,$$
$$\theta_{A1} = 100.034° \quad\quad \theta_{A2} = 145.708°$$

Taking θ_{A1} as θ_A, and θ_{A2} as θ_B, we can find from the equation of k_u that $p_A = 37.854$ mm, $p_B = 47.218$ mm. Similarly we can find from the equation of k_a that $p_{A0} = 46.861$ mm, $p_{B0} = 81.275$ mm. Hence the polar coordinates of the four joints of the four-bar linkage are:

$$A_0(46.861, \quad 100.034°)$$
$$A(37.854, \quad 100.034°)$$
$$B_0(81.275, \quad 145.708°)$$
$$B(47.218, \quad 145.708°)$$

This is the four-bar linkage shown in Fig. 4.11.

Substituting the values of δ, m, l, l_* thus obtained into equation (3.136) yields the coordinates of the Ball point U: $x_U = 85.786$ mm, $y_U = 151.388$ mm.

4.3 Comparison of three methods of guiding an instantaneous motion

We have up to now introduced respectively three methods of guiding an instantaneous motion of a body, namely, (1) the method depicted in Sections 3.7, 3.9, which we may call it the *circling-point curve and centering-point curve method*, (2) the *instantaneous invariants method* mentioned in Section 4.1, and (3) the *polode*

Two other methods of guiding an instantaneous motion of a body

method mentioned in Section 4.2. All three methods should be interchangeable. We are going to find out the relationships among the quantities used in the three methods. For this purpose we shall use the instantaneous invariants as the basis of comparison. In other words, the quantities used in the other two methods shall be expressed in terms of instantaneous invariants, and listed in Table 4.1.

Table 4.1 Correlations among three methods of guiding an instantaneous motion of a body

Method \ Order	3	4	5
Instantaneous invariants	$b_2 = -1$	a_3, b_3	a_3, b_3, a_4, b_4
Circling-point curve and centering-point curve	$\delta = b_2$	$m = \dfrac{3b_2^2}{b_3}$ $l = \dfrac{3b_2^2}{a_3 + 3b_2}$ $l_* = \dfrac{3b_2^2}{a_3}$	m, l, l_* (as in left column) $m' = -\dfrac{3(2b_3^2 - b_4 b_2)}{b_3^3}$ $l' = 3\dfrac{a_4 b_2 - 2a_3 b_3 - 3b_2 b_3}{(a_3 + 3b_2)^2}$ $l'_* = 3\dfrac{a_4 b_2 - 2a_3 b_3}{a_3^2}$
Polode	—	$\rho_f = -(\mathrm{SIGN} b_2)\dfrac{b_2^2}{a_3 + b_2}$ $\rho_c = -(\mathrm{SIGN} b_2)\dfrac{b_2^2}{a_3 + 2b_2}$ $\dfrac{d\delta}{ds} = b_3$	ρ_f, ρ_c (as in left column) $\rho'_f = (\mathrm{SIGN} b_2)\dfrac{2a_3 b_3 + b_2 b_3 - a_4 b_2}{(a_3 + b_2)^2}$ $\rho'_c = (\mathrm{SIGN} b_2)\dfrac{2a_3 b_3 + 2b_2 b_3 - a_4 b_2}{(a_3 + 2b_2)^2}$ $\dfrac{d^2\delta}{ds^2} = -\dfrac{b_4}{b_2}$
Number of instantaneous invariants prescribed	1, (b_2)	2, (a_3, b_3)	4, $(a_3, b_3; a_4, b_4)$

In Table 4.1, the term *order* implies the number of infinitesimally separated positions through which a body is to be guided. For instance, if the *order* is 4, it means that a body is to be guided through four infinitesimally separated positions. As can be seen from the table, among the three methods, the method of instantaneous invariants and the method of circling-point curve and centering-point curve correspond substantially to each other, but the polode method does not completely correspond to these two methods. First of all, b_2 may be considered as a *scale factor*, hence it can be set at -1. In the later higher order guidance, b_2 will not be considered as a decisive quantity. For instance, a third order guidance requires only a b_2. However, if b_2 is considered only as a scale factor, it can also be considered as a *non-prescribed* quantity. From the instantaneous invariants contained in ρ_f, ρ_c listed in Table 4.1, it can be seen that only prescribing ρ_f, ρ_c does not correspond to a fourth order guidance. Just like that mentioned in Section 4.2.5(b), in addition to these, it requires a further prescribed $d\delta/ds$ ($= b_3$) to bring the guidance up to a fourth order. In other words, in a fourth order guidance, a_3, b_3 have to be prescribed. From this point of view, only prescribing ρ_f, ρ_c entails higher than a third order guidance. This is because a_3 is already included in ρ_f, ρ_c, while a third order guidance does not require an a_3. Furthermore, as mentioned in Section 4.2.5(c), prescribing ρ_f, ρ_c, $d\rho_f/ds$, $d\rho_c/ds$ is equivalent to prescribing a_3, b_3, a_4, but to bring it up to a fifth order guidance, a further $d^2\delta/ds^2$ ($= b_4$) has to be prescribed. From this point of view, the case mentioned in Section 4.2.4 that prescribing the four quantities ρ_f, ρ_c, $d\rho_f/ds$ ($= 0$), $d\rho_c/ds$ ($= 0$), is not yet a fifth order guidance, but higher than a fourth order guidance, because a_4 is already included in ρ'_f, ρ'_c.

Exercises

4.1 Show that in using the instantaneous invariants, in the initial position, $d\delta/ds = -b_3/b_2$.

4.2 Verify the coordinates of the Ball point as given in equation (4.18).

4.3 A wheel of radius 24 rolls without slip on the outside of a fixed wheel of

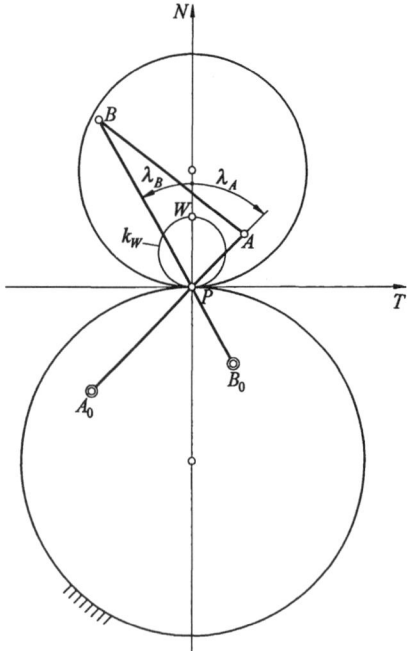

Two other methods of guiding an instantaneous motion of a body

radius 36. Construct a four-bar linkage A_0ABB_0, so that the radius of curvature of its moving polode $\rho_c = +24$, and that of its fixed polode $\rho_f = -36$.

Hint: Taking the sense of PN as shown in the figure as positive, we have then a negative λ_A, a positive λ_B; and $\overline{PA}$, $\overline{PB}$ are positive, while $\overline{PA_0}$, $\overline{PB_0}$ are negative. $\overline{PA}$, $\overline{PA_0}$ should satisfy the Euler-Savary equation (1.14), and so should $\overline{PB}$, $\overline{PB_0}$. Attention should be paid to equation (4.22). In the above two Euler-Savary equations and the two equations (4.23a), (4.23b), there are altogether 6 unknowns: $\overline{PA_0}$, $\overline{PB_0}$, $\overline{PA}$, $\overline{PB}$, λ_A, λ_B. Hence any two of them may be arbitrarily assumed to solve the remaining four. There are ∞^2 solutions.

4.4 In Fig. 4.5, given $\overline{DA_0} = 55.1$, $\overline{DB_0} = 109.1$, $\overline{DA} = 50.2$, $\overline{DB} = 123.2$, $\sphericalangle A_0DA = 15.5°$. Find δ_{ba} and λ_f.
(Ans.: $\delta_{ba} = 139.6945$, $\lambda_f = 37.165°$, $\rho_a = -51.8416$, $\rho_b = -82.4331$)

4.5 In Fig. 4.10, given $\rho_c = 13$, $\rho_f = 15$. Find δ, l, l_*. Assume a value of the angle ε, to find r_1, a, r_2, b.
(Ans.: $\delta = 97.5$, $l = 34.41176$, $l_* = 53.18182$. Assume $\varepsilon = 30°$, $r_1 = 46.05681$, $a = 16.25534$, $r_2 = 65.52971$, $b = 26.33963$)

4.6 Show that in the case $q_A - q_B \cos\varepsilon = 0$ in Section 4.2.4 (b),

$$\tan^2\varepsilon = \frac{l(r_2 - l_*)(r_2 - \rho_f)}{r_2^2 \rho_f}$$

and

$$\tan^2\varepsilon = \frac{l_*[(r_2 - b) - l][(r_2 - b) - \rho_c]}{(r_2 - b)^2 \rho_c}$$

and

$$r_1 = r_2 l_* \sqrt{\frac{l_*}{(2r_2 - l_*)[(l + l_*)r_2 - ll_*]}}$$

4.7 The figure shows a mincing machine similar to that shown in Fig. 1.4. The mincing cutter is of a circular shape with radius r, rolling on a flat plate. Find first l, l_*, δ and k'_u, k'_a, and then take a point C on k'_u, to find the point C_0 on k'_a.
(Ans.: $\rho_c = +r$, $\rho_f = \infty$, $(V = 0)$. Therefore $\delta = r$, $Q = -1$, $l_* = -3r$, $l = 3r/2$, same as Exercise 3.41.)

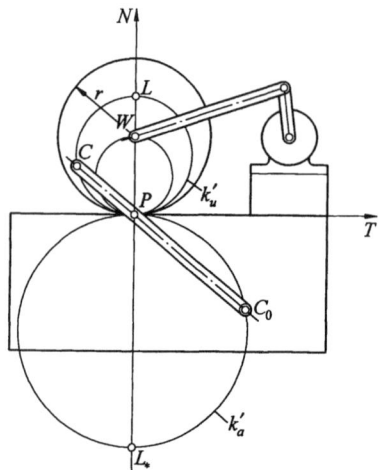

4.8 (Taken from Sieker & Beyer, 1943) The figure shows a flat plate rolling on the outside of a fixed wheel. The radius of the wheel is r. The path of a point on the flat plate is an involute. It is required to construct a four-bar linkage so that the path γ_E of its coupler point E approximates this involute. Assume

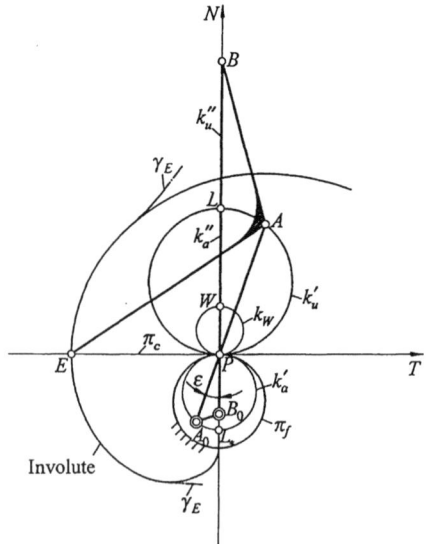

$r_2 = (\overline{PB_0}) = -1.2r$.

Hint: $\rho_c = \infty$, $\rho_f = -r$

(Ans.: $\delta = +r$, $Q = -2$, $l = 3r$, $l_* = -1.5r$, $r_1 = -1.414r$, $a = -4.2426r$, $b = -7.2r$, $\varepsilon = 19.4712°$ ($\tan^2 \varepsilon = 0.125$).)

4.9 Derive the expressions of m', l', l'_* in Table 4.1.

4.10 Derive the expressions of ρ'_f, ρ'_c in Table 4.1.

5
Balance of number of coordinates in synthesis problems

5.1 Kraus's concept of valence

(Kraus, 1952) suggested a concept of valence which reminds us of the balance of valences in a chemical equation before and after the reaction. The valence that we are talking about here is in fact the number of coordinates of a certain geometrical feature. To make it easier to understand, we use from now on the *number of*

Table 5.1 Number of coordinates required for specifying various geometrical features

Specification	Number of coordinates required
Length (distance between two joints or between two points, radius, etc.)	1
Position angle of a straight line	1
Coordination of two angles	1
A fixed point (or joint)	2
Direction of a sliding pair	1
Coordination of a pair of angular displacement and linear displacement	1
Coordination of two angular velocities (a velocity ratio)	1
First position of a coupler point	2
Each consecutive position of a coupler point	1
Tangent (or normal) at a point on the coupler curve	1
Coordination of a coupler point diaplacement with an angular diaplacement of a crank	1
Coordination of a coupler point velocity with an angular velocity (scalar quantity)	1
Pole (finitely separated coupler positions)	2
Instantaneous pole (infinitesimally separated coupler positions)	2
A point on a certain line or curve	1
A three-point contacting circle of curvature (including diameter and position) at a certain point on a coupler curve	2
A four-point contacting circle of curvature (including diameter and position) at a certain point on a coupler curve	3

coordinates in place of *valence*. Table 5.1 shows the number of coordinates required to specify each certain geometrical feature. For example the number of coordinates required to specify a point is two, regardless of which coordinate system we use -- whether it is rectangular (x, y) or polar (r, θ). The same principle applies to other required specifications.

As an example, Fig. 5.1 shows a four-bar linkage with four joints. In other words, the original number of available coordinates is 4×2, or 8. If A_0, B_0 are prescribed, the remaining number of available coordinates for determining the two points A, B is only $8 - 4 = 4$. If we wish to coordinate an angle displacement pair $\Delta\phi : \Delta\psi$ (This symbol represents that, while the crank A_0A rotates through an angle $\Delta\phi$, the crank B_0B should rotate through an angle $\Delta\psi$.), we need one more coordinate, the remaining number of available coordinates being $4 - 1 = 3$. There are therefore ∞^3 solutions.

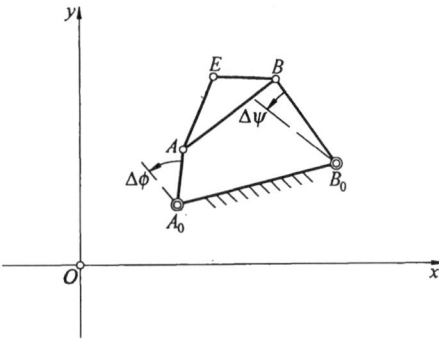

Fig. 5.1. Number of coordinates needed is 10 for five prescribed points A_0, A, B, B_0, E.

Another example is a four-bar linkage A_0ABB_0 together with a coupler point E. The total number of available coordinates of these five points is $5 \times 2 = 10$. If the locations of A_0, B_0 and the initial position E_1 of E are prescribed, these occupy $3 \times 2 = 6$ coordinates, and $10 - 6 = 4$. Hence the number of four-bar linkages which can be synthesized with its coupler point E to meet the specifications is ∞^4, or there are ∞^4 solutions. This result can also be conceived from another point of view: since A_0, B_0, E_1 are prescribed, there are 4 selections in choosing the remaining two points A, B. A further example is to prescribe the four points A_0, B_0, E_1, E_2. To prescribe E_2 needs only one coordinate. The number of possible four-bar linkages with a generating coupler point E is ∞^3. In other words, the four-bar linkage of any one solution is with A_0, B_0 as rotation centres, and is capable of having its coupler point E passing through the two points E_1, E_2. In Exercise 5.1, we pose this question to the reader: why prescribing the first coupler point position E_1 we need two coordinates, while prescribing any one of the subsequent positions $E_2, E_3, \ldots$, we only need one. Bearing this in mind, if we wish to synthesize a four-bar linkage with a coupler point E, the

Balance of number of coordinates in synthesis problems 219

number of available coordinates is 10. If all four points A_0, A, B, B_0 are not prescribed, theoretically *nine* coupler point positions $E_1, E_2, ..., E_9$ can be prescribed.

The concept of number of coordinates will become clearer with the numerical examples in Section 5.3.

5.2 Geometrical constructions for finding a certain unknown joint

In later synthesis problems, we shall frequently have to find certain unknown joints by geometrical constructions. For conciseness and to avoid repetition, we

Table 5.2 Geometrical construction for determining an unknown joint

Kind of construction	Given	To determine	as centre of arc passing through the following three points	
1	A_1E_1, A_2E_2, A_3E_3; B_1	B_0	B_1, B_2, B_3	
2	E_1, E_2, E_3; A_0 and angular displacements ϕ_{12}, ϕ_{13} of crank A_0A_1	A_1	E_1, E_2^1, E_3^1; where $\sphericalangle E_2 A_0 E_2^1 = -\phi_{12}$ $\sphericalangle E_3 A_0 E_3^1 = -\phi_{13}$	
3	A_1E_1, A_2E_2, A_3E_3; B_0	B_1	B_0, B_{02}^1, B_{03}^1; where $\Delta A_2 E_2 B_0 = \Delta A_1 E_1 B_{02}^1$ $\Delta A_3 E_3 B_0 = \Delta A_1 E_1 B_{03}^1$	

group these constructions into three kinds as listed in Table 5.2 and describe them briefly as follows.

The 1st kind of construction requires little explanation. B_0 is the centre of the circle passing through the three points B_1, B_2, B_3.

In the 2nd kind of construction, the points A_0, E_1, E_2, E_3 and two crank rotation angles ϕ_{12}, ϕ_{13} are given. Furthermore, it is specified that, when the crank A_0A rotates through the angle ϕ_{12}, point E is to move from E_1 to E_2 (denoted hereafter as "coordination of $\phi_{12} : E_1 \rightarrow E_2$"); and when A_0A rotates through the angle ϕ_{13}, E is to move from E_1 to E_3 (coordination of $\phi_{13} : E_1 \rightarrow E_3$). Please note that the length of $\overline{A_0A}$ is not known yet, and the positions of A_1, A_2, A_3 are also unknown. The construction is carried out by taking E_2 as a point belonging to A_0A_2 and then rotating A_0A_2 through an angle $-\phi_{12}$ back to A_0A_1. In so doing E_2 is brought to E_2^1. Similarly by taking E_3 as a point belonging to A_0A_3 and then rotating A_0A_3 through an angle $-\phi_{13}$ back to A_0A_1; E_3 is brought to E_3^1. What this construction means is that, although A_0A_1 is unknown yet, the motion of E relative to A_0A_1 can be considered as if it were moving from E_1 through E_2^1 to E_3^1. With the three points E_1, E_2^1, E_3^1 known, the centre of the circle passing through these three points is the point A_1. In fact this inversion concept has already been used in Fig. 3.43.

In the 3rd construction, A_1E_1, A_2E_2, A_3E_3 and B_0 are given. By taking B_0 as a point belonging to A_2E_2 and moving A_2E_2 back to A_1E_1, B_0 is then moved to B_{02}^1. Next by taking B_0 as a point belonging to A_3E_3 and moving A_3E_3 back to A_1E_1, B_0 is then moved to B_{03}^1. The centre of the circle passing through B_0, B_{02}^1, B_{03}^1 is the point B_1. This inversion concept was mentioned in Section 3.4.4. However, the point A_1 was found there from a given A_0 by algebraic method without employing this inversion concept.

5.3 Examples of balancing the number of coordinates

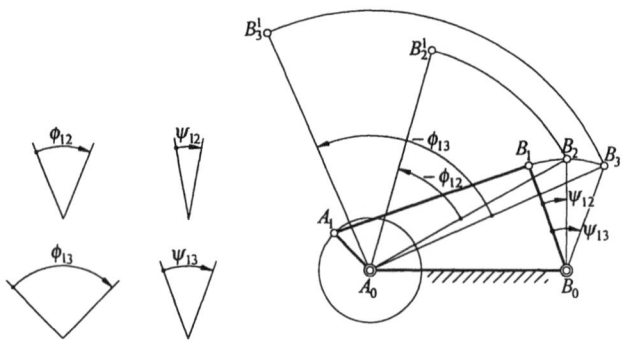

Fig. 5.2.   Example: synthesis of four-bar A_0ABB_0 for prescribed A_0, B_0, and coordinations of angular displacements $\phi_{12}:\psi_{12}$; $\phi_{13}:\psi_{13}$.

Balance of number of coordinates in synthesis problems

In this Section, we shall present a number of numerical examples taken from (Kraus, 1952) to illustrate the concept of balancing the number of coordinates.

Example: In Fig. 5.2, a four-bar linkage A_0ABB_0 is to be synthesized. The conditions are that A_0, B_0 are predetermined, and two angle-pairs $\phi_{12}:\psi_{12}$, $\phi_{13}:\psi_{13}$ are to be coordinated. These are tabulated as follows:

Number of available coordinates of $(A_0,A,B,B_0) = 4 \times 2 = 8$	
Specifications	Number of coordinates needed
prescribed A_0	2
prescribed B_0	2
coordination of $\phi_{12}:\psi_{12}$	1
coordination of $\phi_{13}:\psi_{13}$	1
Total number of coordinates needed	6
Remaining number of available coordinates	$8-6=2$
Possible number of solutions	∞^2

If the coordinates of B_1 are now assigned, the last two available coordinates are used up, and the problem should be solvable.

Given B_1, we determine the two points B_2, B_3 by ψ_{12}, ψ_{13}. According to the 2nd kind of construction shown in Table 5.2, we revolve B_2 about A_0 through angle $-\phi_{12}$ to get B_2^1, and revolve B_3 about A_0 through $-\phi_{13}$ to get B_3^1. Please note that, since ϕ_{12}, ϕ_{13} are both clockwise, $-\phi_{12}$, $-\phi_{13}$ are both counterclockwise. The centre of the circle passing through B_1, B_2^1, B_3^1 is A_1.

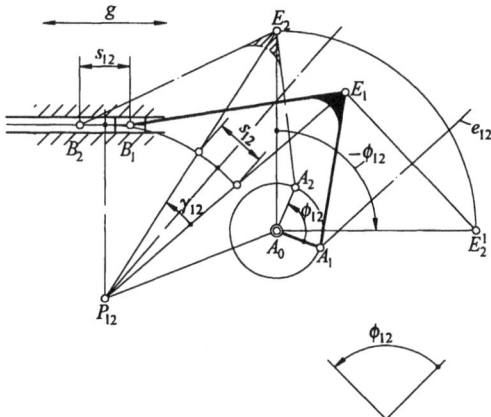

Fig. 5.3. Example: synthesis of linkage A_0ABE for prescribed E_1, E_2, and cooordinations of $\phi_{12}:\widehat{E_1E_2}$ and $\phi_{12}:s_{12}$ for given direction g of s.

The sequence of synthesis of the present example is: A_0, $B_0 \to B_1 \to B_2^1$, $B_3^1 \to A_1$.

A comprehensive discussion on coordinations of angle-pairs will be given in Chapter 7.

Example: In Fig. 5.3, a slider-crank mechanism is to be so synthesized in such a way that, as the crank A_0A rotates through an angle ϕ_{12}, the slider (point B) should travel a distance s_{12} in a certain direction g, and a coupler point E should move at the same time from E_1 to E_2.

To solve this problem, let us examine first the number of available coordinates. Comparing an ordinary four-bar linkage $A_0 ABB_0$ with a slider-crank mechanism $A_0ABB_{0\infty}$, we see the number of available coordinates of the former is 8, while that of the latter is only 7. This is because $B_0 \to \infty$, or because the length of $\overline{B_0B}$ is already prescribed as ∞, the number of available coordinates of B_0 is only 1, being the direction in which B_0 goes to infinity. Alternatively we may say that the sliding direction of the joint B is an available coordinate. Taking also into account the two coordinates of the coupler point E, we have the total number of available coordinates equal to 9. These are tabulated as follows:

number of available coordinates of (A_0, A, B, E) and direction of sliding of joint B	$=$ 9
Specifications number of coordinates needed	
prescribed E_1	2
prescribed E_2	1
prescribed sliding direction (g) of B	1
coordination of ϕ_{12} : $s_{12} (= \overline{B_1B_2})$	1
coordination of ϕ_{12} : $E_1 \to E_2$	1
Total number of coordinates needed	6
Remaining number of available coordinates	$9 - 6 = 3$
Number of possible solutions ∞^3	

The procedure of solution is: Choose any point on the plane as A_0, using up two coordinates. Then find E_2^1 according to the 2nd construction in Table 5.2. Draw the perpendicular bisector e_{12} of $\overline{E_1E_2^1}$, and choose a point on e_{12} as A_1, using up the last available coordinate. The problem should have a unique solution.

Knowing the two positions A_1E_1, A_2E_2 of the coupler, we can determine the pole P_{12} and the rotation angle γ_{12}. Now lay off the length s_{12} in the angle $\sphericalangle E_1P_{12}E_2$, and obtain the perpendicular distance between s_{12} and P_{12}. Take P_{12} as the centre, and rotate s_{12} to the desired direction g. We thus determine the points B_1, B_2 and solve the problem.

The sequence of synthesis of this example is: $A_0 \to E_2^1 \to e_{12} \to A_1 \to P_{12}$, γ_{12}

Balance of number of coordinates in synthesis problems

→ B_1, B_2.

In this occasion we would like to pay some attention to the geometrical features to be prescribed. As shown in the present example, a coordination of $\phi_{12} : E_1 \to E_2$ was prescribed. It could also be an assignment of the locations of E_1, E_2, without ϕ_{12}. However, if s_{12} is prescribed, a coordination of $s_{12} : \phi_{12}$ or $s_{12} : E_1 \to E_2$ must be prescribed at the same time. Otherwise an assignment of s_{12} alone is meaningless. In the previous example, a prescribed coordination of $\phi_{12} : \psi_{12}$ is meaningful, while a prescribed ψ_{12} alone is meaningless. Similarly, it is meaningful to prescribe an angular velocity ratio, but meaningless to prescribe only an angular velocity.

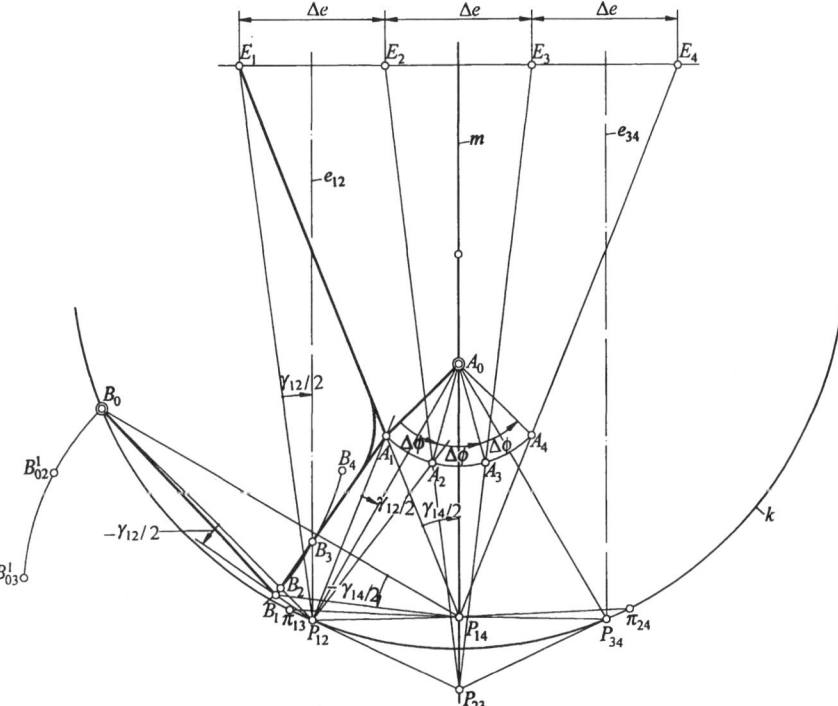

Fig. 5.4. Example: synthesis of A_0ABB_0E for prescribed E_1, E_2, E_3, E_4 and coordinations of $\phi_{12} : \widehat{E_1E_2} \, ; \phi_{23} : \widehat{E_2E_3} \, ; \phi_{34} : \widehat{E_3E_4}$.

Example: Fig. 5.4 shows a path generating mechanism. The path generated by the coupler point E is an approximate straight line. The mechanism to be synthesized is the four-bar linkage A_0ABB_0, with a coupler point E. The number of available coordinates is 10. The specifications are listed as shown:

224   *Kinematics and Design of Planar Mechanisms*

Number of available coordinates $(A_0, A, B, B_0, E) = 5 \times 2 = 10$	
Specifications	Number of coordinates needed
prescribed E_1	2
prescribed E_2	1
prescribed E_3	1
prescribed E_4	1
coordination of $\phi_{12} : E_1 \rightarrow E_2$	1
coordination of $\phi_{23} : E_2 \rightarrow E_3$	1
coordination of $\phi_{34} : E_3 \rightarrow E_4$	1
Total number of coordinates needed	8
Remaining number of available coordinates	$10 - 8 = 2$
Number of possible solutions ∞^2	

In this problem, the four points E_1, E_2, E_3, E_4, in the absence of particular reason, may not be on a straight line. The intervals $\overline{E_1E_2}, \overline{E_2E_3}, \overline{E_3E_4}$ may not be equal, and the angular intervals ϕ_{12}, ϕ_{23}, ϕ_{34} may also not be equal. However, for simplicity, we assume $\overline{E_1E_2} = \overline{E_2E_3} = \overline{E_3E_4} = \Delta e$, and $\phi_{12} = \phi_{23} = \phi_{34} = \Delta \phi$.

Draw the respective perpendicular bisectors e_{12}, m, e_{34} of $\overline{E_1E_2}, \overline{E_2E_3}, \overline{E_3E_4}$. For the coupler, its pole P_{12} must lie on e_{12}, and its pole P_{34} must lie on e_{34}. Its P_{14} and P_{23} are on the line m. We can choose any point on m as A_0, so that one coordinate is used up. Taking line m as the line of symmetry, we can build three angles $\Delta\phi$ at A_0. Although we do not yet know the points A_1, A_2, we have already determined the angle $\sphericalangle A_1A_0A_2 = \Delta\phi$ and its bisector. The intersection of this bisector and e_{12} is P_{12}. Similarly the bisector of the angle $\sphericalangle A_3A_0A_4$ and e_{34} is P_{34}. With the pole P_{12} known, the angle $\sphericalangle E_1P_{12}E_2$ is the angle of rotation γ_{12} of the coupler, and the half of this angle is $\gamma_{12}/2$. Dispose this angle $\gamma_{12}/2$ on both sides of the line $P_{12}A_0$ at P_{12}, and intersect the two lines of $\Delta\phi$ at A_0 to get the two points A_1, A_2. Similarly, we obtain A_3, A_4 by drawing two lines from P_{34}. Hence we are able to determine the length of crank $\overline{A_0A}$. The lines joining A_1E_1, A_4E_4 intersect in P_{14} which lies on the line m. Similarly the lines joining A_2E_2 and A_3E_3 intersect in P_{23}, also lying on the line m. Now the disposition of the four poles P_{12}, P_{14}, P_{23}, P_{34} is just like that shown in Fig. 3.48. Therefore the centre-point curve breaks up into the line m and a circle k. The circle k passes through the four points π_{13}, P_{12}, P_{34}, π_{24}, its centre being on the line m.

Choosing a point on the circle k as B_0, we use up the last available coordinate. With this B_0, we can find the corresponding B_1 according to the method shown in Fig. 3.43(c), by making use of P_{12}, P_{14}. Join $P_{12}B_0$, and build the angle $-\gamma_{12}/2$ on one side of it. Next join $P_{14}B_0$, and build the angle $-\gamma_{14}/2$ on one side of it. These two lines intersect in B_1. Alternatively we can locate the point B_1 by following the 3rd

Balance of number of coordinates in synthesis problems

construction in Table 5.2.

The sequence of synthesis of the present example is: e_{12}, m, e_{34} → A_0 → P_{12},P_{34} → A_1, A_2, A_3, A_4 → P_{14}, P_{23} → m, k → B_0 → B_1.

5.4 The method of point-position reduction

This is a method usually applied to synthesize path generating mechanisms. It appeared in the literature first in (Kraus, 1935). The term *point-position reduction* can be explained by means of the 3rd kind of construction in Table 5.2. Suppose the four points $B_0, B_{02}^1, B_{03}^1, B_{04}^1$ are already found. In general case, these four points are not on a circle, and the point B_1 cannot be found. However, if two of these four points coincide, resulting in three points, there must be a circle passing through these three points, and the point B_1 is obtainable. This is the so-called *point-position reduction*. In order to accomplish this, it is only necessary to let the pole of two

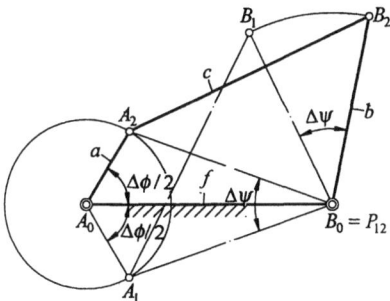

Fig. 5.5. Method of point-position reduction, P_{12} coinciding with B_0.

finitely separated positions coincide with a certain joint. As an example, Fig. 5.5 shows two positions A_1, A_2 of the crank pin A of a crank-rocker, symmetrical with respect to the line of centres A_0B_0. Hence, $\Delta A_1B_0B_1 = \Delta A_2B_0B_2$. A_1B_0 and A_2B_0 are also symmetrical with respect to A_0B_0. ∢$A_1B_0A_2$ = ∢$B_1B_0B_2$ = $\Delta\psi$, and ∢$A_0B_0A_1$ = ∢$A_0B_0A_2$ = $\Delta\psi/2$. For the two positons A_1B_1, A_2B_2 of the coupler AB, its pole P_{12} coincides with the joint B_0.

We can readily illustrate the method of point-position reduction by applying it to the following examples of synthesis.

Example: In Fig. 5.6, it is required to synthesize a path generating mechanism so that its coupler point E passes through the four prescribed points E_1, E_2, E_3, E_4, and while the coupler point E moves from E_1 to E_4, the crank should rotate through an angle ϕ_{14}. We list as before the specifications in a table:

Number of available coordinates $(A_0, A, B, B_0, E) = 5 \times 2 = 10$	
Specifications	Number of coordinates needed
prescribed E_1	2
prescribed E_2, E_3, E_4	3
coordination of $\phi_{14}: E_1 \to E_4$	1
Total number of coordinates needed	6
Remaining number of available coordinates	$10 - 6 = 4$
Number of possible solutions ∞^4	

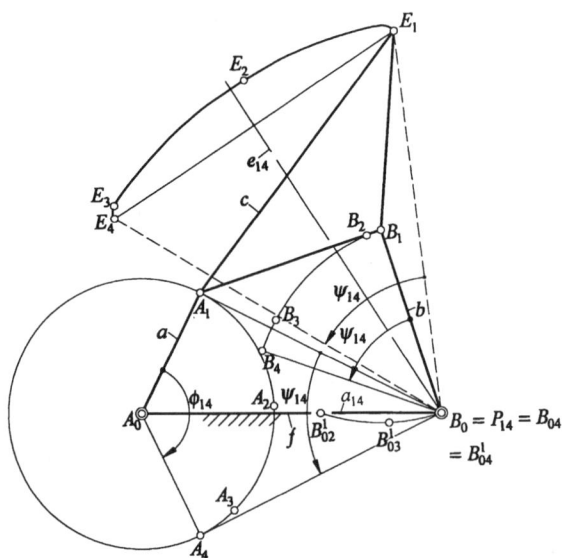

Fig. 5.6. Example: synthesis of A_0ABB_0E for prescribed E_1, E_2, E_3, E_4 and coordination of $\phi_{14} : \widehat{E_1E_4}$.

The application of method of point-position reduction changes the number of possible solutions to ∞^3. The synthesis procedure is as follows: Draw the perpendicular bisector e_{14} of $\overline{E_1E_4}$. We choose a point on e_{14} as B_0, using up one coordinate. And $\sphericalangle E_1B_0E_4 = \psi_{14}$. Choose any length $\overline{AE}$ as radius (using up the second coordinate), and draw circular arcs with E_1, E_4 as centres. Finally choose a length $\overline{AB_0}$ as radius (using up the third coordinate) and draw a circular arc with B_0 as centre, cutting the two former arcs in A_1, A_4. Next draw the bisector a_{14} of the angle $\sphericalangle A_1B_0A_4$; A_0 should lie on a_{14}. However, since we already know ϕ_{14}, we can

Balance of number of coordinates in synthesis problems

then determine A_0 and also determine the crank circle of the joint A. Finally, knowing the length $\overline{AE}$, we can therefore find A_2, A_3.

The last step is to find B_1. According to the 3rd kind of construction in Table 5.2, it can be seen that among the four points $B_0, B_{02}^1, B_{03}^1, B_{04}^1$, the two points B_{04}^1 and B_0 belong both to the first and fourth positions of AE, or it is the pole P_{14}. These four points become three points, the centre of the circle passing through which is the point B_1.

The sequence of synthesis of this example is: $e_{14} \to B_0 \to AE \to AB_0 \to A_1, A_4$, $\to a_{14} \to A_0 \to A_2, A_3 \to B_{02}^1, B_{03}^1 \to B_1$.

The underlying principle of applying the method of point-position reduction is that, as in the last example, in order to gain a simple construction of a circle passing through three points, we have to put up with a decrease of number of solutions from ∞^4 to ∞^3. Nevertheless, ∞^3 is still quite large. Furthermore, although ∞^4 or ∞^3 solutions are numerous, it is quite possible that none of them is satisfactory. Under such circumstances. we have to reassign the prescribed items in order to get new solutions. The main advantage of using the method of point-position reduction is that both centering-point curve and circling-point curve are omitted.

(Hain, 1944) gave very interesting examples in the application of method of point-position reduction. In the following example, we apply point-position reduction twice in synthesizing a backspace mechanism for an old type typewriter.

Example: (taken from Hain, 1944) Fig. 5.7(a) shows a backspace mechanism of a typewriter. As the typist presses down the backspace key on the keyboard, the driving member shown on the lower-left corner in the figure will push upwards, to rotate the crank A_0A in a clockwise sense. The point A travels through A_1, A_2, A_3 to A_4,

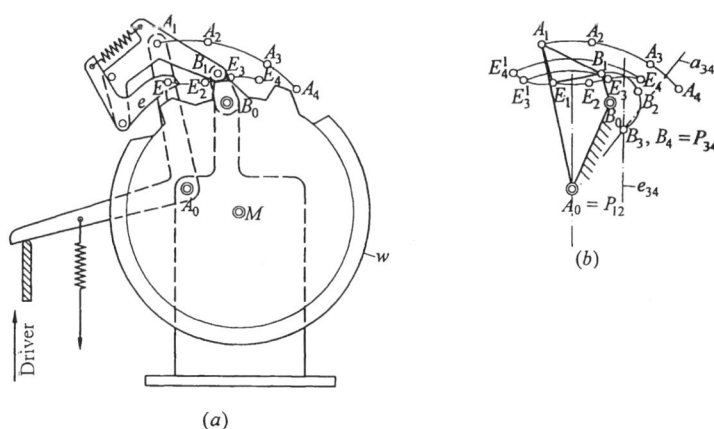

Fig. 5.7. Synthesis of a backspace mechanism of an old typewriter. (a) Prescribed E_1, E_2, E_3, E_4 and coordinations of $\phi_{12} : \widehat{E_1E_2}$; $\phi_{13} : \widehat{E_1E_3}$; $\phi_{14} : \widehat{E_1E_4}$. (b) Solving by method of point-position reduction.

bringing with it a pawl e. A point E on the pawl e engages a tooth of a ratchet wheel w, and pushes the ratchet wheel to rotate also in a clockwise direction. The path of the point E passes through four prescribed positions E_1, E_2, E_3, E_4. When E reaches E_4, the carriage of the typewriter will be brought back one space. The three points E_2, E_3, E_4 lie on a circle with the centre point M of the ratchet w as centre. The point E_1 lies outside this circle, so that as the wheel w rotates, its teeth will keep clear of the point E_1 of the pawl e. Assume the three arcs $\widehat{E_1E_2}$, $\widehat{E_2E_3}$, $\widehat{E_3E_4}$ are of equal length. As the point E travels through $\widehat{E_2E_3}$, the carriage starts to move backwards. In order to overcome the friction resistance within the machine, the arc length $\widehat{A_2A_3}$ through which A travels is made larger than $\widehat{A_3A_4}$, and is of course also larger than $\widehat{A_1A_2}$. Thus the respective corresponding crank angles ϕ_{12}, ϕ_{13}, ϕ_{14} of the crank A_0A are all prescribed. These are listed as before as follows:

Number of available coordinates (A_0, A, B, B_0, E) = 5×2=10	
Speafications	Number of coordinates needed
prescribed E_1	2
prescribed E_2, E_3, E_4	3
coordination of ϕ_{12}:$E_1 \to E_2$	1
coordination of ϕ_{13}:$E_1 \to E_3$	1
coordination of ϕ_{14}:$E_1 \to E_4$	1
Total number of coordinates needed	8
Number of remaining available coordinates	10 − 8=2
Number of possible solutions ∞^2	

The procedure of solution is as follows (please refer to Fig. 5.7(b)): Draw the perpendicular bisector e_{12} of $\overline{E_1E_2}$, and locate the point A_0 on e_{12} by building the angle $\sphericalangle E_1A_0E_2 = \phi_{12}$, then $A_0 = P_{12}$. Next, according to the 2nd kind of construction in Table 5.2, revolve the points E_2, E_3, E_4 respectively through angles $-\phi_{12}$, $-\phi_{13}$, $-\phi_{14}$ to the positions E_2', E_3', E_4' (ϕ_{12}, ϕ_{13}, ϕ_{14} are clockwise, therefore $-\phi_{12}$, $-\phi_{13}$, $-\phi_{14}$ are all counterclockwise). It can be seen that, because of $A_0 = P_{12}$, $E_2' = E_1$. The centre of the circle passing through the three points E_1, E_3', E_4' is A_1. We have applied once point-position reduction. Having got A_1, we can then determine A_2, A_3, A_4 by ϕ_{12}, ϕ_{13}, ϕ_{14}.

Having determined A_1E_1, A_2E_2, A_3E_3, A_4E_4, we can choose a certain point to carry out another point-position reduction. Take for instance the pole P_{34} as B_3 (= B_4). According to the 1st kind of construction in Table 5.2, we can find first the points B_1, B_2, and the centre of the circle passing through the three points B_1, B_2, B_3(= B_4) is B_0.

Balance of number of coordinates in synthesis problems 229

Here, we have applied the point-position reduction the second time.

The synthesis procedure of the present example is: $e_{12} \to A_0 \to E_3^1, E_4^1 \to A_1 \to A_2, A_3, A_4 \to P_{34} \to B_1, B_2 \to B_0$.

In this example, by applying twice point-position reduction, we have reduced the number of possible solutions from ∞^2 to a finite number.

Please note that, as far as synthesizing a single four-bar linkage is concerned, the number of available corrdinates for the four joints A_0, A, B, B_0 is $4 \times 2 = 8$. If non of the joints is prescribed, the number of possible linkages would then be ∞^8. In fact, among all these four-bar linkages some are completely identical in bar lengths and configurations, and some are identical in bar lengths but different in configurations, and some are proportional in bar lengths. These should be considered as duplicated linkages.

Exercises

5.1 In designing a path generating mechanism, positions E_1, E_2, E_3, E_4 of the coupler point E are prescribed. Explain why in prescribing E_1 two coordinates are required, while in prescribing each of E_2, E_3, E_4 only one coordinate is required.

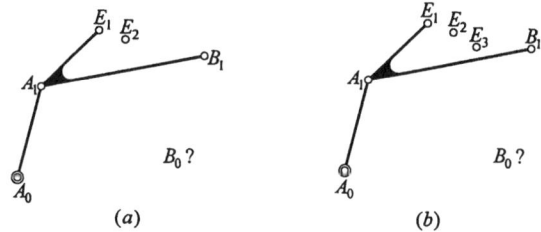

Hint: The reasoning can be grasped by considering the solutions of the problems in the Figures (a), (b).
(a) What is the number of solutions of finding B_0, if A_0, A_1, B_1, E_1, E_2 are prescribed? How can the problem be solved?
(b) What is the number of solutions of finding B_0, if $A_0, A_1, B_1, E_1, E_2, E_3$ are prescribed? How can the problem be solved?
(c) In each of the two problems (a), (b), if B_0 is prescribed, but not B_1, what is the number of solutions, and how can the problem be solved?

5.2 In Section 3.1, it has been mentioned that there are ∞^6 solutions in guiding a body through two finitely separated positions. According to Fig. 3.1, prescribing two finitely separated positions of a body is equivalent to prescribing: $P_{12}, \gamma_{12}/2$. Explain the 6 choices.

5.3 In the example shown in Fig. 5.7, the tabulation has shown that the number of solutions is ∞^2. In this case the location of A_0 has to be on a curve k_{A0}. (Please note that point A_0 selected in Fig. 5.7(b) is such that $\angle E_1 A_0 E_2 = \phi_{12}$, and that this A_0 is only one point on k_{A0}.) What are the two choices?

5.4 In Fig. 5.7(b), suppose the 9 points A_0, A_1, A_2, A_3, A_4; E_1, E_2, E_3, E_4 are prescribed. What is the final choice in finding B_0 and B_1?

5.5 In the example given in Section 5.4 (Fig. 5.6), there were ∞^4 solutions if the method of point position reduction was not used. Explain what these four choices are.

6
Dimensional systhesis of linkages --- path generation problems

In Chapter 5 we gave some examples of synthesis of path generating mechanisms. In the present chapter, we shall introduce more systematically the path generating problems. We shall start from a type of common problems where the mechanism to be synthesized should be able to move its coupler point say E, from E_1 to E_2, and at the same time to coordinate this movement of E with the rotation angles of the rotating links.

6.1 Coordination of two coupler point positions with one crank angle

In this problem one of the rotation centres A_0 of a crank, and two coupler point positions E_1, E_2 are prescribed. It is required to coordinate the rotation angle ϕ_{12} with the movement $E_1 \to E_2$. Because the linkage to be synthesized is not yet a complete mechanism, we are not going to apply the balancing of number of coordinates. Fig. 6.1 shows the solution of this problem. In fact this is a special case of the 2nd kind of construction in Table 5.2. We can revolve as before the point E_2 about A_0 through an angle $-\phi_{12}$ to E_2^1. The perpendicular bisector a_{12}^1 of $\overline{E_1 E_2^1}$ is the locus of the point A_1. Similarly we can revolve E_1 about A_0 through an angle $+\phi_{12}$ to E_1^2. The perpendicular bisector a_{21}^2 of $\overline{E_2 E_1^2}$ is the locus of A_2. If we choose now a certain length a of the crank $\overline{A_0 A}$ as radius and draw an arc with A_0 as centre, the

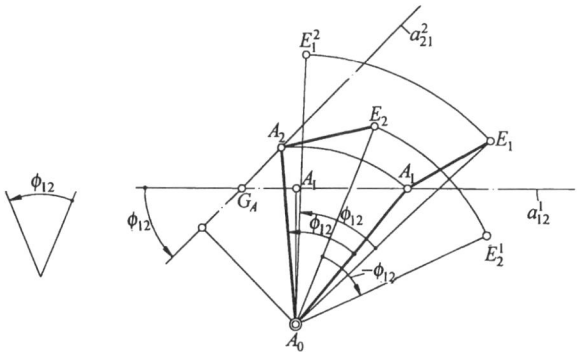

Fig. 6.1. Coordination of ϕ_{12} : $E_1 \to E_2$.

intersections of this arc with a_{12}^1, and a_{21}^2 are respectively A_1 and A_2. Furthermore, a_{12}^1 and a_{21}^2 intersect in the point G_A. If $A_0 A_1 E_1 E_2^1$ is rotated as a rigid body together with the line a_{12}^1 about A_0 through an angle ϕ_{12}, they will reach to the positions $A_0 A_2 E_1^2 E_2$ and a_{21}^2. Hence the perpendicular distance from A_0 to a_{12}^1 is equal to that from A_0 to a_{21}^2. This perpendicular $\overline{A_0 A_I}$ is the lower limit of the length a.

The two links $A_0 A$ and AE as shown in Fig. 6.1, is in general called a *dyad*, as mentioned in Section 1.7.

We shall now deal with the selection of the point A_1 in Fig. 3.43. For a prescribed A_0, the requirement is to coordinate the rotation angle ϕ_{14} of the crank $A_0 A$ and the movement from E_1 to E_4 of the coupler point E. According to the present procedure, E_4 is revolved about A_0 through an angle $-\phi_{14}$ to E_4^1. The perpendicular bisector a_{14}^1 of $\overline{E_1 E_4^1}$ is the locus of A_1. This mechanism is an approximate straight line mechanism. The four points E_1, E_2, E_3, E_4 are exactly on a straight line, but the actual path of E is an approximately straight line.

The example shown in Fig. 5.4 is also an approximate straight-line mechanism. However, the present procedure is not required in that example.

6.2 Coordinations of two coupler point positions with one pair of crank angles

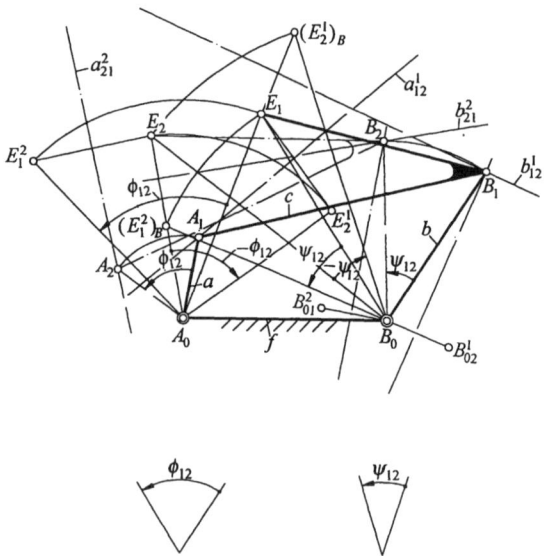

Fig. 6.2. Coordinations of $\phi_{12}: E_1 \rightarrow E_2$, $\psi_{12}: E_1 \rightarrow E_2$.

Dimensional systhesis of linkages --- path generation problems

In Fig. 6.2, the rotation centres A_0, B_0 and the coupler point positions E_1, E_2 are prescribed. It is required to coordinate $\phi_{12} : E_1 \rightarrow E_2$ and $\psi_{12} : E_1 \rightarrow E_2$. The latter coordination can also be considered as to coordinate $\phi_{12} : \psi_{12}$. We tabulate the coordinates as before as follows:

Number of available coordinates of (A_0, A, B, B_0, E) = 5×2=10	
Speifications	Number of coordinates needed
Prescribed A_0	2
Prescribed B_0	2
Prescribed E_1	2
Prescribed E_2	1
Coordination of $\phi_{12} : E_1 \rightarrow E_2$	1
Coordination of $\psi_{12} : E_1 \rightarrow E_2$	1
Total number of coordinates needed	9
Remaining number of available coordinates	10 – 9 = 1
Number of possible solutions	∞^1

The solution is carried out by applying the technique of Fig. 6.1 to each of A_0AE and B_0BE. Revolve first E_1 about A_0 through an angle $+\phi_{12}$ to E_1^2. Draw the perpendicular bisector a_{21}^2 of $\overline{E_2E_1^2}$. Choose a point A_2 on a_{21}^2, thus using up the last coordinate. Having determined A_2, we can determine A_1 correspondingly, thus determining both positions A_1E_1, A_2E_2.

Similarly revolve E_1 about B_0 through an angle $+\psi_{12}$ to $(E_1^2)_B$. Draw the perpendicular bisector b_{21}^2 of $\overline{E_2(E_1^2)_B}$. Then b_{21}^2 is the locus of B_2. Applying the 3rd kind of construction in Table 5.2, we can find B_{01}^2. Draw the perpendicular bisector of $\overline{B_0B_{01}^2}$ to intersect b_{21}^2 in B_2. Alternatively, we can revolve E_2 about B_0 through an angle $-\psi_{12}$ to $(E_2^1)_B$. Draw the perpendicular bisector b_{12}^1 of $\overline{E_1(E_2^1)_B}$, then b_{12}^1 is the locus of B_1. Again applying the 3rd kind of construction in Table 5.2, we can find B_{02}^1. Draw the perpendicular bisector of $\overline{B_0B_{02}^1}$. Its intersection with b_{12}^1 is B_1.

More about coordination of crank angle pairs will be discussed in Chapter 7.

Example: Fig. 6.3 shows a practical example adopted in (Hain, 1942). This is a six-bar linkage. The three rotation centres A_0, B_0, F_0 are all prescribed. The requirements are that as A_0A rotates through an angle ϕ_{12} = 180°, the link B_0B should rotate through ψ_{12}= 110.7°, and at the same time to coordinate a certain coupler point travel $E_1 \rightarrow E_2$. The problem is: if only one four-bar linkage were synthesized to accomplish the task, the transmission angle (please refer to Section 7.5) would be quite poor. The solution is: Draw any line g passing through F_0, and choose any four

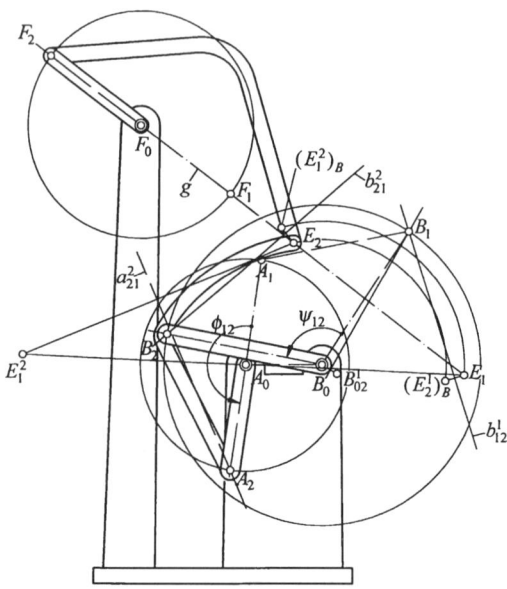

Fig. 6.3. Example: six-bar linkage, for prescribed A_0, B_0, F_0, and coordinations of ϕ_{12}: ψ_{12}: $E_1 \rightarrow E_2$.

points E_1, E_2, F_1, F_2 on g, so that $\overline{E_1E_2} = \overline{F_1F_2}$, where F_1, F_2 are two opposite points on the diameter of the crank circle of F. Now the problem is to coordinate $\phi_{12} : E_1 \rightarrow E_2$, and $\psi_{12} : E_1 \rightarrow E_2$. Following the procedure shown in Fig. 6.2 we can synthesize first the four-bar linkage A_0ABB_0. Next we can build the crank F_0F and coupler EF, so that $\overline{EF} = \overline{E_1F_1} = \overline{E_2F_2}$. If F_0F is taken as the driving crank, the angles $\phi_{12} = 180°$ and $\psi_{12} = 110.7°$ will be coordinated.

6.3 Algebraic method

In the following sections we shall describe the synthesis of path generating mechanisms using the matrix method mentioned in Chapter 3.

6.3.1 Coordinations of three coupler point positions with two crank rotation angles

This problem is in fact the 2nd kind of construction in Table 5.2. Assume A_0, E_1, E_2, E_3, ϕ_{12} and ϕ_{13} are prescribed. First, revolve the point E_2 about A_0 through an angle $-\phi_{12}$ to E_2^1, as indicated in the Table. Denote the coordinates of A_0 by $A_0(x_0, y_0)$,

Dimensional systhesis of linkages --- path generation problems

and those of E_2 and E_2^1 by $E_2(x_2, y_2)$, and $E_2^1(x_2^1, y_2^1)$ respectively. The latter can be written according to equation (3.53) as

$$\begin{bmatrix} x_2^1 \\ y_2^1 \\ 1 \end{bmatrix} = \begin{bmatrix} \cos\phi_{12} & \sin\phi_{12} & x_0(1-\cos\phi_{12}) - y_0 \sin\phi_{12} \\ -\sin\phi_{12} & \cos\phi_{12} & x_0\sin\phi_{12} + y_0(1-\cos\phi_{12}) \\ 0 & 0 & 1 \end{bmatrix} \begin{bmatrix} x_2 \\ y_2 \\ 1 \end{bmatrix} \quad (6.1)$$

Similarly, the coordinates (x_3^1, y_3^1) of E_3^1 can also be written according to equation (6.1).

Denote the coordinates of E_1 by (x_1, y_1). Knowing the three points $E_1(x_1, y_1)$, $E_2^1(x_2^1, y_2^1)$, $E_3^1(x_3^1, y_3^1)$, we can find the coordinates $A_1(x_{A1}, y_{A1})$ of the centre of the circle passing through the three points, using the procedure similar to that mentioned in Section 3.4.3. According to equation (3.46), we can write the equation of the perpendicular bisector a_{12}^1, or the locus of the point A_1, of $\overline{E_1 E_2^1}$ of the 2nd kind of construction in Table 5.2 as

$$a_{12}^1 : (x_1 - x_{A1})^2 + (y_1 - y_{A1})^2 = (x_2^1 - x_{A1})^2 + (y_2^1 - y_{A1})^2 \quad (6.2)$$

Similarly, we can write the equation of the perpendicular bisector a_{13}^1 of $\overline{E_1 E_3^1}$ as

$$a_{13}^1 : (x_1 - x_{A1})^2 + (y_1 - y_{A1})^2 = (x_3^1 - x_{A1})^2 + (y_3^1 - y_{A1})^2 \quad (6.3)$$

Substituting equation (6.1) into equation (6.2), or, substituting the expressions of x_2^1, y_2^1 in terms of x_2, y_2 and ϕ_{12} into equation (6.2) gives

$$a_{12}^1 : G_2' x_{A1} + S_2' y_{A1} + T_2' = 0 \quad (6.4)$$

Similarly, substituting the expressions of x_3^1, y_3^1 in terms of x_3, y_3 and ϕ_{13} into equation (6.3) gives

$$a_{13}^1 : G_3' x_{A1} + S_3' y_{A1} + T_3' = 0 \quad (6.5)$$

The coefficients in equations (6.4) and (6.5) are:

$$\left.\begin{aligned} G_j' &= -x_1 + x_j \cos\phi_{1j} + y_j \sin\phi_{1j} + (x_0 \text{vers}\phi_{1j} - y_0 \sin\phi_{1j}) \quad (j=2,3) \\ S_j' &= -y_1 - x_j \sin\phi_{1j} + y_j \cos\phi_{1j} + (x_0 \sin\phi_{1j} + y_0 \text{vers}\phi_{1j}) \\ T_j' &= x_j(x_0 \text{vers}\phi_{1j} + y_0 \sin\phi_{1j}) + y_j(-x_0 \sin\phi_{1j} + y_0 \text{vers}\phi_{1j}) \\ &\quad - (x_0^2 + y_0^2)\text{vers}\phi_{1j} + (x_1^2 + y_1^2 - x_j^2 - y_j^2)/2 \end{aligned}\right\} \quad (6.6)$$

Each equation of (6.4) and (6.5) is linear in the unknowns x_{A1}, y_{A1}, and all coefficients are constants which can be computed from equation (6.6). Solving these two equations simultaneously, we can find x_{A1}, y_{A1}.

6.3.2 Coordinations of four coupler point positions with three crank rotation angles --- the k_{A0}-curve

This problem is an extension of the previous one. Suppose E_1, E_2, E_3, E_4 and three crank rotation angles ϕ_{12}, ϕ_{13}, ϕ_{14} to be coordinated are prescribed. It is obvious that for an arbitrarily chosen A_0, the four points E_1, E_2^1, E_3^1, E_4^1 may not lie on a circle. The locus of A_0 that can cause these four points on a circle will be called k_{A0}-curve. Now, according to equations (6.4), (6.5) we can also write the equation of the perpendicular bisector a_{14}^1 of $\overline{E_1 E_4^1}$ as

$$a_{14}^1 : G_4' x_{A1} + S_4' y_{A1} + T_4' = 0 \qquad (6.7)$$

The coefficients G_4', S_4', T_4' in equation (6.7) can be obtained by setting $j = 4$ in equation (6.6). Equations (6.4), (6.5), (6.7) are three linear equations in the two unknowns x_{A1}, y_{A1}. In order that they should have nontrivial solution, the determinant of their coefficients must be equal to zero, or

$$k_{A0} : \begin{vmatrix} G_2' & S_2' & T_2' \\ G_3' & S_3' & T_3' \\ G_4' & S_4' & T_4' \end{vmatrix} = 0 \qquad (6.8)$$

The nine elements of equation (6.8) are as those listed in equation (6.6), in which all are known constants, except the coordinates (x_0, y_0) of A_0. This is the equation of the k_{A0}-curve. Because its fourth degree term vanishes identically, it is in fact a cubic equation in x_0, y_0. A comparison between equation (6.6) and the coefficients L_j, Q_j, H_j in equation (3.101) shows that these two equations are just similar, but not identical. Expanding equation (6.8) yields

$$k_{A0}: k_1 y_0^3 + k_2 x_0 y_0^2 + k_3 x_0^2 y_0 + k_4 x_0^3 + k_7 y_0^2 + k_8 x_0 y_0 + k_9 x_0^2 \\ + k_5 y_0 + k_6 x_0 + k_{10} = 0$$

However, as $k_1 = k_3$, $k_2 = k_4$, the above equation can also be written as

$$k_{A0}: (k_1 y_0 + k_2 x_0)(x_0^2 + y_0^2) + k_7 y_0^2 + k_8 x_0 y_0 + k_9 x_0^2 + k_5 y_0 + k_6 x_0 + k_{10} = 0 \qquad (6.9)$$

The coefficients $k_1, k_2, ..., k_{10}$ in equation (6.9) are due to those in G_j', S_j', T_j'. These are listed in Appendix 10. It can be seen from equation (6.9) that k_{A0} is a circular cubic.

The constant term k_{10} in equation (6.9) does not vanish identically. This is because at the time when E_1, E_2, E_3, E_4 and $\phi_{12}, \phi_{13}, \phi_{14}$ are prescribed, the k_{A0}-curve does not necessarily pass, with respect to that original coordinate system, through the origin. For this reason we choose another point O' as the new origin, such that this O' lies on the k_{A0}-curve. As shown in Fig. 6.4, we choose the point O' on the perpendicular bisector of $\overline{E_2 E_1}$ such that $\angle E_2 O' E_1 = -\phi_{12}$. In this way the point E_2^1 coincides with E_1, and the four points E_1, E_2^1, E_3^1, E_4^1 must lie on a circle. Hence

Dimensional systhesis of linkages --- path generation problems

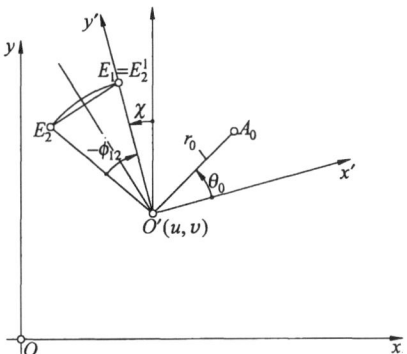

Fig. 6.4.   New coordinate system $x'O'y'$.

O' must be a point on the k_{A0}-curve. This is just the principle of point-position reduction explained in Section 4.4. We take now the point O' as the origin of the new coordinate system (x',y'), and the direction $O'E_1$ as the direction of the y'-axis. Let the coordinates of O' in the original (x, y) coordinate system be (u, v), and the inclination of the line $O'y'$ with respect to Oy be denoted by χ (counterclockwise sense being considered as positive). Then for a point whose coordinates were (x, y) in the original coordination system, we can calculate its coordinates (x', y') in the new coordinate system by

$$\begin{bmatrix} x' \\ y' \\ 1 \end{bmatrix} = \begin{bmatrix} \cos\chi & \sin\chi & -u\cos\chi - v\sin\chi \\ -\sin\chi & \cos\chi & u\sin\chi - v\cos\chi \\ 0 & 0 & 1 \end{bmatrix} \begin{bmatrix} x \\ y \\ 1 \end{bmatrix} \qquad (6.10)$$

After this transformation, we can calculate the coefficients of the equation of the k_{A0}-curve on the basis of the new coordinates of the points. In order to avoid confusion, we denote these new coefficients by $k'_1, k'_2, \ldots$. Now k'_{10} must vanish, hence the equation of k_{A0} is:

$$k_{A0}: (k'_1 y'_0 + k'_2 x'_0)(x'^2_0 + y'^2_0) + k'_7 y'^2_0 + k'_8 x'_0 y'_0 + k'_9 x'^2_0 + k'_5 y'_0 + k'_6 x'_0 = 0 \quad (6.11)$$

It can be seen from equation (6.11) that k_{A0} is a circular cubic. The rectangular coordinates (x'_0, y'_0) of a point A_0 on the k_{A0}-curve can be transformed into polar coordinates by substituting the equations

$$\left. \begin{array}{l} x'_0 = r_0 \cos\theta_0 \\ y'_0 = r_0 \sin\theta_0 \end{array} \right\}$$

or

$$k_{A0}: K_2 r_0^2 + K_1 r_0 + K_0 = 0 \qquad (6.12)$$

The expressions of K_2, K_1, K_0 in equation (6.12) are also listed in Appendix 10, these being functions of θ_0. For a given value of θ_0, substituting it into equation (6.12) will yield two A_0 points. In this way, we can plot out the whole k_{A0}-curve. Choose a point on k_{A0} as A_0, then A_1 is the centre of the circle passing through the four points E_1, E_2^1, E_3^1, E_4^1.

With the four positions A_1E_1, A_2E_2, A_3E_3, A_4E_4 known, we can find the

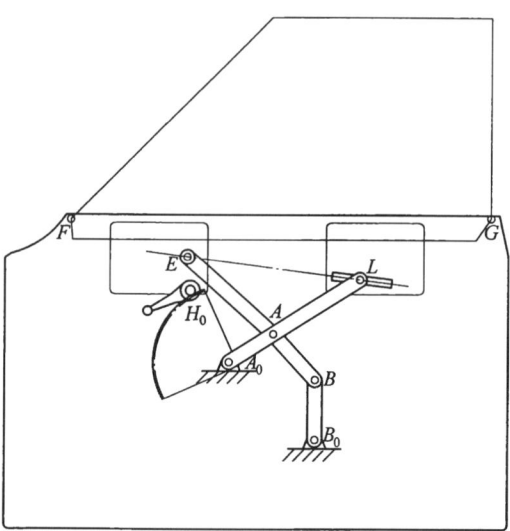

Fig. 6.5. An early manually-operated four-bar window regulator designed by Ford Motor Company.

circle-point curve according to the method mentioned in Section 3.6.6. Choosing a point on this circle-point curve as B_1, we can find B_0 as before.

Example: This example is taken from (Gustavson, 1967). This is a practical application of mechanism design. Fig. 6.5 shows an early manually-operated window regulator in a form of a four-bar linkage designed by Ford Motor Company. To facilitate identification, and to help the reader to grasp the design procedure, we denote the points in the figure by the usual symbols used in this book. In Fig. 6.5, A_0ABB_0 is a four-bar linkage, and E is a coupler point. L is a point on the crank A_0A. The window glass is to be supported on the two points E and L. The point E belongs to the window glass FG, while the elongated slot on which L is lying belongs also to the window glass, the centre line of the slot passing through E. Fig. 6.6 shows the four positions of the window glass through which the latter has to be guided. In other words, the four lines F_1G_1, F_2G_2, F_3G_3, F_4G_4 are prescribed. Hence the four lines E_1L_1, E_2L_2, E_3L_3, E_4L_4 are also predetermined, being denoted respectively by l_1, l_2, l_3, l_4, as shown in Fig. 6.7, while the four points L_1, L_2, L_3, L_4 are not known yet.

The point H_0 is the centre of the handle, which is also the centre of a pinion bound together with the handle. This pinion drives a sector gear that is bound

Dimensional systhesis of linkages --- path generation problems 239

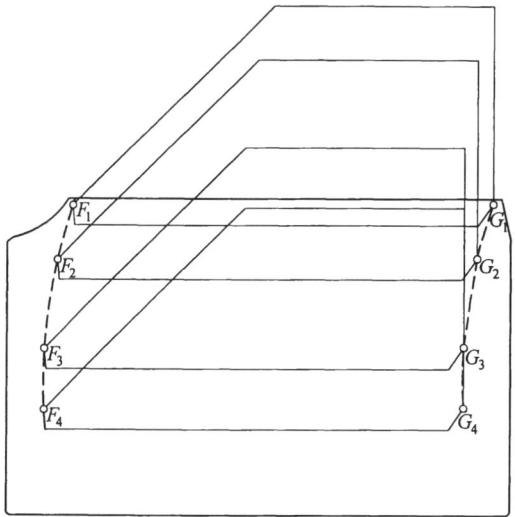

Fig. 6.6. Four window positions planned in Fig. 6.5.

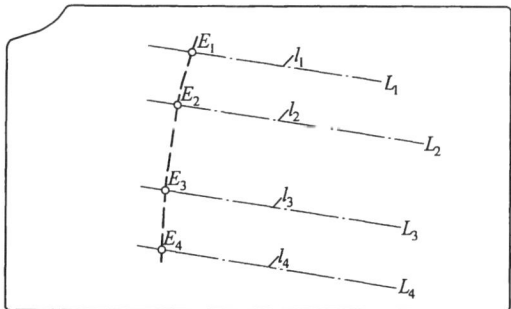

Fig. 6.7. The four positions of EL ($= l$) determined in Fig. 6.6.

together with the crank A_0A. Rotating the handle will drive the crank A_0A and hence the whole four-bar linkage. The locations of H_0 and A_0 are determined by design conditions, and the angle of rotation ϕ_{14} of the crank A_0A is also determined by the number of revolutions of the handle as well as the radius of the sector gear.

Please refer to Fig.6.8. Having determined the locations of A_0, l_1, l_4 and the angle ϕ_{14}, we can calculate the length $\overline{A_0L}$ according to the following equation:

$$\overline{A_0L} = \frac{[a^2 + b^2 + 2ab\cos(\phi_{14} + \beta)]^{1/2}}{\sin(\phi_{14} + \beta)}$$

where a, b are respectively the perpendicular distances from A_0 to l_1, l_4, and β is the angle between l_1 and l_4. In case l_1 and l_4 are parallel, then $\beta = 0$. Draw a circular arc with A_0 as centre, and the calculated length $\overline{A_0L}$ as radius, cutting the lines l_1, l_2, l_3, l_4 respectively in the points L_1, L_2, L_3, L_4, as shown in Fig. 6.9. We can then determine the angles ϕ_{12}, ϕ_{13}. Please note that the angles ϕ_{12}, ϕ_{13}, ϕ_{14} here are all clockwise. We revolve as shown in Fig. 6.1 the points E_2, E_3, E_4 about A_0 through angles $-\phi_{12}$, $-\phi_{13}$, $-\phi_{14}$ respectively to get E_2^1, E_3^1, E_4^1; and erect the respect perpendicular bisectors a_{12}, a_{13}, a_{14} (for simplicity, a_{12} is written here in place of a_{12}^1), as shown in Fig. 6.10(a). Since the point A_0 is not chosen on the k_{A0}-curve according to the procedure mentioned in Section 6.3.2, the three lines a_{12}, a_{13}, a_{14} do not intersect in one point. Fig. 6.10(b) shows an enlarged view of the intersection points of these three lines. The original author used an approximation method. We choose

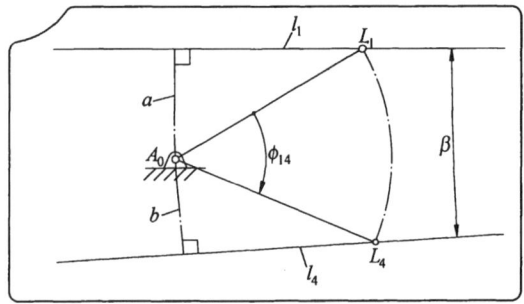

Fig. 6.8.   Determination of length $\overline{A_0L}$ by positions of A_0, l_1, l_4 and two angles β, ϕ_{14}.

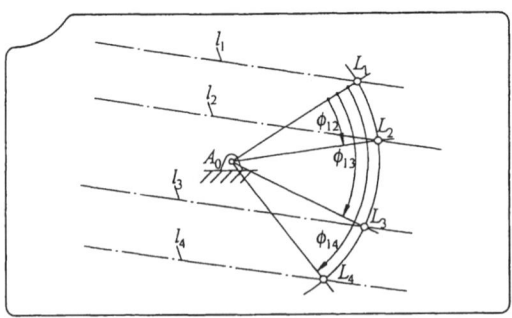

Fig. 6.9.   Determination of points L_1, L_2, L_3, L_4 by length $\overline{A_0L}$ and positions of l_1, l_2, l_3, l_4.

Dimensional systhesis of linkages --- path generation problems   241

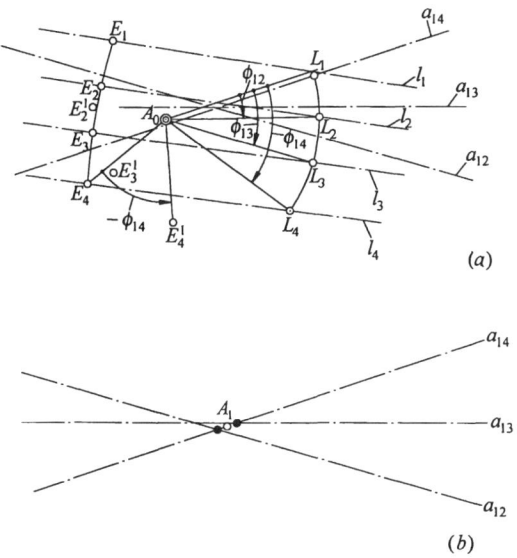

Fig. 6.10. (a) Erecting perpendicular bisectors a_{12}, a_{13}, a_{14} of $\overline{E_1E_2^1}$, $\overline{E_1E_3^1}$, $\overline{E_1E_4^1}$. (b) Determination of A_1 by the three lines a_{12}, a_{13}, a_{14}.

the point A_1 at the middle point (the white dot between two black dots in the figure) between the intersections of a_{14} with a_{12} and a_{13}. Since A_1 is on the line a_{14}, as the crank A_0A_1 of the dyad $A_0A_1E_1$ rotates through an angle ϕ_{14} to the position A_0A_4, the point E_1 will reach precisely the position E_4. However, because A_1 is not on the line a_{12} or a_{13}, as A_0A_1 rotates through an angle ϕ_{12} or ϕ_{13}, the point E can only reach the vicinity of E_2 or E_3, as shown in Fig. 6.11 (an exaggerated view).

Now, knowing the four positions A_1E_1, A_2E_2, A_3E_3, A_4E_4 of the coupler, we can

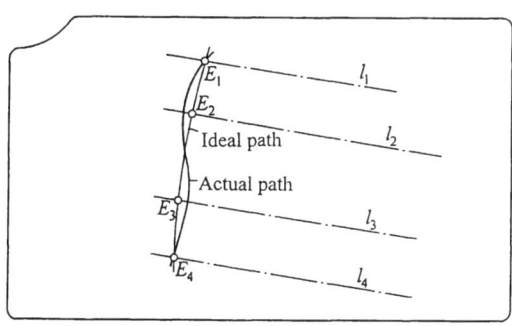

Fig. 6.11. Actual path (enlarged) and ideal path of E.

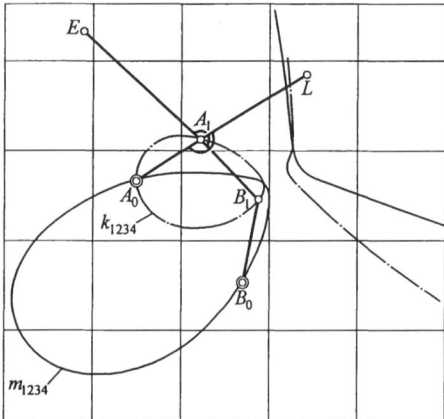

Fig. 6.12. Circle-point curve k_{1234} and centre-point curve m_{1234}, and synthesized four-bar linkage A_0ABB_0.

draw its circle-point curve k_{1234} and centre-point curve m_{1234}. The curve m_{1234} must pass through the point A_0, and k_{1234} must also pass through A_1. Choose a point on m_{1234} as B_0, then we can correspondingly determine the point B_1 on k_{1234}. However, the selection of B_0 should be limited within a certain range according to the design requirements. Fig. 6.12 shows the curves k_{1234} and m_{1234}, as well as the synthesized four-bar linkage $A_0A_1B_1B_0$. Finally, it is necessary to inspect if the deviation between the actually synthesized coupler curve and the ideal path is acceptable. The actual path of E can be computed by means of the later equation (6.22).

The synthesis sequence of this example is: H_0, $A_0 \to L \to E \to A_1E_1, A_2E_2, A_3E_3, A_4E_4 \to k_{1234}$, $m_{1234} \to B_0 \to B_1$.

6.4 Vector method

In this section we shall explain how to apply vector equations to a synthesis problem of a path generating mechanism. It should be pointed out here that, the vector method is not only applicable to path generation, but also to other types of dimensional synthesis problems. What we are going to explain is based on (Freudenstein & Sandor, 1959). To facilitate understanding, a concise introduction of

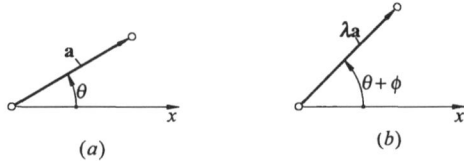

Fig. 6.13. Vectors and complex numbers. (a) $\mathbf{a} = ae^{i\theta}$; (b) $\lambda = e^{i\phi}$, $\lambda\mathbf{a} = ae^{i(\theta+\phi)}$.

Dimensional systhesis of linkages --- path generation problems

vectors and *complex numbers* is warranted. In Fig. 6.13(*a*), **a** is a vector. Its inclination to the real axis (*x*-axis) is θ, counterclockwise sense being considered as positive. We write

$$\mathbf{a} = ae^{i\theta}$$

where *a* is the magnitude of **a**, *e* is the base of natural logarithm, and $i = \sqrt{-1}$. In case *a* is unity, **a** is called a unit vector. Assume $\lambda = e^{i\phi}$ is a unit vector, then the product of λ and **a** is

$$\lambda \mathbf{a} = ae^{i(\theta+\phi)}$$

In other words, multiplying **a** by a unit vector is equivalent to rotating **a** through an angle ϕ, as shown in Fig. 6.13(*b*). Certainly, λ can also be a non-unit vector. In that case $\lambda\mathbf{a}$ is a vector obtained by rotating **a** through an angle ϕ and with its magnitude enlarged (or reduced) by the magnitude of λ. This is the advantage of using exponents of *e* in the multiplication of vectors. As is well-known,

$$e^{i\theta} = \cos\theta + i\sin\theta$$
$$e^{i(\theta+\phi)} = \cos(\theta+\phi) + i\sin(\theta+\phi)$$

This means that by using complex numbers, we can easily decompose a vector equation into two orthogonal parts, namely, a real part and an imaginary part. This is analogous to decomposing a force in mechanics into component F_x and component F_y. The main advantage of applying complex numbers in planar kinematics is that these two components can be handled together in all mathematical operations without causing confusion. Please do not confuse the imaginary part with the imaginary roots of a real number equation.

6.4.1 Coordinations of four coupler point positions with three crank rotations

This problem corresponds to the one treated in Section 6.3.2. Let the prescribed coupler point positions be E_1, E_2, E_3, E_4, and the prescribed angles of crank rotations be ϕ_{12}, ϕ_{13}, ϕ_{14}. We can tabulate as before:

Number of available coordinates $(A_0, A, B, B_0, E) = 5 \times 2 = 10$	
Speifications	Number of coordinates needed
E_1	2
E_2, E_3, E_4	3
$\phi_{12} : E_1 \to E_2$	1
$\phi_{13} : E_1 \to E_3$	1
$\phi_{14} : E_1 \to E_4$	1
Total number of coordinates needed	8
Remaining number of available coordinates	$10 - 8 = 2$
Number of possible solutions	∞^2

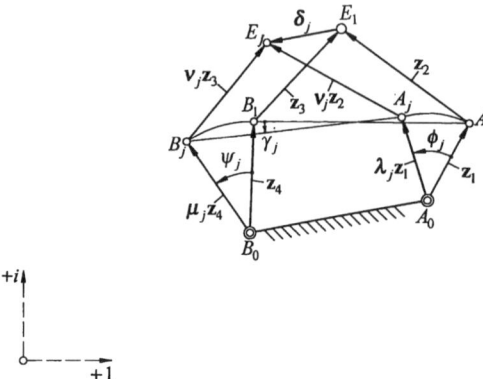

Fig. 6.14. Coordinations of coupler point positions $E_1, ..., E_j$ with crank rotation angles $\phi_{12}, ..., \phi_{1j}$.

Fig. 6.14 shows a given four-bar linkage A_0ABB_0. As this four-bar linkage moves from its first position to the jth position, its coupler AB moves from A_1B_1 to A_jB_j, and the coupler point E moves from E_1 to E_j. Correspondingly, the crank A_0A rotates through an angle ϕ_{1j} and crank B_0B rotates through an angle ψ_{1j}. The angle of rotation of the coupler is the angle between $\Delta A_1B_1E_1$ and $\Delta A_jB_jE_j$, denoted by γ_{1j}. All counterclockwise angles are considered positive. For brevity, the angles $\phi_{1j}, \psi_{1j}, \gamma_{1j}$ are written from now on as ϕ_j, ψ_j, γ_j. For the time being we can neglect the position and orientation of the coordinate axes. Hence the real axis and the imaginary axis are both drawn as dotted lines in the figure. In Fig. 6.14 the vectors are: $\overrightarrow{A_0A_1} = \mathbf{z}_1$, $\overrightarrow{A_1E_1} = \mathbf{z}_2$, $\overrightarrow{B_1E_1} = \mathbf{z}_3$, $\overrightarrow{B_0B_1} = \mathbf{z}_4$, and the vector $\overrightarrow{E_1E_j}$ is denoted by $\boldsymbol{\delta}_j$. The unit vectors are: $\lambda_j = e^{i\phi_j}, \mu_j = e^{i\psi_j}, \nu_j = e^{i\gamma_j}$. As mentioned before, these unit vectors are to be used for rotating another vector. Hence they may be called *rotation coefficients*. Comparing with Fig. 6.13(b), we have

$$\overrightarrow{A_0A_j} = \lambda_j \mathbf{z}_1, \quad \overrightarrow{B_0B_j} = \mu_j \mathbf{z}_4, \quad \overrightarrow{A_jE_j} = \nu_j \mathbf{z}_2, \quad \overrightarrow{B_jE_j} = \nu_j \mathbf{z}_3$$

For the *loop $A_0A_1E_1E_jA_jA_0$*, we have the following *loop equation*

$$\mathbf{z}_1 + \mathbf{z}_2 + \boldsymbol{\delta}_j = \lambda_j \mathbf{z}_1 + \nu_j \mathbf{z}_2$$

or
$$(\nu_j - 1)\mathbf{z}_2 + (\lambda_j - 1)\mathbf{z}_1 = \boldsymbol{\delta}_j \tag{6.13}$$

By setting $j = 2, 3, 4$, equation (6.13) represents in fact three equations:

Dimensional systhesis of linkages --- path generation problems   245

$$\left.\begin{array}{l}(\nu_2-1)z_2+(\lambda_2-1)z_1=\delta_2\\(\nu_3-1)z_2+(\lambda_3-1)z_1=\delta_3\\(\nu_4-1)z_2+(\lambda_4-1)z_1=\delta_4\end{array}\right\} \quad (6.14)$$

Equations (6.14) are three non-homogeneous linear equations in the two unknowns z_1 and z_2. In order that these simultaneous equations should have non-trivial solutions, the determinant of the coefficients must vanish, or

$$\begin{vmatrix} \nu_2-1 & \lambda_2-1 & \delta_2 \\ \nu_3-1 & \lambda_3-1 & \delta_3 \\ \nu_4-1 & \lambda_4-1 & \delta_4 \end{vmatrix} = 0 \quad (6.15)$$

The terms in equation (6.15), except ν_2, ν_3, ν_4, are all known. These three vectors include three scalar unknowns γ_2, γ_3, γ_4. We can therefore assume a value of say, γ_2, to solve for γ_3 and γ_4. Equation (6.15) is a complex equation, corresponding to two scalar equations, hence can be solved for two unknowns. The value of γ_2 should be chosen within a certain limit. If γ_2 is chosen beyond this limit, equation (6.15) may have no solutions. With regard to this limit, please refer to Appendix 11.

Similarly, for the loop $B_0B_1E_1E_jB_jB_0$, we have according to equation (6.13)

$$(\nu_j-1)z_3+(\mu_j-1)z_4=\delta_j \quad (6.16)$$

By setting $j = 2, 3, 4$, equation (6.16) represents also three equations like those in (6.14), including two unknowns z_3 and z_4. In order to have non-trivial solutions, the determinant of the coefficients should vanish. Just like equation (6.15), we have

$$\begin{vmatrix} \nu_2-1 & \mu_2-1 & \delta_2 \\ \nu_3-1 & \mu_3-1 & \delta_3 \\ \nu_4-1 & \mu_4-1 & \delta_4 \end{vmatrix} = 0 \quad (6.17)$$

In this equation, values of ν_2, ν_3, ν_4 have already been found previously, and μ_2, μ_3, μ_4 include three unknowns ψ_2, ψ_3, ψ_4. By assuming a value of ψ_2, we can solve for the values of ψ_3 and ψ_4.

By assuming the values of γ_2 and ψ_2 above, we have thus used up two coordinates.

6.4.2 Coordinations of five coupler point positions with four crank rotations

The five prescribed coupler point positions are E_1. E_2, E_3, E_4, E_5, and the four crank angles prescribed are ϕ_{12}, ϕ_{13}, ϕ_{14}, ϕ_{15}. We tabulate as before.

Number of avaliable coordinates = 5×2 = 10	
Specifications	Number of coordinates needed
E_1	2
E_2, E, E_4, E_5	4
$\phi_{12}: E_1 \to E_2$	1
$\phi_{13}: E_1 \to E_3$	1
$\phi_{14}: E_1 \to E_4$	1
$\phi_{15}: E_1 \to E_5$	1
Total number of coordinates needed	10
Remaining number of available coordinates	10 − 10 = 0
Number of possible solutions: finite	

There are now two sets of equations like equation (6.15):

$$\begin{vmatrix} v_2-1 & \lambda_2-1 & \delta_2 \\ v_3-1 & \lambda_3-1 & \delta_3 \\ v_4-1 & \lambda_4-1 & \delta_4 \end{vmatrix} = 0, \quad \begin{vmatrix} v_2-1 & \lambda_2-1 & \delta_2 \\ v_3-1 & \lambda_3-1 & \delta_3 \\ v_5-1 & \lambda_5-1 & \delta_5 \end{vmatrix} = 0$$

and two sets of equations like equation (6.17):

$$\begin{vmatrix} v_2-1 & \mu_2-1 & \delta_2 \\ v_3-1 & \mu_3-1 & \delta_3 \\ v_4-1 & \mu_4-1 & \delta_4 \end{vmatrix} = 0, \quad \begin{vmatrix} v_2-1 & \mu_2-1 & \delta_2 \\ v_3-1 & \mu_3-1 & \delta_3 \\ v_5-1 & \mu_5-1 & \delta_5 \end{vmatrix} = 0$$

These four vector equations correspond to 8 scalar equations, for solving the eight unknowns $\gamma_2, \gamma_3, \gamma_4, \gamma_5$ and $\psi_2, \psi_3, \psi_4, \psi_5$. In carrying out the computation, we can expand these four equations into a quartic equation in the sole unknown $\tan(\gamma_2/2)$. Solving this equation will yield 12 possible (or impossible) four-bar linkages.

6.5 Algebraic equation of the coupler (point) curve

The algebraic equation of the coupler curve appeared first in (Roberts, 1875). We shall derive this equation and discuss its properties.

6.5.1 Derivation of the coupler curve equation

Fig. 6.15 shows a four-bar linkage A_0ABB_0, with its link lengths denoted as before: $\overline{A_0A} = a, \overline{B_0B} = b, \overline{AB} = c, \overline{A_0B_0} = f$. ABE is the coupler, with $\overline{AE} = m$, $\overline{BE} = n$, and $\angle BEA = \eta$. The point A_0 is now taken as the origin, and A_0B_0 is taken as the direction of the x-axis. The line normal to the x-axis at A_0 is y-axis. Assume in the position shown in the figure, the coordinates of E are $E(x, y)$, and the inclination of the line AE with respect to x-axis is λ. Further let the coordinates of A

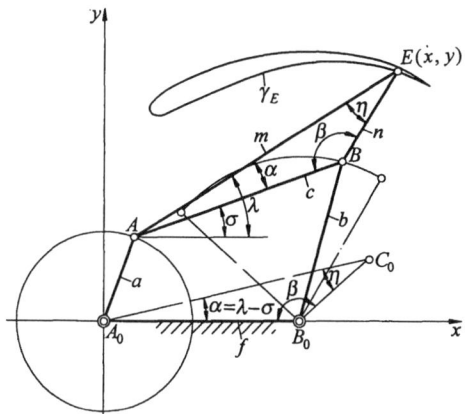

Fig. 6.15. For derivation of equation of the coupler point curve.

be denoted by (x_A, y_A), and those of B by (x_B, y_B). We have then

$$x_A = x - m\cos\lambda \qquad y_A = y - m\sin\lambda \tag{6.18}$$

$$x_B = x - n\cos(\lambda + \eta) \qquad y_B = y - n\sin(\lambda + \eta) \tag{6.19}$$

The crank circles on which the points A and B lie are represented respectively by

$$k_A: x_A^2 + y_A^2 - a^2 = 0 \qquad k_B: (x_B - f)^2 + y_B^2 - b^2 = 0 \tag{6.20}$$

Substituting equations (6.18) and (6.19) into equation (6.20) results in

$$x\cos\lambda + y\sin\lambda = \frac{x^2 + y^2 + m^2 - a^2}{2m}$$

and $\qquad (x - f)\cos(\lambda + \eta) + y\sin(\lambda + \eta) = \dfrac{(x-f)^2 + y^2 + n^2 - b^2}{2n}$ $\qquad$ (6.21a, b)

Eliminating λ from equations (6.21a,b) gives

$$\gamma_E : U^2 + V^2 = W^2 \tag{6.22}$$

where

$$\begin{aligned} U &= n[(x-f)\cos\eta + y\sin\eta](x^2 + y^2 + m^2 - a^2) - mx[(x-f)^2 + y^2 + n^2 - b^2] \\ V &= n[(x-f)\sin\eta - y\cos\eta](x^2 + y^2 + m^2 - a^2) + my[(x-f)^2 + y^2 + n^2 - b^2] \\ W &= 2mn\sin\eta[x(x-f) + y^2 - fy\cot\eta] \end{aligned} \tag{6.23}$$

Equation (6.23) can also be written as

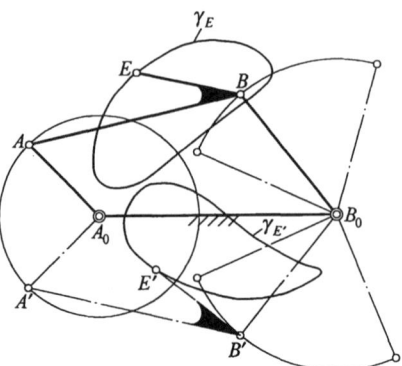

Fig. 6.16. Bicursal coupler point curve γ_E, $\gamma_{E'}$.

$$\begin{aligned}&n^2[(x-f)^2+y^2](x^2+y^2+m^2-a^2)^2\\&-2mn[(x^2+y^2-fx)\cos\eta+fy\sin\eta](x^2+y^2+m^2-a^2)\\&[(x-f)^2+y^2+n^2-b^2]+m^2(x^2+y^2)[(x-f)^2+y^2+n^2-b^2]^2\\&-4m^2n^2[(x^2+y^2-fx)\sin\eta-fy\cos\eta]^2=0\end{aligned} \quad (6.24)$$

As can be seen from equation (6.24), the highest term of the coupler point curve is $(x^2 + y^2)^3$, hence it is a tricircular sextic. Furthermore, the *class* of this curve is 12. Both four-bar linkages A_0ABB_0E and $A_0A'B'B_0E'$ shown in Fig. 6.16 satisfy equation (6.24). This is because both of them have the same dimensions, and the link lengths satisfy *Grashof rule*. In other words, the sum of the longest and shortest links is less than that of the other two links. These two four-bars are separated from each other, i.e. A_0ABB_0 can never reach $A_0A'B'B_0$, and the latter can also never reach the former without dismantling the linkages. Therefore the coupler curves γ_E and $\gamma_{E'}$ are separated from each other. In other words, equation (6.24) represents a *two-part*, or *bicursal* coupler curve. On the other hand, for a *non-Grashof* four-bar, i.e. in which the sum of the lengths of the longest and shortest links is longer than that of the other two links, both four bars can be reached from each other without dismantling, and there is only one four-bar. Both coupler curves γ_E and $\gamma_{E'}$ are connected together to become a *unicursal* curve, as that shown in the later Fig. 6.17. (Please refer to Beyer, 1953,Theorem 50 and Fig. 231.) In this case equation (6.24) represents a *one-part* coupler curve.

As has been mentioned in Section 5.1, if all points A_0, A_1, B_1, B_0 of a four-bar linkage are not specified, with the number of available coordinates 10, nine coupler point positions can be assigned. Let us examine now the number of parameters in equation (6.24). There are altogether six, namely, a, b, f, m, n, η. In other words, if the coordinates of the 6 coupler points $E_1(x_1,y_1)$, $E_2(x_2,y_2)$,...,$E_6(x_6,y_6)$ are specified, 6 unknowns can be found by solving 6 equations. How could we specify 9 coupler

Dimensional systhesis of linkages --- path generation problems 249

point positions? In fact, in equation (6.24), the coordinates of A_0 have already been specified as (0, 0), and those of B_0 have been specified as $(f, 0)$, the remaining number of available coordinates is only 6.

6.5.2 Nodes of a coupler point curve

Equation (6.22) can further be written as

$$F(x,y) \equiv (N\Phi + P\Psi)^2 + (L\Phi + M\Psi)^2 - 4K^2(LP - MN)^2 = 0 \quad (6.25)$$

In Fig. 6.15, let $\angle BAE = \alpha$ and $\angle ABE = \beta$. The notations in equation (6.25) are:

$$L = \sin\alpha[(x-f)\sin\eta - y\cos\eta]$$
$$M = y\sin\beta$$
$$N = \sin\alpha[(x-f)\cos\eta + y\sin\eta]$$
$$P = -x\sin\beta$$
$$\Phi = x^2 + y^2 + m^2 - a^2$$
$$\Psi = (x-f)^2 + y^2 + n - b^2$$
$$K = \frac{n}{\sin\alpha} = \frac{m}{\sin\beta} = \frac{c}{\sin\eta}, \text{ being a constant.}$$

According to the description in Appendix A2.2, we have at a node $\partial F/\partial x = 0$, and $\partial F/\partial y = 0$. Differentiating first $F(x, y)$ partially with respect to x and y gives

$$\frac{\partial F}{\partial x} = 2(N\Phi + P\Psi)\frac{\partial}{\partial x}(N\Phi + P\Psi) + 2(L\Phi + M\Psi)\frac{\partial}{\partial x}(L\Phi + M\Psi)$$
$$- 8K^2(LP - MN)\frac{\partial}{\partial x}(LP - MN) \quad (6.26)$$

$$\frac{\partial F}{\partial y} = 2(N\Phi + P\Psi)\frac{\partial}{\partial y}(N\Phi + P\Psi) + 2(L\Phi + M\Psi)\frac{\partial}{\partial y}(L\Phi + M\Psi)$$
$$- 8K^2(LP - MN)\frac{\partial}{\partial y}(LP - MN) \quad (6.27)$$

Consider now the equation

$$k_F: LP - MN = 0 \quad (6.28)$$

which can also be written as

$$k_F: x(x-f) + y^2 - fy\cot\eta = 0 \quad (6.29)$$

Equation (6.29) represents a circle passing through the two points A_0, B_0, with a circumferential angle η, as shown in Fig. 6.17. This circle is denoted by k_F. At the intersection of the coupler curve γ_E and the circle k_F, we have $LP - MN = 0$. If equation (6.28) is substituted into equation (6.25), we can see that at such intersections

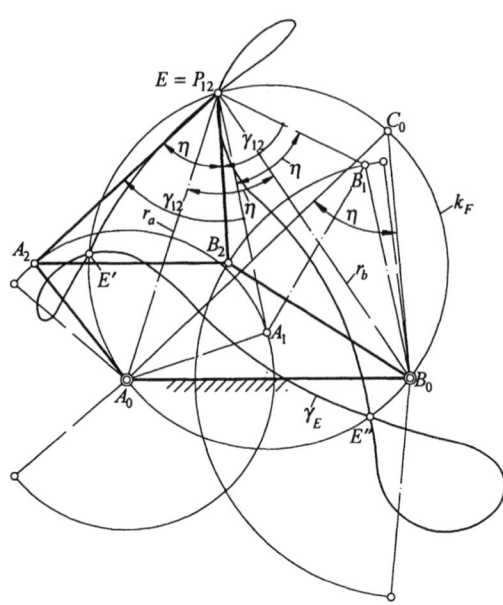

Fig. 6.17. The three nodes of a coupler point curve, r_a, r_b being respectively bisectors of $\sphericalangle A_1EA_2$ and $\sphericalangle B_1EB_2$.

$$N\Phi + P\Psi = 0, \; L\Phi + M\Psi = 0 \qquad (6.30)$$

This is because the squares of these two terms are both positive in equation (6.25). Furthermore, because of equations (6.28) and (6.30), it can be seen from equations (6.26) and (6.27) that at these intersections, $\partial F/\partial x = 0$ and $\partial F/\partial y = 0$. We know that the nodes of a coupler curve lie in its intersections with the circle k_F.

This fact can also be proved from geometry. In Fig. 6.17, the coupler point E is just in a node position. In other words, E is just the pole P_{12} between the two positions A_1B_1 and A_2B_2 of the coupler c. Therefore

$$\sphericalangle A_1EA_2 = \sphericalangle B_1EB_2 = \gamma_{12}$$

where γ_{12} is the angle of rotation of the body c from position 1 to position 2, and

$$\sphericalangle A_1EA_0 = \sphericalangle B_1EB_0 = \gamma_{12}/2$$

therefore

$$\sphericalangle A_0EB_0 = \sphericalangle A_1EB_1 = \eta \qquad (6.31)$$

Hence we know that E lies on the circle k_F.

As an application to synthesis problems, let us consider the given data in (Kraus,

Dimensional systhesis of linkages --- path generation problems

1954), as shown in Fig. 6.18. Suppose the three points A_0, B_0, E_1 are given, and the angle ϕ_{12} is prescribed. In other words, as the crank A_0A_1 rotates through the angle ϕ_{12} to the position A_0A_2, the coupler point E should return to the position E_1. In this case E_1 should be counted as two positions, i.e. E_1 and E_2 coincide together. We tabulate as before as follows:

Number of available coordinates = 5 × 2 = 10	
Specifications	Number of coordinates needed
A_0, B_0, E_1	6
$E_2 (= E_1)$	1
$\phi_{12} : E_1 \to E_2$	1
Total number of coordinates needed	8
Remaining number of available coordinates	10 – 8 = 2
Number of possible solutions	∞^2

Join A_0E_1 and B_0E_1. Build at A_0 the angle ϕ_{12} symmetrically on both sides of A_0E_1. Choose an arbitrary crank length a, and draw the crank circle of A, then the two points A_1, A_2 can be determined. Join the lines E_1A_1 and E_1A_2, then the angle $\sphericalangle A_1E_1A_2$ is γ_{12}. Build at E_1 the angle γ_{12} symmetrically on both sides of E_1B_0, and choose an arbitrary length b to draw the crank circle of B, then the two points B_1, B_2 can be determined. The four-bar linkage $A_0A_1B_1B_0$ is thus completed. The above two choices of a and b used up the last two available coordinates.

If only the positions of the four points A_0, A_1, B_1, B_0 are prescribed, we know

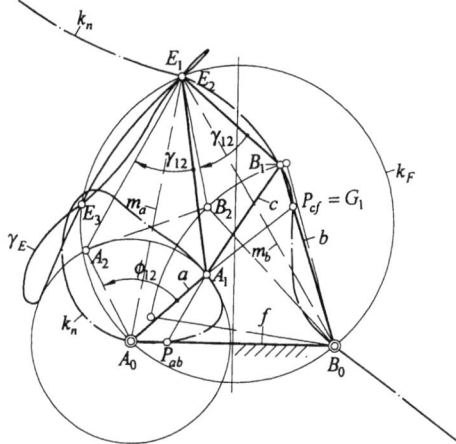

Fig. 6.18. Synthesis of four-bar linkage $A_0A_1B_1B_0$ and determination of curve k_n for prescribed A_0, B_0, E_1 ($= E_2$) and coordination of $\phi_{12} : E_1 \to E_2$.

from equation (6.31) that in this position of the four-bar linkage, the angles subtended by $\overline{A_1B_1}$ and $\overline{A_0B_0}$ at a coupler point on the body c are equal, if the coupler point is just at a node of its path. However, this angle is no longer equal to the constant η, and the circle k_F is also not a constant circle. This occasion reminds us of the methods of constructing a centre-point curve mentioned in Section 3.6.2. According to those methods, we only need to replace an opposite-pole quadrilateral by $A_0A_1B_1B_0$ in order to locate the locus of the node point. This locus is denoted by k_n, as shown in Fig. 6.18. Just like the centre-point curve, the curve k_n is also a circular cubic, passing through the six points A_0, B_0, A_1, B_1, P_{ab}, P_{cf}.

A coupler point curve can at most have three real nodes, therefore the *genus* of a general coupler point curve is 1 (please refer to Section A2.4).

6.5.3 Singular foci of a coupler point curve

According to what is described in Section A2.5 of Appendix 2, since a coupler point curve is a tricircular sextic, it should have three singular foci. To find these three singular foci, let us find first the intersections of the straight line

$$y = \pm ix + h \quad (i = \sqrt{-1}) \tag{6.32}$$

with the coupler curve γ_E. Substituting equation (6.32) into equation (6.24) we see that, because the coupler curve has a triple root at each of the circular points (I, J), the terms of 6th, 5th and 4th degrees in equation (6.24) vanish identically; and therefore that equation (6.24) reduces to a cubic equation. If the straight line of equation (6.32) is an asymptote of the original coupler curve at (I, J), the coefficient of the third degree term of this cubic should be equal to zero, hence the values of h can be determined. It can be shown that the three real singular foci are at the points A_0, B_0, C_0 in Fig. 6.15, and $\triangle A_0B_0C_0 \sim \triangle ABE$ (please refer to Exercise 6.4), and the point C_0 lies on k_F. We have therefore the following theorem:

Theorem 22 The coupler point curve is a tricircular sextic, its intersections with the circle k_F passing through the three singular foci A_0, B_0, C_0 being its three nodes, and $\triangle A_0B_0C_0 \sim \triangle ABE$.

6.6 Roberts-Chebyshev theorem

This theorem appeared first in (Roberts, 1875), and subsequently in (Chebyshev, 1879). It is generally called *Roberts-Chebyshev Theorem*. The derivations used by these two authors, however, were not the same. What we are going to present here is the concise result, rather than a rigorous mathematical treatment.

6.6.1 The content of Roberts-Chebyshev theorem

Suppose in Fig. 6.19, an original four-bar linkage A_0ABB_0 is given. E is a coupler point on its coupler ABE. Draw parallelograms A_0AEA' and B_0BEB', and $\triangle A'EC \sim$

Dimensional systhesis of linkages --- path generation problems

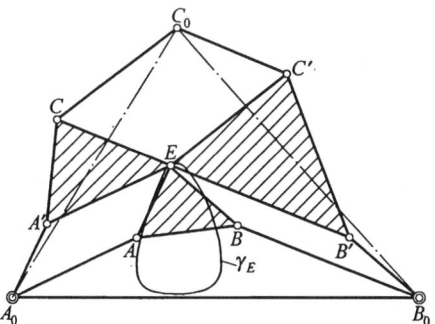

Fig. 6.19. Roberts-Chebyshev theorem: triple generation of coupler curve γ_E.

$\Delta EB'C' \sim \Delta ABE$. Finally complete the parallelogram $CEC'C_0$. Then $\Delta A_0 B_0 C_0 \sim \Delta ABE$. Fixing C_0 will result in, besides the original four-bar linkage $A_0 ABB_0$, two other four-bar linkages $A_0 A'CC_0$ and $B_0 B'C'C_0$. For the whole mechanism, the number of links is $n = 10$, and the number of joints is $j = 14$, where each of the points A_0, B_0, C_0, E is counted as two joints. If we apply Grübler criterion to this mechanism, we will find its degrees of freedom to be $3(n-1) - 2j = -1$. In other words, this mechanism would not be movable. However, if the three couplers are dismantled at the joint E, it can be shown that the three coupler point curves γ_E generated by each coupler point E of the three four-bar linkages are identical. This is the *Roberts-Chebyshev theorem*:

Theorem 23 The three four-bar linkages $A_0 ABB_0$, $A_0 A'CC_0$ and $B_0 B'C'C_0$ have a

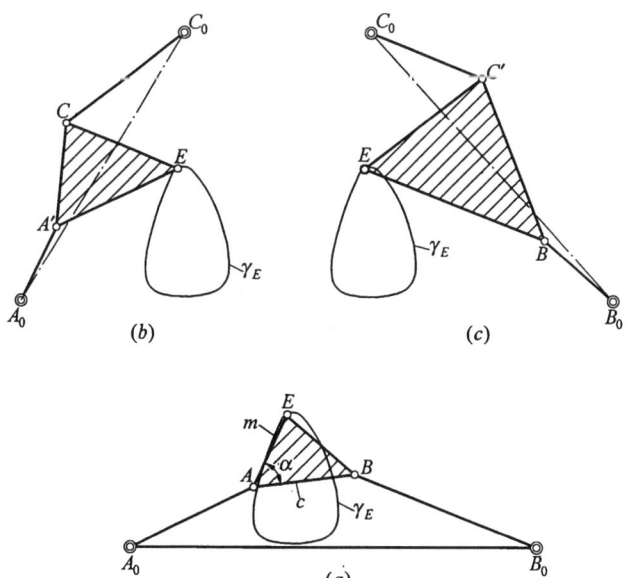

Fig. 6.20. Three cognate linkages (a)(b)(c).

common coupler point E, where A_0, B_0, C_0 are fixed and $\triangle ABE \sim \triangle A'EC \sim \triangle EB'C' \sim \triangle A_0B_0C_0$. The coupler point curves generated by the point E on each of these three four-bar linkages are identical.

Fig. 6.20 $(a)(b)(c)$ shows separately the three four-bar linkages of Fig. 6.19 with their respective coupler curves generated by E. It can be seen that the three γ_E's are identical. In practical applications, if, for instance, we find it impossible to mount the fixed joint B_0 in Fig. 6.20(a), while still requiring the same coupler point curve γ_E, we may use the four-bar linkage $A_0A'CC_0$ of Fig. 6.20(b) instead of the original four-bar. Also, if the transmission angle (please refer to Section 7.5) of the original four-bar linkage is not satisfactory, we may also replace it by the one with a better transmission angle from the other two four-bar linkages. These three four-bar linkages are called *cognate linkages*. In other words, each four-bar linkage has two other four-bar linkages as its cognate linkages.

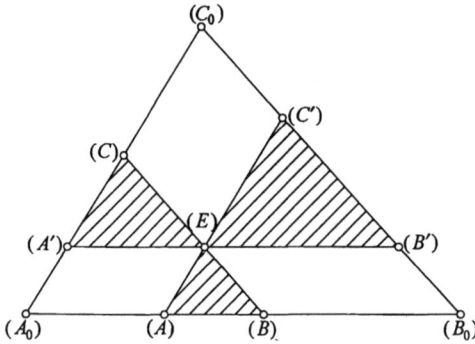

Fig. 6.21. Cayley diagram of cognate linkages.

In practical applications, it is expedient to carry out the procedure of finding out the cognate linkages as that suggested by (Cayley, 1876). Suppose a four-bar linkage as shown in Fig. 6.20(a) is given, and it is required to find out its two other cognate linkages. We may draw a diagram as shown in Fig. 6.21. Draw first the lengths $\overline{A_0A}$, $\overline{AB}$, $\overline{BB_0}$ on a straight line $(A_0)(A)(B)(B_0)$, construct the $\triangle(A)(B)(E)$, and draw parallel lines to complete the whole diagram. Finally lay the points (A_0), (B_0) of Fig.6.21 respectively at the positions A_0, B_0 of Fig. 6.19, build $\triangle A_0B_0C_0 \sim \triangle ABE$, and lay the point (C_0) at C_0. The whole mechanism is thus completed.

6.6.2 A simple proof of Roberts-Chybeshev theorem

We shall prove by simple vector relations that the coupler point curves generated by both points E in Fig. 6.20(a) and 6.20(b) are identical. This proof originates in (Pflieger-Haertel, 1944), but the first simple proof was by (Hart, 1883).

Dimensional systhesis of linkages --- path generation problems 255

The mathematical tools used in the proof below are not quite the same as those of (Hart, 1883), though the mathematical implication is the same. Let

$$z = \overrightarrow{A_0 E}$$

and

$$v = \frac{\overrightarrow{AE}}{\overrightarrow{AB}} = \frac{m}{c} e^{i\alpha} \quad \text{(a constant)}$$

We have then in Fig. 6.20(a)

$$z = \overrightarrow{A_0 A} + v \overrightarrow{AB} \tag{6.33}$$

The vector v can be considered as an operator. In other words, applying v to a vector is equivalent to rotate that vector through an angle α, and then magnify (or reduce) it by a factor m/c. Referring to Fig. 6.19, we have now

$$\overrightarrow{AB} = \overrightarrow{A_0 B_0} + \overrightarrow{B_0 B} - \overrightarrow{A_0 A}$$

Therefore

$$v\overrightarrow{AB} = v\overrightarrow{A_0 B_0} + v\overrightarrow{B_0 B} - v\overrightarrow{A_0 A}$$
$$= \overrightarrow{A_0 C_0} + \overrightarrow{C'E} - \overrightarrow{A'C}$$
$$= \overrightarrow{A_0 C_0} + \overrightarrow{C_0 C} + \overrightarrow{CA'} \tag{6.34}$$

Note further that $\overrightarrow{A_0 A} = \overrightarrow{KE}$. Substituting equation (6.34) into equation (6.33) gives

$$z = \overrightarrow{A_0 C_0} + \overrightarrow{C_0 C} + \overrightarrow{CA'} + \overrightarrow{A'E}$$

This means that the curve γ_E in Fig. 6.20(a) is identical with the γ_E in Fig. 6.20(b). Similarly it can be proved that the γ_E in Fig. 6.20(c) is identical with these two γ_E's.

6.6.3 Special cases

(a) Fig. 6.22 shows a four-bar linkage $A_0 ABB_0$, the coupler point E of which lies on the line AB, i.e. $\triangle ABE$ becomes a straight line. In this case $\triangle A_0 B_0 C_0$ becomes also a straight line, and we know from Fig. 6.21 that both $\triangle EB'C'$ and $\triangle A'EC$ become straight lines. However, $\overrightarrow{A_0 B_0} / \overrightarrow{A_0 C_0} = \overrightarrow{EB'} / \overrightarrow{EC'} = \overrightarrow{A'E} / \overrightarrow{A'C} = \overrightarrow{AB} / \overrightarrow{AE}$ and $A_0 AEA'$, $B_0 BEB'$, $C_0 CEC'$ remain parallograms.

(b) Fig. 6.23 shows the cognate linkages of a slider-crank mechanism $A_0 AB$. In this case the point B_0 lies at infinity, but in a definite direction which is the direction normal to the sliding direction of B. The diagram in Fig. 6.21 becomes therefore now a diagram as shown in Fig. 6.24. Because (B_0) lies at infinity in the right hand direction, therefore all the points (C_0), (C'), (B') lie at infinity. In Fig. 6.23, $\angle BA_0 C$ = $\angle EA'C = \alpha$. The point C moves on a straight line. The sliding guide at C can therefore be removed without affecting the straight line motion of C.

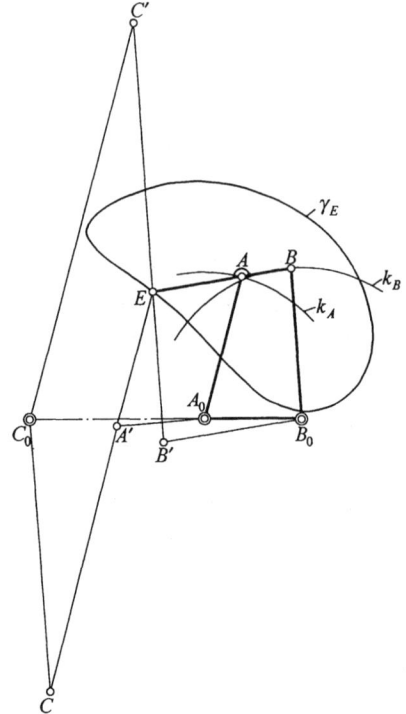

Fig. 6.22. Roberts-Chebyshev theorem: points A, B, E on a straight line.

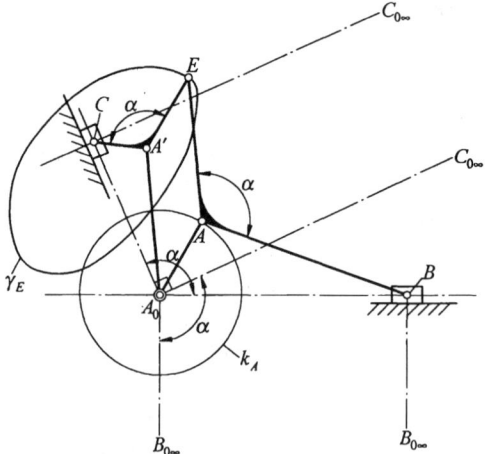

Fig. 6.23. Roberts-Chebyshev theorem: slider-crank mechanism.

Dimensional sythesis of linkages --- path generation problems 257

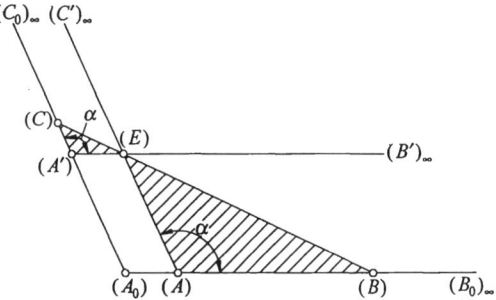

Fig. 6.24. Diagram of the cognate linkages in Fig. 6.23.

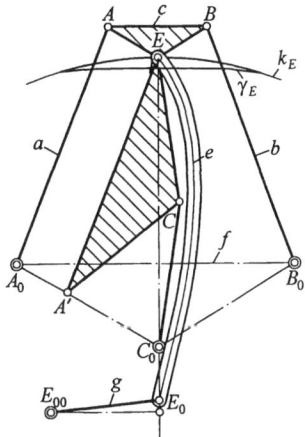

Fig. 6.25. Cognate linkage $A_0A'CC_0$ of isosceles four-bar A_0ABB_0, and dwell mechanism.

(c) Fig. 6.25 shows an isosceles four-bar linkage A_0ABB_0. Use has been made of the centre of curvature E_0 of the path γ_E of the coupler point E in its middle position by connecting a dyad e, g, to form a dwell mechanism, so that during the motion of the linkage the link g may have a temporary standstill. As the transmission with this mechanism is rather poor, no matter whether A_0A or B_0B is chosen as the driving crank; hence we replace the original linkage with one of its cognate linkages $A_0A'CC_0E$, as shown in Fig. 6.26. Using the link A_0A' as the driving crank improves the transmission.

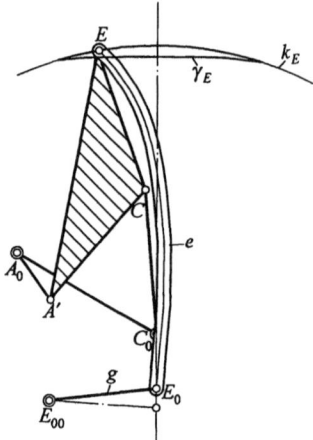

Fig. 6.26. Replacement of the original four-bar linkage A_0ABB_0E in Fig. 6.25 with $A_0A'CC_0E$.

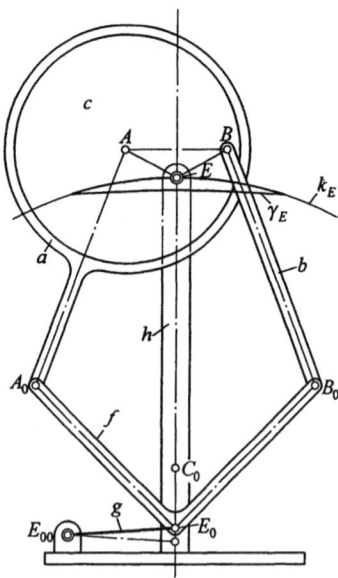

Fig. 6.27. Six-bar dwell mechanism derived from Fig. 6.25.

Dimensional systhesis of linkages --- path generation problems 259

Fig. 6.27 shows a mechanism derived from that shown in Fig. 6.25, taken from (Rauh, 1951, p.79). Here the link e is omitted, while links a, c, b, f remain unchanged. However, f is no longer fixed. The point E is now fixed, and c is taken as the driving member, while the original centre of rotation E_{00} of the link g is still a fixed centre. Please note that the frame h, being a link connecting the two points E and E_{00}, was not present in the original drawing Fig. 6.25. The path γ_E of the point E in Fig. 6.27 is drawn on f, not on h. This transformation retains the dwell motion of g.

6.6.4 Cognate linkages of multiple-bar linkages

According to the definition of (Dijksman, 1976), the three Roberts cognates shown in Fig. 6.20 may be called *curve cognates*, because the three four-bar linkages can generate identical coupler point curve γ_E. The four-bar linkage A_0ABB_0 in Fig. 6.20(*a*) may be called the *source mechanism* or *source linkage*, because the other two mechanisms shown in Fig. 6.20(*b*), (*c*) are derived from it. For a source four-bar linkage, there are only two uniquely determined curve cognates.

For a source linkage with six or more bars, there are, besides the curve cognates, the so called *coupler cognates* and *function cognates*. If the motion of the coupler of a derived linkage is identical with the motion of the coupler of the source linkage, the derived linkage is called a coupler cognate. If the functional relationship $\psi = \psi(\phi)$ (please refer to Chapter 7) represented by the two cranks (rotating about fixed centres) of a derived linkage is the same as the functional relationship $\psi = \psi(\phi)$ represented by the two cranks of the source linkage, the derived linkage is called a function cognate.

As is well-known, there are five types of six-bar linkages, namely: Watt-I, with two fixed centres of rotation; Watt-II, with three fixed centres of rotation; Stephenson-I and Stephenson-II, both with two fixed centres of rotation; and Stephenson-III, with three fixed centres of rotation (please refer to Dijksman, 1976, p.157). Investigations on six-bar curve cognates derived from a Watt-I source linkage include (Soni, 1970) and (Dijksman, 1971a). These cognates are also Watt-I six-bar linkages among which there are also coupler cognates. In general, the number of cognate linkages can be infinite. Based on a Stephenson-III six-bar source linkage, (Rischen, 1962) developed from Roberts-Chebyshev theorem 6 six-bar cognate linkages[§], and (Roth, 1965) derived six- to eight-bar cognates by means of skew pantograph and geared five-bar linkages. (Soni, 1971a, b) derived from eight-bar source linkages other eight-bar coupler cognate linkages. A systematic description of all curve-cognates, coupler-cognates and function cognates derived from Watt and Stephenson types six-bar source linkages can be found in (Dijksman, 1976). In the above literature, an operation of *stretch rotation* (please refer to the later Section 6.12) was applied frequently which should be quite interesting. Readers are recommended to refer to them, and we shall not go into detail here.

Finally, (Dijksman, 1971b) proved that the well-known Peaucellier inversor and

---

[§] Among the 8 cognates claimed by Rischen only 6 are six-bar linkages, as shown by (Dijksman, 1971a).

Hart inversor are in fact cognates, although the former is an eight-bar, and the latter is a six-bar linkage.

6.7 R_M-curve as a special case of coupler point curve

Since a general coupler point curve is a tricircular sextic, and the R_M-curve mentioned in Section 3.3.4 is also a tricircular sextic, the question arises: is there any relation between these two curves? (Alt, 1921) considered the R_M-curve as a special coupler point curve. A comparison between equation (6.24) and equation (3.28) shows that if the points A_0, B_0 and f in Fig. 6.15 are considered respectively as P_{12}, P_{23} and s in Fig. 3.22, and if in Fig. 6.15

$$\left.\begin{array}{l} m = a = \dfrac{R_{123}}{2\sin\alpha} \\ n = b = \dfrac{R_{123}}{2\sin\beta} \\ c = \dfrac{R_{123}\sin\eta}{2\sin\alpha\sin\beta} \end{array}\right\} \quad (6.35)$$

then equation (6.24) becomes equation (3.28) (please see Fig. 6.28). The symbol R_{123} in equation (6.35) stands for R_A in equation (3.28). In order to avoid confusion, we

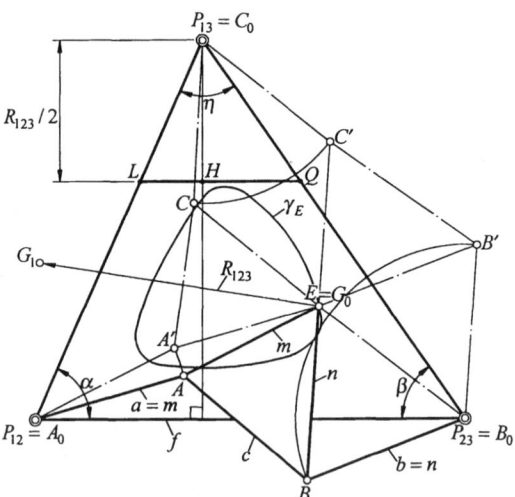

Fig. 6.28. R_M-curve as a coupler point curve.

now rewrite A_0 in Fig. 3.22 as G_0, and A_1 as G_1. The R_{123} is the radius from G_0 to G_1, G_2, or G_3, and α, β, η are the three apex angles of the poletriangle $\Delta P_{12}P_{23}P_{13}$, as

Dimensional systhesis of linkages --- path generation problems 261

shown in Figs. 3.22 and 6.28.

The three cognate linkages in Fig. 6.19 take now the form as shown in Fig. 6.28. C_0 is P_{13}. Because of $m = a$, $n = b$, the three parallelograms become rhombuses. Let us repeat from the beginning. We have first the poletriangle $\Delta P_{12}P_{23}P_{13}$ which is fixed. For a prescribed R_{123}, we can either calculate the lengths $m\ (=a)$, $n\ (=b)$ and c from equations (6.35), or find these three lengths from simple geometric relations in the figure. Draw a line from P_{13} perpendicular to $P_{12}P_{23}$. Lay off on this altitude the distance $\overline{P_{13}H} = R_{123}/2$. Draw a line passing through H and parallel to $P_{12}P_{23}$, meeting the two sides of the poletriangle in L and Q. We have then according to equations (6.35) $\Delta LQP_{13} \equiv \Delta ABE$. For variable values of R_{123}, these three four-bar linkages are also variable. The path γ_E generated by the coupler point E is just the locus R_M of the centre point C_0 for the prescribed radius R_{123}.

The question is now, is there any upper limit or lower limit for R_{123}? Of course there is no upper limit for R_{123}, for $R_{123} \to \infty$ when the three homologous points G_1, G_2, G_3 lie on a straight line. On the other hand, there is a lower limit for R_{123}. This lower limit is reached when A_0, A, B, B_0 in Fig. 6.28 are on a straight line, or

$$a + c + b = f$$

as shown in Fig. 6.29. The four-bar linkage becomes now a locked chain, being not

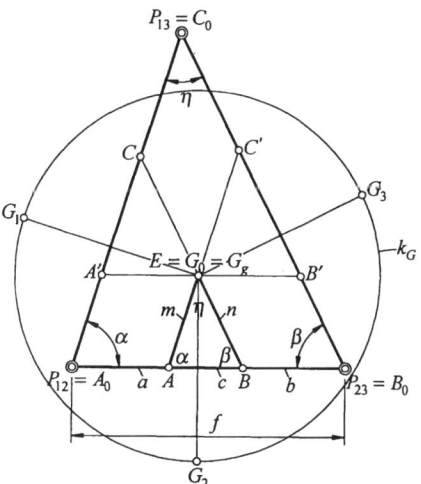

Fig. 6.29. Lower limit of R_{123}.

movable. However, the relation $\Delta ABE \approx \Delta P_{12}P_{23}P_{13}$ still holds. The coupler point curve γ_E generated by the point E shrinks into a point, which is just the intersection point of the bisectors of the three apex angles of the poletriangle $\Delta P_{12}P_{23}P_{13}$, or its incentre. The three rhombuses still exist. From the geometrical relations it can be seen that the conditions $m = a$, $n = b$, $a + c + b = f$ are satisfied. The two points A_g and A_0 corresponding to Fig. 3.12 combine into one point, or $G_g = G_0$, which is E. The three homologous points G_1, G_2, G_3 can be found as mirror images of G_g with respect to the three sides, as mentioned before. The circle k_G passing through G_1, G_2, G_3 is the circle with minimum radius R_{123}. The diameter of this circle is equal to twice the diameter of the inscribed circle of the poletriangle $\Delta P_{12}P_{23}P_{13}$. We have therefore the following theorem:

Theorem 24 For three finitely separated positions of a body, there exists one and only one circle with minimum radius passing through three homologous points, the centre of which is at the incentre of the poletriangle, and its diameter being equal to twice the diameter of the inscribed circle of the poletriangle.

Similarly, it can be proved that the R^1-curve is also a coupler point curve. As mentioned in Section 3.3.4, R_M and R^1 are disposed symmetrically with respect to the line $P_{12}P_{13}$.

6.8 Transition curve

As has been mentioned in Section 6.5.2 (please refer to Fig. 6.18), when a coupler point E is at a node of its path, the locus k_n of E is a circular cubic. Now the question is, for a coupler point E as that shown in Fig. 6.17, if two nodes of its path combine into one, what is the locus of E? This problem was first investigated by (Müller, 1889a, b, 1891b). Now please refer to Fig. 6.30. According to Fig. 6.17, if E is just at a node of its path, then

$$\sphericalangle A_0EB_0 = \sphericalangle AEB = \eta$$

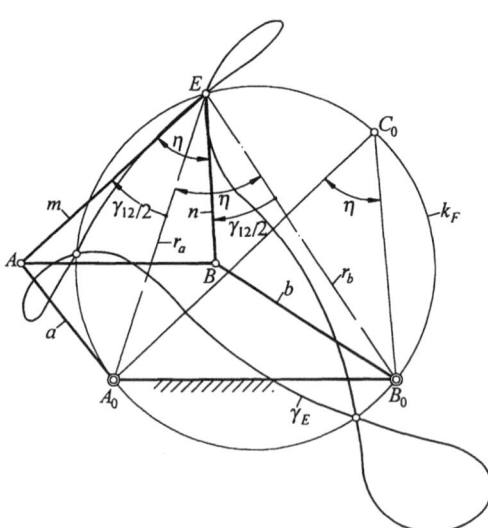

Fig. 6.30. A partial view of Fig. 6.17, coupler point E being at a node of its path.

Dimensional systhesis of linkages --- path generation problems

or
$$\sphericalangle A_0EA = \sphericalangle B_0EB = \gamma_{12}/2$$

Let $r_a = \overline{A_0E}$, $r_b = \overline{B_0E}$, hence

$$\cos \sphericalangle A_0EA = \frac{r_a^2 + m^2 - a^2}{2r_a m}$$

$$\cos \sphericalangle B_0EB = \frac{r_b^2 + n^2 - b^2}{2r_b n}$$

Therefore

$$\frac{r_a^2 + m^2 - a^2}{r_a m} = \frac{r_b^2 + n^2 - b^2}{r_b n}$$

or

$$m\frac{r_a}{r_b} + m(n^2 - b^2)\frac{r_a}{r_b}\frac{1}{r_b^2} - n\left(\frac{r_a}{r_b}\right)^2 - n(m^2 - a^2)\frac{1}{r_b^2} = 0 \tag{6.36}$$

where r_a, r_b, m, n are all variables. Among these variables there exist the following relationships:

$$f^2 = r_a^2 + r_b^2 - 2r_a r_b \cos\eta$$

or

$$\frac{f^2}{r_b^2} = \left(\frac{r_a}{r_b}\right)^2 + 1 - 2\frac{r_a}{r_b}\cos\eta \tag{6.37}$$

and

$$\cos\eta = (m^2 + n^2 - c^2)/(2mn) \tag{6.38}$$

Let $r_a/r_b = u$. Eliminating $\cos\eta$ from equations (6.37) and (6.38), and substituting the relation of equation (6.36) into it results in the following cubic equation in the unknown u

$$nm^2(n^2 - b^2)u^3 - m[n^2 f^2 + n^2(m^2 - a^2) + (n^2 - b^2)(m^2 + n^2 - c^2)]u^2$$
$$+ n[m^2 f^2 + m^2(n^2 - b^2) + (m^2 - a^2)(m^2 + n^2 - c^2)]u$$
$$- mn^2(m^2 - a^2) = 0 \tag{6.39}$$

As is well-known, the roots of a cubic equation can be detected by means of its discriminant. Let this discriminant be denoted by Δ. A cubic equation must have one real root, while the other two roots are determined by Δ. If $\Delta > 0$, these two roots are imaginary; if $\Delta = 0$, these two roots are real and become a double root; if $\Delta < 0$, these two roots are also real but different. Hence equating the discriminant of equation (6.39) to zero gives

$$\Delta = 27m^4n^4(n^2-b^2)^2(m^2-a^2)^2$$
$$+4n^2(n^2-b^2)(2m^2n^2+m^4-a^2n^2-l^2m^2+a^2c^2)^3$$
$$+4m^2(m^2-a^2)(n^4+2m^2n^2-l^2n^2-b^2m^2+b^2c^2)^3$$
$$-18m^2n^2(n^2-b^2)(m^2-a^2)(n^4+2m^2n^2-l^2n^2-b^2m^2+b^2c^2)$$
$$(2m^2n^2+m^4-a^2n^2-l^2m^2+a^2c^2)$$
$$-(n^4+2m^2n^2-l^2n^2-b^2m^2+b^2c^2)^2(2m^2n^2+m^4-a^2n^2-l^2m^2+a^2c^2)^2=0 \quad (6.40)$$

where $l^2 = a^2 + b^2 + c^2 - f^2$ is a constant.

In equation (6.40), only m, n are two variables. Take the middle point O of $\overline{AB}$ as the origin, and OB as the x-axis. Then the coordinates of $E(x, y)$ satisfy the following two equations:

$$\left. \begin{array}{l} m^2 = (x+c/2)^2 + y^2 \\ n^2 = (x-c/2)^2 + y^2 \end{array} \right\} \quad (6.41)$$

Substituting equations (6.41) into equation (6.40) yields an equation of 10th degree in the variables x, y, the highest term of which includes the term $x^2(x^2 + y^2)^4$. This means that $\Delta = 0$ is a quadricircular curve of 10th degree. This is the Müller's *transition curve*, denoted here by $\ddot{u}$. This curve is drawn on the coupler plane. For any coupler point on one side of this curve, its path exhibits three real nodes; for a coupler point on the other side of this curve, its path exhibits only one real node but two imaginary nodes; and for a coupler point on this curve, its path exhibits, besides one real node, another real node which is a combination of other two real nodes; in other words, the coupler point curve becomes self-tangent at this point. Please refer to Fig. 6.31 for a

(a) (b)

Fig. 6.31. (a) Two distinct real nodes. (b) Combination of two nodes, the curve becoming self-tangent.

clear understanding. Fig. (a) shows the case with two distinct nodes, and Fig. (b) shows the case with a combination of two nodes; i.e. where the curve becomes self-tangent.

A query may arise: the curve k_n shown in Fig. 6.18 is only a cubic; why should the equation (6.40) of the $\ddot{u}$-curve be one of 10th degree? This is because the positions of A_0, A_1, B_1, B_0 in Fig. 6.18 are fixed, while the positions of A and B in Fig. 6.30 are variable.

Fig. 6.32 shows an example taken from (Müller, 1889b). A_0ABB_0 is a symmetrical four-bar linkage, in which $a = b = f$, $c = 2a$. Apart from the straight lines,

Dimensional systhesis of linkages --- path generation problems

the solid curve is the transition curve $\ddot{u}$. Because of $b^2 = a^2$, $c^2 = 4a^2$, $l^2 = 5a^2$, therefore equation (6.40) is simplified to

$$\ddot{u}:\ 3840a^4y^6 + 16a^2y^4[88(x^2+y^2)^2 - 428a^2(x^2+y^2) + 97a^4]$$
$$+ 8(x^2+y^2-a^2)^2y^2[-18(x^2+y^2)^2 - 232a^2(x^2+y^2) + 43a^4]$$
$$+ (x^2+y^2-a^2)^3[36(x^2+y^2)^2 + 349a^2(x^2+y^2) - a^4] = 0 \qquad (6.42)$$

This $\ddot{u}$-curve exhibits two cusps at S, S', whose coordinates are respectively (0,

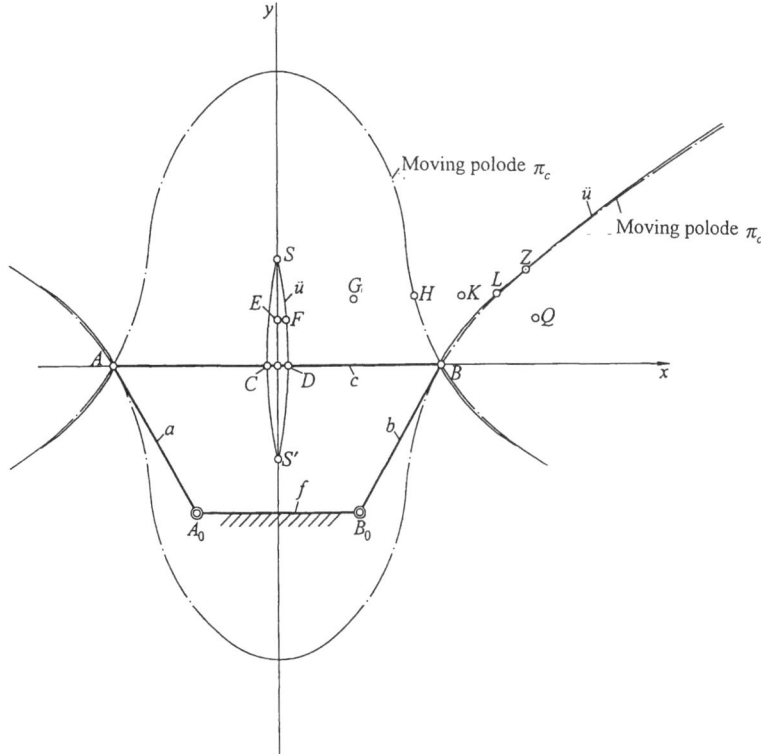

Fig. 6.32. Transition curve $\ddot{u}$ (solid curve), and moving polode π_c (chain line).

$a/\sqrt{3}$), (0, $-a/\sqrt{3}$). There are four points A, C, D, B on the x-axis, among which the coordinates of D, C are respectively ($\pm 0.05352a$, 0). The $\ddot{u}$-curve divides the xy-plane into several districts. We take now only the first quadrant into consideration. This is divided into three regions, namely, (1) left hand side of the curve DS, (2) the area between DS and BL, and (3) the right hand side of BL. For illustration purposes, let us consider now in the three regions the coupler points E, F, G, H, K, L, and Q.

E (on the left hand side of DS). Its coupler point curve exhibits three nodes, as shown in Fig. 6.33.

F (on the curve DS). Its coupler point curve exhibits one node and one self-tangent point, as shown in Fig. 6.34.

The three points G, H, K lie in the same region. For a coupler point in this region, its coupler point curve exhibits only one real node, and two imaginary double points. However, as the point H lies just on the moving polode π_c (please note that π_c is represented by the chain line in the figure and that both $ü$ and π_c are curves on the coupler c) which divides further this region into two parts, namely:

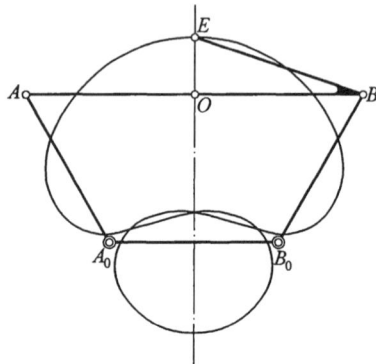

Fig. 6.33. Path of the coupler point E in Fig. 6.32.

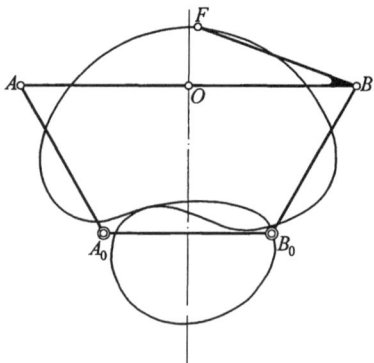

Fig. 6.34. Path of the coupler point F in Fig. 6.32.

G (on the left hand side of π_c). Its coupler point curve exhibits a real node, and two conjugate imaginary double points.

H (on π_c). Its coupler point curve exhibits a cusp, and two conjugate imaginary double points.

Dimensional systhesis of linkages --- path generation problems

K (on the right hand side of π_c). Its coupler point curve exhibits one isolated point (please refer to Appendix 2), and two conjugate imaginary double points.

Finally we come to the two points L and Q. L lies just on $\ddot{u}$, while Q lies on the right hand side of $\ddot{u}$. The curves $\ddot{u}$ and π_c are tangent to each other at Z.

L (on $\ddot{u}$). Its coupler point curve exhibits an isolated point and a self-tangent point.

Q (on the right hand side of $\ddot{u}$). Its coupler point curve exhibits two isolated points and one node.

An isolated point can be counted as a real node. Hence any one of the above coupler curves exhibits three nodes, real or imaginary. In fact, the moving polode π_c shown in Fig. 6.32 possesses in passing through the point B a branch quite close to the branch BL of $\ddot{u}$, which can hardly be observed from the figure. For a coupler point on that branch of π_c, its coupler point curve exhibits certainly a cusp.

As mentioned in Section 3.2.2, the moving polode π_c is a bicircular curve of 8th degree, having at each of the points A, B a quadruple point. As mentioned before, the transition curve $\ddot{u}$ is a quadricircular curve of 10th degree, having at each of the points A, B a double point. The two curves π_c and $\ddot{u}$ intersect in 80 points, real and imaginary. Each of the points A, B is counted as 8 intersections, and each of the imaginary points I, J at infinity is also counted as 8 intersections. Hence there are altogether 32 intersections at the four points A, B, I, J. Among the remaining 48 intersections there are 12 tangent points between π_c and $\ddot{u}$ which should therefore be counted as 24 intersections. The coupler point curves of such coupler points exhibit a cusp in a beak shape as that shown in Fig. 3.66. For each of the last 24 intersections between π_c and $\ddot{u}$ the coupler point curve exhibits a cusp and a self-tangent point (please refer to Müller, 1889b, 1903; Beyer, 1953, No. 98).

For a slider-crank mechanism as that shown in Fig. 6.35, as suggested by (Müller, 1891a), we may let first $f - b = e$ in equation (6.40), and then let $f \to \infty$, and $b \to \infty$ (please note that $l^2/b \to -2e$), and then simplify equation (6.40) into

$$m^2(m^2 - c^2)^2[m^2 e^2 + (m^2 - a^2)(m^2 - c^2)] = 0 \qquad (6.43)$$

Equation (6.43) indicates that the transition curve $\ddot{u}$ breaks up now into:

(1) $m^2 = 0$, i.e. two *isotropic lines* (please refer to Section A2.5) passing through A.

(2) $m^2 e^2 + (m^2 - a^2)(m^2 - c^2) = 0$, i.e. two circles with A as the centre, and with

$$\begin{matrix} m_1 \\ m_2 \end{matrix} = \frac{1}{\sqrt{2}}\left[(a^2 + c^2 - e^2) \pm \sqrt{(a^2 + c^2 - e^2)^2 - 4a^2 c^2}\right]^{1/2} \qquad (6.44)$$

as the radii, denoted respectively by $\ddot{u}_1$ and $\ddot{u}_2$.

(3) $(m^2 - c^2)^2 = 0$, i.e. a circle in duplicate with A as the centre, and c as the radius, denoted by $\ddot{u}_3$.

The values of m_1, m_2 in equation (6.44) are real only when $c > a + e$ or $c < a - e$. The example shown in Fig. 6.35 is drawn with $a = 2$, $c = 9$ and $e = 3$. Take a point E on $\ddot{u}$ as the generating coupler point, and the coupler point curve becomes two oval

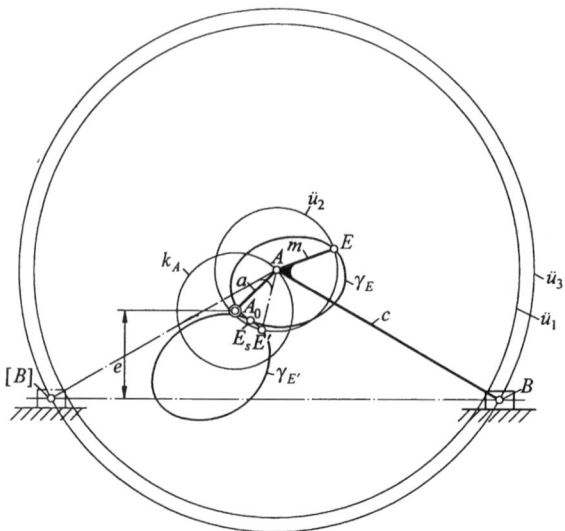

Fig. 6.35. Transition curve $\ddot{u}_1$, $\ddot{u}_2$, $\ddot{u}_3$ of the coupler of a slider-crank, where $a = 2$, $c = 9$, $e = 3$, coupler curves γ_E and $\gamma_{E'}$ being tangent at E_s.

curves γ_E, $\gamma_{E'}$. However, these two curves are not generated by the same slider-crank. The curve on the lower left hand side is generated by the linkage $A_0A[B]E'$, where $\sphericalangle [B]AE' = \sphericalangle BAE$. The two curves γ_E and $\gamma_{E'}$ are tangent to each other at E_s.

(Lichtenheldt, 1936) showed how to apply the transition curve to a shaper, the coupler points of whose swinging block linkage (inversion of the slider-crank) generated coupler point curves.

6.9 Generation of ellipses

As mentioned in Section 6.5.1, a coupler point curve is a sextic. It seems to be quite straightforward to generate, by means of a four-bar linkage, a quadratic curve such as an ellipse. But the fact is not so. Consider, for instance, a given four-bar linkage. For any four positions of the linkage, we can find for the four corresponding positions of the coupler a circle-point curve k_{1234} according to Section 3.6.3. Choose any point E_1 on this circle-point curve. This point E_1 lies of course together with its three homologous points E_2, E_3, E_4 on a circle. As the four-bar linkage moves from its first position to its fourth position, although the coupler curve γ_E of E_1 passes precisely through E_2, E_3, E_4, we are not sure if this path γ_E does osculate the circle passing through the points E_1, E_2, E_3, E_4. Similar consideration applies to the generation of ellipses. The following descriptions are based mainly on (Beyer, 1953, Chap. IX).

Dimensional sythesis of linkages --- path generation problems

6.9.1 Basic geometrical concept

As is well-known, an ellipse is determined by five points. This means that, for any five points on a plane, an ellipse can be drawn passing through these five points. Please refer to Section A12.1. In other words, if five positions of a body are given, then any point E_1 on the first position of this body, must lie together with its four homologous points E_2, E_3. E_4. E_5 on an ellipse (or other conic sections).

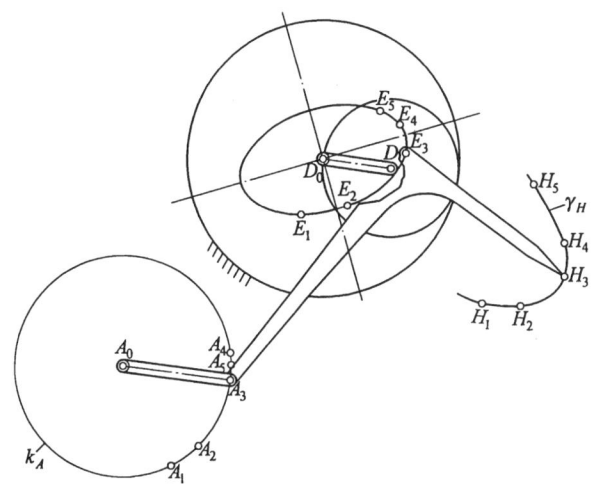

Fig. 6.36. Generation of a given curve γ_H by an ellipse.

Fig. 6.36 shows a given curve γ_H. Choose on this curve any five points H_1, H_2, H_3, H_4, H_5. Choose any fixed point as A_0, and any radius $\overline{A_0 A}$. Draw the crank circle k_A, and draw arcs with any radius $\overline{HA}$ and centres at H_1, ..., H_5, cutting k_A in the five points A_1, ..., A_5. Now the five positions $A_1 H_1$, ..., $A_5 H_5$ of the body AH are all known. Choose any point E on this body, then the positions of the five homologous points E_1, ..., E_5 are also known. In case the conic section passing through the five points E_1, ..., E_5 is an ellipse, it is feasible to make use of a Cardan circle-pair to generate this ellipse. We can thus obtain a mechanism for generating a curve approximating the given γ_H.

Please note that, the linkage $A_0 AEH$ here is just a *dyad*, not yet a four-bar linkage. Furthermore, the sequence of the A's is such that A_5 lies between A_3 and A_4, therefore $D_0 D$ is chosen as the driving crank, not $A_0 A$.

6.9.2 Six finitely separated positions of a body --- the conic section point curve

We shall not consider here how to guide a body through six finitely separated positions. Rather, we shall consider a somewhat different problem. We know that for five finitely separated positions of a body, any point on the body always lies together with its other four homologous points on a conic (e.g. an ellipse). We now wish to find, for six finitely separated positions of a body, the locus of such point on the body such that all its six homologous points lie on a conic.

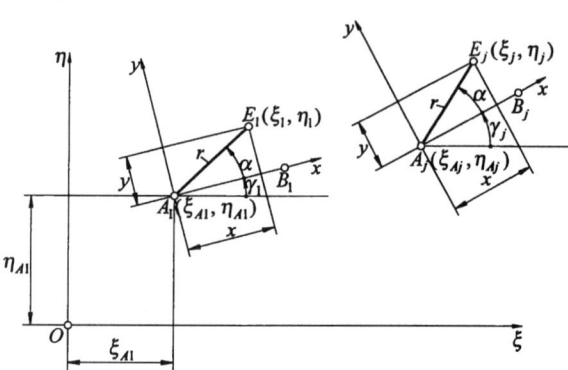

Fig. 6.37. Fixed coordinate system (ξ, η) and moving coodinate system (x, y).

In Fig.6.37, with respect to a fixed coordinate system $\xi\eta$, there is a moving body, on which there is a reference point A and a reference line AB. The jth position of the body is specified by the coordinates (ξ_{Aj}, η_{Aj}) of A_j and the orientation angle γ_j of the line A_jB_j. Let E be a point on this moving body, and the coordinates of E with respect to a coordinate system xy embedded in the body be (x, y), or, in polar coordinates (r, α)

$$\left. \begin{array}{l} x = r\cos\alpha \\ y = r\sin\alpha \end{array} \right\}$$

In the jth position of the moving body, the coordinates of E_j with respect to the fixed coordinate system $\xi\eta$ are

$$\left. \begin{array}{l} \xi_j = \xi_{Aj} + x\cos\gamma_j - y\sin\gamma_j \\ \eta_j = \eta_{Aj} + x\sin\gamma_j + y\cos\gamma_j \end{array} \right\} \quad (6.45)$$

Please note that, equations (6.45) are exactly the same as equation (4.1). Suppose the points $E_1, ..., E_6$ lie on an ellipse on the fixed plane, and the equation of the ellipse is

$$a_{11}\xi^2 + 2a_{12}\xi\eta + a_{22}\eta^2 + 2a_{13}\xi + 2a_{23}\eta + 1 = 0 \quad (6.46)$$

The coordinates of the six points E_j should satisfy equation (6.46). Thus a set of six linear equations in the five unknowns $a_{11}, a_{12}, a_{22}, a_{13}, a_{23}$ is formed. The determinant of the coefficients of these six equations must vanish, or

Dimensional systhesis of linkages --- path generation problems

$$k_{E1}: \begin{vmatrix} \xi_1^2 & \xi_1\eta_1 & \eta_1^2 & \xi_1 & \eta_1 & 1 \\ \xi_2^2 & \xi_2\eta_2 & \eta_2^2 & \xi_2 & \eta_2 & 1 \\ \xi_3^2 & \xi_3\eta_3 & \eta_3^2 & \xi_3 & \eta_3 & 1 \\ \xi_4^2 & \xi_4\eta_4 & \eta_4^2 & \xi_4 & \eta_4 & 1 \\ \xi_5^2 & \xi_5\eta_5 & \eta_5^2 & \xi_5 & \eta_5 & 1 \\ \xi_6^2 & \xi_6\eta_6 & \eta_6^2 & \xi_6 & \eta_6 & 1 \end{vmatrix} = 0 \qquad (6.47)$$

Substituting equation (6.45) into equation (6.47) results in an equation of 8th degree in the variables x, y, or the equation of the locus of the point E_1(or that of $E_2, ..., $ or E_6) on the moving plane, denoted by k_{E1}. This locus may be called the *conic section point curve*. The main conclusion is (Beyer, 1939):

Theorem 25 The conic section point curve k_{E1} on the first position of the moving body passes through the five poles $P_{12}, P_{13}, P_{14}, P_{15}, P_{16}$ and the ten image poles $P_{23}^1, P_{24}^1, P_{25}^1, P_{26}^1, P_{34}^1, P_{35}^1, P_{36}^1, P_{45}^1, P_{46}^1$ and P_{56}^1.

In general, the equation of this curve of 8th degree is quite complicated. However, In special cases, it may break up into straight lines and a circle.

6.9.3 A special case of conic section point curve

With respect to Fig. 6.37, suppose the points A_1, A_j ($j = 2, 3, 4, 5, 6$) lie on a straight line, and among the 6 positions of the body AB, each two positions are parallel, i.e. $A_1B_1 \| A_6B_6$, $A_2B_2 \| A_5B_5$, $A_3B_3 \| A_4B_4$, as shown in Fig. 6.38. In this case the coefficients of the 8th, 7th and 6th degree terms in equation (6.47) vanish identically, and the remaining equation of a curve of 5th degree breaks up into three straight lines and a circle. Let the origin be at A_1, and let the line A_1A_j be the ξ-axis. Hence $\eta_{A1} = \eta_{Aj} = 0$, and $\gamma_1 = \gamma_6, \gamma_2 = \gamma_5, \gamma_3 = \gamma_4$. Equation (6.47) now reduces to

$$[\sin(\alpha+\gamma_1)-\sin(\alpha+\gamma_2)][\sin(\alpha+\gamma_2)-\sin(\alpha+\gamma_3)][\sin(\alpha+\gamma_3)-\sin(\alpha+\gamma_1)]$$

$$\left[4r\sin\frac{\gamma_2-\gamma_1}{2}\sin\frac{\gamma_3-\gamma_2}{2}\sin\frac{\gamma_1-\gamma_3}{2}-(h_2+h_5-h_6)\cos\left(\alpha+\frac{\gamma_3+\gamma_2}{2}\right)\sin\frac{\gamma_3-\gamma_2}{2}\right.$$

$$\left.+(h_3+h_4-h_2-h_5)\cos\left(\alpha+\frac{\gamma_1+\gamma_2}{2}\right)\sin\frac{\gamma_1-\gamma_2}{2}\right]=0 \qquad (6.48)$$

where $h_j = \overline{A_1A_j}$ ($j = 2, 3, 4, 5, 6$). Each of the first three factors on the left hand side of equation (6.48) represents a straight line. These three factors may also be written as

$$\cos\left(\alpha+\frac{\gamma_1+\gamma_2}{2}\right)\cos\left(\alpha+\frac{\gamma_2+\gamma_3}{2}\right)\cos\left(\alpha+\frac{\gamma_3+\gamma_1}{2}\right)=0 \qquad (6.49)$$

All three lines pass through the origin A_1, denoted respectively by g_{12}, g_{23} and g_{13}. Their inclinations with respect to the polar line are respectively

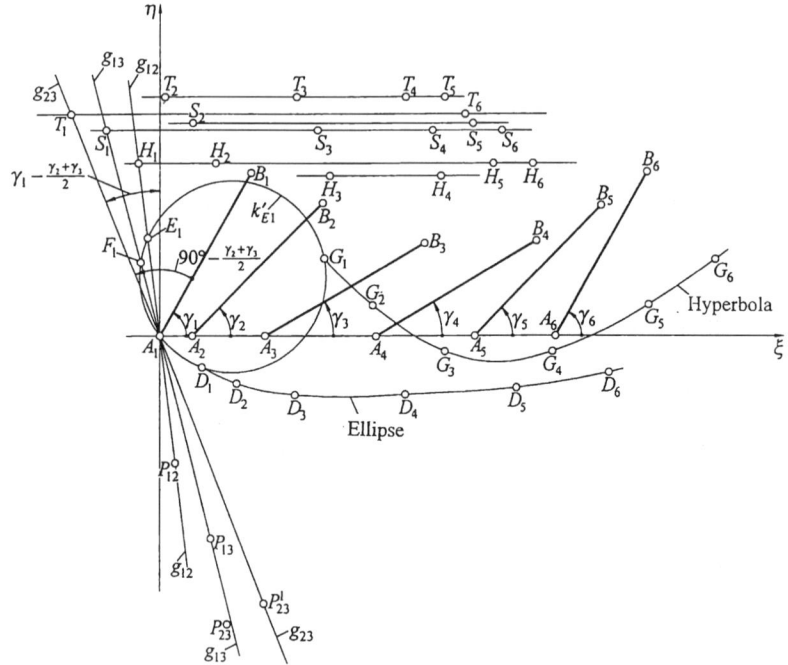

Fig. 6.38. Special case of k_{E1}: k_{E1} breaking up into lines g_{12}, g_{23}, g_{13} and circle k'_{E1}.

$$90° - \frac{\gamma_1 + \gamma_2}{2}, \quad 90° - \frac{\gamma_2 + \gamma_3}{2}, \quad 90° - \frac{\gamma_3 + \gamma_1}{2}$$

Please note that, by the definition of the polar coordinates (r, α) in Fig. 6.37, the polar line is $A_1 B_1$, and not $A_1 \xi$. The line g_{12} passes through P_{12}, g_{13} passes through P_{13}, and g_{23} passes through P^1_{23}.

The fourth factor of equation (6.48) represents a circle passing through A_1, denoted by k_{E1}'. This circle k_{E1}' intersects g_{12} in E_1, and g_{13} in F_1, and

$$\overline{A_1 E_1} = \frac{h_2 + h_5 - h_6}{4 \sin \frac{\gamma_2 - \gamma_1}{2}}, \quad \overline{A_1 F_1} = \frac{h_3 + h_4 - h_6}{4 \sin \frac{\gamma_3 - \gamma_1}{2}} \qquad (6.50a, b)$$

Hence the circle k'_{E1} is determined by the three points A_1, E_1, F_1.

Since k_{E1} breaks up into straight lines g_{12}, g_{23}, g_{13} and the circle k'_{E1}, take any point on one of the three lines, then this point must lie together with its five homologous points on a conic section. Take, for instance, the three points H_1, S_1, T_1 in Fig. 6.38. The points H_1, H_2, H_5, H_6 lie on a straight line, and H_3, H_4 lie on another

Dimensional systhesis of linkages --- path generation problems

line. The points S_1, S_3, S_4, S_6 and S_2, S_5 lie respectively on two lines. The points T_1, T_6 and T_2, T_3, T_4, T_5 lie also respectively on two lines. All these lines are parallel to the line A_1A_6. In other words, each of these conics has broken up into two straight lines. They are therefore of no use in generating an approximately elliptical coupler curve.

For a point such as D_1 or G_1 on the circle k'_{E1}, we see that the homologous points D_1, D_2, ..., D_6 lie on an ellipse, and that G_1, G_2, ..., G_6 lie on a hyperbola. Hence if we wish to generate an approximately hyperbolical coupler curve, we have to choose a point such as G_1; and if we wish to generate an approximately elliptical coupler curve, we have to choose a point such as D_1.

Fig. 6.39 shows a practical application of the present special case. B_0BA is a slider-crank mechanism. The points A_1, A_2, ..., A_6 lie precisely on a straight line B_0A_1. On the crank circle of B, make the points B_6, B_5, B_4 respectively symmetrical to B_1, B_2, B_3, with respect to the vertical line passing through B_0 and normal to B_0A. Hence the pair B_1, B_6 is of the same altitude. Similarly, each of the two pairs, B_2, B_5 and B_3, B_4, is also of the same altitude. Hence $A_1B_1 \| A_6B_6$, $A_2B_2 \| A_5B_5$, $A_3B_3 \| A_4B_4$. We omit here the three lines g_{12}, g_{23}, g_{13} in Fig. 6.38, because they are useless. The circle k_{E1}', or the last factor in equation (6.48), is simplified to a circle with $\overline{A_1B_1}$ as its diameter. Take any point on this circle as E_1, then E_1 should lie together with its five homologous points E_2, ..., E_6 on an ellipse. The curve γ_E shown in the figure is the path of E. It can be seen that γ_E osculates well the ellipse passing through E_1, ..., E_6.

To generate the ellipse passing through E_1, ..., E_6, we have to use a double-slider mechanism. The cross guide is bound to the fixed link f, as shown in Fig. 6.40(a). By means of the formulae in Appendix 12, we can compute the orientation of the axes, the location of the centre point of the ellipse, and the magnitudes of its semi-major

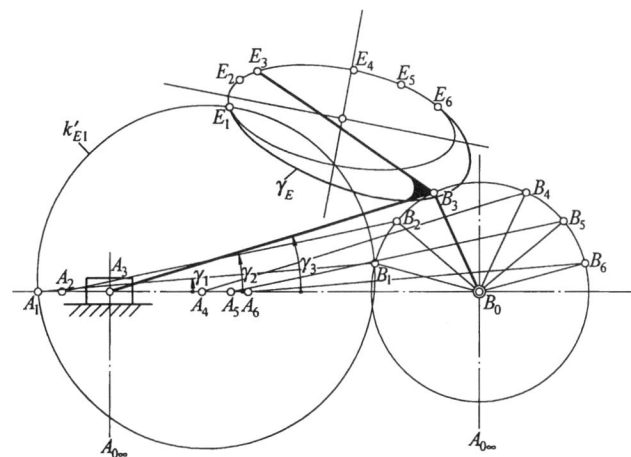

Fig. 6.39. Conic section point curve for six positions of the coupler of a slider-crank, circle k'_{E1} with $\overline{A_1B_1}$ as diameter being a part of this curve.

axis a_e and semi-minor axis b_e. We then determine the position of the cross guide by the axes of the ellipse. We can also find the inclination angle t of the coupler link GH of the double-slider with respect to the cross guide by means of equation (A12.4). With this mechanism, the motion of the point E no longer follows the path γ_E, but

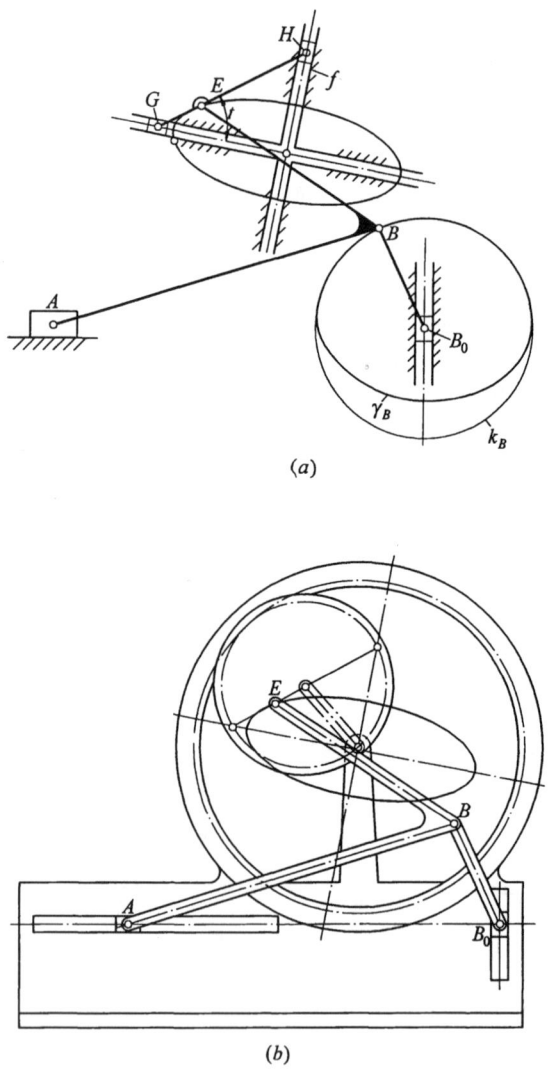

Fig. 6.40. Ellipse generating mechanism for Fig. 6.39. (a) Double-slider. (b) Cardan circle pair.

Dimensional systhesis of linkages --- path generation problems

along the elliptical path. If the relative positions of the three points A, B, E remain unchanged, we have to remove the crank B_0B. The point B shall then move along the path γ_B. The curve γ_B osculates along the upper part of the crank circle of B. If the crank B_0B is reassembled, the joint B_0 may not be fixed, but may move along a vertical guide. As the point B moves along the upper part $B_1 \to B_6$ shown in Fig. 6.39, the point B_0 remains nearly standstill. As B moves to the lower part of γ_B, the point B_0 moves slightly upwards, and then comes down to its original place. We have therefore constructed a dwell mechanism.

A more feasible mechanism for generating an exact ellipse is by using a Cardan circle pair, somewhat like the one shown in Fig. 1.20 or Fig. 1.21, with an internal gear transmission as shown in Fig. 6.40(b). Horizontal and vertical guides and sliders are made respectively at A and B_0, to enable A and B_0 to move respectively in horizontal and vertical directions.

6.9.4 An ellipse tangent to a given curve in four points

The derivations in the present section and Section 6.9.6 are based on (Beyer, 1938a). Let $y = y(x)$ be a given curve. Choose any point (x_1, y_1) on this curve. We want to find an ellipse which should be tangent to the curve at (x_1, y_1) in four points, as shown in Fig. 6.41. In other words, with prescribed orientation of the axes of the osculating ellipse, we want to find the equation of the ellipse which should have four infinitesimally separated points in common with the given curve at (x_1, y_1). This is the so-called kind-I osculating ellipse by Beyer. It is applied in the old fashion chuck

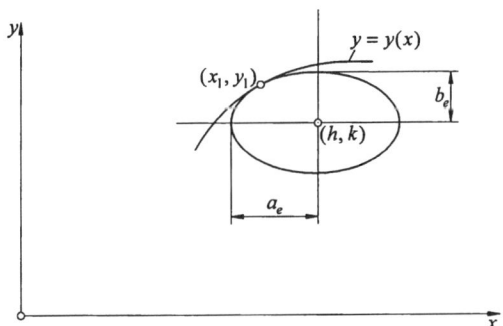

Fig. 6.41. Osculating ellipse of the kind-I.

for a lathe to replace a *logarithmic spiral* (*Archimedes spiral*, or *equiangular spiral*). The condition that the ellipse and the curve $y(x)$ should be tangent in four-point at (x_1, y_1) is that they should both have the same y, y', y'' and y''' at (x_1, y_1) ((')denotes the differential coefficient of a certain function with respect to x). Suppose the equation of this osculating ellipse is

$$b_e^2(x-h)^2 + a_e^2(y-k)^2 = a_e^2 b_e^2 \qquad (6.51)$$

Differentiating equation (6.51) three times with respect to x gives

$$\left.\begin{array}{l}b_e^2(x-h)^2 + a_e^2(y-k)y' = 0\\ b_e^2 + a_e^2[y'^2 + (y-k)y''] = 0\\ 3y'y'' + (y-k)y''' = 0\end{array}\right\} \quad (6.52a,b,c)$$

Consider equations (6.51), (6.52a,b,c) as four simultaneous equations in the four unknowns $a_e^2, b_e^2, x-h, y-k$. The solution of these equations is

$$\left.\begin{array}{l}h = x_1 - \dfrac{3y'y''}{3y''^2 - y'y'''}\\ k = y_1 + \dfrac{3y'y''}{y'''}\\ a_e^2 = \dfrac{27y'y''^4}{y'''(3y''^2 - y'y''')^2}\\ b_e^2 = \dfrac{27y'^2y''^4}{y''^2(3y''^2 - y'y''')}\end{array}\right\} \quad (6.53a,b,c,d)$$

If we wish to replace a given curve $y = y(x)$ at a certain point (x_1, y_1) by an osculating ellipse, we find first the values of y', y'', y''' of the curve at (x_1, y_1), substitute them into the four equations (6.53a,b,c,d), and therefore readily find the determining magnitudes h, k, a_e, b_e for the ellipse.

For a special case, in which the chosen point (x_1, y_1) is at the top point or at the bottom point of the ellipse, the characteristics of the ellipse ensure not only $y' = 0$, but also $y''' = 0$. Hence the original curve $y = y(x)$ to be simulated should also satisfy the conditions of $y' = 0$ and $y''' = 0$ at the point (x_1, y_1). However, since in the fraction terms in equations (6.53b,c,d) both numerators and denominators become zero, these fraction terms are indeterminate. We can make use of the well-known method in differential calculus by letting the numerator and denominator approach to their limits separately to find these expressions, namely,

$$\left.\begin{array}{l}h = x_1\\ k = y_1 + \dfrac{3y''^2}{y''''}\\ a_e^2 = \dfrac{3y''}{y''''}\\ b_e^2 = \dfrac{9y''^4}{y''''^2}\end{array}\right\} \quad (6.54a,b,c,d)$$

6.9.5 Dwell mechanism by means of osculating ellipse

In Fig. 6.40 (a, b), we made use of an elliptical path to construct a dwell mechanism. Here, we make use of a four-point tangent ellipse based on the method mentioned in Section 6.9.4 to construct a dwell mechanism. Fig. 6.42 shows an

Dimensional systhesis of linkages --- path generation problems

epicyclic gear train, in which π_c is a small wheel rolling on the inside of a fixed, large wheel π_f. Let the radius of π_c be r, and that of π_f be $3r$. The path of a point E on the circle of π_c is a hypocycloid $E_1E_2E_3$. The parametric equations of the part E_1E_2 of this hypocycloid are

$$\begin{aligned} x &= r(\sin 2\alpha + 2\sin\alpha) \\ y &= r(3 + \cos 2\alpha - 2\cos\alpha) \end{aligned} \right\} \qquad (6.55a,b)$$

In these equations, the origin O of the xy-coordinate system is taken at the lowest point of π_f, and x-axis is the tangent of π_f. Let the centre of π_c be denoted by D, and that of π_f by D_0. The arm length $s = \overline{D_0D}$, and $\sphericalangle OD_0D = \alpha$. Differentiating equations (6.55a,b) with respect to α yields $y' = \dfrac{dy}{d\alpha} \Big/ \dfrac{dx}{d\alpha}$, $y'' = \dfrac{dy'}{d\alpha} \Big/ \dfrac{dx}{d\alpha}$, $\cdots$. Evaluating these differential coefficients at $E_m(x, y) = E_m(0, 2r)$, we have

$$y' = 0,\ y'' = -\frac{1}{8r},\ y''' = 0,\ y'''' = -\frac{3}{128r^3}$$

Substituting these values into equations (6.54 a, b, c, d), we have

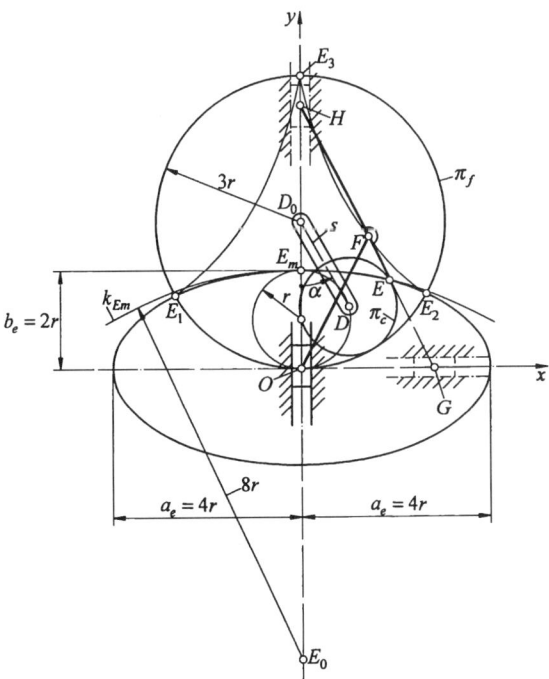

Fig. 6.42. Replacement of hypocycloid with a four-point tangent ellipse, radius of small wheel = r, radius of large wheel = $3r$.

$$h = 0, k = 0, a_e = 4r, b_e = 2r$$

These four values determine the equation of the osculating ellipse that is tangent to the hypocycloid in four points at E_m. As mentioned before, this ellipse can readily be generated by the point E on the coupler HEG of a double-slider mechanism, where $\overline{HE} = a_e$, $\overline{EG} = b_e$. It can also be generated by an isosceles slider-crank mechanism

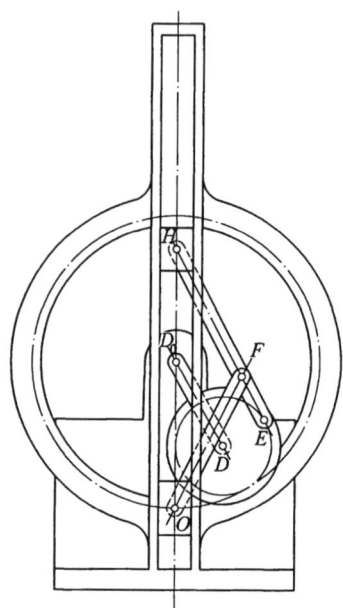

Fig. 6.43. Internal gear mechanism of Fig. 6.42, with sliders mounted at O, E.

OFH where $\overline{OF} = \overline{FH} = (a_e + b_e)/2$. Certainly, if we wish to retain both Cardan circle-pair π_c, π_f and the isosceles slider-crank, we have to make at O a slider and guide, to allow the joint O to slide vertically; otherwise the whole mechanism would become an immovable locked chain. Bearing these considerations in mind, we show in Fig.6.43 a practical construction of the mechanism.

The radius of curvature of the osculating ellipse at E_m is equal to $a_e^2/b_e = 8r$, which is also the radius of curvature of the hypocycloid at E_m, as to be expected. For as long as two curves are tangent to each other in three points, they must have a common radius of curvature, particularly in the present case where the two curves are in a four-point contact.

A link i may be connected from E to E_0. Let $\overline{EE_0} = 8r$, and take the point E_0 on the y-axis. The link i drives further a rocker l as shown in Fig.6.44. The point E_{30} is the other extreme position of E_0 when E moves to the position E_3, and also lies on the

Dimensional sythesis of linkages --- path generation problems 279

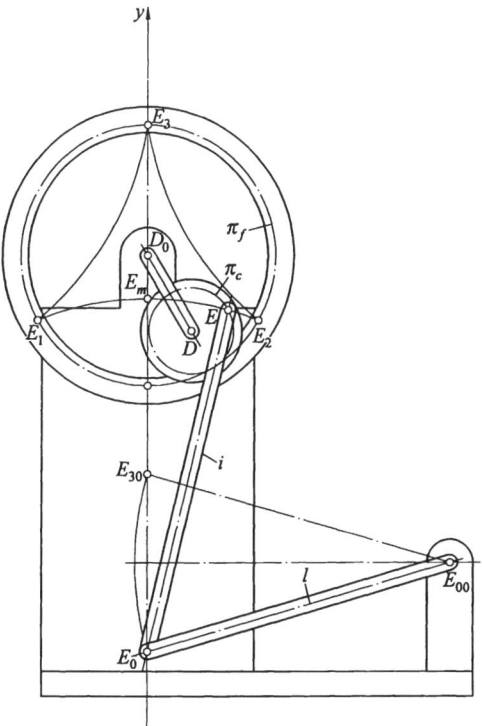

Fig. 6.44. Dwell mechanism constructed according to Fig. 6.42, $\overline{EE_0} = 8r$.

y-axis. The pivot point E_{00} of the lever l is taken on the perpendicular bisector of $\overline{E_0 E_{30}}$. In this way, as π_c rolls on the inside of π_f, and as the point E moves from E_1 to E_2, the link l remains approximately stationary to form a dwell mechanism.

6.9.6 An ellipse tangent to a given curve in five points

Let $y = y(x)$ be a given curve. Suppose we wish to construct an ellipse to be tangent to this curve at (x_1, y_1) in five points. We shall derive the conditional equations. We transform first the origin of the coordinates to the point (x_1, y_1), so that the required ellipse will be tangent to the given curve in five points at the origin. In other words, both the curve and the ellipse should have common values of y', y'', y''', y'''' at the origin. Since the ellipse is passing through the origin, its constant term should vanish. We may therefore assume the equation of the ellipse to be

$$a_{11}x^2 + 2a_{12}xy + a_{22}y^2 + 2a_{13}x + 2y = 0 \qquad (6.56)$$

Differentiating equation (6.56) four times with respect to x gives a set of four simultaneous equations in the unknowns a_{11}, a_{12}, a_{22}, a_{13}. Solving these equations yields

$$\left.\begin{aligned} a_{11} &= (4y'^2 y'''^2 + 6y'y''^2 y''' - 3y'^2 y''y'''' - 9y''^4)/(9y''^3) \\ a_{12} &= (3y'y''y'''' - 3y''^2 y''' - 4y'y'''^2)/(9y''^3) \\ a_{22} &= (4y'''^2 - 3y''y'''')/(9y''^3) \\ a_{13} &= -y' \end{aligned}\right\} \quad (6.57a,b,c,d)$$

Substituting the values of y', y'', y''', y'''' of the original curve $y = y(x)$ at the new origin into equations (6.57a, b, c, d), we can find the four coefficients a_{11}, a_{12}, a_{22}, a_{13}. Hence, we are able to determine the osculating ellipse. We can determine the equations of the major and minor axes of the ellipse using equations (A12.3a, b). In these equations, it does not matter whether the constant term of the equation (A12.1) of the ellipse is equal to 1, the values of p, q, u, v in equation (A12.2) remaining the same. Therefore, it does not matter whether the ellipse passes through the origin, equations (A12.3a, b) are always valid.

The osculating ellipse found by this method is different from the one discussed in Section 6.9.4, since the directions of the major and minor axes of the ellipse are not prescribed. This is what Beyer calls the osculating ellipse of kind-II. It was applied in non-circular wheel or teeth of a milling cutter to replace the logarithmic spiral (Beyer, 1950). Please refer to Exercise 6.8.

6.10 Generating coupler curves with cusps

6.10.1 Generating coupler curves with two cusps

We have discussed in Section 6.5.2 how to synthesize a four-bar linkage to generate a node of a coupler curve. In Fig. 6.18, if the intersection point P_{cf} of A_0A_1 and B_0B_1 is taken as a coupler point G_1, or if the point G_1 is just the velocity pole of the coupler c (please note that G_1 is not a point on B_0B_1), then according to Section 3.7.4, the path of G_1 here becomes a cusp. Therefore, in order to generate a coupler curve with a cusp, the coupler point in question should become a velocity pole in that position, or at the intersection of the centre lines of the two cranks A_0A and B_0B. In order to generate a coupler curve with two cusps, the coupler point in question should become the velocity pole in both prescribed positions.

In Fig. 6.45, suppose the three points A_0, E_1, E_2 are prescribed. In particular, E_1, E_2 are the prescribed positions of the cusps of the coupler curve. It is required to synthesize the four-bar linkage A_0ABB_0. The synthesis is as follows: Join A_0E_1 and A_0E_2 to determine ϕ_{12}. Erect the bisector of ϕ_{12} and the perpendicular bisector e_{12} of $\overline{E_1E_2}$, to get the intersection B_0. The angle $\angle E_1B_0E_2$ is then ψ_{12}. Dispose the angle ψ_{12} symmetrically on both sides of A_0B_0. The intersections with A_0E_1 and A_0E_2 give the points A_1, A_2 respectively. Take any point on B_0E_1 as B_1. The synthesis is thus completed. What we could choose was only the pole B_1 on B_0E_1, or B_2 on B_0E_2. There

Dimensional systhesis of linkages --- path generation problems 281

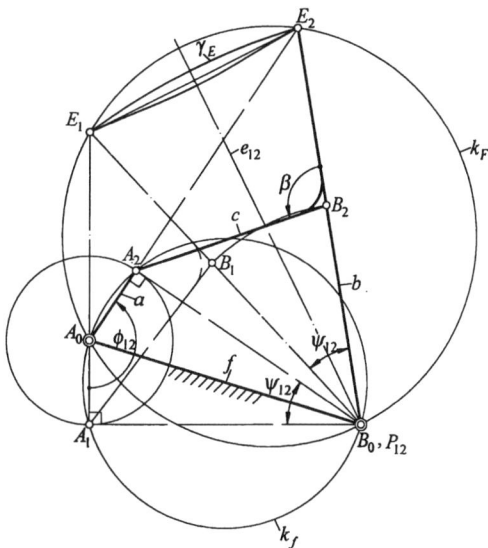

Fig. 6.45. Coupler point curve with two cusps: construction of A_0ABB_0 for prescribed A_0, E_1, E_2.

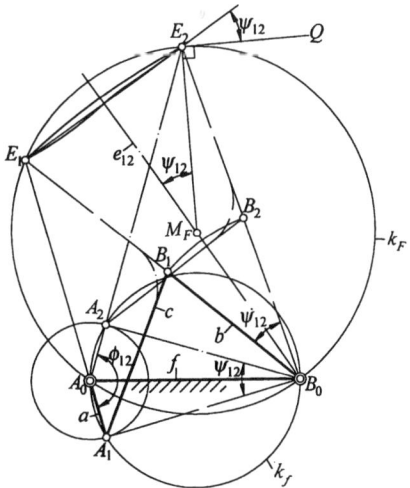

Fig. 6.46. Coupler point curve with two cusps: construction of A_0ABB_0 for prescribed E_1, E_2, ϕ_{12}.

are therefore ∞^1 solutions. We can now see that E_1 is just the velocity pole of A_1B_1, and E_2 is just the velocity pole of A_2B_2. For the whole triangles $\triangle A_1B_0E_1$ and $\triangle A_2B_0E_2$, B_0 is the pole P_{12} between these two positions of the body AE. Therefore the angle between A_1E_1 and A_2E_2 is ψ_{12}, or $\sphericalangle E_1B_0E_2$ is its angle of rotation. Hence we have $\phi_{12} + \psi_{12} = 180°$, and $\phi_{12}/2 + \psi_{12}/2 = 90°$. In other words, if a circle k_f is drawn with $\overline{A_0B_0}$ as its diameter to intersect A_0E_1 and A_0E_2, the points A_1, A_2 can also be located. A circle like k_f, the angle subtended by the two end points on whose diameter at any point on the half circle is always 90°, is commonly called a *Thales circle*.

Since a cusp belongs to a kind of nodes, the points E_1, E_2 are on the circle k_F passing through A_0, B_0. Another kind of problem is that shown in Fig. 6.46. The positions E_1, E_2 and angle ϕ_{12} are prescribed. It is required to synthesize the four-bar linkage. The synthesis is as follows: Join $\overline{E_1E_2}$ and erect its perpendicular bisector e_{12}. According to Fig. 6.45, we can see that the circumferential angle subtended by $\overline{E_1E_2}$ on circle k_F is ψ_{12}, and $\psi_{12} = 180° - \phi_{12}$. Hence we draw a line E_2Q at E_2 in Fig. 6.46, such that it builds an angle ψ_{12} with E_1E_2, and then draw a perpendicular to E_2Q, to intersect e_{12} in M_F. M_F is the centre of the circle k_F, and the circle k_F is thus determined. The lower intersection of e_{12} and k_F is B_0. It can be seen that $\sphericalangle E_1B_0E_2 = \psi_{12}$. Now choose any point on k_F as A_0, and any point on B_0E_1 as B_1. The synthesis is thus completed. There are ∞^2 solutions.

Please note that, although the E_1, E_2 in Figs. 6.45 and 6.46 are the extreme positions of E, they do not correspond to the extreme positions of B.

According to the above geometrical method, for a given A_0ABB_0 in Fig. 6.45, to locate the point E on the coupler c (=AB) that can generate two cusps, the procedure

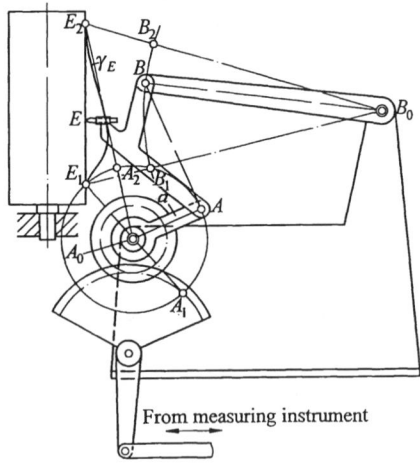

Fig. 6.47. Application of coupler curve with two cusps and with an approximately straight line portion for guiding a recorder pen (Hain, 1941 a,b).

Dimensional systhesis of linkages --- path generation problems

is quite simple. It is as follows: Draw the circle k_f with $\overline{A_0B_0}$ as the diameter to cut the crank circle of point A in the two points A_1, A_2. Find the two points B_1, B_2 corresponding to A_1, A_2. The intersection point of A_0A_1 and B_0B_1 is E_1, and that of A_0A_2 and B_0B_2 is E_2. Check: the four points E_1, E_2, A_0, B_0 should lie on the same circle k_F.

Fig. 6.47 shows an example of application given by (Hain, 1941a, b). A_0ABB_0 is a four-bar linkage, with E as its coupler point, being made in the form of a recording pen. The coupler curve γ_E of E exhibits two cusps at E_1, E_2. Between these two points the path γ_E is approximately a straight line. The position of the point E on this straight line is determined by the "quantity" from a measuring instrument through a rotating crank a ($= A_0A$). Outside the straight line, the recording pen and the point E are lifted up from the recording drum.

Fig. 6.48 shows another example of application of a coupler curve with two cusps (Hain, 1941a, b; 1961, Fig.12-28). The purpose was to reduce the noise in a typewriter mechanism. D_0DGG_0 is a four-bar linkage. If a downward force is applied at one end of the lever d, as shown in the figure, the follower link g will reach its extreme position. Upon releasing the lever d, the link g will be pulled back by a spring s_2 and will reach its another extreme position as shown in the figure. In each extreme position, a certain *stop* should be provided to limit the motion of the link g. Hence noise occurs whenever the link g hits the stop. Cushions made of rubber or felt were provided to reduce the noise. However, the noise returned as the cushions wore away. In order to overcome this defect, cushions were replaced by a mechanism designed for reducing noise. In Fig. 6.48, another four-bar linkage A_0ABB_0 is so constructed that its coupler curve γ_E forms cusps at E_1, E_2. A link j is connected between c and g, to enable the positions E_1, E_2 just to correspond to the two extreme

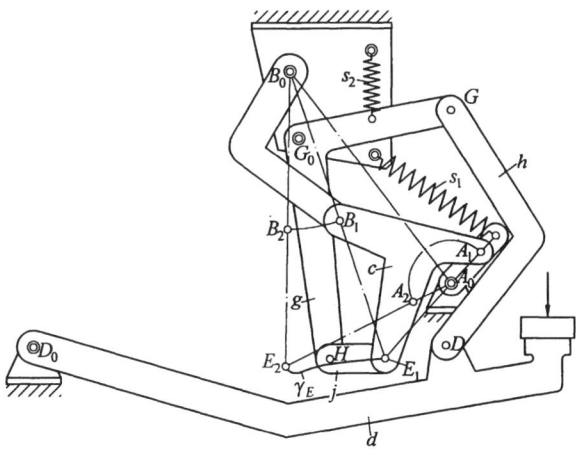

Fig. 6.48. Application of coupler curve with two cusps to reduce noice (Hain, 1961).

positions of g. Since the point E must move backwards as it reaches E_1 or E_2, g cannot move beyond its extreme positions. The four-bar linkage A_0ABB_0 thus serves as a replacement for hard stops.

6.10.2 Generating coupler curves with three cusps

(Mayer, 1938) first investigated the mathematics of generating coupler curve with three cusps. Subsequently, (Beyer, 1938b) digested the concepts from an engineering point of view. However, as the materials involved some geometrical concepts which may not be readily applicable to engineers, in the following, we shall introduce only the major concepts and set a few simple geometrical relationships as Exercises 6.13, 6.14, 6.15 for familiarity purposes.

In the case of the coupler point E of a four-bar linkage generating a coupler curve with three cusps E_1, E_2, E_3, we know from Fig.6.17 that the three nodes of the coupler curve become three cusps, but they should still lie on the circle k_F. In other words, the five points A_0, B_0, E_1, E_2, E_3 must lie on the same circle k_F. Suppose the positions of the three points E_1, E_2, E_3 are prescribed, the circle k_F is then determined, as shown in Fig. 6.49. Join $\overline{E_1E_3}$ and erect its perpendicular bisector e_{13}, intersectiong the circle k_F in the two points A_0, A_0'. Join $\overline{E_1E_2}$ and erect its

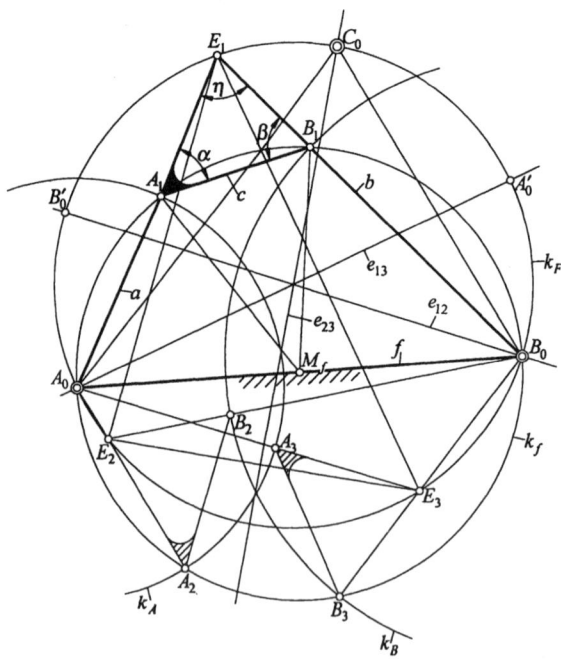

Fig. 6.49. Coupler point curve with three cusps.

Dimensional systhesis of linkages --- path generation problems   285

perpendicular bisector e_{12}, intersecting the circle k_F in the two points B_0, B_0'. Take now A_0, B_0 as two fixed centres of rotation and draw the circle k_f with $\overline{A_0B_0}$ as its diameter, intersecting A_0E_1, B_0E_1 respectively in A_1, B_1. The four-bar $A_0A_1B_1B_0$ together with its coupler triangle $A_1B_1E_1$ is one of the required linkages.

Knowing A_1, B_1, we can determine the respective crank circles k_A, k_B of the two points A, B. Join A_0E_2 to intersect k_A in A_2, and join B_0E_2 to intersect k_B in B_2. Similarly, the intersections of A_0E_3, B_0E_3 and k_A, k_B respectively are A_3, B_3. It can be shown by simple geometric relations that $\triangle A_1B_1E_1 \equiv \triangle A_2B_2E_2 \equiv \triangle A_3B_3E_3$. Next join $\overline{E_2E_3}$, and erect its perpendicular bisector e_{23}, cutting the circle k_F in C_0. It can be proved that $\sphericalangle A_0B_0E_1 = \sphericalangle B_1A_1E_1 = \sphericalangle B_0A_0C_0$, $\sphericalangle B_0A_0E_1 = \sphericalangle A_1B_1E_1 = \sphericalangle A_0B_0C_0$, $\triangle A_1B_1E_1 \sim \triangle B_0A_0E_1 \sim \triangle A_0B_0C_0$. Hence C_0 is just the singular focus of the coupler curve in Fig. 6.17.

Since e_{12}, e_{13} intersect the circle k_F further in B_0' and A_0', there are altogether four possibilities of selecting the fixed link, namely: A_0B_0, $A_0'B_0'$, $A_0'B_0$ and A_0B_0'. The perpendicular bisectors selected each time, besides the lines e_{12}, e_{13} mentioned above, can also be e_{12}, e_{23} or e_{13}, e_{23}. There are therefore altogether 12 possible four-bar linkages. Of course, these linkages include some non-movable mechanisms, of which we cannot reach the cusp positions without dismantling and reassembling the mechanism. The mechanism shown in Fig. 6.49 is a double rocker. However, the linkage can reach the three cusp positions on the coupler curve through one continuous motion.

For generating three cusps by means of a six-bar linkage, please refer to (Hwang & Chang, 1989); and for generating four cusps by six-bar linkages, please refer to (Chang & Hwang, 1994).

6.11 Generating symmetrical coupler curves

The first literature on this subject was the book (Rauh, 1951, p.91). However, it considered only a special case of symmetrical coupler curves (namely, $\beta = 180°$, please see following sections). The next literature was (Meyer zur Capellen, 1956a), which is the basis for generating all symmetrical coupler-curves with non-symmetrical four-bar linkages. This is also what we are going to discuss.

6.11.1 Proof of the principle

In Fig. 6.50, A_0ABB_0 is a four-bar linkage, in which

$$\overline{AB} = \overline{B_0B} = \overline{BE} = b \tag{6.58}$$

The path of the coupler point E of a four-bar linkage satisfying the conditions (6.58) is a symmetrical curve. The proof is as follows: because of the conditions (6.58), the three points E, A, B_0 always lie on a circle with B as centre, and $\overline{B_0B}(=b)$ as the

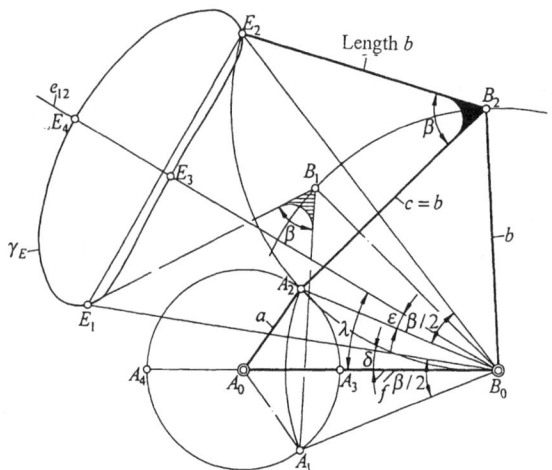

Fig. 6.50. Four-bar linkage for generating symmetrical coupler curves: $\overline{AB} = \overline{B_0B} = \overline{BE}$, $\sphericalangle AB_0E = \beta/2 =$ constant.

radius. Therefore

$$\sphericalangle AB_0E \equiv \frac{1}{2} \sphericalangle ABE = \frac{\beta}{2}$$

where β is a constant angle. Consider two positions A_1 and A_2 of A, symmetrically disposed with respect to A_0B_0. The two configurations $A_1B_0B_1E_1$ and $A_2B_0B_2E_2$ are congruent; for if we rotate the former configuration about B_0 through an angle $\sphericalangle A_1B_0A_2$, we obtain the latter. Erect the perpendicular bisector e_{12} of $\overline{E_1E_2}$, and denote the angle between e_{12} and A_0B_0 by λ. Further let $\sphericalangle A_0B_0E_1 = \delta$, and denote the angle between e_{12} and A_2B_0 by ε. Now because E_1, E_2 are symmetrical with respect to e_{12}, we have

$$\lambda - \delta = \beta/2 - \varepsilon \tag{6.59}$$

Also because A_1B_0, A_2B_0 are symmetrical with respect to A_0B_0, we have

$$\lambda - \varepsilon = \beta/2 - \delta \tag{6.60}$$

A comparison between equations (6.59) and (6.60) shows that

$$\lambda = \beta/2 \tag{6.61}$$

This means that the inclination of the line e_{12} with respect to A_0B_0 is a constant, or e_{12} is an invariable line. Hence e_{12} is the *axis of symmetry*, inclining at a constant angle of $\beta/2$ to A_0B_0. As A_1 and A_2 coincide, i.e., A takes the position A_3 or A_4, the corresponding position of E is at E_3 or E_4, both being on the line e_{12}.

6.11.2 Simple proof of the condition (6.58)

In Fig. 6.51(a), A_0ABB_0 is a symmetrical four-bar linkage, the well-know Roberts mechanism. $\triangle AEB$ is an isosceles coupler. It is obvious that the path γ_E of the point E is a symmetrical coupler curve. Now draw a diagram in Fig. 6.51(b) similar to Fig. 6.21. According to Roberts-Chebyshev theorem in Section 6.6, we can construct two other cognate linkages. In Fig. 6.51(a) only one cognate linkage is

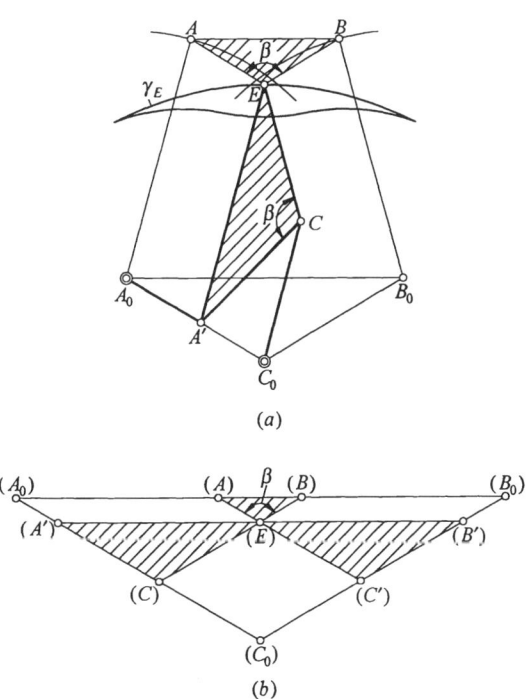

Fig. 6.51. (a) Roberts symmetrical mechanism, path of E being symmetrical.
(b) Cognate linkages of the linkage in (a).

drawn, which is $A_0A'CC_0$. Obviously, in this mechanism $\overline{CC_0} = \overline{CA'} = \overline{CE}$. This is just the four-bar linkage shown in Fig. 6.50. Hence, we have proved equation (6.58), as originally derived by (Meyer zur Capellen, 1956a).

Please note that the range of the angle β in Fig. 6.50 is $0° \leq \beta \leq 360°$.

6.11.3 Antuma's triangular nomogram

For a variety of generating symmetrical coupler curves, (Antuma, 1972)

288   *Kinematics and Design of Planar Mechanisms*

proposed a crank-rocker nomogram as shown in Fig. 6.52, for practical designers. In this nomogram the angles β_1, β_2 are two parameters, as defined in Fig. 6.53. β_1 is the angle between B_0B_4 and A_0B_0 when the point A is at A_4. β_2 is the angle between B_0B_3 and the vertical line when A is at A_3. β_1, β_2 are taken respectively as the abscissa and ordinate of the nomogram in Fig. 6.52. After having selected β_1, β_2, various types of symmetrical coupler curves can be generated, depending on the value of β selected, as will be explained in detail as follows:

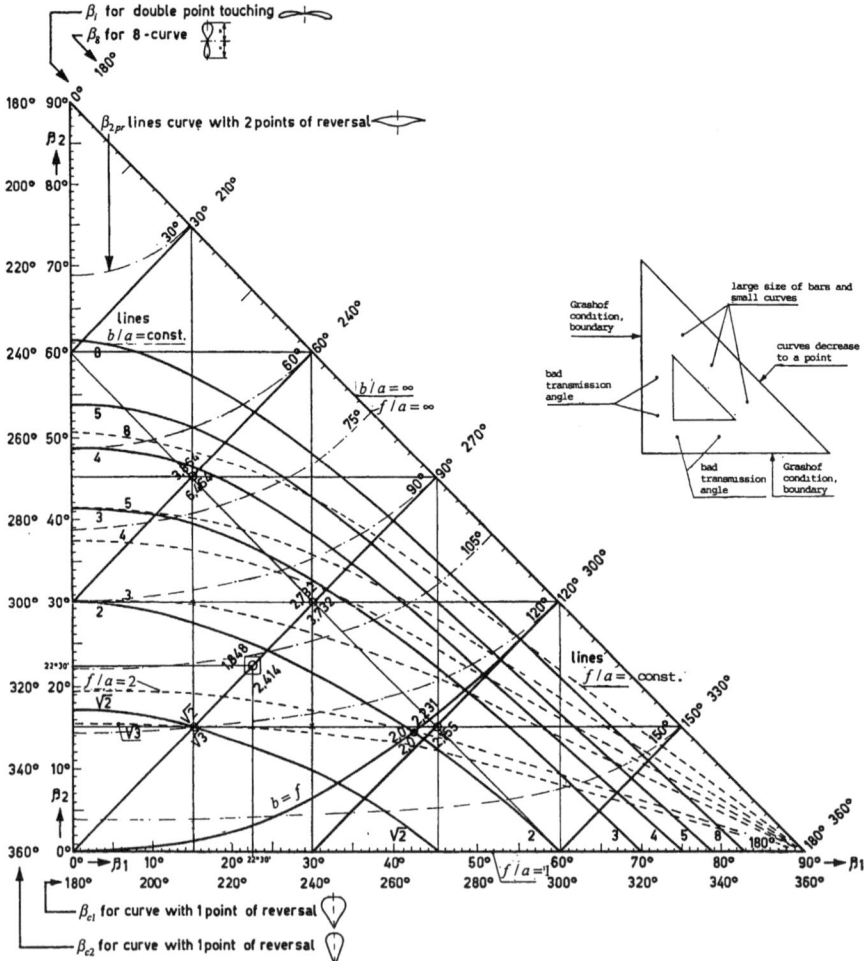

Fig. 6.52. Antuma nomogram for crank-rocker mechanisms generating symmetrical coupler curves.

This figure is reproduced with kind permission of IMechE, London, UK.

Dimensional systhesis of linkages --- path generation problems

(a) Generating a coupler curve with an internally touching node

It can seen from Fig. 6.53, as mentioned before, that the positions of the coupler point are E_3, E_4, coinciding together, and lying on the vertical line passing through

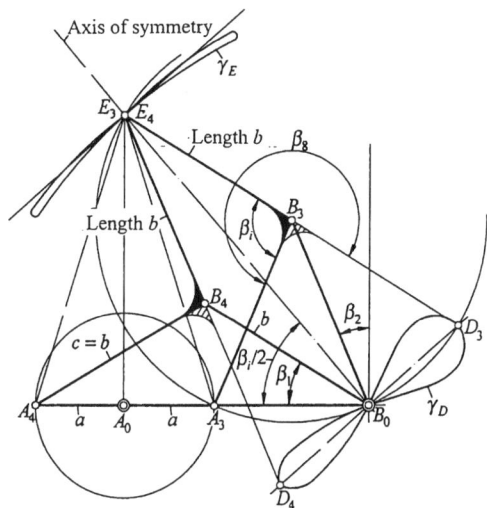

Fig. 6.53. Definitions of β_1, β_2 in Fig. 6.52.

A_0, but still corresponding respectively to A_3, A_4. Moreover, the axis of symmetry is just the bisector of $\angle B_3B_0B_4$. The value β_i of the angle β is therefore

$$\beta_i/2 - \beta_1 = 90° - (\beta_i/2 + \beta_2)$$

or

$$\beta_i = 90° + \beta_1 - \beta_2$$

This is the pencil of lines inclining at an angle of 45°, being all normal to the hypotenuse of the right isosceles triangle. The tangents of the caupler curve at E_3, E_4 are normal to the axis of symmetry, because the instantaneous velocity pole of the coupler is just at B_0, and the axis of symmetry is normal to the volocities of E_3, E_4.

(b) Generating a figure 8 coupler curve

Suppose the coupler point is taken at D, where D is the point lying on the circle with B as centre, and with $\overline{B_0B}$ as radius, and opposite to E. The locations of D_3, D_4 corresponding to A_3, A_4 are shown in Fig. 6.53, and $\overline{B_0D_3} = \overline{B_0D_4}$, and D_3, B_0, D_4 are collinear. The value β_8 of β should therefore be

$$\beta_8 = 180° + \beta_i = 270° + \beta_1 - \beta_2$$

Please note that the shape of the coupler is $\Delta A_3 B_3 D_3$ or $\Delta A_4 B_4 D_4$. In Fig 6.52 the pencil of straight lines is identical with that for β_i. Simply adding 180° to the respective β_i values changes those into β_8 values.

(c) Symmetrical coupler curves with one cusp

Again as mentioned in Section 3.7.4, if the coupler point becomes just the velocity pole, the path of this point exhibits a cusp. Hence as shown in Fig. 6.54(a), the coupler point F_4 coinciding with B_0 is taken as the generating point when A is at A_4. The path of F_4 forms a cusp. Certainly the value β_{c1} of the angle β depends on β_1 or

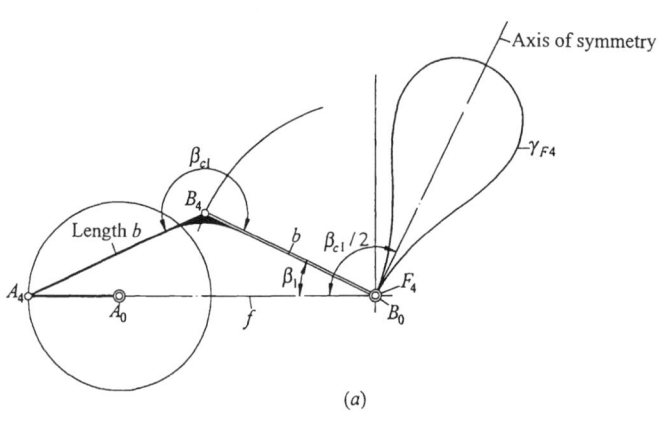

(a)

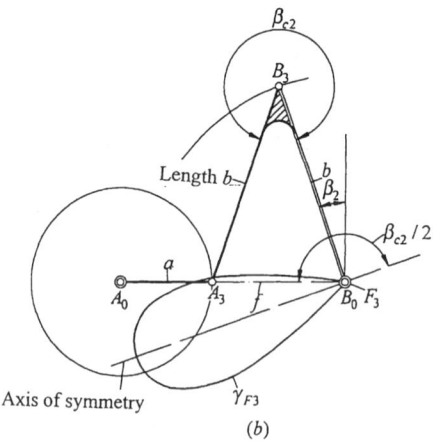

(b)

Fig. 6.54. Angle β for symmetrical coupler curve with one cusp. (a) β_{c1}. (b) β_{c2}.

Dimensional systhesis of linkages --- path generation problems

$$\beta_{c1} = 180°+2\beta_1$$

These are indicated on the bottom line of the nomogram in Fig. 6.52. For example, for $\beta_1 = 40°$, $\beta_{c1} = 260°$.

Similarly, as shown in Fig. 6.54(b), the coupler point F_3 coinciding with B_0 is taken as the generating point when A is at A_3. The path of F_3 also forms a cusp. This value β_{c2} of β depends only on β_2, or

$$\beta_{c2} = 360°-2\beta_2$$

These are those indicated on the left vertical line of the nomogram. For example, for $\beta_2 = 30°$, $\beta_{c2} = 300°$.

(d) Relations among $b/a, f/b, f/a$ and β_1, β_2

For given values of β_1, β_2, the three ratios $b/a, f/b$ and f/a are defined. It can be seen respectively from Figs. 6.54(a)(b) that

$$\frac{f+a}{2b} = \cos\beta_1 \tag{6.62}$$

and

$$\frac{f-a}{2b} = \sin\beta_2 \tag{6.63}$$

From equations (6.62) (6.63) we get

$$\frac{b}{a} = \cos\beta_1 - \sin\beta_2$$

$$\frac{f}{b} = \cos\beta_1 + \sin\beta_2$$

$$\frac{f}{a} = \frac{\cos\beta_1 + \sin\beta_2}{\cos\beta_1 - \sin\beta_2}$$

Referring to Fig. 6.52, we see that b/a = constant is represented by the pencil of solid curves, concave downward, while f/b = 1 is represented by the sole concave upward solid curve. f/a = constant is represented by the pencil of concave downward dotted curves.

(e) Symmetrical coupler curves with two cusps

In the present case, the conditions to be satisfied by the four-bar linkages, besides the above equations (6.62) and (6.63), include furthermore the relation shown in Fig. 6.45. Combining Fig. 6.50 and 6.45, or imagining that $\overline{A_2B_2}$ and $\overline{B_0B_2}$ in Fig. 6.45 were equal, we have

$$\overline{A_0B_0} = 2b\cos\frac{\beta}{2}$$

Because of

$$\overline{A_2B_0}^2 = f^2 - a^2 = (f+a)(f-a)$$

we have, by substituting equations (6.62), (6.63) into the above two equations, the value β_{2pr} of β for generating symmetrical coupler curves with two cusps:

$$\beta_{2pr} = 2\cos^{-1}\sqrt{\cos\beta_1 \sin\beta_2}$$

β_{2pr} = constant is represented by the pencil of concave upward chain curves in Fig. 6.52.

(f) Symmetrical coupler curves with two nodes

To generate a symmetrical coupler curve with two nodes, the procedure can be carried out as follows. Assume first values of β_1, β_2; or values of b/a, f/a, then a point can be determined in the nomogram Fig. 6.52. Draw a perpendicular from this point to the hypotenuse of the isosceles right triangle and read off the value of β_i. Next observe the β_{2pr} curve which passes through this point and read off the value of β_{2pr}. Any value of β between these β_i and β_{2pr} values is the required one. For example, let $\beta_1 = 30°$, $\beta_2 = 36°$; we have then $\beta_i = 84°$ and $\beta_{2pr} \approx 91°$. Therefore any value of β within the range $84° < \beta < 91°$ will result in a coupler curve with two nodes of the shape $\infty\infty$.

Finally, there are some further features to be mentioned:

(i) The value β_g is just the average between β_{c1} and β_{c2}. For example, if $\beta_1 = 30°$, then $\beta_{c1} = 240°$; and if $\beta_2 = 36°$, then $\beta_{c2} = 288°$; but for $\beta_1 = 30°$, $\beta_2 = 36°$ the angle $\beta_g = 264°$.

(ii) The two legs of the isosceles right triangle represent two extreme Grashof conditions $\beta_1 = 0$ and $\beta_2 = 0$. On the hypotenuse, the coupler curve shrinks into a point. For points rather close to the hypotenuse large bar sizes and small coupler curves will result. Preferable ranges of β_1, β_2 are $\beta_1 \geq 15°$, $\beta_2 \geq 15°$, $\beta_1 + \beta_2 \leq 60°$. This is the region within the smaller isosceles right triangle in the insert on the right hand side of the nomogram in Fig. 6.52.

(Antuma, 1978) extended this triangular nomogram further, and developed altogether 27 different nomograms, including a great deal of coupler curve characteristics, for instance, those with two tangents touching in 3 separate points, and those with two symmtrical Ball points, etc. The reader may refer to this work.

6.11.4 Generating symmetrical coupler curves by six-bar linkages

There are two principles which (Antuma, 1972) used to generate symmtrical coupler curves with six-bar linkages. These will be explained as follows:

(a) In Fig. 6.55, A_0ABB_0 is again a four-bar linkage like that in Fig. 6.50, in which $c = b$. However, in the coupler triangle $\triangle ABE$, $\overline{BE} \neq \overline{BA}$. In this case the path of E is of course not a symmetrical curve.

In fact, the sequence of construction of this figure is : build an arbitrary triangle

Dimensional systhesis of linkages --- path generation problems

$\triangle B_0 BD$ on $B_0 B$. Let $\angle B_0 AB = \varepsilon$, and draw the line s at B_0 to build the angle ε with $B_0 D$. Choose a point F on s so that $\overline{DF} = \overline{B_0 D} = e$. Finally complete the parallelogram $BDFE$, and the point E is thus determined. The triangle $\triangle ABE$ thus constructed is congruent with $\triangle BB_0 D$. This can be realized by rotating the latter about B_0 through an angle of 2ε to the chain line position $\triangle B'B_0 D'$.

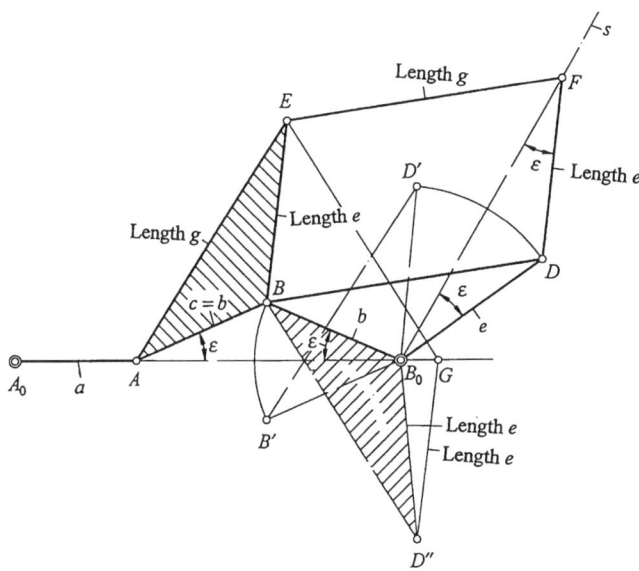

Fig. 6.55. Principle of symmetrical coupler curve generated by six-bar linkage.

Since $\triangle B_0 DF \sim \triangle B_0 BA$, and $B_0 D$, $B_0 B$ belong to the same body and rotate together, therefore the path generated by the point F is similar to the path of A, in a proportion $\overline{B_0 D} / \overline{B_0 B} = e/b$. Now the path of A is a circle, therefore the path of F is also a circle.

Another mechanism is obtained by folding $\triangle BB_0 D$ along $B_0 B$ to $\triangle BB_0 D''$ as shown in the figure, and then completing the parallelogram $D''BEG$. The point G should lie on line $A_0 B_0$. Now if the crank $A_0 A$ is removed, then as long as the path of A is a non-circular but symmetrical curve with respect to $A_0 B_0$, the path of G is also a symmetrical curve. This is because:

(i) If the whole mechanism is locked up and rotated about B_0 through equal but opposite angles, the two positions taken by the point A are symmetrical with respect to $A_0 B_0$, and the two positions taken by point G are also symmetrical with respect to $A_0 B_0$.

(ii) Let A be moved along the line $A_0 B_0$, then G will also be moved along $A_0 B_0$, because if the angle of rotation of AB is $+\Delta\varepsilon$, then the angle of rotation of B_0B, B_0D'' is $-\Delta\varepsilon$, and the angle of rotation of $D''G$ is $+\Delta\varepsilon$. The point G keeps on the line A_0B_0.

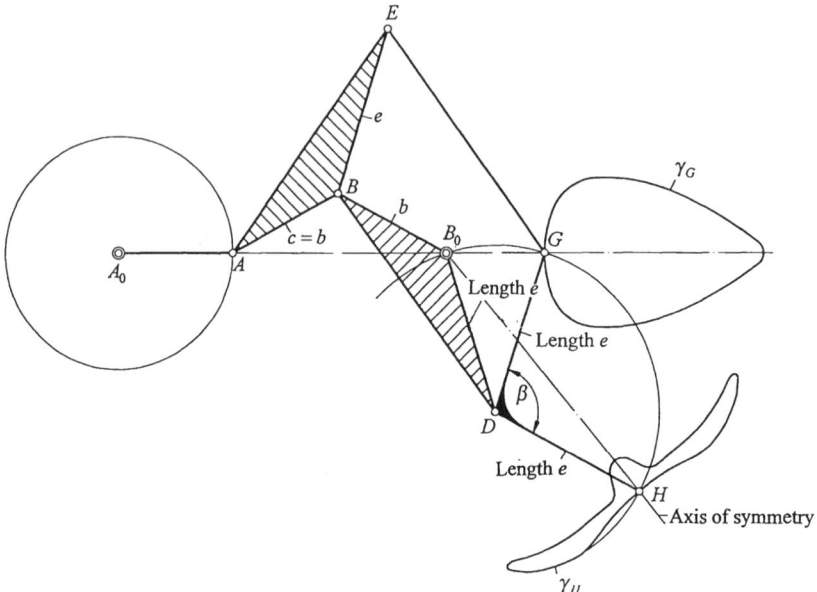

Fig. 6.56. Symmetrical coupler curve generated by a Watt-I six-bar linkage.

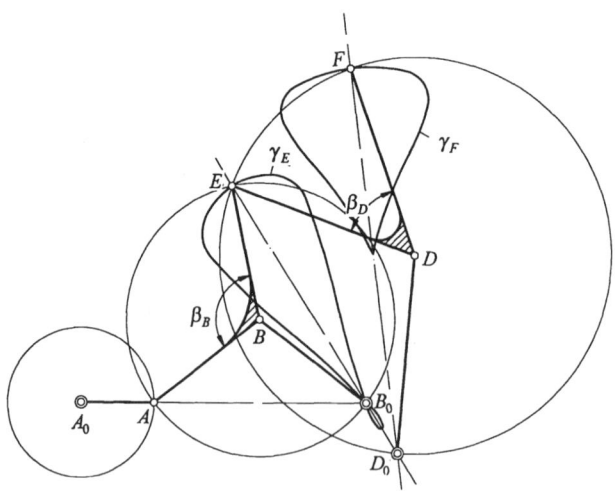

Fig. 6.57. Symmetrical coupler curve generated by a Stephenson-III six-bar linkage.

Dimensional systhesis of linkages --- path generation problems

(b) In Fig. 6.50, if the path of A were not a circle, but a symmetrical curve with respect to A_0B_0, the path of E would also be a symmetrical curve.

According to the above two principles (a) (b), we can get two six-bar symmetrical coupler curve mechanisms as shown in Figs. 6.56 and 6.57. The mechanism in Fig. 6.56 is a Watt-I type six-bar linkage like that shown in Fig. 6.55, the path γ_G of its coupler point G being a symmetrical curve with A_0B_0 as the axis of symmetry. $GDHB_0$ is a dyad correspoinding to $ABEB_0$ in Fig. 6.50. The only difference is that the point G does not like point A moving on a circle, but on a symmetrical curve. Here $\overline{DH} = \overline{DB_0} = \overline{DG}$, and $\angle GDH = \beta$ is a fixed angle. Therefore the path γ_H of H is also a symmetrical curve, its axis of symmetry passing through point B_0. The point H may be any point on the circle with D as centre, and $\overline{DB_0} = \overline{DG}$ as the radius.

Fig. 6.57 shows a Stephenson-III type six-bar linkage, in which A_0ABEB_0 is a symmetrical coupler curve four-bar similar to that shown in Fig. 6.50. The axis of symmetry of the path γ_E of the point E is EB_0. Choose any point D_0 on EB_0, and any point D on the prependicular bisector of $\overline{ED_0}$. Draw a circle with D as centre and $\overline{DE}$ as its radius, and take any point F on this circle. $EDFD_0$ is then another symmetrical coupler curve linkage corresponding to $ABEB_0$. The only difference is that E does not like point A moving on a circle, but on a symmetrical curve. Therefore the path γ_F of F is also a symmetrical curve, its axis of symmetry being FD_0.

6.12 Higher order path curvature

The methods of guiding a body through two, three, four and five infinitesimally separated positions discussed in Chapter 3 and the principle of instantaneous invariants as well as the polode method in Chapter 4, were mainly concerned with a body to be guided as a whole. In the present section, the analysis will be concerned with the motion characteristics of the path generated by a moving point on the body. For instance, we mentioned frequently in the preceding sections how to replace a path of a moving point by the circle of curvature of that path. That technique may be considered as a first step of simulation. In order to understand in more detail the characteristics of a certain path or curve, higher orders of path curvature have to be taken into consideration. The following description is taken from (Freudenstein, 1965).

In Fig. 6.58, let the centre of curvature of a given curve or given path at A be A_0, and the radius of curvature be ρ. After an infinitesimal displacement, A moves to A', with T_A as the tangent of its path, and A_0 moves to A_0'. The path of A_0 is called the evolute. The centre of curvature of the evolute is at A_{00}, and A_{00} moves to A_{00}'. The path of A_{00} is the evolute of the evolute, its centre of curvature being at A_{000}. It can be seen from the figure that the lines AA_0, A_0A_{00}, $A_{00}A_{000}$ move respectively to

Kinematics and Design of Planar Mechanisms

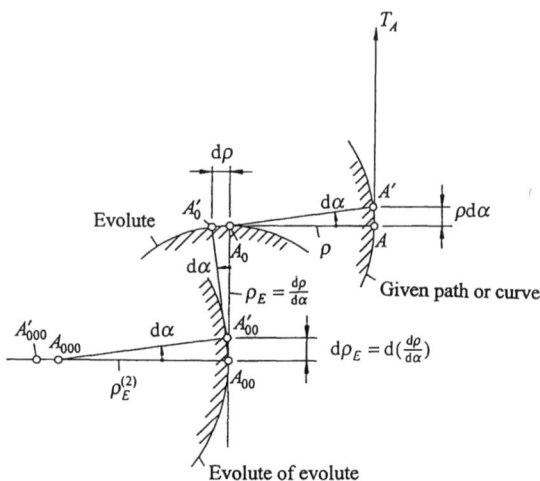

Fig. 6.58. Higher order characteristics of the path of a moving point.

$A'A_0'$, $A_0'A_{00}'$, $A_{00}'A_{000}'$. The angular displacements of these lines are all equal, being denoted by $d\alpha$. Let $\rho, \rho_E, \rho_E^{(2)}$ be respective radii of higher order curvatures, or

$$\rho = \overline{A_0 A}$$

$$\rho_E = \overline{A_{00} A_0} = \frac{d\rho}{d\alpha}$$

$$\rho_E^{(2)} = \overline{A_{000} A_{00}} = \frac{d^2\rho}{d\alpha^2}$$

$$\ldots\ldots\ldots$$

$$\ldots\ldots\ldots$$

$$\rho_E^{(n)} = \frac{d^n\rho}{d\alpha^n}$$

Suppose a curve is given, and the planar motion of a moving body is also given. It is required to find out such points on the moving body, that its path can be used to approximate the given curve at a certain point, i.e. to simulate the given curve. Let us consider first, what are the points, the radius of curvature of the path of which is equal to the radius of curvature of the given curve at a certain point? We assert that, the radius of curvature of the path of any point on the moving body can be made equal to the radius of curvature of the given curve at a certain point by enlarging or reducing the scale (the so-called *stretch*) of the plane of the moving body and by rotating the orientation (the so-called *rotation*) of this plane. This means that the path of any point on the moving body can be made to be tangent to the given curve at

Dimensional systhesis of linkages --- path generation problems

a certain point in three points. Hence it only makes sense to simulate the given curve by an approximate path of a moving point at least in a four-point tangent. In other words, we start from searching the path of a moving point which has the same value of ρ_E / ρ as that of the given curve. This value is called the *characteristic number*, and is denoted by λ_1, or

$$\lambda_1 = \frac{\rho_E}{\rho} = \frac{1}{\rho}\left(\frac{d\rho}{d\alpha}\right) \tag{6.64}$$

In case a moving point having a five-point tangent with the given curve is to be sought, then, besides the same value of λ_1, both curves should have the same value of the characteristic number $\rho_E^{(2)} / \rho$, denoted by λ_2, or

$$\lambda_2 = \frac{\rho_E^{(2)}}{\rho} = \frac{1}{\rho}\left(\frac{d^2\rho}{d\alpha^2}\right) \tag{6.65}$$

In the following, we shall show how to find the λ_1 and λ_2 values for a given curve. Compare Fig. 6.58 with Figs. 3.56 (b)(c), and take equation (3.108) into consideration (for simplicity purposes, θ_A is written here as θ)

$$\rho = \frac{p^2}{\delta \sin\theta - p} \qquad [(3.108)]$$

where $p = \overline{PA}$, δ is the diameter of the inflection circle and, as defined before, $\delta = -1/\gamma'$. The expression of $d\rho/ds$ was given in equation (3.109). ds is the infinitesimal distance along the fixed polode, as that shown in Fig. 3.56 (b) (please note that ds is not the $\overline{AA'}$ shown in Fig. 6.58). It can be seen from that figure that

$$\overline{AA'} = \rho d\alpha = -p d\gamma (\text{since } d\gamma \text{ is negative})$$
$$= -p\frac{d\gamma}{ds} ds = \frac{p}{\delta} ds$$

Therefore
$$ds = \frac{\rho\delta}{p} d\alpha$$

or
$$\frac{ds}{d\alpha} = \frac{p\delta}{\delta\sin\theta - p} \tag{6.66}$$

Now substituting, as before, equations (3.110), (3.111), (3.112) into equation (3.109), and then substituting the relation (6.66) into it, we get

$$\frac{d\rho}{d\alpha} = \frac{d\rho}{ds} \cdot \frac{ds}{d\alpha} = \frac{3p^3\delta}{(\delta\sin\theta - p)^3}\left(\frac{\cos\theta}{l} + \frac{\sin\theta}{m} - \frac{\sin\theta\cos\theta}{p}\right) \tag{6.67}$$

Substituting equations (6.67) and (3.108) into equation (6.64) gives

$$\lambda_1 = \frac{1}{\rho}\left(\frac{d\rho}{d\alpha}\right) = \frac{3p\delta^2}{(\delta\sin\theta - p)^2}\left(\frac{\cos\theta}{l} + \frac{\sin\theta}{m} - \frac{\sin\theta\cos\theta}{p}\right) \qquad (6.68)$$

For a given path or curve, equation (6.68) can be used for computing the value of λ_1 at a certain point. On the other hand, if the value of λ_1 is prescribed, equation (6.68) serves as the polar coordinate equation of the locus of the point whose path satisfies this ratio $\lambda_1 = \rho_E / \rho$. Equation (6.68) can be transformed into a quadratic equation in the unknown t by replacing p by $p = t\sin\theta$:

$$\lambda_1: \; g_2 t^2 + g_1 t + g_0 = 0 \qquad (6.69)$$

where $g_2 = \lambda_1 \tan\theta$

$g_1 = -2\lambda_1 \delta \tan\theta - 3\dfrac{\delta^2}{m}\tan\theta - 3\dfrac{\delta^2}{l}$

$g_0 = \lambda_1 \delta^2 \tan\theta + 3\delta^2$

To transform into rectangular coordinates, taking the poletangent as the x-axis, and the polenormal as the y-axis, or making the substitutions $p\cos\theta = x$, $p\sin\theta = y$, we can rewrite equation (6.68) into

$$\lambda_1: \; \lambda_1 lm(x^2 + y^2 - \delta y)^2 - 3\delta^2[(x^2 + y^2)(ly + mx) - lmxy] = 0 \qquad (6.70)$$

Equation (6.70) is the rectangular coordinate equation of the locus of all points satisfying the prescribed λ_1-value, and is called the λ_1-curve, or the *quartic of derivative curvature*. This is a bicircular quartic with double points at the circular points I and J (please refer to Appendix A1.2), and also a node at the origin. The first bracket in equation (6.70) represents the inflection circle, and the second bracket represents the circling-point curve. Hence the λ_1-curve passes through the intersection of inflection circle and the circling-point curve, or the Ball point.

The solid curve shown in Fig. 6.59 represents the λ_1-curve for a body performing a motion for $\lambda_1=1/4$. The so-called cardioid motion is a kinematic inversion of the Cardan circle pair as that shown in Fig. 3.67. In this inversion the smaller circle is fixed, while the larger circle rolls without slip on the outside of the smaller circle, as shown in the latter Fig. 6.61. In this case $\rho_c = 2\rho_f$. The path of any point on the body bound together with the larger circle is a *cardioid*. In this case $1/l = 0$, $1/m = 0$ (please refer to Dijksman, 1976, 3.5.3), and the λ_1-curve becomes

$$\lambda_1(x^2 + y^2 - \delta y)^2 + 3\delta^2 xy = 0$$

This equation is transformed into polar coordinate form by substituting $x = p\cos\theta$, $y = p\sin\theta$:

$$\lambda_1(p - \delta\sin\theta)^2 + 3\delta^2 \sin\theta\cos\theta = 0$$

Dimensional systhesis of linkages --- path generation problems

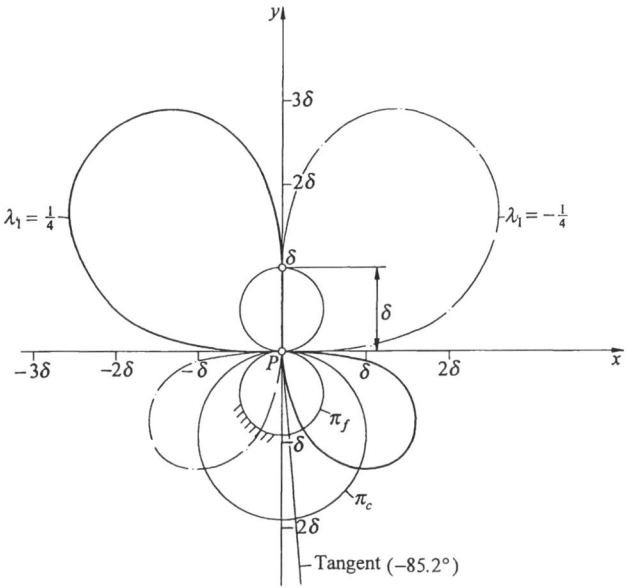

Fig. 6.59. λ_1-curve for $\lambda_1 = 1/4$ and $\lambda_1 = -1/4$ when a body is performing a cardioid motion ($\rho_c = 2\rho_f$).

or
$$p = \delta\left(\sin\theta + \sqrt{-\frac{3}{2\lambda_1}\sin 2\theta}\right) \tag{6.71}$$

so that the λ_1-curve in Fig. 6.59 can be drawn. The equations of the tangents at the circular points are $y = 0$ (the x-axis) and $\lambda_1 y + 3x = 0$, or $y + 12x = 0$. Please note that no curve exists within the interval $0 < y < +\delta$ on the y-axis. Both π_c, π_f are symmetrical with respect to the y-axis in Fig. 6.59. The question is, why is the λ_1-curve (silid line) not symmetrical with respert to the y-axis ? It can be seen from equation (6.71), for $\lambda_1 = -1/4$, the λ_1-curve is shown as the chain line in Fig. 6.59. Hence if the λ_1-value is taken as $\lambda_1^2 = 1/16$, the whole λ_1-curve will be the complete curve including both solid and chain lines.

For special cases of the λ_1-curve, please refer to (Freudenstein, 1965).

We come to see how to find the λ_2-value of a given curve. Equation (6.68) can be rewritten as

$$\lambda_1(\delta\sin\theta - p)^2 = 3p\delta^2 K \tag{6.72}$$

where

$$K \equiv \frac{\cos\theta}{l} + \frac{\sin\theta}{m} - \frac{\sin\theta\cos\theta}{p}$$

which is the circling-point curve in equation (3.116). Differentiating equation (6.72) with respect to α, and noting that $d\lambda_1/d\alpha = \lambda_2 - \lambda_1^2$, and taking equations (3.124), (3.125) and (1.15) into consideration, we get

$$(\lambda_2 - \lambda_1^2)(\delta\sin\theta - p)^2 + 2\lambda_1 p\delta\left[\frac{d\delta}{ds}\sin\theta + \delta\cos\theta\left(\frac{\sin\theta}{p} - \frac{1}{\rho_c}\right) + \cos\theta\right]$$

$$= \frac{3p\delta^3}{(\delta\sin\theta - p)}\left[p\frac{dK}{ds} - K\left(\cos\theta + \frac{6p}{m}\right)\right]$$

(6.73)

In equation (6.73), $\rho_c = \overline{PM}$ is the radius of curvature of the moving polode at the point P, and

$$\frac{dK}{ds} = -\left(\frac{m'}{m^2}\sin\theta + \frac{l'}{l^2}\cos\theta + \frac{\sin\theta\cos^2\theta}{p^2}\right) + \theta'\left(\frac{\cos\theta}{m} - \frac{\sin\theta}{l} - \frac{\cos^2\theta - \sin^2\theta}{p}\right)$$

(6.74)

In equation (6.74), ($'$) represents the differential coefficient with respect to s, and θ' can be taken from equations (3.111) and (3.125), or

$$\theta' = \frac{\sin\theta}{p} - \frac{1}{\rho_c}$$

(6.75)

Replacing p by $p = t\sin\theta$ in equation (6.73), and substituting expressions of K in equation (6.72) and θ' in equation (6.75) into it, we get

$$\tan^2\theta(\lambda_2 - \lambda_1^2)(\delta - t)^3 + 2\lambda_1\delta^2(\delta - t)\left[-3\frac{t}{m}\tan^2\theta + \left(1 - \frac{t}{\rho_f}\right)\tan\theta\right]$$

$$= 3\delta^3\left\{-\frac{m'}{m^2}t^2\tan^2\theta - \frac{l'}{l^2}t^2\tan\theta - 1 + \left(1 - \frac{t}{\rho_c}\right)\left[\frac{t}{m}\tan\theta + \left(1 - \frac{t}{l}\right)\tan^2\theta - 1\right]\right\}$$

$$- \lambda_1\delta(\delta - t)^2\left(\tan\theta + \frac{6t}{m}\tan^2\theta\right) = 0$$

(6.76)

where $\rho_f = \overline{PM_0}$ is the radius of curvature of the fixed polode at the point P. Arranging equation (6.76) into a cubic polynomial in t gives

$$h_3 t^3 + h_2 t^2 + h_1 t + h_0 = 0$$

(6.77)

where the coefficients are

$$h_3 = \tan^2\theta\left[-(\lambda_2 - \lambda_1^2) + \frac{6\lambda_1\delta}{m}\right]$$

Dimensional systhesis of linkages --- path generation problems

$$h_2 = \delta \tan^2\theta \left[3(\lambda_2 - \lambda_1^2) - \frac{6\lambda_1\delta}{m} + 3\delta^2\left(\frac{m'}{m^2} - \frac{1}{l\rho_c}\right)\right]$$

$$+ \delta^2 \tan\theta \left[\lambda_1\left(\frac{1}{\rho_c} + \frac{1}{\rho_f}\right) + 3\delta\left(\frac{l'}{l^2} + \frac{1}{m\rho_c}\right)\right]$$

$$h_1 = \delta^2 \tan^2\theta \left[-3(\lambda_2 - \lambda_1^2) + 3\delta\left(\frac{1}{\rho_c} + \frac{1}{l}\right)\right]$$

$$-\delta^3 \tan\theta \left[2\lambda_1\left(\frac{1}{\rho_f} + \frac{2}{\delta}\right) + \frac{3}{m}\right] - \frac{3\delta^3}{\rho_c}$$

$$h_0 = \delta^3[(\lambda_2 - \lambda_1^2 - 3)\tan^2\theta + 3\lambda_1 \tan\theta + 6]$$

For a given path, the values of δ, m, l, m', l' of the moving body at a certain point are known. Equation (6.77) can be used for calculating the λ_2-value at that point.

Conversely, if the values of λ_1, λ_2 are prescribed, it is possible to find out at any instant on a body performing any plane motion, such points that satisfy the required λ_1, λ_2 values. In other words, it is possible to simulate a given curve at any point by the path of a moving point in a five-point tangent. It should be noted, however, that the kinematic data of the curve to be simulated should not be confused with those of the simulating path. Now setting equations (6.69) and (6.77) simultaneously. These are two simultaneous equations in the unknowns t and $\tan\theta$. Eliminating t by the dialytic method yields

$$\begin{vmatrix} 0 & 0 & g_2 & g_1 & g_0 \\ 0 & g_2 & g_1 & g_0 & 0 \\ g_2 & g_1 & g_0 & 0 & 0 \\ h_3 & h_2 & h_1 & h_0 & 0 \\ 0 & h_3 & h_2 & h_1 & h_0 \end{vmatrix} = 0 \qquad (6.78)$$

Equation (6.78) may be further rewritten as

$$\begin{vmatrix} 0 & 0 & g_2 & g_1 & g_0 \\ 0 & g_2 & g_1 & g_0 & 0 \\ g_2 & 0 & g_0 & 0 & 0 \\ h_3 & -\frac{g_1}{g_2}h_3 + h_2 & h_1 & h_0 & 0 \\ 0 & h_3 & h_2 & h_1 & h_0 \end{vmatrix} = 0$$

It can be seen that, in the first column of this determinant, both g_2 and h_3 have the same factor $\tan\theta$, and in the second column, g_2, $-(g_1/g_2)h_3 + h_2$ and h_3 also have the same factor, therefore $\tan^2\theta$ can be factorized from the determinant. Hence equation (6.78), being a 7th degree equation in $\tan\theta$, is reduced to a 5th degree equation in $\tan\theta$. In other words, in general cases, for prescribed λ_1,λ_2, it is possible to find out five points on the moving plane, whose paths are tangent to the given curve at a certain point in five points.

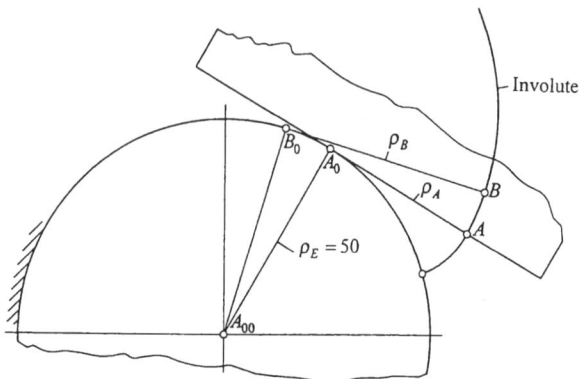

Fig. 6.60. Example: given curve being an involute, and $\lambda_1 = 9/(4\sqrt{3})$ at point A on the curve.

Example: A straight flat plate, rolls without slip on the outside of a fixed cylinder. The path generated by a point A on the flat plate is an involute, as shown in Fig. 6.60. Assume the radius of the cylinder is 50 mm. It is required to simulate this involute in the vicinity of point A. Assume it is prescribed at A that

$$\lambda_1 = \frac{\overline{A_{00}A_0}}{\overline{A_0A}} = \frac{\rho_E}{\rho_A} = \frac{9}{4\sqrt{3}}$$

Since ρ_E is constantly 50 mm, the position of the point A is at $\overline{A_0A}(=\rho_A) = 50\cdot 4\sqrt{3}/9 = 200\sqrt{3}/9$ mm$(= 38.490$ mm$)$. Hence at the point A, the values λ_1,λ_2 of the involute are respectively: $\lambda_1 = 9/(4\sqrt{3})$ (prescribed), $\lambda_2 = 0$ (since $\rho_E^{(2)} = 0$).

Suppose the simulating motion is the cardioid motion in Fig. 6.59. As shown in Fig. 6.61, the fixed polode π_f is a circle with a radius of 10 mm, or $\rho_f = -10$ mm. The moving polode π_c is a circle with a radius of 20 mm, or $\rho_c = -20$ mm. In this case the inflection circle is a circle equal to π_f, but on the opposite side of the poletangent, or $\delta = +10$ mm. The aim of the present problem is to find out on the moving body π_c such points whose paths are tangent to the path of A in Fig.6.60 in five points, or whose characteristic numbers are

Dimensional systhesis of linkages --- path generation problems 303

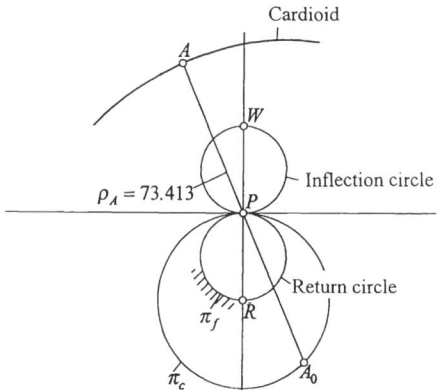

Fig. 6.61. Example: determination of point A on the cardioid, so that the cardioid is tangent to the involute at point A in Fig. 6.60 in five points.

$\lambda_1 = 9/(4\sqrt{3})$, $\lambda_2 = 0$. In other words, we may substitute these values of λ_1, λ_2 into equation (6.78), to solve for the roots $\tan\theta$.

For simplicity purposes, let $\tan\theta = z$. Because of proportionality, we may assume $\delta = 1$, then $\rho_f = -1/2$, and $\rho_c = -1$. As mentioned before, in this cardioid motion, we have $1/m = 0$, $1/l = 0$. The coefficients of equation (6.69) are:

$$g_2 = \lambda_1 z, \quad g_1 = -2\lambda_1 z, \quad g_0 = \lambda_1 z + 3$$

and the coefficients of equation (6.77) are

$$h_3 = \lambda_1^2 z^2, \ h_2 = -3\lambda_1^2 z^2 - 3\lambda_1 z, \ h_1 = 3\lambda_1^2 z^2 - 3z^2 + 3,$$
$$h_0 = -z^2(\lambda_1^2 + 3) + 3\lambda_1 z + 6$$

Substitute these values into equation (6.78). It can be seen that one of the solutions is $\lambda_1 z = -3$, or $z = \tan\theta_A = (-3/9)(4\sqrt{3})$, which is $\theta_A = -66.58678°$. The corresponding value of p can be found from the λ_1-curve. Substituting the present values of θ_A and λ_1 into equation (6.71), we obtain (in Fig. 6.61 we take arbitrarily $\delta = 20$ mm) $p = -36.7065$ mm. This is the point A shown in Fig. 6.61, the centre of curvature A_0 of whose path is just on the circle π_c. The radius of curvature of the path of A is $\rho_A = 2|p| = 73.413$mm.

To simulate the involute in Fig. 6.60 by the cardioid in Fig. 6.61, the plane of Fig. 6.61 should first be rotated until the $A_0 A$ lines in both figures coincide with each other, and then Fig. 6.61 is reduced by the proportion $\dfrac{200\sqrt{3}}{9}/73.4130 = 0.5243$ (the so-called stretch-rotation). Both curves are thereafter tangent to each other in five points.

In a special case, if $\lambda_1 = 0$, $\lambda_2 = 0$, or in case ρ is constant, the points on the moving plane satisfying the conditions $\lambda_1 = 0$, $\lambda_2 = 0$ are the well-known Burmester points. As mentioned is Section 3.8, there are four Burmester points. This is why the points on the moving plane satisfying prescribed non-zero values of λ_1, λ_2 are called *generalized Burmester points* by Freudenstein and there are five such points.

(Veldkamp, 1967) interpreted the meanings of λ_1 and λ_2 by means of instantaneous invariants (see Section 4.1) as well as from a point of view of differential geometry. In spite of the conciseness of these interpretations, mathematical equations may not easily be handled by engineers. Readers may refer to it directly.

Exercises

6.1 The coordinates of the points and the angles shown in the figure are
$E_1(54.7, 62.6)$
$E_2(45.6, 63.5)$, $\phi_{12} = 19.2°$
$E_3(31.0, 60.7)$, $\phi_{12} = 45.2°$
$E_4(19.0, 51.8)$, $\phi_{14} = 70.7°$

In order that the four points E_1, E_2^1, E_3^1, E_4^1 should lie on a circle, the locus of the centre point A_0 is the k_{A0}-curve. Find the equation of k_{A0} and draw the curve.

What is the number or solutions of synthesizing a four-bar linkage to generate a coupler curve passing through E_1, E_2, E_3, E_4 and at the same time to coordinate the three crank angles ϕ_{12}, ϕ_{13}, ϕ_{14} ? What are the choices?

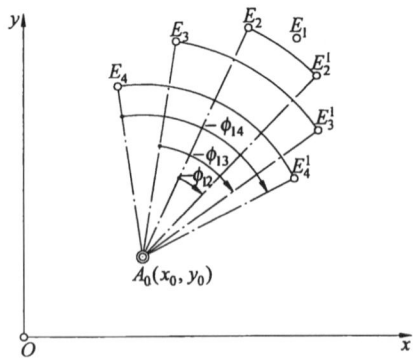

Dimensional systhesis of linkages --- path generation problems

6.2 Similar to Exercise 6.1, but the coordinates of the points and angles are:
$E_1(29, 89)$,
$E_2(17, 92)$, $\phi_{12} = 20°$
$E_3(6, 90)$, $\phi_{13} = 40°$
$E_4(-4, 85)$, $\phi_{14} = 60°$
Construct a four-bar linkage to fit the four coupler point positions and to coordinate the crank angles.

6.3 For the problem of coordinating four coupler point positions with three crank angles as that mentioned in Section 6.4.1, how can it be solved by a geometrical method?

6.4 Show that the coupler point curve as that given by equation (6.24), possesses three asymptotes at each of the two circular points I, J. These three pairs of asymptotes intersect respectively in three real points A_0, B_0, C_0, the three singular foci.
Hint: Please refer to Fig. 6.15. The coordinates of these three points are: $A_0(0,0)$, $B_0(f, 0)$, and

$$x_{C0} = \frac{\sin\beta\cos\alpha}{\sin\eta} f$$

$$y_{C0} = \frac{\sin\beta\sin\alpha}{\sin\eta} f$$

6.5 A given four-bar linkage A_0ABB_0 is shown in Fig. 6.19, in which $\overline{A_0B_0} = f = 100$, $\overline{A_0A} = a = 25$, $\overline{AB} = c = 36$, $\overline{B_0B} = b = 52$, $\overline{AE} = 32$, $\overline{BE} = 45$.
Find the other two cognate linkages.

6.6 By means of a vector equation similar to equation (6.33), show that the coupler curves γ_E in Figs. 6.20(c) and 6.20(a) are identical.

6.7 The coordinates of five coupler point positions are given: $E_1(24, 36)$, $E_2(16, 50)$, $E_3(22.2, 60)$, $E_4(37.6, 70)$, $E_5(58.2, 70)$.
  (a) Find the equation of the ellipse passing through these five points and the equations of the major and minor axes of this ellipse.
  (b) Synthesize a four-bar linkage so that its coupler curve can approximate this ellipse. What is the number of solutions? Do any satisfactory solutions exist?
  (c) Construct a Cardan circle-pair to generate this ellipse.

6.8 (Taken from Beyer, 1950) A logarithmic spiral is given as shown in Figure (a), the polar coordinate equation of which is

$$r = ce^{k\theta}$$

where $r = c$ at $\theta = 0$. The angle β between the radial line and the tangent at a point on the spiral is constant, or $\tan\beta = r/r' = 1/k$. Assume $c = 30$, $\beta = 56°$.
Find an ellipse which is tangent to the spiral at E in four points as shown in Figure (b).

(Ans.:Transfer the origin to the point E, then the equation of the ellipse becomes $-0.031577378x^2 - 0.001846023xy - 0.0203395y^2 - 1.482560969x + y = 0$).

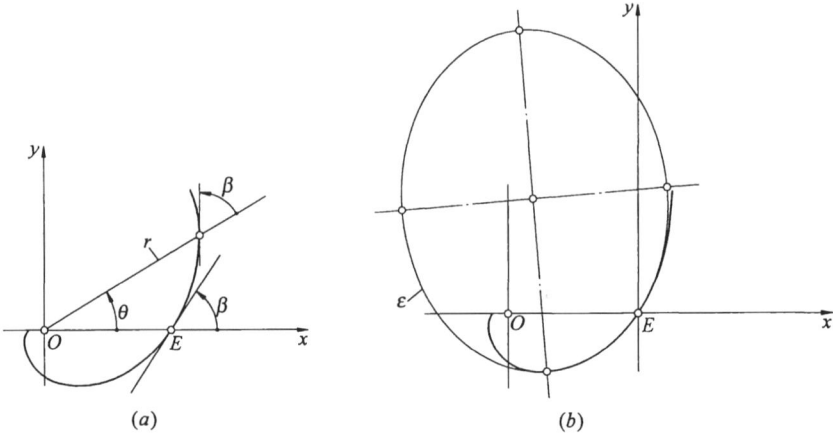

(a) Logarithmic spiral $r = ce^{k\theta}$, $\tan \beta = r/r' = 1/k$.
(b) Ellipse ε in a four-point tangent to the logarithmic spiral at E.

6.9 Show that the radius of the circle passing through the three points E_1, E_m, E_2 in Fig. 6.42 is $7r$. Hence if it is taken that $\overline{EE_0} = 7r$, a dwell mechanism can be constructed even more directly.

6.10 The positions of A_0, E_1, E_2 are prescribed as shown. Construct a four-bar linkage such that the coupler curve of its coupler point E exhibits a cusp at E_1 and E_2 respectively.

6.11 Show, by means of Kraus's concept of balancing the number of coordinates as mentioned in Section 5.3, that the number of solutions of the problem shown in Fig. 6.45 is ∞^1.

Dimensional systhesis of linkages --- path generation problems

6.12 Show that in the construction method of Fig. 6.49 as described in Section 6.10.2, $\overline{A_1E_1} = \overline{A_2E_2}$ and $\not\triangleleft A_1B_1E_1 = \not\triangleleft A_1A_0B_0$.

6.13 Show that in Fig. 6.49,

$$a^2 + b^2 + c^2 + \frac{2abc}{f} = f^2$$

6.14 Find the two other cognate linkages of the four-bar linkage $A_0A_1B_1B_0$ including the coupler triangle $\Delta A_1B_1E_1$ shown in Fig. 6.49.

6.15 Discuss the case in which the coupler curve triangle $\Delta E_1E_2E_3$ in Fig. 6.49 becomes an isosceles triangle (i.e. $\overline{E_1E_2} = \overline{E_1E_3}$), as well as becomes an equilateral triangle. (Please refer to Hain, 1961, Figs. 12-30 and 12-31.)

6.16 Design a four-bar linkage to generate a symmetrical coupler curve. The two parameters as defined in Fig. 6.53 are : $\beta_1 = 40°$, and $\beta_2 = 30°$.
(a) The symmetrical coupler curve exhibits two cusps.
(b) The symmetrical coupler curve exhibits an 8 figure.
(Ans.: (a) $\beta_{2pr} = 103.53°$; (b) $\beta_8 = 280°$)

6.17 Construct, according to the method of Fig. 6.56, a Watt-I type six-bar symmetrical coupler curve generating mechanism.

6.18 Construct, according to the method of Fig. 6.57, a Stephenson-III type six-bar symmetrical coupler curve generating mechanism.

6.19 A given curve is an involute as that shown in Fig. 6.60. At the point B of the involute $\rho_B = \overline{B_0B} = \rho_E = 50$, or $\lambda_1 = 1$. Find the points on the plane of a wheel rolling on a flat plate as shown in the following figure, the path of which should be in a 5-point contact with the involute of Fig. 6.60 at the point B.

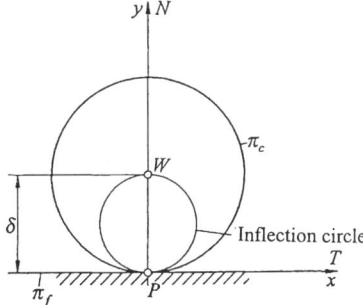

(Ans.: Two points can be found. Their polar coordinates are, with respect to the polar line Px in the figure, on the assumption $\delta = 1$, $E(1.969200285, 26.98346545°)$, $F(1.147330101, 41.2461307°.)$)

7
Synthesis of function generators

7.1 Function generator and its applications

A mechanism, which keeps a certain functional relationship between the angles of rotation of the input crank and those of the output crank is called a function generator. Fig. 7.1 shows a four-bar function generator. The angle of rotation $\Delta\phi$ of

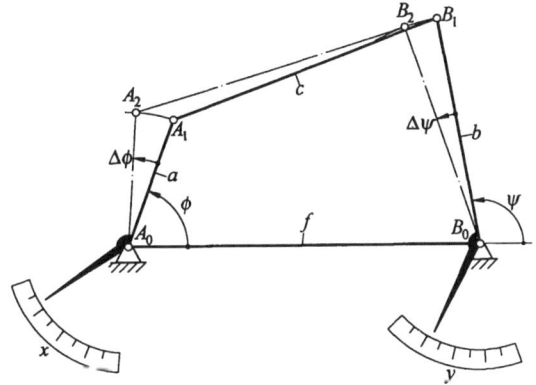

Fig. 7.1. Function generator.

the link a represents Δx, and the angle of rotation $\Delta\psi$ of the link b represents Δy. This mechanism is capable of generating a function $\psi = \psi(\phi)$ which is a close approximation to the prescribed functional relationship $y = y(x)$.

In practical applications, as long as the angles of rotation of two rotating links should keep a certain functional relationship, a function generator is needed. Fig. 7.2 shows an example taken from (Kracke, 1970). Suppose the input quantity for a measuring instrument is x_M, and the output quantity is y_M. This quantity y_M measured by the instrument, however, cannot readily exhibit x_M, but $y_M = k_M x_M^n$, where k_M is a proportional factor. In order to show x_M, the measured y_M has to be put into a function generator. For the function generator, the input quantity is y_M, and the output quantity is y_G, and $y_G = k_G k_M^{1/n} x_M$. In other words, it is only necessary to adjust the value of the proportional factor $k_G k_M^{1/n}$, and the

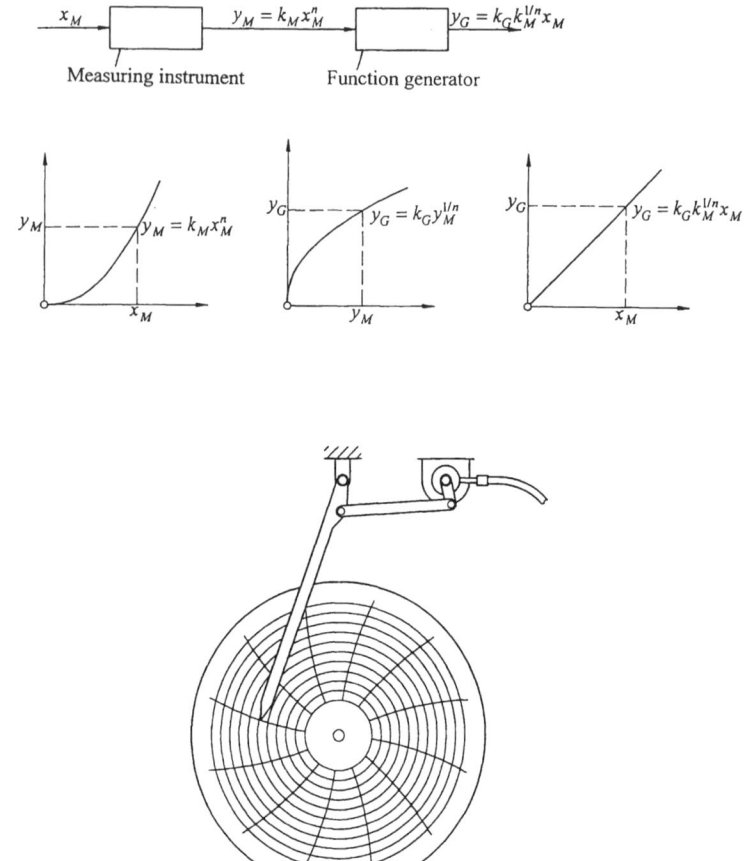

Fig. 7.2. Function generator as a linearizer (Kracke, 1970).

output quantity y_G of the function generator can be used to represent the x_M value to be measured. Hence the function generator is a device to recover the linearization of the measured quantity by means of an inverse function.

Fig. 7.3 shows another example taken from (Kracke, 1970). In this figure a steering mechanism of a container transport is shown. As the centre line A_0A of the tractor makes an angle $\Delta\phi$ relative to the centre line A_0B_0 of the semi-trailer, the line B_0B normal to the rear axle of the semi-trailer has to turn through an angle $\Delta\psi$ relative to A_0B_0, in order that the axes of all wheels may possibly intersect in one point. A_0ABB_0 should therfore be a function generator.

Synthesis of function generators

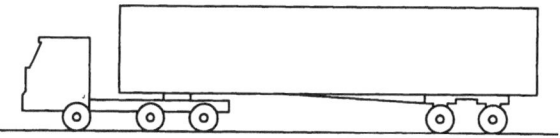

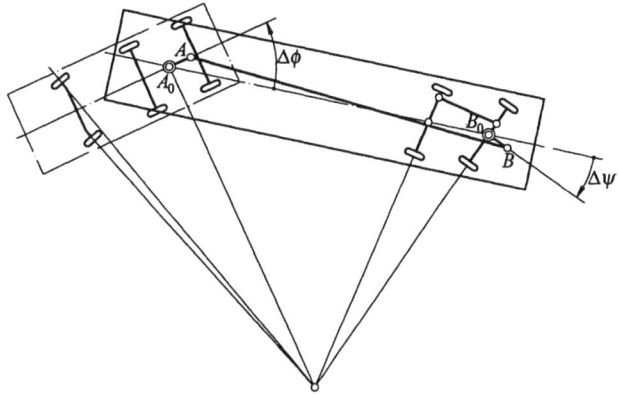

Fig. 7.3. Function generator applied on semi-trailer steering in container transport (Kracke, 1970).

In general cases, to convert a given functional relationship $y = y(x)$ into a functional relationship $\psi = \psi(\phi)$, certain scale factors exist between $\Delta\phi$ and Δx, and between $\Delta\psi$ and Δy, or

$$\left. \begin{aligned} M_\phi &= \frac{\Delta\phi}{\Delta x} \text{in rad / unit of } x \\ M_\psi &= \frac{\Delta\psi}{\Delta y} \text{in rad / unit of } y \end{aligned} \right\} \qquad (7.1)$$

Please note that, ϕ is not necessarily equal to $M_\phi x$, and ψ is not necessarily equal to $M_\psi y$. In other words, ϕ is not necessarily equal to zero when $x = 0$, and ψ is also not necessarily equal to zero when $y = 0$.

In Fig. 7.4, the solid curve represents a prescribed functional relationship $\psi(\phi)$. Select on this curve sereval points P_1, P_2, P_3, P_4. These points are called *precision points*. Suppose it is required to synthesize a function generator, that as the rotating link a rotates through angles $\Delta\phi_{12}$, $\Delta\phi_{13}$, $\Delta\phi_{14}$, the rotating link b should rotate respectively through angles $\Delta\psi_{12}$, $\Delta\psi_{13}$, $\Delta\psi_{14}$, as shown in Fig. 7.5. The generated functional relationship of the synthesized mechanism will somewhat like

312   *Kinematics and Design of Planar Mechanisms*

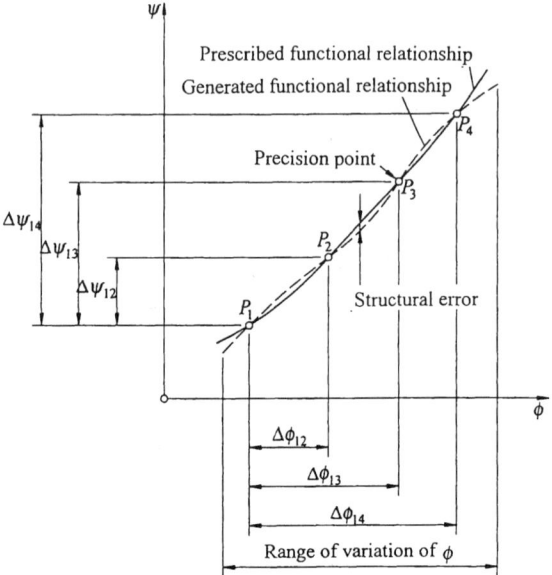

Fig. 7.4.   Functional relationship $\psi(\phi)$.

the dotted curve shown in Fig. 7.4. Certainly this dotted curve passes through the points P_1, P_2, P_3, P_4. Please note that, the origin of the coordinate system corresponding to the dotted curve is not the original origin shown in Fig. 7.4. In general the origin of the coordinate system is determined once the mechanism is synthesized. In Fig. 7.5, from now on, the initial position angle ϕ_1 of the link a will be denoted by ϕ_0, and the initial position angle ψ_1 of link b will be denoted by ψ_0.

The vertical distance between the solid and dotted curves in Fig. 7.4 is called the *structural error*.

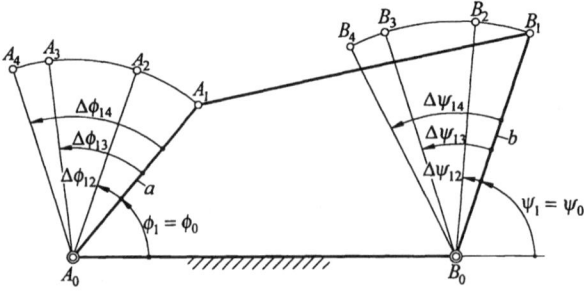

Fig. 7.5.   Coordinations of crank rotation pairs: $\Delta\phi_{12}{:}\Delta\psi_{12}$, $\Delta\phi_{13}{:}\Delta\psi_{13}$, $\Delta\phi_{14}{:}\Delta\psi_{14}$.

Synthesis of function generators

Just like the cases of body guidance problems depicted in Section 3.10, problems of coordinations of angular rotations of the two rotating links can also be distinguished into cases of finitely separated relative positions and infinitesimally separated relative positions. The case shown in Fig. 7.4, may be called a case of four finitely separated relative positions $P_1-P_2-P_3-P_4$.

7.2 Coordination of finitely separated angular displacements --- geometrical method

7.2.1 Coordination of a single angle-pair $\Delta\phi_{12} : \Delta\psi_{12}$ (P_1-P_2) and the relative pole

In Fig. 7.6, two bodies a and b rotate respectively about fixed centres A_0 and B_0. It is required to connect a and b by a coupler, such that as a rotates through an angle $\Delta\phi_{12}$, b should rotate through an angle $\Delta\psi_{12}$. We may draw an arbitrary line p_1 on body a, and an arbitrary line q_1 on body b. The requirement is: as p_1 rotates through $\Delta\phi_{12}$, q_1 should rotate through $\Delta\psi_{12}$. For clarity purposes, without loss of generality, both p_1 and q_1 are drawn in line with A_0B_0. Moreover, p_1 is drawn from A_0 towards the right, and q_1 is also drawn from B_0 towards the right. As p_1 rotates through $\Delta\phi_{12}$ to p_2, q_1 is rotated through $\Delta\psi_{12}$ to q_2. Imagine that body a were fixed, i.e. consider p_2, q_2 bound together with A_0B_0 and p_2 rotated back to p_1. Thus B_0 would be moved to B_{02}^1, and q_2 to q_2^1. In other words, in this

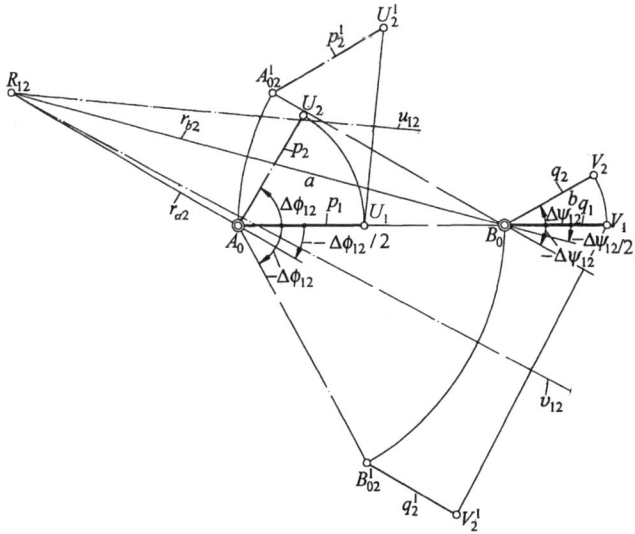

Fig. 7.6. Relative pole R_{12} for coordinating a single angle-pair ϕ_2: ψ_2.

relative motion the body a is considered motionless, while the line q_1 on b is moved to q_2^1. According to the method of locating the pole in Section 3.1, to locate the pole for this relative motion, we erect first the perpendicular bisector r_{a2} of $\overline{B_0 B_{02}^1}$, and then choose any point V_1 on q_1 and get its corresponding point V_2^1 on q_2^1, and erect the perpendicular bisector v_{12} of $\overline{V_1 V_2^1}$. The intersection point of r_{a2} and v_{12} is the relative pole R_{12} of the motion of b relative to a. Please note that, V_1 is any point chosen on q_1. The line v_{12} varies, depending on the position of V_1 chosen. However, the line v_{12} always intersects r_{a2} in the same R_{12} whatever v_{12} may be.

Similarly, we may consider body b as motionless, and a moves relative to b. Imagine as before that p_2, q_2 were bound together with $A_0 B_0$ but q_2 were moved back to q_1. In this relative motion A_0 is moved to A_{02}^1, and p_2 to p_2^1. In other words, b is considered fixed, while the line p_1 on a is moved to p_2^1. To locate the relative pole for this relative motion, we erect the perpendicular bisector r_{b2} of $\overline{A_0 A_{02}^1}$, and choose any point U_1 on p_1 and get its corresponding point U_2^1 on p_2^1. The perpendicular bisector of $\overline{U_1 U_2^1}$ is u_{12}. The intersection point of r_{b2} and u_{12} is the same relative pole R_{12}.

It is clear that, in order to locate R_{12}, all the points and lines V_1, V_2^1, v_{12} and U_1, U_2^1, u_{12} can be omitted. Simply draw a line r_{a2} from A_0, and a line r_{b2} from B_0; the intersection of r_{a2} and r_{b2} is R_{12}, while r_{a2} can be considered as a line obtained by rotating $A_0 B_0$ about A_0 through an angle $-\Delta\phi_{12}/2$, and r_{b2} as a line obtained by rotating $A_0 B_0$ about B_0 through an angle $-\Delta\psi_{12}/2$, as shown in Fig. 7.7. We have the following throem:

Theorem 26 To coordinate a pair of rotation angles $\Delta\phi_{12} : \Delta\psi_{12}$, rotate the line of centres $A_0 B_0$ about A_0 through an angle $-\Delta\phi_{12}/2$ to get r_{a2}, and then rotate it about B_0 through an angle $-\Delta\psi_{12}/2$ to get r_{b2}. The intersection of r_{a2} and r_{b2} is the relative pole R_{12}.

After having located R_{12}, it can be seen that $\sphericalangle A_0 R_{12} B_0 = (\Delta\phi_{12} - \Delta\psi_{12})/2$. Suppose in Fig. 7.7, a four-bar linkage $A_0 A_1 B_1 B_0$ is so synthesized that it is capable of coordinating the prescribed angle-pair $\Delta\phi_{12} : \Delta\psi_{12}$. It is clear that any two opposite sides of the quadrilateral $A_0 A_1 B_1 B_0$ subtend equal angles at R_{12}, or

$$\left. \begin{array}{l} \sphericalangle A_0 R_{12} A_1 = \sphericalangle B_0 R_{12} B_1 \\ \sphericalangle A_0 R_{12} B_0 = \sphericalangle A_1 R_{12} B_1 = (\Delta\phi_{12} - \Delta\psi_{12})/2 \end{array} \right\} \quad (7.2a,b)$$

Synthesis of function generators

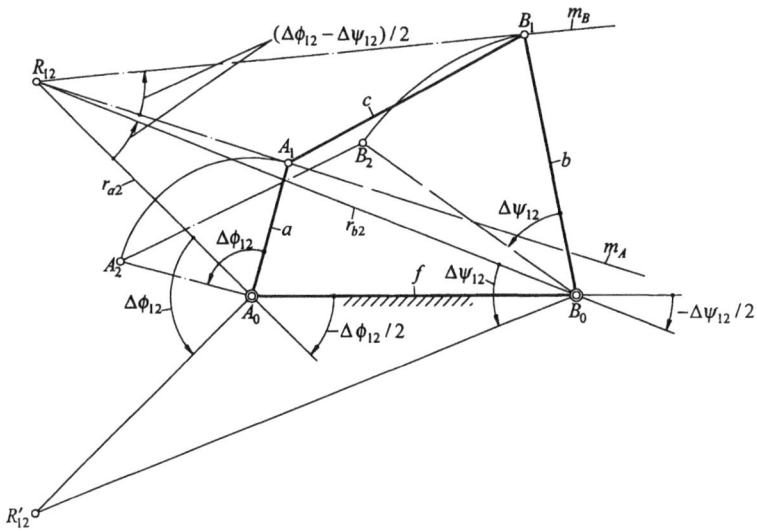

Fig. 7.7. Synthesis of four-bar linkage A_0ABB_0 to coordinate ϕ_2: ψ_2, (P_1-P_2).

This is a case just like that shown in Fig. 3.1 as mentioned in Theorem 7 between the quadrilateral $B_0 A_0 A_1 B_1$ and the pole P_{12}. Hence in Fig. 7.7, suppose the four-bar $A_0 A_1 B_1 B_0$ has not been found yet; we can rotate both lines r_{a2}, r_{b2}, keeping their relative angle $(\Delta\phi_{12} - \Delta\psi_{12})/2$ unchanged, to any position m_A, m_B. Choose any point A_1 on m_A, and any point B_1 on m_B, and join A_1B_1, then $A_0A_1B_1B_0$ is the first position of the required four-bar linkage. Rotate A_0A_1 through an angle $\Delta\phi_{12}$ to A_0A_2; the angle through which the link B_0B_1 rotates to B_0B_2 must be $\Delta\psi_{12}$. Since there are altogether three choices, there are ∞^3 solutions.

This synthesis can be analyzed from the point of view of balancing the number of coordinates mentioned in Section 5.1. The number of available coordinates for the four joints is 8. Prescribed A_0, B_0 and coordination of the angle-pair $\Delta\phi_{12} : \Delta\psi_{12}$ use up 5 coordinates. The remaining number of available coordinates is $8 - 5 = 3$, hence there are ∞^3 possible solutions.

As A_1, B_1 can be any points chosen respectively on m_A, m_B, a question arises: what will happen if both A_1, B_1 are chosen at R_{12}? In this case the length of the coupler reduces to zero, and the mechanism $A_0R_{12}B_0$ becomes a rigid triangle, being certainly not movable. However, if the triangle is dismantled at R_{12}, and reassembled at R'_{12} as shown in Fig. 7.7, it can be seen that the angle of rotation from A_0R_{12} to $A_0R'_{12}$ is still $\Delta\phi_{12}$, and that from B_0R_{12} to $B_0R'_{12}$ is still $\Delta\psi_{12}$, the requirement of coordinating $\Delta\phi_{12}:\Delta\psi_{12}$ being still satisfied. This means that if the points A_1, B_1 are chosen too close to R_{12}, or the coupler length chosen is too short, it is possible that the rotating links may not rotate continuously through the respective angles $\Delta\phi_{12}$ and $\Delta\psi_{12}$.

7.2.2 Coordination of a pair of angular and linear displacements $\Delta\phi_{12}$: Δs_{12} --- synthesis of a slider-crank function generator

Suppose it is required to synthesize a slider-crank mechanism, such that as the crank A_0A rotates through an angle $\Delta\phi_{12}$, the slider should travel through a distance Δs_{12} (a leftward sense of Δs_{12} is considered as positive). In Fig. 7.8,

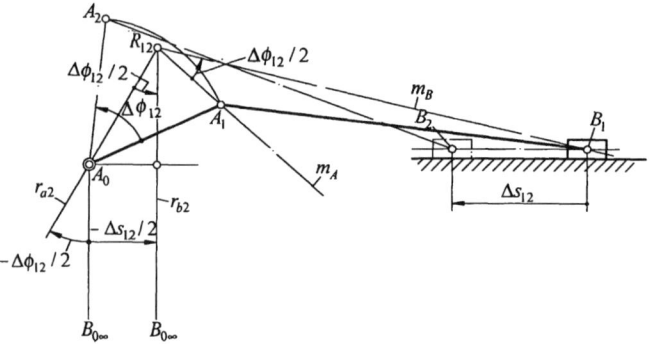

Fig. 7.8. Synthesis of slider-crank $A_0A_1B_1$ to coordinate $\Delta\phi_{12}:\Delta s_{12}$.

imagine according to Fig. 7.7, that B_0 goes to infinity, in a direction normal to the direction of sliding of the slider. According to theorem 26, rotate the line $A_0B_{0\infty}$ about A_0 through $-\Delta\phi_{12}/2$ to r_{a2}, and draw a vertical line r_{b2} from A_0 at a distance of $-\Delta s_{12}/2$ (rightwards) in the direction of sliding of B. The intersection of r_{a2}, r_{b2} is still the relative pole R_{12}. In this case $\sphericalangle A_0R_{12}B_{0\infty}=\Delta\phi_{12}/2$ (because of $\Delta\psi_{12}=0$). Rotate as before both lines r_{a2}, r_{b2}, keeping their relative angle $\Delta\phi_{12}/2$ unchanged, to any position m_A, m_B. Choose any point A_1 on m_A, and any point B_1 on m_B, and join A_1B_1. The first position $A_0A_1B_1$ of the slider-crank is thus determined. Rotate A_0A_1 through an angle $\Delta\phi_{12}$ to A_0A_2; the distance through which B_1 travels to B_2 must be Δs_{12}. There are again three choices, hence there are ∞^3 solutions.

Synthesis of function generators

7.2.3 Coordinations of two pairs of crank rotations $\Delta\phi_{12} : \Delta\psi_{12}$; $\Delta\phi_{13} : \Delta\psi_{13}$ $(P_1-P_2-P_3)$

This case, referring to Fig. 7.4, can be represented by $P_1-P_2-P_3$. For brevity reasons, we write from now on the symbols ϕ_2, ϕ_3, ... in place of $\Delta\phi_{12}$, $\Delta\phi_{13}$,...; and ψ_2, ψ_3, ... in place of $\Delta\psi_{12}$, $\Delta\psi_{13}$,... respectively. In this case, besides the relative pole R_{12} located for the coordination of $\phi_2 : \psi_2$, the relative pole R_{13} for the coordination of $\phi_3 : \psi_3$ can be located by a similar method according to theorem 26, as shown in Fig. 7.9. $\sphericalangle A_0R_{13}B_0 = (\phi_3 - \psi_3)/2$. Rotate first the two lines r_{a2}, r_{b2} about R_{12}, keeping their relative angle $(\phi_2 - \psi_2)/2$ unchanged, to any position m_A, m_B; and

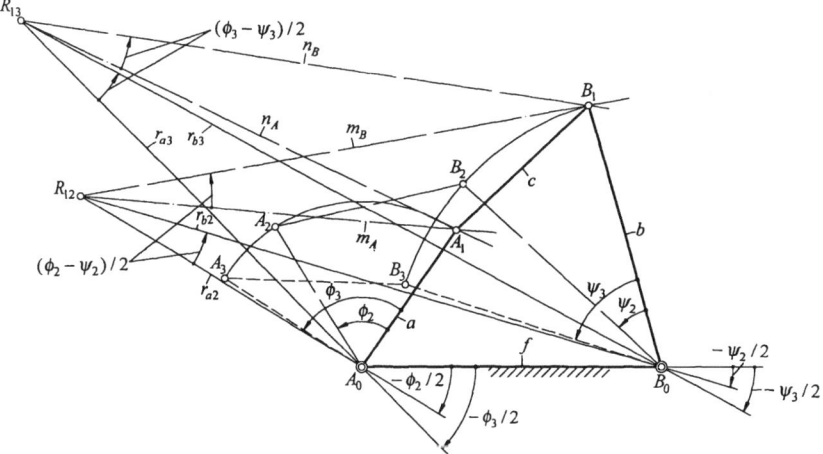

Fig. 7.9. Synthesis of four-bar linkage $A_0A_1B_1B_0$ to coordinate $\phi_2: \psi_2$, $\phi_3: \psi_3$, $(P_1-P_2-P_3)$.

rotate both lines r_{a3}, r_{b3}, about R_{13}, also keeping their relative angle $(\phi_3 - \psi_3)/2$ unchanged, to any position n_A, n_B. The intersection between m_A, n_A is A_1 and the intersection between m_B, n_B is B_1. The synthesized $A_0A_1B_1B_0$ is the first position of the required four-bar linkage. Rotate A_0A_1 through an angle ϕ_2 to A_0A_2, then the angle through which B_0B_1 rotates to B_0B_2 must be ψ_2; and rotate A_0A_1 through an angle ϕ_3 to A_0A_3, then the angle through which B_0B_1 rotates to B_0B_3 must be ψ_3. There are two choices, hence there are ∞^2 solutions.

Analyzing by the principle of balancing coordinates, we see the number of available coordinates is still 8. Prescribed A_0, B_0 and coordinations of $\phi_2: \psi_2$, $\phi_3: \psi_3$ use up 6 coordinates. The remaining number of available coordinates is $8 - 6 = 2$, hence ∞^2 possible solutions.

To coordinate angular and linear displacements $\Delta\phi_{12} : \Delta s_{12}$; $\Delta\phi_{13} : \Delta s_{13}$, the procedure can be carried out according to Fig. 7.8 and by analogy with the present method. Find first the relative poles R_{12}, R_{13}, and then rotate the two lines r_{a2}, r_{b2} and the two lines r_{a3}, r_{b3}, to locate their corresponding intersections A_1, B_1. The problem is thus solved.

7.2.4 Coordinations of three pairs of crank rotations $\phi_2 : \psi_2$; $\phi_3 : \psi_3$; $\phi_4 : \psi_4$ (P_1–P_2–P_3–P_4)

In case three pairs of crank rotations are to be coordinated, three relative poles R_{12}, R_{13}, R_{14} can be located as before; rotate both lines r_{a2}, r_{b2}, both lines r_{a3}, r_{b3}, and both lines r_{a4}, r_{b4}, keeping the relative angles between each pair of lines unchanged, as shown in Fig. 7.10. Suppose r_{a2}, r_{b2} are rotated to m_A, m_B ; r_{a3}, r_{b3} to n_A , n_B ; and r_{a4}, r_{b4} to l_A, l_B ; and m_A, n_A, l_A are concurrent at a point A_1. The three lines m_B, n_B, l_B may not intersect in one common point. In order that m_B , n_B , l_B should also intersect in one common point, the point A_1 may not be chosen arbitrarily on the plane.

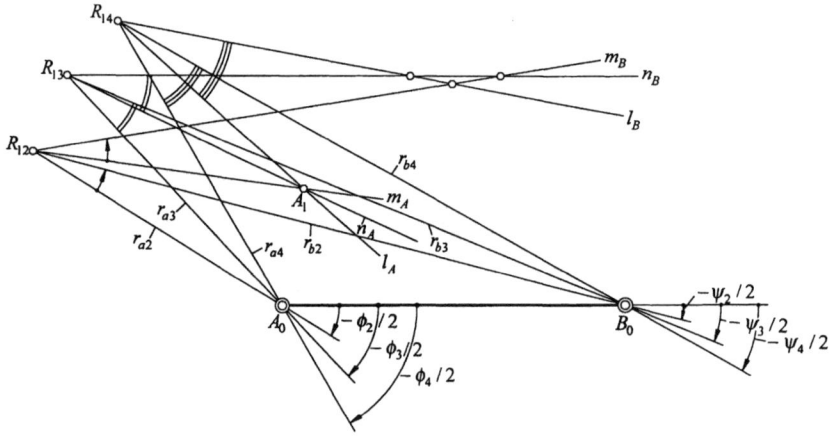

Fig. 7.10. In coordinating three angle-pairs $\phi_2 : \psi_2$, $\phi_3 : \psi_3$, $\phi_4 : \psi_4$, as m_A, n_A, l_A intersect in one point A_1; m_B, n_B, l_B may not intersect in one point.

Analyzing by balancing the number of coordinates, we see that for prescribed A_0, B_0 and coordinations of three pairs of angular displacements, the remaining number of available coordinates is $8 - 7 = 1$, which means ∞^1 possible solutions. This indicates that the point A_1 should lie on a locus (a curve). This locus can be interpreted by means of Fig. 7.6. We can find the three positions q_2^1, q_3^1, q_4^1 (not shown in the figure) of b relative to a, and then find the relative centre-point curve for the four positions q_1, q_2^1, q_3^1, q_4^1 which is the locus on which A_1 should lie. The relative circle-point curve is the locus on which B_1 should lie.

In the past the method for finding the relative centre-point curve was still based on the method mentioned in Section 3.6.2, by finding out first one of the opposite-pole quadrilaterals from the six relative poles R_{12}, R_{13}, R_{14}, $(R_{23})_a$, $(R_{24})_a$,

Synthesis of function generators

$(R_{34})_a$. For instance, R_{12}, R_{14}, $(R_{23})_a$, $(R_{34})_a$, where $(R_{23})_a$ is the pole between q_2^1 and q_3^1, and $(R_{34})_a$ is the pole between q_3^1 and q_4^1. The reason for the subscript (a) here is, because these poles do not like R_{12}, R_{13} or R_{14}, which are unique with respect to a or b, but $(R_{23})_a$ and $(R_{23})_b$ are two different points. As mentioned before, $(R_{23})_a$ is the pole between q_2^1 and q_3^1, being a rotation pole with respect to p_1, while $(R_{23})_b$ is the pole between p_2^1 and p_3^1, being a rotation pole with respect to q_1. Thus confusion is liable to arise and would result in mistakes.

Similarly, to find the locus of B_1, or the relative circle-point curve, one of the opposite-pole quadrilaterals from the six relative poles R_{12}, R_{13}, R_{14}, $(R_{23})_b$, $(R_{24})_b$, $(R_{34})_b$ had to be found out for the four positions p_1, p_2^1, p_3^1, p_4^1 (not shown in Fig. 7.6). The locus of B_1 may also be called the relative centre-point curve, and the locus of A_1 the relative circle-point curve. However, A_1 is usually called the relative centre-point, and B_1 the relative circle-point.

There is a certain correspondence between points on the relative centre-point curve and points on the relative circle-point curve, as has been tabulated by (Alt, 1936) (please refer to Beyer, 1953, equation (141)).

We are not going to make use of the construction based on opposite-pole quadrilaterals. As mentioned in Section 3.11, the relative centre-point curve or relative circle-point curve should be able to be completely determined by the three basic relative poles R_{12}, R_{13}, R_{14} and the six angles in $(\phi_2 - \psi_2)/2$, $(\phi_3 - \psi_3)/2$, $(\phi_4 - \psi_4)/2$. This is the technique we are going to introduce in Section 7.7, using the algebraic method.

To coordinate four angular displacement pairs $\phi_2: \psi_2$; $\phi_3: \psi_3$; $\phi_4: \psi_4$; $\phi_5: \psi_5$, or the case $(P_1 - P_2 - P_3 - P_4 - P_5)$, it is again adequate to use the algebraic method. This will be explained in Section 7.7.3.

7.3 Order type synthesis --- geometrical method

The so-called order type synthesis implies the methods of coordination of prescribed values of $d\psi/d\phi, d^2\psi/d\phi^2, \ldots$ Coordination of differential coefficients higher than $d^2\psi/d\phi^2$ is called higher order synthesis.

7.3.1 Conversion of differential coefficients in an order type synthesis

There is a certain rule for the scale factors M_ϕ, M_ψ in equations (7.1) in the conversion of differential coefficients in an order type synthesis. Let a given function be $y = y(x)$. Its succesive differential coefficients are $y' = dy/dx$, $y'' = d^2y/dx^2$, $y''' = d^3y/dx^3$, $y'''' = d^4y/dx^4$. In the synthesized mechanism, x is represented by ϕ, and y is represented by ψ, or $y = y(x)$ is represented by $\psi = \psi(\phi)$. The successive differential coefficients of the function ψ are $\psi' = d\psi/d\phi$, $\psi'' = d^2\psi/d\phi^2$, $\psi''' = d^3\psi/d\phi^3$,

$\psi'''' = d^4\psi / d\phi^4$. The corresponding differential coefficients should be converted from each other as follows:

$$\left. \begin{aligned} \psi' &= \frac{M_\psi}{M_\phi} y' \\ \psi'' &= \frac{M_\psi}{M_\phi^2} y'' \\ \psi''' &= \frac{M_\psi}{M_\phi^3} y''' \\ \psi'''' &= \frac{M_\psi}{M_\phi^4} y'''' \end{aligned} \right\} \quad (7.3)$$

If it is required to synthesize a mechanism to approximately generate a given function $y = y(x)$ in the vicinity of a certain point $x = x_m$, the values of y', y'', y''', y'''' at x_m have to be calculated first, and then converted into values of $\psi', \psi'', \psi''', \psi''''$ respectively by equations (7.3) before applying the synthesis methods. Theoretically, if we could synthesize a mechanism such that all values of its infinite differential coefficients $\psi', \psi'', \ldots$ match the required values, this mechanism would then be able to completely reproduce the given function $y = y(x)$. This can be seen clearly from the Taylor expansion of the function $y = y(x)$ at the point $x = x_m$:

$$y(x_m + \Delta x) = y(x_m) + \frac{\Delta x}{1!} y'(x_m) + \frac{\Delta x^2}{2!} y''(x_m) + \cdots$$

However, we know that it is only possible to match as high as y''''.

7.3.2 Matching a single angular velocity ratio (P_1P_2)

Suppose the two points P_1, P_2 in Fig. 7.4 approach each other indefinitely and

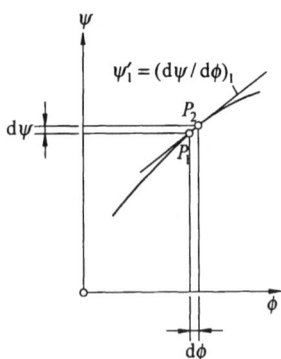

Fig. 7.11. $\psi(\phi)$ relationship in the second order synthesis (P_1P_2).

Synthesis of function generators

become one point. This case is represented by (P_1P_2), and is called a *second order synthesis*. Now both $\Delta\phi_{12}$, $\Delta\psi_{12}$ approach infinitesimals, and may be written as $d\phi$, $d\psi$, as shown in Fig. 7.11. The ratio $d\psi/d\phi$ is not indefinite, but approaches a definite value, because

$$\frac{d\psi}{d\phi} = \frac{d\psi/dt}{d\phi/dt} = \frac{\dot\psi}{\dot\phi} = \frac{\omega_b}{\omega_a} \qquad (7.4)$$

This means that matching a prescribed value of $d\psi/d\phi$ is equivalent to matching a certain angular velocity ratio between the rotating links a, b. In this case since both $\Delta\phi_{12}$, $\Delta\psi_{12}$ approach infinitesimals, the two lines r_{a2}, r_{b2} combine into one line, coinciding with A_0B_0. In other words, the relative pole R_{12} falls on the line A_0B_0, but its position is definite. Considering first the distance $\overline{R_{12}A_0}$ in Fig. 7.7, we have

$$f \sin\frac{\Delta\psi_{12}}{2} = \overline{R_{12}A_0} \sin\frac{\Delta\phi_{12} - \Delta\psi_{12}}{2}$$

As $\Delta\phi_{12} \to d\phi$, $\Delta\psi_{12} \to d\psi$, the above equation becomes

$$f d\psi = \overline{R_{12}A_0}(d\phi - d\psi)$$

Replacing $d\psi/d\phi$ by ψ', we get

$$\overline{R_{12}A_0} = \frac{\psi'}{1-\psi'} f \qquad (7.5)$$

or

$$\psi' = \frac{\overline{R_{12}A_0}}{\overline{R_{12}A_0} + f}$$

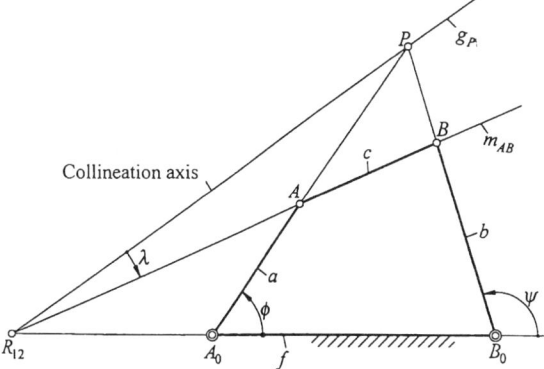

Fig. 7.12.   Third order synthesis, solved by determining R_{12} and angle λ.

or

$$\psi' = \frac{\omega_b}{\omega_a} = \frac{\overline{R_{12}A_0}}{\overline{R_{12}B_0}} \qquad (7.6)$$

In Fig. 7.12, R_{12} is the D_{AB} in Figs. 1.17 and 1.22, being the relative pole between a and b, or P_{ab}. Equation (7.6) is the well-known relation that the angular velocity ratio between ω_b and ω_a is equal to the ratio of the distances from R_{12} ($=P_{ab}$) to A_0 and B_0.

In the synthesis process, the point R_{12} is uniquely determined by ψ'. In the present case, since the two lines m_A, m_B in Fig. 7.7 combine into one single line in Fig. 7.12, the line is denoted by m_{AB}. Rotate m_{AB} about R_{12} to any position, and choose any two points on it, one as A and the other as B. The four-bar linkage A_0ABB_0 is the required mechanism. In this problem of matching a single angular velocity ratio, there are ∞^3 possible solutions.

7.3.3 Matching ψ' and ψ'' ($P_1P_2P_3$)

In the present case, the three points $P_1P_2P_3$ in Fig. 7.4 approch one another indefinitely, as shown in Fig. 7.13. Now the values of $\psi' = d\psi / d\phi$,

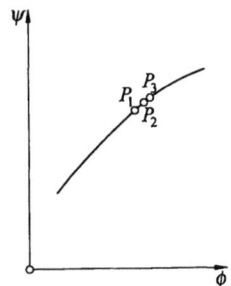

Fig. 7.13. $\psi(\phi)$ relationship in the third order synthesis ($P_1P_2P_3$).

$\psi'' = d^2\psi / d\phi^2$ are prescribed, and it is required to synthesize a four-bar linkage to match them. This case is represented by $P_1P_2P_3$, and is called a *third order synthesis*.

As mentioned in Section 7.3.2, the location of R_{12} is uniquely determined by the prescribed value of ψ' according to equation (7.5), as shown in Fig. 7.12. As to ψ'', we have already derived in equation (2.16) its relation with respect to the angle λ in Fig. 1.17, where ψ' was written as i_{ba}. Rewrite equation (2.16) in accordance with present notation:

$$\cot \lambda = -\frac{\psi''}{\psi'(1-\psi')} \qquad (7.7)$$

As mentioned before, R_{12} is the point D_{AB} in Fig. 1.17, and λ is $\angle AD_{AB}P$ in Fig. 1.17,

Synthesis of function generators

or ∢ *KJP* in Fig. 2.7(a). This angle is considered to be clockwise in Fig. 2.7(a), hence is negative. In both Figs. 1.17 and 7.12, λ is also counted from the collineation axis, being negative for a clockwise sense. Now let the collineation axis $R_{12}P$ be denoted by g_P. The procedure of solution is: in Fig. 7.12, keep the angle λ unchanged, and rotate both lines g_P and m_{AB} about R_{12} to any position. Then choose any point on g_P as P. Join PA_0, PB_0, intersecting respectively the line m_{AB} in A and B. The thus synthesized four-bar linkage A_0ABB_0 possesses both prescribed ψ' and prescribed ψ''. There are altogether ∞^2 solutions.

Example: It is required to synthesize a four-bar linkage to match prescribed $\psi' = 0.453450$, $\psi'' = 0.349066$.

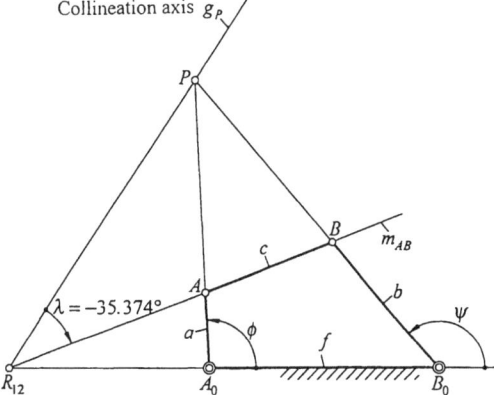

Fig. 7.14. Example: synthesis of four bar linkage A_0ABB_0 for prescribed $\psi'= 0.453450$, $\psi''= 0.349066$.

Solution: In Fig. 7.14, locate first the two points A_0, B_0, and then calculate $\overline{R_{12}A_0} = 0.82966f$ according to equation (7.5) to locate the point R_{12}. Next calculate $\lambda = -35.374°$ according to equation (7.7). Lay the collineation axis g_P and m_{AB} in any position, and choose any point P on g_P. Join PA_0, PB_0 to intersect m_{AB} in the points A, B.

7.3.4 Matching prescribed ψ', ψ'' and ψ''' ($P_1P_2P_3P_4$) --- the Carter-Hall circle

In this problem, the values of $\psi' = d\psi/d\phi$, $\psi'' = d^2\psi/d\phi^2$ and $\psi''' = d^3\psi/d\phi^3$ are prescribed, and it is required to synthesize the four-bar linkage. In this case, the four points P_1, P_2, P_3, P_4 in Fig. 7.4 approach one another indefinitely, hence the case can be represented by $P_1P_2P_3P_4$, or be called a *fourth order synthesis*.

As mentioned before, the location of point R_{12} in Fig. 7.15 is determined by the

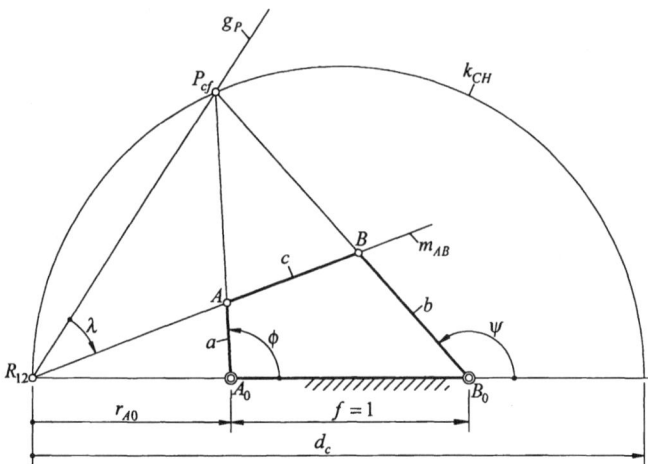

Fig. 7.15. The Carter-Hall circle k_{CH}.

prescribed value of ψ' according to equation (7.5). From now on, $\overline{R_{12}A_0}$ will be denoted by r_{A0}. The angle λ is also detenimed by ψ', ψ'' according to equation (7.7). Both lines g_P and m_{AB} can be placed in any position. After having determined the position of these two lines, there is only one point on the line g_P which can be chosen as P_{cf} such that the value of ψ''' of the synthesized four-bar linkage matches the prescribed ψ'''. The locus of such P_{cf} on the g_P line-pencil to match a certain prescribed value of ψ''', as discovered by (Carter, 1957), is a curve which, as subsequently proved by (Hall, 1957), is a circle. This circle is commonly called *Carter-Hall circle*, and is denoted here by k_{CH}, as shown in Fig. 7.15. The centre of k_{CH} lies on A_0B_0, and k_{CH} passes through R_{12}, its diameter being dependent on ψ'''. Let the diameter of k_{CH} be denoted by d_c, then

$$d_c = \frac{3f[\psi'^2(1-\psi')^2 + \psi''^2]}{(1-\psi')[\psi'(1+\psi')(1-\psi')^2 + 3\psi''^2 + (1-\psi')\psi''']} \quad (7.8)$$

The derivation of equation (7.8) is given in Appendix A13. It can be seen from equation (A13.9) that, as both d_c and r_{A0} are originating from R_{12}, they have the same sign convention, i.e. d_c is positive if it lies on the right hand side of R_{12}, and negative if it lies on the left hand side of R_{12}.

Having determined R_{12} and the circle k_{CH}, we can place the two lines g_P and m_{AB} in any position. The intersection of g_P and k_{CH} is P ($=P_{cf}$). Join PA_0 and PB_0, to intersect m_{AB} respectively in A and B. The problem is thus solved. There are ∞^1 solutions.

If it is required to match further a prescribed value of ψ'''', or to do a *fifth order*

Synthesis of function generators

synthesis $P_1P_2P_3P_4P_5$, a specific point P on the Carter-Hall circle has to be found, such that the synthesized four-bar linkage A_0ABB_0 should exhibit the prescribed value of ψ''''. However, the equations based on the present method would be quite complicated. We would therefore not derive these equations further, but will come back to this point in Section 7.6.3.

7.4 Error of a function generator

A function generating mechanism synthesized to match prescribed values of ψ', ψ'',... means that it exhibits such values of ψ', ψ'',... in the position of synthesis. Let the values of x, y at which the function $y = y(x)$ is to be generated be denoted respectively by x_m, y_m. As the value of x moves an amount Δx away from x_m, or $\Delta x = x - x_m$, the driving link should rotate through an angle $\Delta \phi = M_\phi \Delta x$. Due to the construction of the function generator, the driven link is then rotated through an angle $\Delta \psi$ which, as displayed on the indicating dial, shows the change of the value of y. However, the ideal change of the angle ψ should be $M_\psi \Delta y$, where Δy is the theoretical change of y due to Δx. The difference between the acutual $\Delta \psi$ and the ideal $M_\psi \Delta y$ is the error of the function generator, or

$$\varepsilon = |\Delta \psi| - |M_\psi \Delta y| \tag{7.9}$$

Equation (7.9) indicates that the error ε is the magnitude difference between $|\Delta \psi|$ and $|M_\psi \Delta y|$, not an algebraic difference. It shows that if ε is positive, the actual swinging angle of the follower link B_0B is too large, and if ε is negative, the swinging angle of B_0B is too small. The value of $\Delta \psi$ can be calculated by the following formula:

$$\Delta \psi = \psi(\phi_m + \Delta \phi) - \psi(\phi_m) = \psi[\phi_m + M_\phi(x - x_m)] - \psi_m \tag{7.10}$$

The functional value of $\psi(\phi)$ can be calculated by equation (2.5).

7.5 Transmission angle

The concept of the *transmission angle* was first proposed by (Alt, 1932c). According to the initial definition, it is defined as shown in Fig. 7.16(a). As a force is applied by a link AB on a point B of the follower link, the direction of the absolute motion of B is along t_a; while the direction of the motion of B relative to A is along t_r. The angle μ between t_r and t_a is called the transmission angle. For a four-bar linkage as that shown in Fig. 7.16(b) in which A_0A is the driving link and B_0B is the driven link, this angle is just $\mu = \sphericalangle ABB_0$. For a crank-rocker as shown in Fig. 7.17, in its inner frame position, i.e. when A is at A_1, the transmission angle $\sphericalangle A_1B_1B_0 = \mu_{min}$ is a minimum; and in its outer frame position, i.e. when A is at A_2, the transmission angle $\sphericalangle A_2B_2B_0 = \mu_{max}$ is a maximum. In all other positions of the four-bar linkage, the transmission angle μ lies between μ_{max} and μ_{min}.

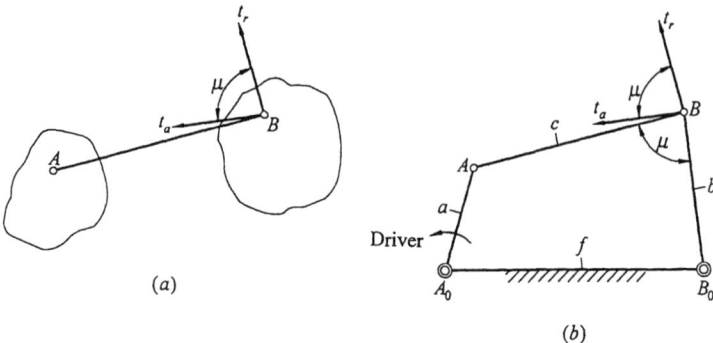

Fig. 7.16. (a) Definition of transmmision angle μ. (b) Transmission angle μ in a four-bar linkage A_0ABB_0.

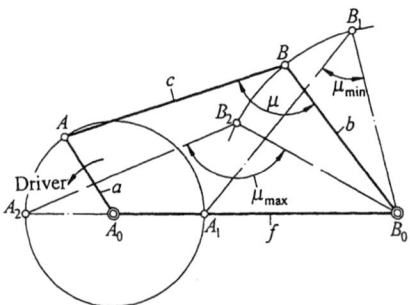

Fig. 7.17. μ_{max} and μ_{min} of a four-bar linkage.

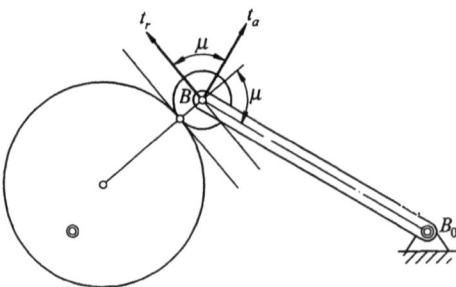

Fig. 7.18. Transmission angle in a cam mechanism.

Synthesis of function generators

For a cam mechanism as shown in Fig. 7.18, the angle μ between t_r and t_a is equal to the angle between the follower link B_0B and the normal of the cam profile at the point of contact. For a cam mechanism with a sliding follower, Fig. 7.19, the transmission angle μ and the pressure angle θ are complementary angles. In a gear mechanism, Fig. 7.20, no relation exists between the transmission angle μ and pressure angle θ. For a Stephenson-III type six-bar mechanism as that shown in Fig. 7.21, the transmission angle μ should be the angle between the connecting link AB and the line $P_{cf}B$. (Wu & Yan, 1987) has pointed out that, for a six-bar linkage, it is sometimes inadequate to judge the transmission *performance* by the transmission

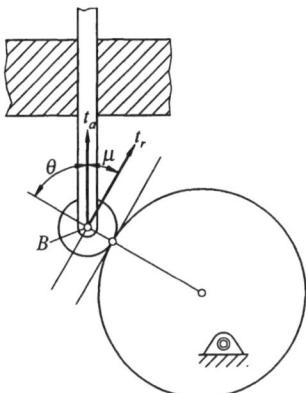

Fig. 7.19. Transmission angle in a cam mechanism.

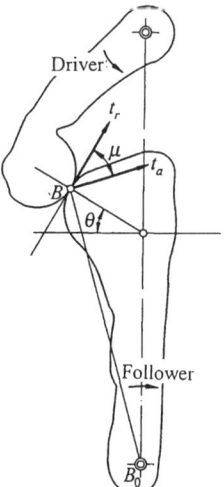

Fig. 7.20. Transmission angle μ in a gear transmission.

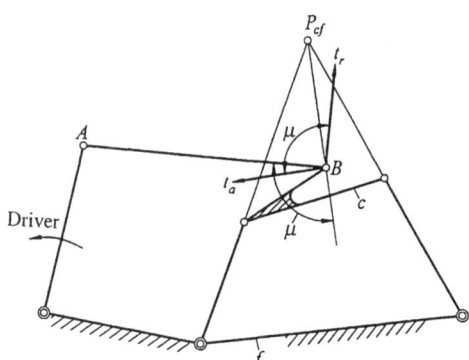

Fig. 7.21. Transmission angle μ in a Stephenson-III type six-bar linkage.

angle.

We come back to the four-bar linkage. In Fig. 7.16(b), the optimum value of μ is 90°. However, the transmission angle changes during the movement of the mechanism, hence it is desirable to keep $|\mu - 90°|$ to a minimum. For instance, the effect of $\mu = 80°$ and $\mu = 100°$ are the same. In the synthesis of a four-bar linkage, it is frequently required to find out among the ∞ possible solutions the one with the optimum transmission angle, i.e. to do the optimization of the transmission angle. However, since $\sin \mu = \sin(180° - \mu)$, it would be preferable to optimize $\sin \mu$, or even to maximize $\sin^2 \mu$, or minimize $\cos^2 \mu$, rather than to optimize $|\mu - 90°|$.

Please note that, in Fig. 7.16(b), if A_0A is the driver, then $\angle A_0AB$ has nothing to do with the transmission performance. $\angle A_0AB$ is the transmission angle only when B_0B is the driving link.

7.6 Simplified geometric method for higher order synthesis

This method is derived from the simplified acceleration analysis of a four-bar linkage as mentioned in Section 2.3. It is originally developed from the analysis of the angular acceleration α_b, hence the second angluar acceleration $\dot{\alpha}_b$ of the follower link b (Chiang, 1971) (Chiang, Pennestrì & Chung, 1989). This method can be compared with the method mentioned in Section 4.2.2, and also the method of (Lakshminarayana, 1971).

Carrying out the geometrical acceleration analysis of a four-bar linkage shown in Fig. 2.7(a) in a reverse order, we can synthesize the linkage to match prescribed values of ψ', ψ''. In Fig. 2.7(a), please note that P is the instantaneous velocity pole of the coupler CB of the four-bar linkage A_0CBB_0, but not the velocity pole of AB (= c). The relationships between the differential coefficients ψ', ψ'',..., and the

Synthesis of function generators

angular velocity, angular acceleration,..., of the link b are as follows (assume ω_a is constant):

$$\left.\begin{aligned}\omega_b &= \frac{d\psi}{dt} = \frac{d\psi}{d\phi} \cdot \frac{d\phi}{dt} = \psi' \omega_a \\ \alpha_b &= \frac{d\omega_b}{dt} = \psi'' \omega_a^2 \\ \dot{\alpha}_b &= \psi''' \omega_a^3 \end{aligned}\right\} \quad (7.11)$$

Therefore matching a prescribed value of ψ' is equivalent to matching a certain angular velocity ratio, and matching a prescribed value of ψ'' is equivalent to matching a certain ratio of α_b and ω_a^2, etc.

7.6.1 Matching ψ' and ψ'', third order synthesis ($P_1P_2P_3$)

It has been explained in Section 2.3 that for $\omega_a = 1$, in Fig. 2.7(a), v_B is represented by $\overline{PB}$, therefore

$$\frac{\overline{PB}}{\overline{B_0B}} = \frac{\overline{KP}}{\overline{B_0P}} = \frac{\omega_b}{\omega_a} = \psi' \quad (7.12)$$

Also it has been proved in equation (2.10) that $a_B^t = \overline{JK}\omega_a^2$, hence

$$\frac{\overline{JK}}{\overline{B_0B}} = \psi'' \quad (7.13)$$

First draw a vertical line $\overline{B_0B} = 1$ unit length in Fig. 7.22. Then according to the

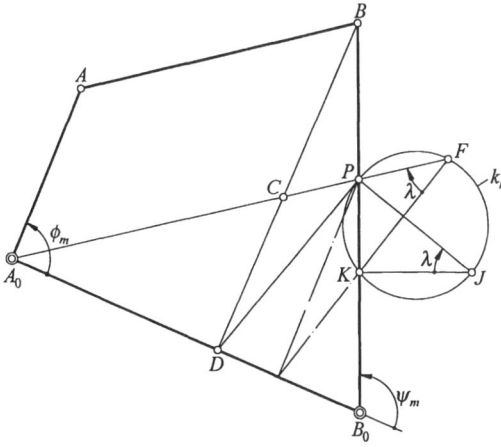

Fig. 7.22. Simplified third order synthesis for matching prescribed ψ', ψ''.

prescribed ψ' the points P and K can be determined on $\overline{B_0B}$ by equation (7.12). For example if $b = \overline{B_0B}$ is taken as 100 mm, and $\psi' = 0.4$, then $\overline{PB} = 40$ mm, $\overline{KP} = 24$ mm. Secondly, the point J can be determined according to equation (7.13). For example, if $\psi'' = 0.3$, then $\overline{JK} = 30$ mm, directing leftwards (since ψ'' is positive). The triangle $\triangle PKJ$ and its circumcircle k_r are thus uniquely determined.

The following procedure is to choose any point D on the plane and draw the following lines:

(1) join DB_0, DP and DB;
(2) draw $KF \| DP$, intersecting the circle k_r in F;
(3) join FP, intersecting DB and B_0D respectively in C and A_0;
(4) complete the parallelogram A_0CBA.

The four-bar linkage A_0ABB_0 is the required mechanism. As there are ∞^2 ways in choosing the point D, there are ∞^2 solutions.

7.6.2 Matching ψ', ψ'' and ψ''', fourth order synthesis ($P_1P_2P_3P_4$)

Since for the four-bar linkage A_0ABB_0 in Fig. 7.15, the corresponding positions of the input angle ϕ and output angle ψ for prescribed values of ψ', ψ'', ψ''' are such that the locus of P_{cf} is a circle, the Carter-Hall circle k_{CH}, with a diameter d_c as given in equation (7.8), we may apply this relationship to the four-bar B_0A_0CB in Fig. 7.23, by considering $\angle BB_0A_0 = \beta$ as the input angle, and the angle v shown in the figure

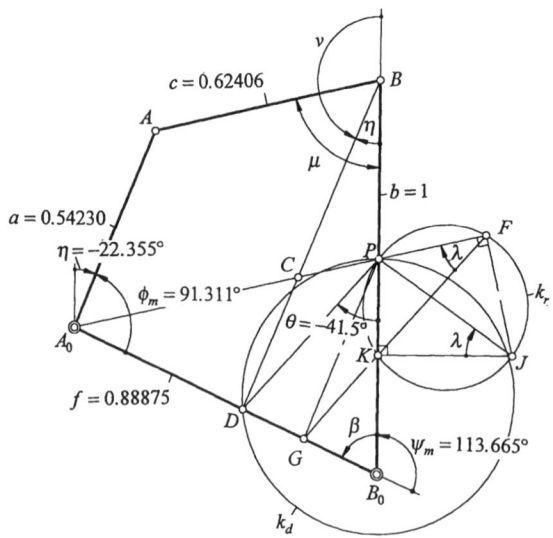

Fig. 7.23. Simplified fourth order synthesis for matching prescribed ψ', ψ'', ψ'''.

Synthesis of function generators

as the output angle. The point D here corresponds to P_{cf} in Fig. 7.15. The locus of D should also be a circle. The centre of this circle lies on B_0B, and the circle passes through the points P, D. Denote this circle by k_d, and its diameter by d_d. Using equation (7.8) as a formula, we may write

$$d_d = \frac{3b[v'^2(1-v')^2 + v''^2]}{(1-v')[v'(1+v')(1-v')^2 + 3v''^2 + (1-v')v''']} \quad (7.14)$$

where $v' = dv/d\beta$, $v'' = d^2v/d\beta^2$, $v''' = d^3v/d\beta^3$. Converting these three differential coefficients in terms of ψ', ψ'', ψ''', we see that

$$\beta + \psi = 180°$$
$$v = \beta + \phi$$

hence $\quad d\beta = -d\psi$, $d\phi/d\beta = -1/\psi'$

$$\left.\begin{array}{l} v' = 1 - 1/\psi' \\ v'' = -\psi''/\psi'^3 \\ v''' = (-3\psi''^2/\psi' + \psi''')\psi'^4 \end{array}\right\} \quad (7.15a,b,c)$$

Substituting equations (7.15 a,b,c) into equation (7.14), we get

$$d_d = \frac{3b[\psi'^2(1-\psi')^2 + \psi''^2]}{\psi'(1-\psi')(1-2\psi') + \psi'''} \quad (7.16)$$

The sign convention of d_d is similar to that of d_c. As the circle k_d in Fig. 7.23 corresponds to the circle k_{CH} in Fig. 7.15, the line segment in Fig. 7.23 corresponding to r_{A0} in Fig. 7.15 is $\overline{PB_0}$. Therefore d_d has the same sign convention as $\overline{PB_0}$. In other words, for a circle k_d below P the sign of d_d is negative, for k_d above P, d_d is positive.

Please note that, if a circle k_d is drawn in Fig. 7.22 passing through P, D, then the value of ψ''' can be calculated from equation (7.16) according to this point D. Conversely, in matching prescribed values of ψ', ψ'', ψ''', the value of d_d has to be calculated according to equation (7.16), and the circle k_d has to be drawn according to its sign. Finally choose any point on k_d as D, and follow the procedure mentioned in Section 7.6.1 to complete the synthesis.

Although the present method is called a geometric method, all dimensions are calculated according to geometrical relations. In Fig. 7.23, $\lambda = \sphericalangle KJP$, $\eta = v - 180°$, $\theta = \sphericalangle B_0PD$, where η represents the orientation of A_0A with respect to the vertical line B_0B. Please note that in Fig. 7.23, λ, η, θ are all negative. Take θ as the parameter in selecting D on k_d. In other words, select a value of θ, and do the following calculations:

$$\lambda = \tan^{-1}\left[-\frac{\psi'(1-\psi')}{\psi''}\right] \qquad (7.17)$$

$$\psi_m = \tan^{-1}\frac{\sin\theta\cos\theta}{-\frac{b(1-\psi')}{d_d} - \cos^2\theta} \qquad (7.18)$$

$$\eta = \tan^{-1}\frac{\sin\theta\cos\theta}{-\frac{b\psi'}{d_d} + \cos^2\theta} \qquad (7.19)$$

The lengths of the links are:

$$a = \frac{b\psi'\sin(\lambda+\theta)}{\sin(\lambda+\theta-\eta)} \qquad (7.20)$$

If $a < 0$, then replace η by $\eta - 180°$, and recalculate a.

$$f = \frac{a\sin(\lambda+\theta-\eta) - b\sin(\lambda+\theta)}{\sin(\lambda+\theta+\psi_m)} \qquad (7.21)$$

If $f < 0$, then replace ψ_m by $\psi_m + 180°$, and recalculate f.

$$\phi_m = \psi_m + \eta \qquad (7.22)$$

$$c = \left|\frac{-a\sin\phi_m + b\sin\psi_m}{\sin(\lambda+\theta+\psi_m)}\right| \qquad (7.23)$$

In the above equations, ϕ_m, ψ_m are the respective values of ϕ, ψ in the design position. If $b = 1$ is assumed in the first place, all calculations can be simplified. According to the above equations, the corresponding computer program is listed in Appendix 14 for easy reference. The following example will clearly illustrate the synthesis procedure. In case a fourth order synthesis of a slider-crank function generator is required, please refer to (Chiang, Pennestri & Chung, 1989), the corresponding computer program being listed in Appendix 15.

Example: It is required to synthsize a four-bar linkage to match prescribed $\psi' = 0.453450$, $\psi'' = 0.349066$, $\psi''' = -0.806133$.

Solution: Take $b = 1$. Find first by equations (7.16), (7.17)

$$d_d = -0.702124, \quad \lambda = -35.3743°$$

Hence k_d lies below the point P. Choosing $\theta = -41.5°$, we get dimensions of the linkage:

Synthesis of function generators

$a = 0.54230$, $b = 1$, $c = 0.62406$, $f = 0.88875$,
$\psi_m = 113.665°$, $\eta = -22.355°$, $\phi_m = 91.311°$

The synthesized four-bar linkage is that shown in Fig. 7.23. Five significant figures are taken in the results. This is because if insufficient significant figures were taken, it would be difficult to find out the correct value of the error. In particular in the synthesis process of a function generator, because of the error propagation due to multiple calculating operations, the input data should be accurate to six to seven significant figures, if the results are to be accurate to five to six significant figures.

7.6.3 Matching ψ', ψ'', ψ''' and ψ'''', fifth order synthesis ($P_1P_2P_3P_4P_5$)

In Fig. 7.23, the four-bar linkage synthesized from any point D chosen on the circle k_d will match the prescribed ψ', ψ'', ψ'''. However, there are only a finite number of D's by which the four-bar linkage thus synthesized will match a certain prescribed value of ψ''''. Let us find these points. Assume first $b = 1$. Rewrite equation (7.16) in the form

$$d_d = \frac{Y}{X} \tag{7.24}$$

that is

$$\left. \begin{array}{l} Y = 3[\psi'^2(1-\psi')^2 + \psi''^2] \\ X = \psi'(1-\psi')(1-2\psi') + \psi''' \end{array} \right\} \tag{7.25 a,b}$$

Furthermore, we see from Fig. 7.23 that

$$d_d = -\frac{|PD|}{\cos\theta} \tag{7.26}$$

The four link lengths a, b, c, f of the four-bar linkage in Fig. 7.23 are considered as constants, and d_d can be considered as function of ϕ only. Differentiating equations (7.24) and (7.26) respectively with respect to ϕ, and comparing the corresponding coefficients, we get the following cubic equation in the unknown $\tan\theta$ for matching a prescribed value of ψ'''':

$$a_3 \tan^3\theta + a_2 \tan^2\theta + a_1 \tan\theta + a_0 = 0 \tag{7.27}$$

where

$$\left. \begin{array}{l} a_3 = \psi'' \\ a_2 = 3\psi'(1-\psi') - \dfrac{Y}{X}(1-2\psi') \\ a_1 = 5\psi'' - \dfrac{Y}{X^2}\{[1-6\psi'(1-\psi')]\psi'' + \psi''''\} \\ a_0 = \left(\dfrac{Y}{X} - \psi'\right)\left(1 + \dfrac{Y}{X} - \psi'\right) \end{array} \right\} \tag{7.28}$$

For the derivation of equations (7.28) please refer to (Chiang, 1973). Equation (7.27) can have one, or three real roots. Therefore at least one four-bar linkage can be synthesized to match the four prescribed values of differential coefficients ψ', ψ'', ψ''', ψ''''.

Example: It is required to carry out a fourth order synthesis of a four-bar linkage to generate the function $y = \log_{10} x$. Assume the design position is at $x_m = 4$, and the range of x is $2 \leq x \leq 9$. The scale factors are assumed to be:

$$M_\phi = \Delta\phi/\Delta x = 10°/\text{unit of } x = 0.17453293 \text{ rad}/\text{unit of } x$$
$$M_\psi = \Delta\psi/\Delta y = -60°/\text{unit of } y = -1.04719755 \text{ rad}/\text{unit of } y$$

Solution: The following differential coefficients are calculated according to equations (7.3):

$$\psi' = -0.6514417;$$
$$\psi'' = 0.9331215;$$
$$\psi''' = -2.6731963;$$
$$\psi'''' = 11.4872149.$$

Set again $b = 1$. The following values are calculated according to equations (7.25 a,b), as well as equations (7.16), (7.17) and (7.28):

$$Y = 6.0843002;$$
$$X = -5.1506797;$$
$$d_d = -1.1812616;$$
$$\lambda = +49.0629°;$$
$$a_3 = 0.9331215;$$
$$a_2 = -0.5071461;$$
$$a_1 = 0.4357468;$$
$$a_0 = -0.2491108.$$

There is only one real root of equation (7.27), i.e. $\tan\theta = 0.56035$, or $\theta = 29.264°$. This angle will be denoted by θ_0. The synthesized four-bar linkage is shown in Fig.7.24. Its dimensions calculated by corresponding equations are:

$$\psi_m = 213.801°, \quad \eta = -116.170°, \quad \phi_m = 97.632°,$$
$$a = 2.549, \quad b = 1, \quad c = 3.327, \quad f = 1.746.$$

The transmission angle of this synthesized four-bar linkage is 48.214° at $x = 2$, and 41.448° at $x = 9$. Its error curve is shown in Fig. 7.25. It can be seen that the error ε in the vicinity of x_m is indeed quite small. However, the error increases rapidly at $x = 9$, being even worse than the error of the four-bar linkage synthesized for $\theta = 27°$, as shown by the chain curve in the figure. Another drawback of this four-bar linkage is that its link lengths are not in proper proportions. It reminds us to consider if it is worthwhile to match precisely the prescribed ψ'''', or if we can find another point D

Synthesis of function generators

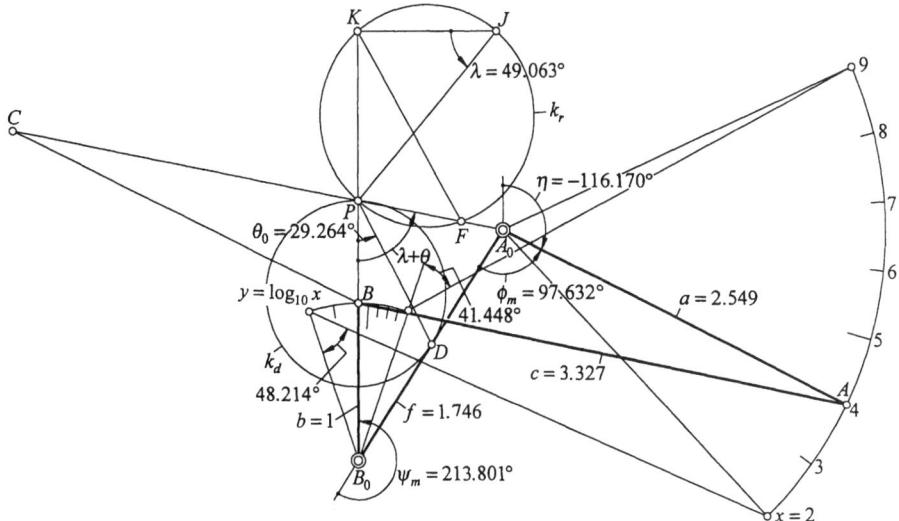

Fig. 7.24. Example: fifth order synthesis of function generator to generate function $y = \log_{10}x$, $\theta = 29.26417°$.

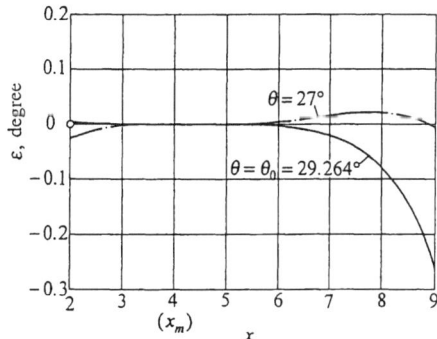

Fig. 7.25. Example in page 334: error curves for $\theta = \theta_0 = 29.264°$ and $\theta = 27°$.

on k_d by varying θ slightly away from θ_0, such that the transmission angle, the error curve and the link proportions of the synthesized four-bar linkage are acceptable. It has been found that, if $\theta > \theta_0$ is taken, the four-bar linkage thus synthesized is not satisfactory, and also that only values of θ within the range $4° \leq \theta \leq \theta_0$ are useful. Fig. 7.26 shows the variations of the error ε at $x = 2$ and $x = 9$ as functions of θ. The scale of the right hand portion of the figure in the range $24° \leq \theta \leq 30°$ is enlarged for easy reading. Try $\theta = 25.5°$. The synthesized four-bar function generator is shown

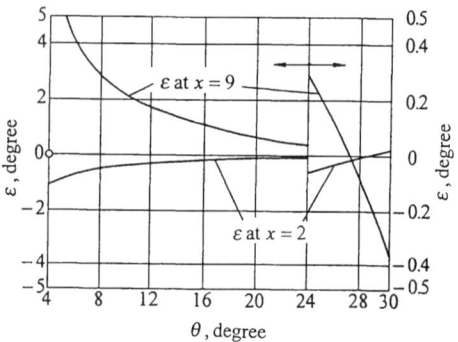

Fig. 7.26.   Example in page 334: error curves at $x = 2$ and $x = 9$ as functions of θ.

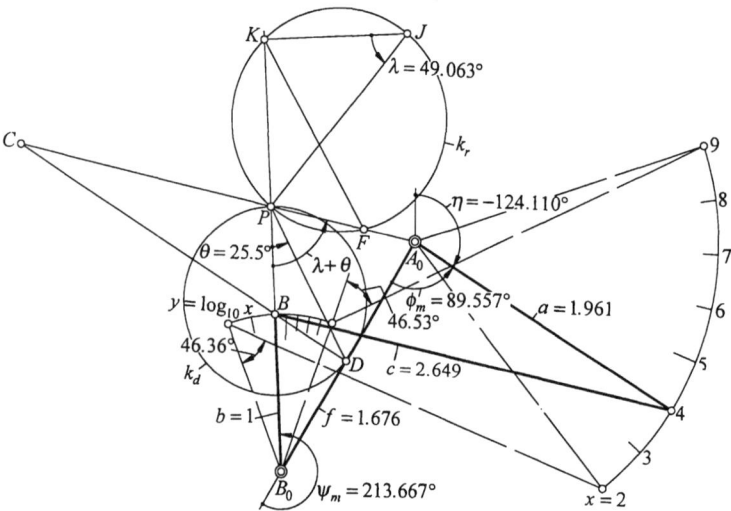

Fig. 7.27.   Example: improved solution of preceding example, $\theta = 25.5°$.

in Fig. 7.27, its dimensions being:

$\psi_m = 213.667°$, $\eta = -124.110°$, $\phi_m = 89.557°$,
$a = 1.961$, $b = 1$, $c = 2.649$, $f = 1.676$.

The errors of this mechanism at $x = 2$ and $x = 9$ are respectively $-0.041°$ and $0.153°$. The transmission angles at both ends of x are respectively $46.36°$ and $46.53°$. The link proportions are better than those of the linkage shown in Fig. 7.24.

Synthesis of function generators

7.7 Algebraic methods

The algebraic methods for synthesizing function generators include those using loop equations (Freudenstein, 1955; Sieker, 1956) which involve also order type synthesis, and those using relative displacement matrices between output link and input link (Suh & Radcliffe, 1967). However, the present algebraic method is developed from the geometrical concept depicted in Section 7.2. Crank-rockers and double rockers synthesized by algebraic methods (Chen, 1969 a,b) can easily be treated as function generators.

7.7.1 Basic equations

Fig. 7.28 is taken partly from Fig. 7.7. Suppose it is required to coordinate a pair of angular displacements $\phi_j : \psi_j$. The relative pole R_{1j} can be determined according to Theorem 26. As has been mentioned in Section 7.2.1, and also indicated in

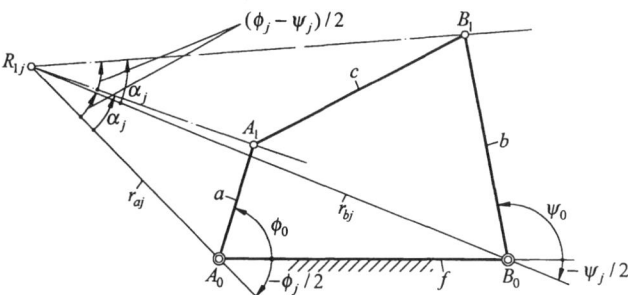

Fig. 7.28. Configuration for derivation of basic equations of a four-bar function generator.

equation (7.2 a), $\overline{A_0 A_1}$ and $\overline{B_0 B_1}$ subtend equal angles at R_{1j}. Let this angle be denoted by α_j, and

$$r_{aj} = \overline{R_{1j} A_0}, \quad r_{bj} = \overline{R_{1j} B_0}$$

$$a = \overline{A_0 A_1}, \quad b = \overline{B_0 B_1}$$

We can formulate the above condition in equation form. From Fig. 7.28 we have

$$r_{aj} = \frac{\sin\frac{\psi_j}{2}}{\sin\frac{\phi_j - \psi_j}{2}} f$$

$$r_{bj} = \frac{\sin\frac{\phi_j}{2}}{\sin\frac{\phi_j - \psi_j}{2}} f$$

(7.29 a,b)

Both r_{aj}, r_{bj} are constants. For simplicity reasons, assume $f = 1$. Denote values of ϕ, ψ in the design position by ϕ_0, ψ_0 respectively. From $\Delta R_{1j}A_0A_1$ and $\Delta R_{1j}B_0B_1$ we have

$$\cot\alpha_j = \frac{r_{aj} + a\cos(\phi_0 + \phi_j/2)}{a\sin(\phi_0 + \phi_j/2)}$$

$$\cot\alpha_j = \frac{r_{bj} + b\cos(\psi_0 + \psi_j/2)}{b\sin(\psi_0 + \psi_j/2)}$$

(7.30 a,b)

Eliminating $\cot\alpha_j$ from equations (7.30 a,b), we get

$$r_{aj}\sin\left(\psi_0 + \frac{\psi_j}{2}\right)\frac{1}{a} - r_{bj}\sin\left(\phi_0 + \frac{\phi_j}{2}\right)\frac{1}{b} = \sin\left[\left(\phi_0 + \frac{\phi_j}{2}\right) - \left(\psi_0 + \frac{\psi_j}{2}\right)\right] \quad (7.31)$$

Let $t = \tan\phi_0$, $q = \tan\psi_0$, $X = 1/(a\cos\phi_0)$, $Y = 1/(b\cos\psi_0)$. Equation (7.31) can be rewritten as

$$r_{aj}\left(q\cos\frac{\psi_j}{2} + \sin\frac{\psi_j}{2}\right)X - r_{bj}\left(t\cos\frac{\phi_j}{2} + \sin\frac{\phi_j}{2}\right)Y$$

$$= \left(t\cos\frac{\phi_j}{2} + \sin\frac{\phi_j}{2}\right)\left(\cos\frac{\psi_j}{2} - q\sin\frac{\psi_j}{2}\right)$$

$$- \left(\cos\frac{\phi_j}{2} - t\sin\frac{\phi_j}{2}\right)\left(q\cos\frac{\psi_j}{2} + \sin\frac{\psi_j}{2}\right) \quad (7.32)$$

For simplicity reasons, let $\xi_j = \phi_j/2$, $\varepsilon_j = \psi_j/2$; and use the symbols $S_{\xi j} = \sin\xi_j$, $C_{\xi j} = \cos\xi_j$, $S_{\varepsilon j} = \sin\varepsilon_j$, $C_{\varepsilon j} = \cos\varepsilon_j$. Equation (7.32) can further be written as

$$-r_{bj}(tC_{\xi j} + S_{\xi j})Y + [r_{aj}C_{\varepsilon j}X - t\sin(\xi_j - \varepsilon_j) + \cos(\xi_j - \varepsilon_j)]q$$
$$= -r_{aj}S_{\varepsilon j}X + t\cos(\xi_j - \varepsilon_j) + \sin(\xi_j - \varepsilon_j) \quad (7.33)$$

Equation (7.33) is the basic equation used for function generator synthesis. In this equation there are altogether four unknowns t, q, X, Y. In the present arrangement of equation (7.33), this is a linear equation with respect to the two unknowns Y, q.

Synthesis of function generators

Certainly it can also be arranged as another linear equation with respect to any two of the four unknowns.

7.7.2 Coordinations of three angular displacement pairs --- four finitely separated relative positions (P_1–P_2–P_3–P_4)

Since the problems of coordinations of one or two angular displacement pairs are rather simple, we shall start from the coordinations of three angular pairs, i.e. to coordinate $\phi_2:\psi_2$; $\phi_3:\psi_3$; $\phi_4:\psi_4$. Let $j = 2, 3, 4$ in equation (7.33), we get three linear equations in the unknowns Y, q. The determinant of the coefficients of these equations should vanish identically if they have a non-trivial solution, or

$$\begin{vmatrix} -r_{b2}(tC_{\xi 2}+S_{\xi 2}), r_{a2}C_{\varepsilon 2}X - t\sin(\xi_2-\varepsilon_2)+\cos(\xi_2-\varepsilon_2), -r_{a2}S_{\varepsilon 2}X + t\cos(\xi_2-\varepsilon_2)+\sin(\xi_2-\varepsilon_2) \\ -r_{b3}(tC_{\xi 3}+S_{\xi 3}), r_{a3}C_{\varepsilon 3}X - t\sin(\xi_3-\varepsilon_3)+\cos(\xi_3-\varepsilon_3), -r_{a3}S_{\varepsilon 3}X + t\cos(\xi_3-\varepsilon_3)+\sin(\xi_3-\varepsilon_3) \\ -r_{b4}(tC_{\xi 4}+S_{\xi 4}), r_{a4}C_{\varepsilon 4}X - t\sin(\xi_4-\varepsilon_4)+\cos(\xi_4-\varepsilon_4), -r_{a4}S_{\varepsilon 4}X + t\cos(\xi_4-\varepsilon_4)+\sin(\xi_4-\varepsilon_4) \end{vmatrix} = 0$$

(7.34)

Equation (7.34) contains only two unknowns X and t. It can be expanded into a quadratic equation in the unknown X:

$$k_{A1(1234)} = a_0 X^2 + a_1 X + a_2(1+t^2) = 0 \tag{7.35}$$

Substituting $X = 1/(a\cos\phi_0)$ into equation (7.35), and noting that $1 + t^2 = 1/\cos^2\phi_0$, it can be transformed into

$$k_{A1(1234)} = a_2 a^2 + (a_1\cos\phi_0)a + a_0 = 0 \tag{7.36}$$

The three coefficients in equation (7.36) are:

$$\left.\begin{array}{l} a_2 = u_1 t + u_2 \\ a_1 = u_7 t^2 + u_8 t + u_9 \\ a_0 = u_5 t + u_6 \end{array}\right\} \tag{7.37}$$

a_2, a_1, a_0 are all functions of t, or functions of ϕ_0. The seven $u_1, \ldots, u_9$ are constants, depending on the values of ϕ_j, ψ_j. Their expressions are listed in equations (A16.1) in Appendix 16. For a given value of ϕ_0, or a value of $t = \tan\phi_0$, the three coefficients a_2, a_1, a_0 can be calculated, and substituted into equation (7.36), then two values of a can be found, hence two points on the curve $k_{A1(1234)}$ can be located. The complete curve $k_{A1(1234)}$ found in this way is the locus of A_1, or the relative centre-point curve of the motion of b relative to a. This curve passes of course through A_0.

Corresponding to each set of (a, ϕ_0), or to each point A_1, the point B_1 is uniquely determined. This is found by converting the known (a, ϕ_0) values into (X, t) values and substituting these into the three equations of (7.33) to find the (Y, q) values, or

$$Y = \frac{\Delta_Y}{\Delta}, q = \frac{\Delta_q}{\Delta}$$

$$\Delta_Y = \left| -r_{aj} S_{\varepsilon j} X + t\cos(\xi_j - \varepsilon_j) + \sin(\xi_j - \varepsilon_j), r_{aj} C_{\varepsilon j} X - t\sin(\xi_j - \varepsilon_j) + \cos(\xi_j - \varepsilon_j), 1 \right|$$

$$\Delta_q = \left| -r_{bj}(tC_{\xi j} + S_{\xi j}), -r_{aj} S_{\varepsilon j} X + t\cos(\xi_j - \varepsilon_j) + \sin(\xi_j - \varepsilon_j), 1 \right|$$

$$\Delta = \left| -r_{bj}(tC_{\xi j} + S_{\xi j}), r_{aj} C_{\varepsilon j} X - t\sin(\xi_j - \varepsilon_j) + \cos(\xi_j - \varepsilon_j), 1 \right|$$

(7.38)

The expanded forms of equation (7.38) are listed as equations (A16.2) in Appendix 16. Having found Y, q, the values of b and ψ_0 can be obtained from $q = \tan \psi_0$, $Y = 1/(b \cos \psi_0)$, or the point B_1 can be located. The locus of B_1, the curve $k_{B1(1234)}$ is the relative circle-point curve of the motion of b relative to a. This curve passes of course through B_0.

The length $c = \overline{AB}$ of the coupler can be calculated from equation (2.1), or (assume $f = 1$):

$$c = \left\{ a^2 + b^2 + 1 - 2[a\cos\phi_0 - b\cos\psi_0 + ab\cos(\phi_0 - \psi_0)] \right\}^{1/2} \quad (7.39)$$

Example: It is required to coordinate three angle-pairs: $\phi_2 = -90°$, $\psi_2 = 60°$; $\phi_3 = -54°$, $\psi_3 = 50°$; $\phi_4 = -30°$, $\psi_4 = 36°$. Find the relative centre-point curve $k_{A1(1234)}$ and relative circle-point curve $k_{B1(1234)}$, and choose a four-bar linkage A_0ABB_0 to complete the function generator.

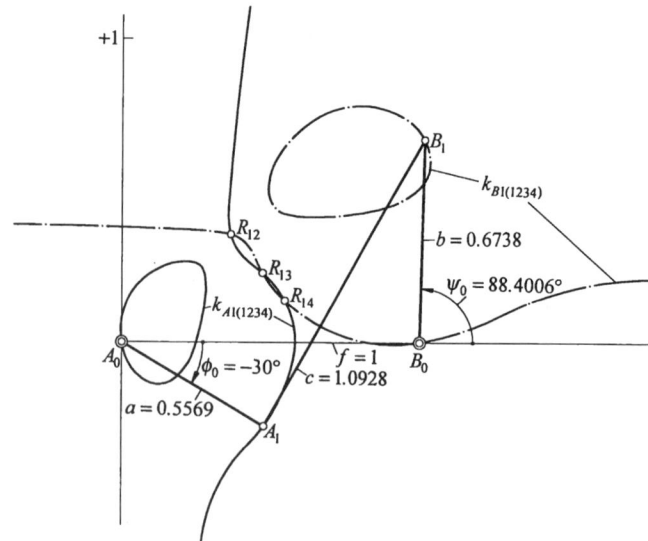

Fig. 7.29. Example: relative centre-point curve $k_{A1(1234)}$ and relative circle-point curve $k_{B1(1234)}$ for coordinating three angle-pairs $\phi_2 = -90°$, $\psi_2 = 60°$; $\phi_3 = -54°$, $\psi_3 = 50°$; $\phi_4 = -30°$, $\psi_4 = 36°$.

Synthesis of function generators

Solution: Take $f = 1$. The curves $k_{A1(1234)}$ and $k_{B1(1234)}$ are plotted as shown in Fig. 7.29. Assuming $\phi_0 = -30°$, we get $a = 0.5569$. The corresponding point B_1 is : $\psi_0 = 88.4006°$, $b = 0.6738$, and $c = 1.0928$. It can be seen that both curves pass through the three relative poles R_{12}, R_{13}, R_{14}.

7.7.3 Coordinations of four angle-pairs---five finitely separated relative positions (P_1–P_2–P_3–P_4–P_5)

This is to coordinate $\phi_2 : \psi_2$; $\phi_3 : \psi_3$; $\phi_4 : \psi_4$; $\phi_5 : \psi_5$. This case is similar to the case of guiding a body through five finitely separated positions and may be considered as a combination of two four-finitely separated positions (P_1–P_2–P_3–P_4) and (P_1–P_2–P_3–P_5). Suppose we have for the former the equation (7.36)

$$k_{A1(1234)} : a_2 a^2 + (a_1 \cos\phi_0) a + a_0 = 0 \qquad [(7.36)]$$

Similarly, for (P_1-P_2-P_3-P_5), we can write

$$k_{A1(1235)} : b_2 a^2 + (b_1 \cos\phi_0) a + b_0 = 0 \qquad (7.40)$$

The three coefficients in equation (7.40) are

$$\left. \begin{array}{l} b_2 = v_1 t + v_2 \\ b_1 = v_7 t^2 + v_8 t + v_9 \\ b_0 = v_5 t + v_6 \end{array} \right\} \qquad (7.41)$$

The seven $v_1, \ldots, v_9$ in equation (7.41) can be calculated by analogy with $u_1, \ldots, u_9$ from Appendix 16, simply by setting $j = 2, 3, 5$. Eliminating a from the simultaneous equations (7.36) and (7.40), and noting that $\cos^2\phi_0 = 1/(1 + t^2)$, we get

$$\frac{1}{\cos^2\phi_0} \begin{vmatrix} a_2 & a_0 \\ b_2 & b_0 \end{vmatrix}^2 + \begin{vmatrix} a_2 & a_1 \\ b_2 & b_1 \end{vmatrix} \begin{vmatrix} a_0 & a_1 \\ b_0 & b_1 \end{vmatrix} = 0 \qquad (7.42)$$

Expanding equation (7.42) yields a sextic equation in the unknown t:

$$f_6 t^6 + f_5 t^5 + f_4 t^4 + f_3 t^3 + f_2 t^2 + f_1 t + f_0 = 0 \qquad (7.43)$$

Equation (7.42) is of the same pattern as that of equation (3.141), hence the equations in Appendix 8 can be used as formulae. However, in order to avoid confusion, expansion of equation (7.42) is listed as equations given in Section A16.2. Equation (7.43) corresponds to the relative Burmester centre point equation of guiding a body through five finitely separated positions. Among the roots of equation (7.43), there are the three known roots corresponding to R_{12}, R_{13}, $(R_{23})_a$. The relative pole $(R_{23})_a$ has already been mentioned is Section 7.2.4, being the relative pole between the 2nd and 3rd positions of b relative to the 1st position of a. Denote these three roots respectively by t_1, t_2, t_3, or

$$t_1 = \tan(-\phi_2/2)$$
$$t_2 = \tan(-\phi_3/2)$$
$$t_3 = \tan[-(\phi_2+\phi_3)/2]$$

All these three points R_{12}, R_{13}, $(R_{23})_a$ are not relative Burmester centre points. After deleting these three roots, equation (7.43) can be reduced to a cubic equation:

$$f_6 t^3 + f_{20} t^2 + f_{10} t + f_{00} = 0 \tag{7.44}$$

The coefficients of equation (7.44) are given in equations (A16.4) in Appendix 16. Since A_0 is one of the four relative Burmester centres, and equations (7.36), (7.40) do not include the origin A_0, the three roots of equation (7.44) correspond to the other three Burmester centres. They are either three real roots, or one real root and two conjugate imaginary roots. Hence the number of possible four-bar linkages to coordinate four angle-pairs is 6, or 1.

The value of a corresponding to each real root t_0 of equation (7.44) can be calculated by

$$a = -\begin{vmatrix} a_2 & a_0 \\ b_2 & b_0 \end{vmatrix} \bigg/ \begin{vmatrix} a_2 & a_1\cos\phi_0 \\ b_2 & b_1\cos\phi_0 \end{vmatrix} \tag{7.45}$$

The expanded form of equation (7.45) is given in equation (A16.5) of Appendix 16. Values of (b, ψ_0) and c corresponding to each set of (a, ϕ_0) values can be calculated as in the case of coordinating three angle-pairs as mentioned in Section 7.7.2.

Example: It is required to generate the function $y = \log_{10} x$. Assumed ranges of the variables are $1 \leq x \leq 10$ and $0 \leq y \leq 1$. The range of $\Delta\phi$ is taken within $-90°$, and that of $\Delta\psi$ is taken within $+90°$. Hence the scale factors are

$$M_\phi = \Delta\phi/\Delta x = -10°/\text{unit of } x$$

$$M_\psi = \Delta\psi/\Delta y = 90°/\text{unit of } y$$

Choose five precision points at $x = 1, 3, 5.5, 7.75, 10$. Values of ϕ_j, ψ_j ($j = 2, 3, 4, 5$) are as follows:

x	Δx_j	ϕ_j	ψ_j
1			
3	2.00	−20.00°	+42.9409°
5.5	4.50	−45.00°	+66.6326°
7.75	6.75	−67.50°	+80.0372°
10	9.00	−90.00°	+90.0000°

In this case there is only one real root of equation (7.44), i.e. $a = 0.6732$ (again assume $f = 1$), $\phi_0 = -49.7557°$. The corresponding B_1 point is : $b = 0.4807$, $\psi_0 = 72.6042°$. Length of coupler $c = 1.2034$. The mechanism is shown in Fig. 7.30.

Synthesis of function generators

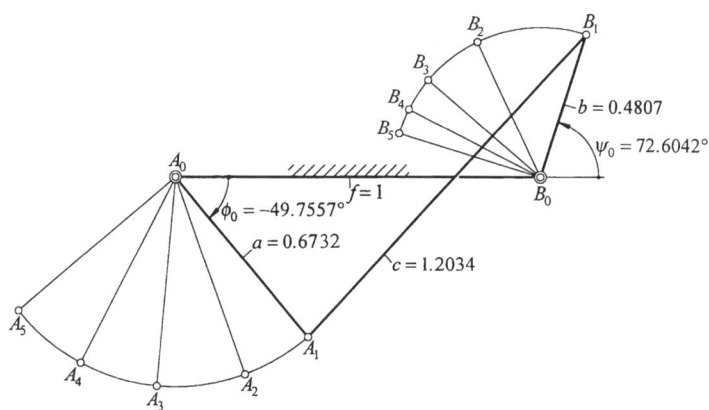

Fig. 7.30. Example: coordinations of four angle pairs $\phi_2 = -20°$, $\psi_2 = 42.9409°$; $\phi_3 = -45°$, $\psi_3 = 66.6326°$; $\phi_4 = -67.50°$, $\psi_4 = 80.0372°$; $\phi_5 = -90.00°$, $\psi_5 = 90.0000°$.

7.7.4 Intermediate cases of four relative positions

For infinitesimally separated relative positions between the two rotating links a, b, such as (P_1P_2) shown in Fig. 7.11 and $(P_1P_2P_3)$ shown in Fig. 7.13, the function generator synthesis problems can all easily be solved by the geometrical method mentioned in Section 7.3, hence we shall not repeat the algebraic methods here. What we are going to explain are such intermediate cases of four relative positions.

(a) $P_1P_2-P_3-P_4$

The $\psi(\phi)$ relationship is as shown in Fig. 7.31, in which $\phi_2 \to d\phi$, $\psi_2 \to d\psi$. For J

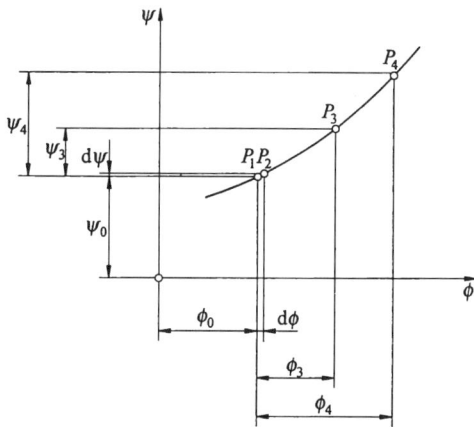

Fig. 7.31. $\psi(\phi)$ relationship in the $(P_1P_2-P_3-P_4)$ synthesis.

= 2, equation (7.31) is simplified into

$$\frac{r_{a2}}{a}\sin\psi_0 - \frac{r_{b2}}{b}\sin\phi_0 = \sin(\phi_0 - \psi_0) \tag{7.46}$$

In this case the position of R_{12} is as that shown in Fig. 7.12. We have already equation (7.5)

$$r_{a2} = r_{A0} = \overline{R_{12}A_0} = \frac{\psi'}{1-\psi'}f \qquad [(7.5)]$$

and also

$$r_{b2} = \frac{1}{1-\psi'}f \tag{7.47}$$

Assume again $f = 1$. Equation (7.46) can be written as

$$\frac{\psi'}{a}\sin\psi_0 - \frac{1}{b}\sin\phi_0 = (1-\psi')\sin(\phi_0 - \psi_0) \tag{7.48}$$

Use as before the symbols $X = 1/(a\cos\phi_0)$, $Y = 1/(b\cos\psi_0)$, $t = \tan\phi_0$, $q = \tan\psi_0$. Equation (7.48) can further be written as

$$-tY + (\psi'X + 1 - \psi')q = (1-\psi')t \tag{7.49}$$

Hence the equation of the relative centre-point curve $k_{A1(1234)}$ (the locus of A_1) corresponding to equation (7.34) now takes the form

$$\begin{vmatrix} -t & \psi'X + 1 - \psi' & (1-\psi')t \\ -r_{b3}(tC_{\xi3} + S_{\xi3}), & r_{a3}C_{\xi3}X - t\sin(\xi_3 - \varepsilon_3) + \cos(\xi_3 - \varepsilon_3), & -r_{a3}S_{\xi3}X + t\cos(\xi_3 - \varepsilon_3) + \sin(\xi_3 - \varepsilon_3) \\ -r_{b4}(tC_{\xi4} + S_{\xi4}), & r_{a4}C_{\xi4}X - t\sin(\xi_4 - \varepsilon_4) + \cos(\xi_4 - \varepsilon_4), & -r_{a4}S_{\xi4}X + t\cos(\xi_4 - \varepsilon_4) + \sin(\xi_4 - \varepsilon_4) \end{vmatrix}$$
$$= 0 \tag{7.50}$$

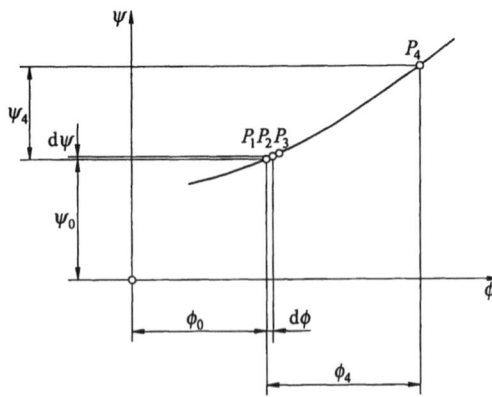

Fig. 7.32. $\psi(\phi)$ relationship in the $(P_1P_2P_3-P_4)$ synthesis.

Synthesis of function generators

(b) $P_1P_2P_3-P_4$

The $\psi(\phi)$ relationship is as shown in Fig. 7.32. For $j = 2$, we have equation (7.49). For $j = 3$, we have to differentiate equation (7.46) with respect to ϕ. However, it should be noted that, ϕ_0, ψ_0 should first be changed to ϕ, ψ respectively, and equations (7.5) and (7.47) should be substituted into equation (7.46) before differentiation. After the differentiation ϕ, ψ should be changed back to ϕ_0, ψ_0. We get then

$$(\psi''q + \psi'^2)X - Y = -\psi''(t-q) + (1-\psi')^2(1+tq) \tag{7.51}$$

Arranging equation (7.51) into a linear form in the two unknowns Y, q yields

$$-Y + [\psi''X - (1-\psi')^2 t - \psi'']q = -\psi'^2 X - \psi''t + (1-\psi')^2 \tag{7.52}$$

Hence the equation of the relative centre-point curve $k_{A1(1234)}$ (the locus of A_1) corresponding to equation (7.34) now takes the form

$$\begin{vmatrix} -t & \psi'X + 1 - \psi' & (1-\psi')t \\ -1 & \psi''X - (1-\psi')^2 t - \psi'' & -\psi'^2 X - \psi''t + (1-\psi')^2 \\ -r_{b4}(tC_{\xi 4} + S_{\xi 4}), & r_{a4}C_{\varepsilon 4}X - t\sin(\xi_4 - \varepsilon_4) + \cos(\xi_4 - \varepsilon_4), & -r_{a4}S_{\varepsilon 4}X + t\cos(\xi_4 - \varepsilon_4) + \sin(\xi_4 - \varepsilon_4) \end{vmatrix}$$
$$= 0 \tag{7.53}$$

The values of ψ', ψ'' are those at P_1.

(c) $P_1P_2-P_3P_4$

The $\psi(\phi)$ relationship in this case is as shown in Fig. 7.33. It is required that the function generator should

(1) match an angular velocity ratio $(d\psi/d\phi)_1 = \psi_1'$ at P_1;

(2) coordinate an angle pair ϕ_3: ψ_3; and

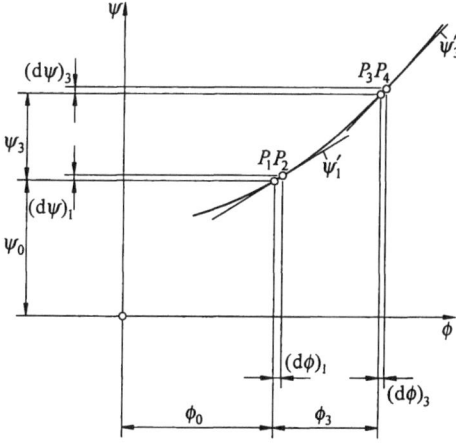

Fig. 7.33. $\psi(\phi)$ relationship in the $(P_1P_2-P_3P_4)$ synthesis.

(3) match another angular velocity ratio $(d\psi/d\phi)_3 = \psi'_3$ at P_3.
For $j = 2$, we have equation (7.49), but ψ' has to be replaced by ψ'_1. For $j = 3$, we have equation (7.33). For $j = 4$, we may write by analogy with equation (7.48)

$$\frac{\psi'_3}{a}\sin(\psi_0 + \psi_3) - \frac{1}{b}\sin(\phi_0 + \phi_3) = (1 - \psi'_3)\sin[(\phi_0 + \phi_3) - (\psi_0 + \psi_3)] \qquad (7.54)$$

Equation (7.54) can be rewritten as

$$-(tC_{\phi 3} + S_{\phi 3})Y + \{\psi'_3 C_{\psi 3} X + (1 - \psi'_3)[(tC_{\phi 3} + S_{\phi 3})S_{\psi 3} + (C_{\phi 3} - tS_{\phi 3})C_{\psi 3}]\}q$$
$$= -\psi'_3 S_{\psi 3} X + (1 - \psi'_3)[(tC_{\phi 3} + S_{\phi 3})C_{\psi 3} - (C_{\phi 3} - tS_{\phi 3})S_{\psi 3}] \qquad (7.55)$$

where $S_{\phi 3} = \sin\phi_3$, $C_{\phi 3} = \cos\phi_3$, $\cdots$ etc. Hence the equation of the relative centre-point curve $k_{A1(1234)}$ (the locus of A_1) corresponding to equation (7.34) now takes the form

$$\begin{vmatrix} -t & \psi'_1 X + 1 - \psi'_1 & (1 - \psi'_1)t \\ -r_{b3}(tC_{\xi 3} + S_{\xi 3}), & r_{a3}C_{\varepsilon 3}X - t\sin(\xi_3 - \varepsilon_3) + \cos(\xi_3 - \varepsilon_3), & -r_{a3}S_{\varepsilon 3}X + t\cos(\xi_3 - \varepsilon_3) + \sin(\xi_3 - \varepsilon_3) \\ -(tC_{\phi 3} + S_{\phi 3}), & \psi'_3 C_{\psi 3} X + (1 - \psi'_3)[(tC_{\phi 3} + S_{\phi 3})S_{\psi 3} & -\psi'_3 S_{\psi 3} X + (1 - \psi'_3)[(tC_{\phi 3} + S_{\phi 3})C_{\psi 3} \\ & + (C_{\phi 3} - tS_{\phi 3})C_{\psi 3}], & -(C_{\phi 3} - tS_{\phi 3})S_{\psi 3}] \end{vmatrix}$$
$$= 0 \qquad (7.56)$$

Equation (7.56) is the basic equation for synthesizing crank-rockers, double rockers and double-cranks.

7.7.5 Synthesis of crank-rockers

(a) Synthesis of crank-rockers considered as synthesis of function generators

As shown in Fig. 7.34, the two positions P_1, P_2 are infinitesimally separated in the outer dead-centre position, and the velocity ratio $(d\psi/d\phi)_1 = \psi'_1 = 0$. In the

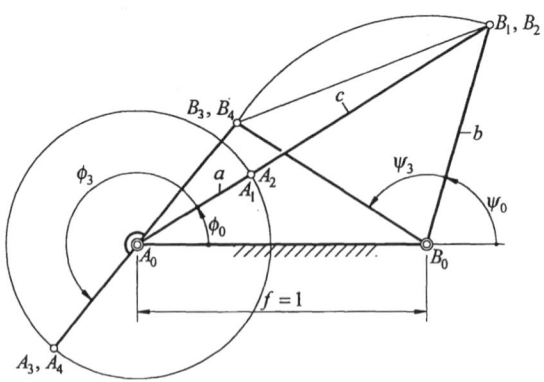

Fig. 7.34. Inner dead-centre position and outer dead-centre position of a crank-rocker.

Synthesis of function generators

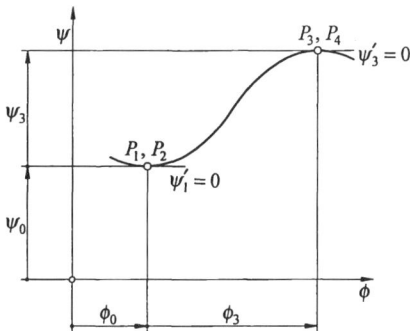

Fig. 7.35. $\psi(\phi)$ relationship of a crank-rocker.

inner dead-centre position, the two positions P_3, P_4 are infinitesimally saperated, and the velocity ratio $(d\psi/d\phi)_3 = \psi_3' = 0$. The $\psi(\phi)$ diagram in this case is shown in Fig. 7.35. It is clear from this diagram, that this is a special case of synthesizing a P_1P_2–P_3P_4 function generator. In other words, $\psi_1' = 0$, $\psi_3' = 0$, and both values of ϕ_3, ψ_3 are prescribed. In equation (7.56), setting $\psi_1' = 0$, $\psi_3' = 0$, we get

$$\begin{vmatrix} -t & 1 & t \\ -r_{b3}(tC_{\xi 3} + S_{\xi 3}), & r_{a3}C_{\varepsilon 3}X - t\sin(\xi_3 - \varepsilon_3) + \cos(\xi_3 - \varepsilon_3), & -r_{a3}S_{\varepsilon 3}X + t\cos(\xi_3 - \varepsilon_3) + \sin(\xi_3 - \varepsilon_3) \\ -(tC_{\phi 3} + S_{\phi 3}), & (tC_{\phi 3} + S_{\phi 3})S_{\psi 3} + (C_{\phi 3} - tS_{\phi 3})C_{\psi 3}, & (tC_{\phi 3} + S_{\phi 3})C_{\psi 3} - (C_{\phi 3} - tS_{\phi 3})S_{\psi 3} \end{vmatrix}$$
$= 0$
(7.57)

Expanding equation (7.57) results in an equation of the locus of $A_1(a, \phi_0)$, or the relative centre-point curve k_{A1} for the motion of b relative to a:

$$k_{A1}: \sin\left(\phi_0 + \frac{\phi_3}{2}\right)\left[a\sin\frac{\phi_3 - \psi_3}{2} + \sin\frac{\psi_3}{2}\cos\left(\phi_0 + \frac{\phi_3}{2}\right)\right] = 0 \quad (7.58)$$

Equation (7.58) indicates that the present k_{A1}-curve breaks up into a straight line and a circle, as shown in Fig. 7.36. This result is identical with that of (Alt, 1925). The first factor in equation (7.58) represents the straight line A_0R_{13}, while the second factor represents the circle with $\overline{A_0R_{13}}$ as its diameter which can also be written as

$$a = -\frac{\sin\frac{\psi_3}{2}}{\sin\frac{\phi_3 - \psi_3}{2}}\cos\left(\phi_0 + \frac{\phi_3}{2}\right) \quad (7.59)$$

Equations (7.58) and (7.59) have also been derived by (Chiang, 1986a) by vector method.

Similarly, if the three equations (7.49), (7.33) and (7.54) are written as

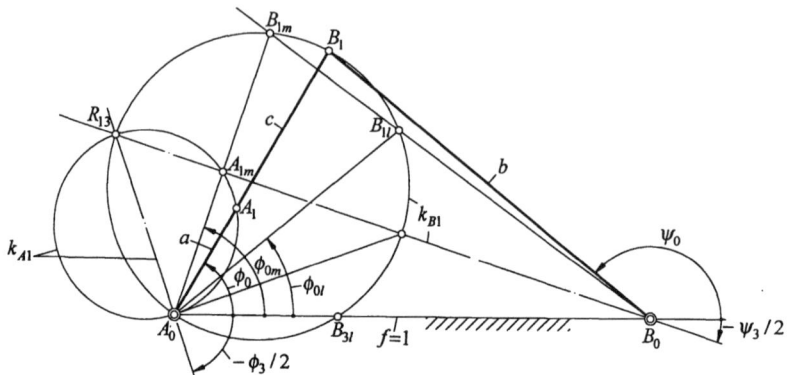

Fig. 7.36. Loci k_{A1} and k_{B1} of crank-rockers for ϕ_3 in the lower range.

simultaneous equations in the two unknowns X, t, and $\psi_1' = 0$, $\psi_3' = 0$ are substituted and the determinant of the coefficients of the three equations are set to zero, then the expansion of this determinant is the following equation of the locus of $B_1(b, \psi_0)$, or the relative circle-point curve for the motion of b relative to a:

$$k_{B1}: \sin\left(\psi_0 + \frac{\psi_3}{2}\right)[b^2 \sin(\phi_3 - \psi_3) + b\sin(\psi_0 + \phi_3) \\ - b\sin(\psi_0 - \phi_3 + \psi_3) + \sin\phi_3] = 0 \qquad (7.60)$$

Equation (7.60) indicates that the present k_{B1}-curve also breaks up into a straight line and a circle. The first factor in equation (7.60) represents the straight line B_0R_{13} in Fig. 7.36, while the second factor represents the circle passing through A_0, R_{13} with its centre lying on B_0R_{13}. If the origin of the polar coordinate system is changed from B_0 to A_0, then this factor represents the locus of $B_1(a+c, \phi_0)$, being

$$a + c = \frac{\sin(\phi_0 + \phi_3 - \psi_3) - \sin(\phi_0 + \phi_3)}{\sin(\phi_3 - \psi_3)} \qquad (7.61)$$

For a given value of ϕ_0, the length $a + c$ can be calculated from equation (7.61), and the length b of the rocker is

$$b = [1 + (a+c)^2 - 2(a+c)\cos\phi_0]^{1/2} \qquad (7.62)$$

The position angle ψ_0 of the rocker b in the design position is

$$\cos\psi_0 = [(a+c)^2 - 1 - b^2]/(2b) \qquad (7.63)$$

The four equations (7.59), (7.61), (7.62), (7.63) are those used for synthesizing crank-rockers. The sole parameter is ϕ_0.

Synthesis of function generators

(b) Available ranges of ψ_3, ϕ_3 and ϕ_0

First of all, it can be seen from Fig. 7.34 that, since the rocker b should never swing twice across the line of centres A_0B_0, we may reasonably set

$$0° \leq \psi_3 \leq 180°$$

Let us now examine the ranges of ϕ_3 and ϕ_0. In Fig. 7.36, draw a line from A_0 normal to B_0R_{13}, intersecting it in A_{1m}. Extend this normal to B_{1m}, so that $\overline{A_0A_{1m}} = \overline{A_{1m}B_{1m}}$, then B_{1m} lies on circle k_{B1}. Join B_0B_{1m}, to intersect k_{B1} in B_{1l}. Choose any angle ϕ_0, and draw a straight line from A_0, intersecting the cincle k_{A1} in A_1, and the circle k_{B1} in B_1. Denote $\angle B_0A_0A_{1m}$ by ϕ_{0m}. It can be seen from the figure that

$$\phi_{0m} = 90° - \frac{\psi_3}{2} \tag{7.64}$$

Take the point A_{1m} as a transition point. In other words, for R_{13} on the lefthand side of A_{1m}, ϕ_3 is within its lower range, and for R_{13} on the righthand side of A_{1m}, ϕ_3 is within its upper range. We shall discuss these separately in the following.

(i) lower range, $\phi_3 / 2 < 180° - \phi_{0m}$

This is the case shown in Fig. 7.36. R_{13} lies now on the left of A_{1m}. Consider first the range of ϕ_0. If the angle ϕ_0 chosen is such that B_1 lies on the left of B_{1m}, then c would be smaller than a, and the four-bar linkage would become a double rocker, no longer a crank-rocker. Therefore B_{1m} is the lefthand limit of B_1. On the other hand, if B_1 is chosen at B_{1l}, then as the rocker b swings through an angle ψ_3, the point B falls just on the point B_{3l}, which is the intersection point of the circle k_{B1} with the line of centres A_0B_0. If the point B_1 is chosen further on the righthand side of B_{1l}, then as the rocker b swings through an angle ψ_3, the point B would fall below the line of centres A_0B_0. Therefore B_{1l} is the righthand limit of B_1. Hence the range of B_1 is between B_{1m} and B_{1l}. In other words, the upper limit of ϕ_0 is ϕ_{0m}, and its lower limit is at ϕ_{0l}. It can be seen from the relation of circumferential angle that

$$\angle B_0A_0B_{1l} = \phi_{0l} = 180° - \phi_3 \tag{7.65}$$

However, as ϕ_{0l} should never be negative, therefore in case $\phi_3 > 180°$, the lower limit of ϕ_0 is $0°$.

For the range of ϕ_3, the upper limit of ϕ_3 is of course

$$\frac{\phi_3}{2} \leq 180° - \phi_{0m}$$

As to the lower limit of ϕ_3, it can be seen from Fig. 7.36 that as ϕ_3 decreases, the point R_{13} moves away from A_{1m} toward the left, and the circle k_{B1} moves also along the line B_0R_{13} toward the left. The available range $\overarc{B_{1m}B_{1l}}$ decreases correspondingly, until the length of this arc reduces to zero, or the circle k_{B1} becomes tangent to the line B_0B_{1m}. This is the lower limit of ϕ_3. In this limit the points B_{3l} and

A_0 also coincide. To find this lower limit of ϕ_3, it is expedient to observe the equation of the circle k_{B1}, or the second factor of equation (7.60). In this equation the two variables are ψ_0 and b. The condition that b should have a double root for $\psi_0 = 0$ (the points of intersection between the line A_0B_0 and the circle k_{B1}) gives

$$\phi_{3\min} = \tan^{-1} \frac{S_{\psi 3}}{C_{\psi 3} - 1} = 180° - \phi_{0m} \qquad (7.66)$$

The range of ϕ_0 can be determined once ϕ_3 has been chosen.

(ii) As the point R_{12} coincides with A_{1m}, the circle k_{B1} breaks further up into the straight line A_0B_{1m} and the line at infinity. In other words, the whole k_{B1}-curve breaks up into three straight lines.

(iii) upper range, $\phi_3 / 2 > 180° - \phi_{0m}$

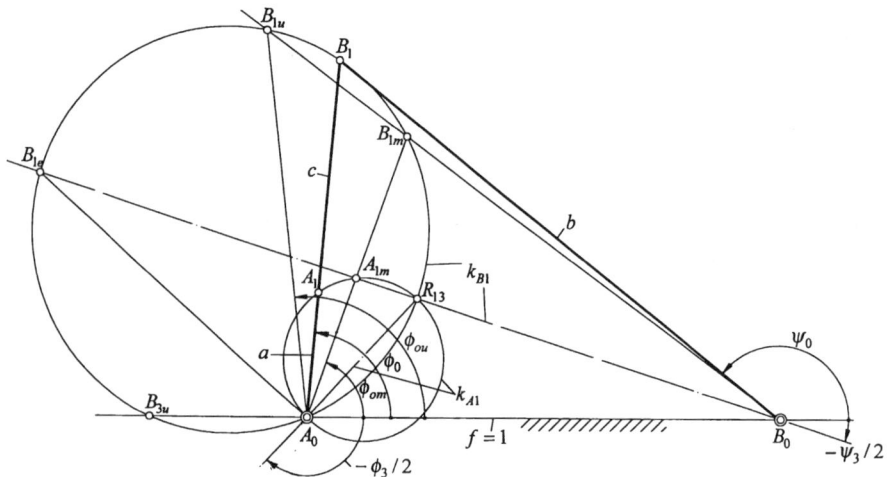

Fig. 7.37. Loci k_{A1} and k_{B1} of crank-rockers for ϕ_3 in the upper range.

In this case the point R_{13} is on the righhand side of A_{1m}, as shown in Fig. 7.37. The circle k_{B1} intersects the line B_0B_{1m} in another point B_{1u}, and also the line A_0B_0 in another point B_{3u}. Now the range of B_1 is between B_{1m} and B_{1u}. Therefore the lower limit of ϕ_0 is ϕ_{0m}, and its upper limit ϕ_{0u} is the angle corresponding to B_{1u}. Again from the relation of circumferential angle we have

$$\sphericalangle B_0A_0B_{1u} = \phi_{0u} = 360° - \phi_3 \qquad (7.67)$$

However, as ϕ_{0u} should never be larger than 180°, in case $\phi_3 < 180°$, the upper limit of ϕ_0 is 180°.

For the range of ϕ_3, the lower limit of ϕ_3 is of course $\phi_3 / 2 > 180° - \phi_{0m}$. As to the upper limit of ϕ_3, it can be seen from Fig. 7.37 that as ϕ_3 increases, the point R_{13}

Synthesis of function generators

moves away from A_{1m} toward the right. The available range $\widehat{B_{1m}B_{1u}}$ decreases correspondingly, until the two points B_{1m}, B_{1u} coincide, or the two intersection points B_{3u} and A_0 between k_{B1} and A_0B_0 also coincide. By analogy with equation (7.66), the condition that b should have a double root gives

$$\phi_{3\max} = \tan^{-1}\frac{S_{\psi 3}}{C_{\psi 3} - 1} + 180° = 360° - \phi_{0m} \tag{7.68}$$

For clarity purposes, ranges of ϕ_3, ϕ_0 as mentioned above are listed in Table 7.1.

Table 7.1 Ranges of ϕ_3 and ϕ_0

	$\phi_{0m} = 90° - \dfrac{\psi_3}{2}$		
	lower range		upper range
Range of $\phi_3 / 2$	$\dfrac{180° - \phi_{0m}}{2}$	$180° - \phi_{0m}$	$180° - \dfrac{\phi_{0m}}{2}$
Range of ϕ_0	$\begin{array}{c}180° - \phi_3\\ (\text{or } 0°, \text{ if } \phi_3 > 180°)\end{array}$	ϕ_{0m}	$\begin{array}{c}360° - \phi_3\\ (\text{or } 180°, \text{ if } \phi_3 < 180°)\end{array}$

(c) Optimization of transmission angle

As mentioned in Section 7.5, in Fig. 7.16(b) the transmission angle of a four-bar linkage A_0ABB_0 is $\mu = \angle ABB_0$. As shown in Fig. 7.17, $\mu_{\max}$ and $\mu_{\min}$ occur respectively when the crank AB and line of centres A_0B_0 are collinear. For brevity, $\mu_{\min}$ will be denoted by μ_i, and $\mu_{\max}$ by μ_a. The problem now is: how can we find out from a family of four-bar linkages the optimum one whose $|\mu_a - 90°|$ or $|90° - \mu_i|$ is the smallest in the whole family? In the problem of synthesizing crank-rockers mentioned above, we have already found the family of ∞^1 solutions. The question is, how to find from these ∞ four-bar linkages the one with an optimum transmission angle? To answer this question, (Volmer & Jensen, 1962) has devised charts for practical designers. However, we shall introduce the Lagrange multiplier method proposed by (Freudenstein & Primrose, 1972) to yield more accurate design data. In the first place, we see that the minimization of $|\mu_a - 90°|$ or $|90° - \mu_i|$ can be replaced by maximization of $\sin \mu_a$ or $\sin \mu_i$. In other words, the transmission quality can be represented by the *largeness* of $\sin \mu_a$ or $\sin \mu_i$, rather than by the *smallness* of $|\mu_a - 90°|$ or $|90° - \mu_i|$. Applying law of cosine to $\Delta A_1B_1B_0$ and $\Delta A_2B_2B_0$ in Fig. 7.17 gives

$$\Psi(a,b,c) = \frac{\cos^2 \mu_i}{\cos^2 \mu_a} = \left[\frac{b^2 + c^2 - (1 \mp a)^2}{2bc}\right]^2 \tag{7.69}$$

In equation (7.69), $f = 1$ has been assumed, and the $(-)$ sign refers to μ_i, while $(+)$

sign refers to μ_a. "Optimizing" μ is equivalent to "maximizing" $\sin \mu_i$ or $\sin \mu_a$, or to "minimizing" $\cos^2 \mu_i$ or $\cos^2 \mu_a$, or the function $\Psi(a,b,c)$.

In equation (7.69), a, b, c are three variables. Among them there are two conditions to be satisfied. Eliminating ϕ_0 between equations (7.59) and (7.61) gives

$$\Psi_1(a,c) = \sin^2\left(\frac{\phi_3 - \psi_3}{2}\right) a^2 + \cos^2\left(\frac{\phi_3 - \psi_3}{2}\right) c^2 - \sin^2 \frac{\psi_3}{2} = 0 \quad (7.70)$$

Next, from $\Delta B_1 A_0 B_3$ and $\Delta B_1 B_0 B_3$ in Fig. 7.34 we have

$$\Psi_2(a,b,c) = \left(\sin^2 \frac{\phi_3}{2}\right) a^2 + \left(\cos^2 \frac{\phi_3}{2}\right) c^2 - \left(\sin^2 \frac{\psi_3}{2}\right) b^2 = 0 \quad (7.71)$$

The problem of "minimizing" the function $\Psi(a,b,c)$ is: for given values of ϕ_3, ψ_3, to find a set of values of a, b, c which, subject to the constraints of equations (7.70) and (7.71), will render the function $\Psi(a,b,c)$, representing either $\cos^2\mu_i$ or $\cos^2\mu_a$, a minimum. The principle of Lagrange multipliers may be described briefly as follows: the set of values of the three independent variables (a,b,c) which will render the function $\Psi(a,b,c)$ an extreme value, will also render the function

$$F(a,b,c) = \Psi + \lambda_1 \Psi_1 + \lambda_2 \Psi_2$$

an extreme value, where λ_1 and λ_2 are two unknown constants, called *Lagrange multipliers*. Since at an extreme value of F, $\partial F/\partial a = 0$, $\partial F/\partial b = 0$, $\partial F/\partial c = 0$, therefore

$$\left. \begin{array}{l} \dfrac{\partial \Psi}{\partial a} + \lambda_1 \dfrac{\partial \Psi_1}{\partial a} + \lambda_2 \dfrac{\partial \Psi_2}{\partial a} = 0 \\[4pt] \dfrac{\partial \Psi}{\partial b} + \lambda_1 \dfrac{\partial \Psi_1}{\partial b} + \lambda_2 \dfrac{\partial \Psi_2}{\partial b} = 0 \\[4pt] \dfrac{\partial \Psi}{\partial c} + \lambda_1 \dfrac{\partial \Psi_1}{\partial c} + \lambda_2 \dfrac{\partial \Psi_2}{\partial c} = 0 \end{array} \right\} \quad (7.72)$$

Equations (7.72) are three simultaneous linear equations in the two unknowns λ_1, λ_2. In order to have non-trivial solutions, the determinant of the coefficients of these equations must vanish identically, hence we have

$$\begin{vmatrix} \dfrac{\partial \Psi}{\partial a}, & \dfrac{\partial \Psi_1}{\partial a}, & \dfrac{\partial \Psi_2}{\partial a} \\[4pt] \dfrac{\partial \Psi}{\partial b}, & \dfrac{\partial \Psi_1}{\partial b}, & \dfrac{\partial \Psi_2}{\partial b} \\[4pt] \dfrac{\partial \Psi}{\partial c}, & \dfrac{\partial \Psi_1}{\partial c}, & \dfrac{\partial \Psi_2}{\partial c} \end{vmatrix} = 0 \quad (7.73)$$

The set of required values of a, b, c and the parameter ϕ_0 for the extreme of $\Psi(a,b,c)$ are determined by the four equations (7.59), (7.61), (7.62) and (7.73). However, in order to facilitate computation, it is expedient to replace the three independent variables a, b, c in equation (7.73) by a^2, b^2, c^2 respectively. It can be written

Synthesis of function generators

$$\frac{\Psi_{3i}(a,b,c)}{\Psi_{3a}(a,b,c)} = \begin{vmatrix} \frac{\partial\Psi}{\partial(a^2)}, & \frac{\partial\Psi_1}{\partial(a^2)}, & \frac{\partial\Psi_2}{\partial(a^2)} \\ \frac{\partial\Psi}{\partial(b^2)}, & \frac{\partial\Psi_1}{\partial(b^2)}, & \frac{\partial\Psi_2}{\partial(b^2)} \\ \frac{\partial\Psi}{\partial(c^2)}, & \frac{\partial\Psi_1}{\partial(c^2)}, & \frac{\partial\Psi_2}{\partial(c^2)} \end{vmatrix} = \begin{vmatrix} \frac{\partial\Psi}{\partial(a^2)}, & S^2_{\eta/2}, S^2_{\phi/2} \\ \frac{\partial\Psi}{\partial(b^2)}, & 0, -S^2_{\psi/2} \\ \frac{\partial\Psi}{\partial(c^2)}, & C^2_{\eta/2}, C^2_{\phi/2} \end{vmatrix} = 0 \qquad (7.74)$$

In equation (7.74),

$$\frac{\partial\Psi}{\partial(a^2)} = \frac{\partial\Psi}{\partial a}\frac{da}{d(a^2)} = \pm\frac{(1\mp a)}{a}[b^2 + c^2 - (1\mp a)^2]\frac{1}{2b^2c^2}$$

$$\frac{\partial\Psi}{\partial(b^2)} = \frac{\partial\Psi}{\partial b}\frac{db}{d(b^2)} = \frac{1}{2b^2}\{2b^2 - [b^2 + c^2 - (1\mp a)^2]\}[b^2 + c^2 - (1\mp a)^2]\frac{1}{2b^2c^2}$$

$$\frac{\partial\Psi}{\partial(c^2)} = \frac{\partial\Psi}{\partial c}\frac{dc}{d(c^2)} = \frac{1}{2c^2}\{2c^2 - [b^2 + c^2 - (1\mp a)^2]\}[b^2 + c^2 - (1\mp a)^2]\frac{1}{2b^2c^2}$$

$$S_{\eta/2} = \sin\frac{\phi_3 - \psi_3}{2}, \quad S_{\phi/2} = \sin\frac{\phi_3}{2}, \quad C_{\phi/2} = \cos\frac{\phi_3}{2},$$

$$C_{\eta/2} = \cos\frac{\phi_3 - \psi_3}{2}, \quad S_{\psi/2} = \sin\frac{\psi_3}{2}.$$

Equation (7.73) is then replaced by equation (7.74). Please note the signs ($\pm$) and ($\mp$) in these equations, the upper signs corresponding to Ψ_{3i}, and the lower signs corresponding to Ψ_{3a}. The simplest way of solving these equations is setting various values to ϕ_0, and calculating a, c, b values from equations (7.59), (7.61) and (7.62), and substituting them into equation (7.74) to see when the determinant changes sign. The procedure can easily be understood from the following example.

Example: Given $\phi_3 = 216.5°$, $\psi_3 = 85°$. It is required to synthesize the crank-rockers and to find out the one with optimum transmission angle.

Solution: According to equation (7.64), we have $\phi_{0m} = 90° - 85°/2 = 47.5°$, hence ϕ_3 falls in the lower range. The lower limit of ϕ_0 is $0°$, and the upper limit is ϕ_{0m}. The available range of ϕ_0 is therefore $0° \leq \phi_0 \leq 47.5°$. Fig. 7.38 shows the variations of $\cos\mu_i$, $\cos^2\mu_i$; $\cos\mu_a$, $\cos^2\mu_a$ of all possible crank-rockers synthesized within the ϕ_0-range. It can be seen that as $\cos\mu_a$ passes across the zero-line, $\cos^2\mu_a$ also becomes zero, and it exhibits a "mininum". However, $\cos\mu_a$ is not a real "minimum". The minimum of $\cos^2\mu_i$ is a real "minimum". To locate this minimum, we can follow the procedure mentioned above by assigning various values to ϕ_0, and calculating the values a, c, b from equations (7.59), (7.61), and (7.62), and then substituting them into equation (7.74) to watch when it changes sign. In this problem the minimum of $\cos\mu_i$ is 0.9119, or a maximum $\mu_i = 24.224°$ occuring at $\phi_0 = 33.73°$. The dimensions of the synthesized crank-rocker are (for $f = 1$) : $a = 0.584$, $c = 1.013$, $b = 0.945$, $\psi_0 = 69.70°$.

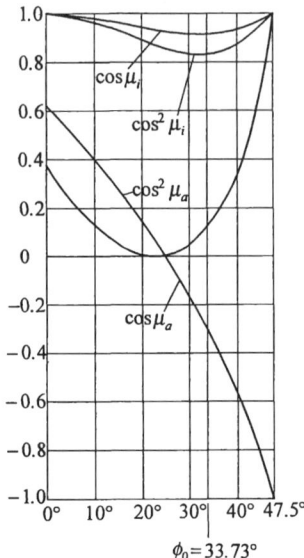

Fig. 7.38.   Example: $\cos \mu_{min}$ occurs at $\phi = 33.73°$ for prescribed $\phi_3 = 216.5°$, $\psi_3 = 85°$, or $\phi_{om} = 47.5°$.

7.7.6 Synthesis of double-rockers

The problem of synthesizing double rockers can also be treated as a problem of synthesizing function generators. In other words, this is again a P_1P_2–P_3P_4 problem, and the locus k_{A1} of A_1 and the locus k_{B1} of B_1 are again cubic curves and can be treated as quadratic equations. Hence equation (7.56) still applies.

Figs. 7.39 (a)-(d) show four types of double rockers and their respective $\psi(\phi)$ diagrams. Fig. 7.39 (a) shows a Grashof double rocker which means that the coupler c is capable of making complete revolutions. Those shown in Figs. 7.39 (b),(c),(d) are all non-Grashof double rockers with (b) internal rocking angles, (c) external rocking angles and (d) overlapping rocking angles. It is clear from these $\psi(\phi)$ diagrams that each one is a closed curve, having four vertices corresponding respectively to the four extreme positions of the two rockers. Each vertex represents two infinitesimally separated positions, being denoted by two numerials, just as in Fig. 7.35. However, the angular velocity ratios are now: $\psi_1' = 0$, $\psi_5' = 0$, and $1/\psi_3' = 0, 1/\psi_7' = 0$. We have only to take four relative positions P_1P_2–P_3P_4, and the double rocker can be synthesized. In the following we shall consider two different types of requirements.

(a) ϕ_3, ψ_3 and f are prescribed

Assume as before $f = 1$. Substituting $\psi_1' = 0$, and $1/\psi_3' = 0$ into equation (7.56) yields the locus of the point A_1:

Synthesis of function generators

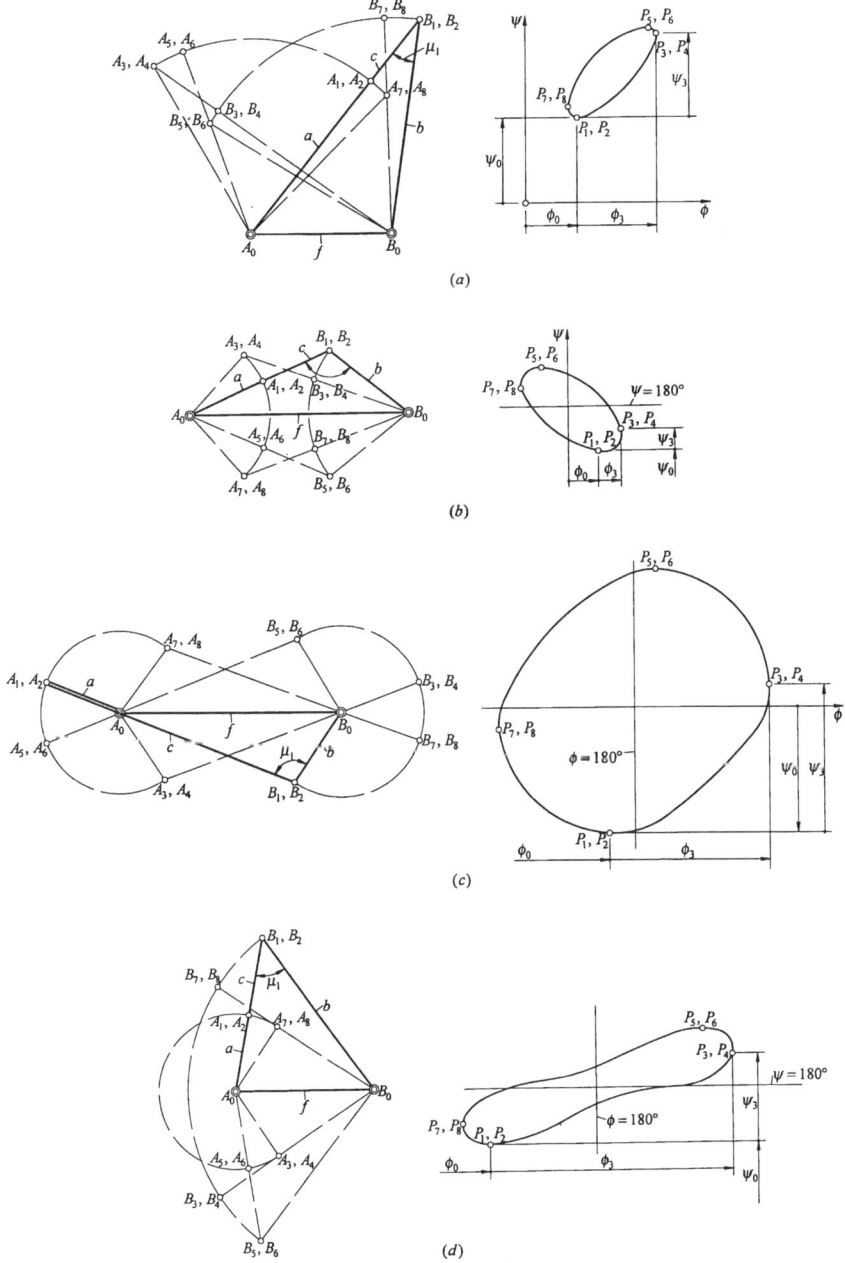

Fig. 7.39. Four types of double-rockers.

$$k_{A1}: M_2 a^2 + M_1 a + M_0 = 0 \tag{7.75}$$

where

$$M_2 = S'_{\phi 0}(1 - C_\eta) + S_\eta[-C_{\phi 0} + \cos(\phi_0 + \phi_3)]$$
$$M_1 = S_{\phi 0}[\cos(\phi_0 + \psi_3) + \cos(\phi_0 + \eta) - 2\cos(\phi_0 + \phi_3)]$$
$$\quad + \sin(\phi_0 + \psi_3)[-C_{\phi 0} + \cos(\phi_0 + \phi_3)]$$
$$M_0 = S_{\phi 0}(1 - C_\psi)$$

where the symbols are as before: $S_{\phi 0} = \sin\phi_0$, $C_{\phi 0} = \cos\phi_0$, and $\eta = \phi_3 - \psi_3$, $S_\eta = \sin\eta$, $C_\eta = \cos\eta$, $C_\psi = \cos\psi_3$.

Similarly, equations (7.49), (7.33) and (7.54) can also be written as three non-homogeneous linear equations in the two unknowns X, t and set $\psi'_1 = 0, 1/\psi'_3 = 0$. Setting the determinant of the coefficients of these three equations equal to zero, and developing, we get the following equation of the locus of B_1 (b, ψ_0), or the relative circle-point curve for the motion of b relative to a:

$$k_{B1}: N_2 b^2 + N_1 b + N_0 = 0 \tag{7.76}$$

where

$$N_2 = S_\eta C_{\psi 0} + \sin(\psi_0 + \psi_3) - \sin(\psi_0 + \phi_3)$$
$$N_1 = -S_\phi - C_{\psi 0} \sin(\psi_0 - \eta) - \sin(\psi_0 + \psi_3)[\cos(\psi_0 + \phi_3) - 2C_{\psi 0}]$$
$$N_0 = \sin(\psi_0 + \psi_3)(1 - C_\phi)$$

where, as before, $C_{\psi 0} = \cos\psi_0$, $S_\phi = \sin\phi_3$, $C_\phi = \cos\phi_3$.

Example: Given $\phi_3 = 38°$, $\psi_3 = 28°$. The k_{A1}- and k_{B1}- curves are found as

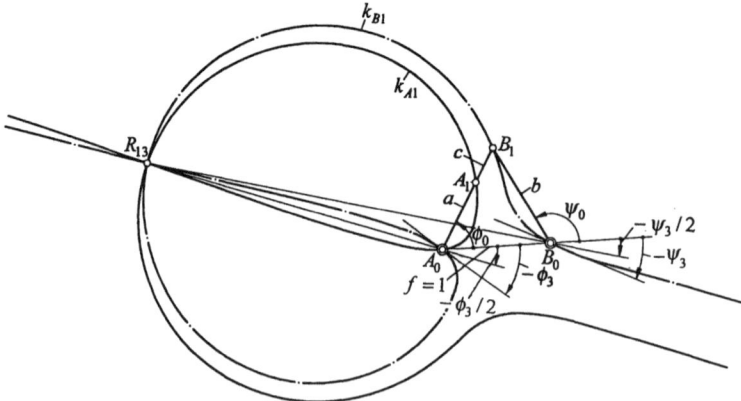

Fig. 7.40. Example: loci k_{A1} and k_{B1} of double-rockers for prescribed $\phi_3 = 38°$, $\psi_3 = 28°$.

Synthesis of function generators

shown in Fig. 7.40.

For a single value of ϕ_0, equation (7.75) yields two a's, or two A_1-points. Since the points A_1 and points B_1 are in a one-to-one correspondence, it is sometimes convenient to calculate $B_1(b, \psi_0)$ directly from $A_1(a, \phi_0)$ by the following equations. From equations (7.49) and (7.54) we can get

$$\tan \psi_0 = \frac{a \sin(\phi_0 + \eta) + \sin \psi_3}{a \cos(\phi_0 + \eta) - \cos \psi_3} \quad (7.77)$$

$$b = \frac{\sin \phi_0}{\sin(\psi_0 - \phi_0)} \quad (7.78)$$

As b is assumed to be positive, in case the computed b by equation (7.78) is negative, ψ_0 should be replaced by $180° + \psi_0$ to recalculate b. For detailed discussion on synthesis of double rockers please refer to (Chiang, 1986b).

7.7.7 Synthesis of double cranks (drag-links)

The double crank mechanism, or drag-link mechanism, is an inversion of the crank-rocker mechanism, by fixing the shortest link. The problems of synthesizing drag-links can, in general, be classified into two kinds. The first kind belongs to the P_1-P_2 type, as shown in Fig. 7.41. The mechainsm is to be synthesized between the stretched position of crank a and frame f and the folded position of a and f. The second kind of problem is to synthesize the drag-links between positions of unity velocity ratio. We begin now with the first kind of problem.

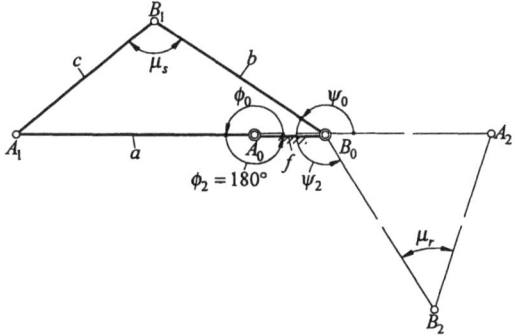

Fig. 7.41. Double-crank in a (P_1-P_2) synthesis.

(*a*) First kind of problem

This sort of problem could also be treated as synthesis of function generators. However, in plane kinematics the problem is rather simple. (Hain, 1957) has solved this problem by the trial and error method, and provided charts for designers. We shall solve it by simple algebraic means. As shown in Fig. 7.41, the transmission angle is a maximum when a and f are in a stretched position, which is denoted by μ_s; and it is a minimum when a and f are in a folded position, which is denoted by μ_r. In order to get an optimum transmission angle, the following condition has to be satisfied

$$\mu_s - 90° = 90° - \mu_r = \Delta\mu \tag{7.79}$$

Equation (7.79) is related to an inversion of the drag-link. We shall come back to this point in the following (b) section. From equation (7.79) it is obvious that

$$\cos\mu_s = -\cos\mu_r$$

Hence if $\Delta\mu$ is given, both μ_s, μ_r are known. Furthermove, from $\Delta B_0 A_1 B_1$ and $\Delta B_0 A_2 B_2$ we have

$$\left. \begin{array}{l} (a+f)^2 = b^2 + c^2 - 2bc\cos\mu_s \\ (a-f)^2 = b^2 + c^2 - 2bc\cos\mu_r \end{array} \right\} \tag{7.80}$$
$$\tag{7.81}$$

Adding equations (7.80) and (7.81) gives

$$a^2 + f^2 = b^2 + c^2 \tag{7.82}$$

and subtracting equation (7.81) from equation (7.80) gives

$$bc\cos\mu_r - af = 0 \tag{7.83}$$

Assume $f = 1$. Under the two conditions (7.82) and (7.83), for every known value of a, there exists a quadratic equation of c (or b), or two sets of values (a, b, c). For each set of (a, b, c) values, the corresponding value of ψ_2 can be found. In case ψ_2 is prescribed, the required solutions can be found by varying a. The procedure is as follows.

Referring to Fig. 7.41, eliminating b from equations (7.82) and (7.83) gives

$$c^2 = \frac{1}{2}\left[(1+a^2) \pm \sqrt{(1+a^2)^2 - 4\left(\frac{a}{\cos(90°-\Delta\mu)}\right)^2}\right] \tag{7.84}$$

From equation (7.82) we have

$$b = \sqrt{f^2 + a^2 - c^2} = \sqrt{1 + a^2 - c^2}$$

With the values $a, c, b, (f = 1)$ known, the value of ψ_2 can be calculated from the equation

$$\psi_2 = 180° + \cos^{-1}\frac{b^2 + (a+1)^2 - c^2}{2b(a+1)} - \cos^{-1}\frac{b^2 + (a-1)^2 - c^2}{2b(a-1)} \tag{7.85}$$

Example: Given $\Delta\mu = 61°$ and $\psi_2 = 112°$. It is required to synthesize the

Synthesis of function generators

drag-link mechanism.

Solution: From $\Delta\mu = 61°$, we have $\mu_s = 151°$, $\mu_r = 29°$. Fig. 7.42 shows the corresponding ψ_2-a curve. This curve includes two parts. The upper part corresponds to the (−) sign in equation (7.84), while the lower part corresponds to the (+) sign in equation (7.84). Both parts start from $a = 1.6977$, where $\psi_2 = 118.53°$. Since $\psi_2 = 112°$ is given, the solution is on the lower part of the curve, which gives $a = 1.7058$, $\psi_2 = 111.992°$, $b = 1.3497$, $c = 1.4450$.

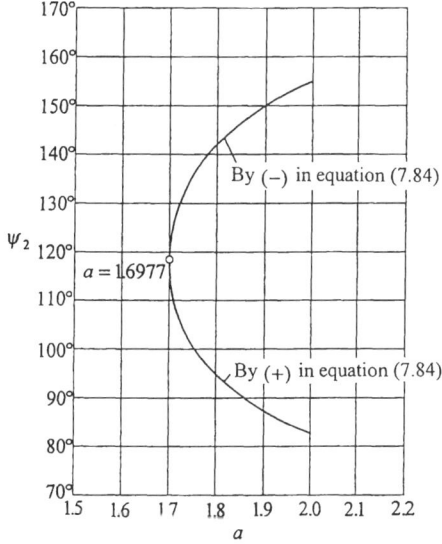

Fig. 7.42. Example: ψ_2-a relationship for a prescribed $\Delta\mu = 61°$.

Example: (taken from Hain, 1957). Fig. 7.43 shows a Geneva intermittent mechanism driven by a drag-link A_0ABB_0. The requirements for the drag-link before synthesis are : given $\Delta\mu = 40°$, i.e. $\mu_s = 130°$, $\mu_r = 50°$ and $\psi_2 = 100°$. The synthesized dimensions are: $f = \overline{A_0B_0} = 1$, $\psi_2 = 99.99°$, $a = 2.6000$, $b = 1.4148$, $c = 2.4000$. Within one cycle of motion of this drag-link, we have found a portion of satisfactory angular displacements coordination such that as a rotates through an angle $\Delta\phi = 180°$, b rotates through just an angle $\Delta\psi = 90°$, as shown by the two positions $A_0A_IB_IB_0$ and $A_0A_{II}B_{II}B_0$ in the figure. In this way the link b can be bound rigidly with the driving wheel B_0C of the Geneva mechanism to drive the latter.

(*b*) Second kind of problem

The $\psi(\phi)$ diagram of this kind of problem is as shown in Fig. 7.44. In this case ϕ_3, ψ_3 and $\psi_1' = (d\psi/d\phi)_1 = 1$, $\psi_3' = (d\psi/d\phi)_3 = 1$ are prescribed. This problem has been solved by (Tsai, 1983a), and a contrast was made between the

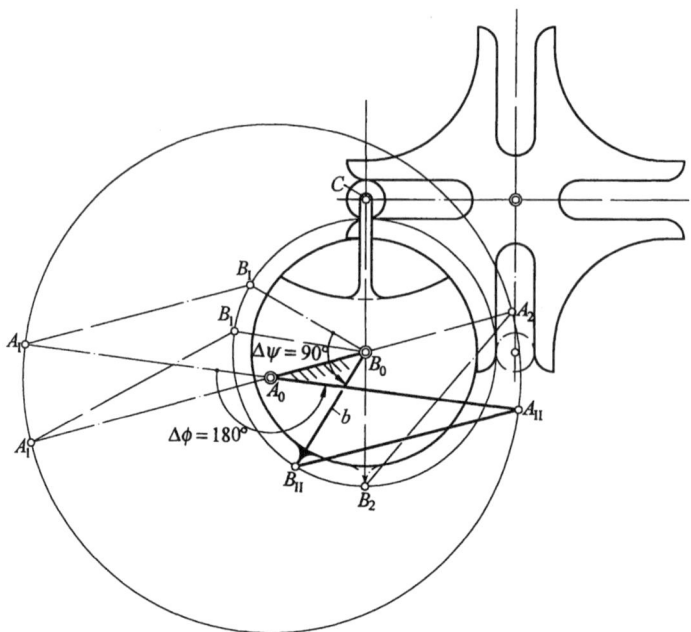

Fig. 7.43. Example: double crank synthesized for prescribed $\Delta\mu = 40°$, $\psi_2 = 100°$, to drive a Geneva mechanism.

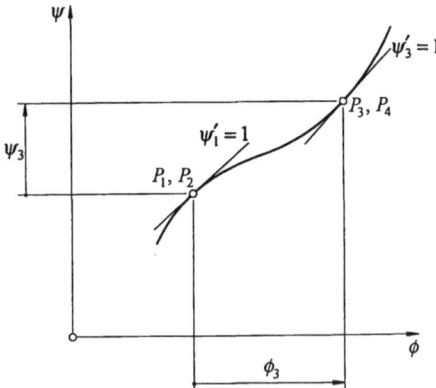

Fig. 7.44. Prescribed $\psi'_1 = 1$, $\psi'_3 = 1$ in a $(P_1P_2-P_3P_4)$ synthesis.

synthesized drag-link and its corresponding crank-rocker. Fig. 7.45 shows a drag-link in its two unity velocity ratio positions. In such positions, as $A_0A_1B_1B_0$ or $A_0A_3B_3B_0$,

Synthesis of function generators

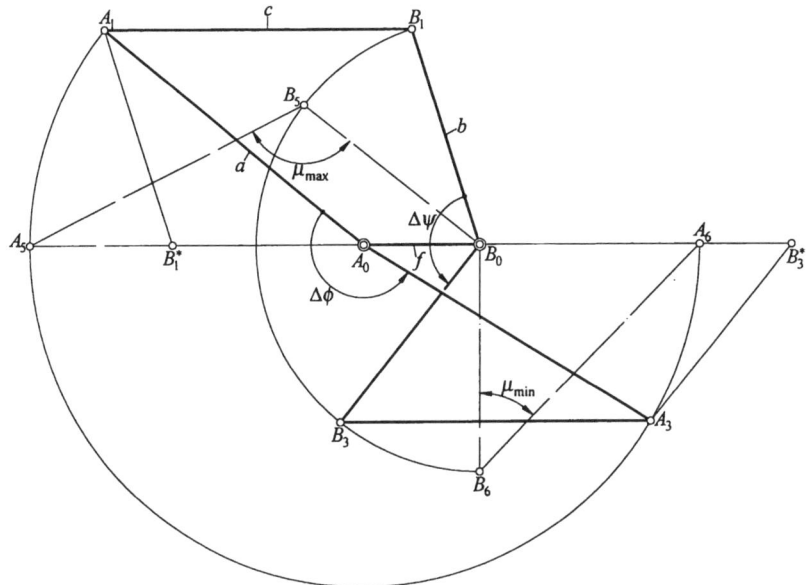

Fig. 7.45. Synthesis of a double-crank between $\psi'_1 = 1$ and $\psi'_3 = 1$.

the coupler A_1B_1 or A_3B_3 is parallel to A_0B_0. Assume as before that both A_0A and B_0B are rotating in a counterclockwise sense. After passing through the position $A_0A_1B_1B_0$, link a always rotates faster than b, and after passing through the position $A_0A_3B_3B_0$, link a always rotates slower than b. This is why (Tsai, 1983b) stressed that a drag-link should be synthesized between two unity velocity ratio positions, in order to fully utilize its capability. To avoid confusion with the symbols ϕ_2, ψ_2 in Fig. 7.41, the symbols ϕ_3, ψ_3 in Fig. 7.45 are replaced by $\Delta\phi$, $\Delta\psi$ respectively. According to the hint given by (Freudenstein, 1983), parallelograms $B_0 B_1 A_1 B_1^*$ and $B_0 B_3 A_3 B_3^*$ are constructed in Fig. 7.45. It is clear that A_0, B_0, B_1^* are collinear, and A_0, B_0, B_3^* are also collinear. In the four-bar linkage $A_0 B_0 B_3^* A_3$, imagine that A_0A_3 were the fixed link. This is a crank-rocker mechanism in its outer dead-centre position and with A_0B_0 as its crank. Similarly, in the four-bar linkage $A_0 B_0 B_1^* A_1$, imagine A_0A_1 were the fixed link. This is a crank-rocker in its inner dead-centre position and again with A_0B_0 as its crank. In Fig. 7.46 these two four-bars are redrawn. Imagine that $A_0 B_0 B_1^* A_1$ were locked together and rotated about A_0 through an angle $\Delta\phi(=\phi_3)$ until A_0A_1 and A_0A_3 coincide, to the position $A_0 (B_0)^3 (B_1^*)^3 A_3$. A comparison of this figure with Fig. 7.34 shows that for this crank-rocker, as the crank A_0B_0 rotates through an angle $\Delta\phi$, the angle through wihch

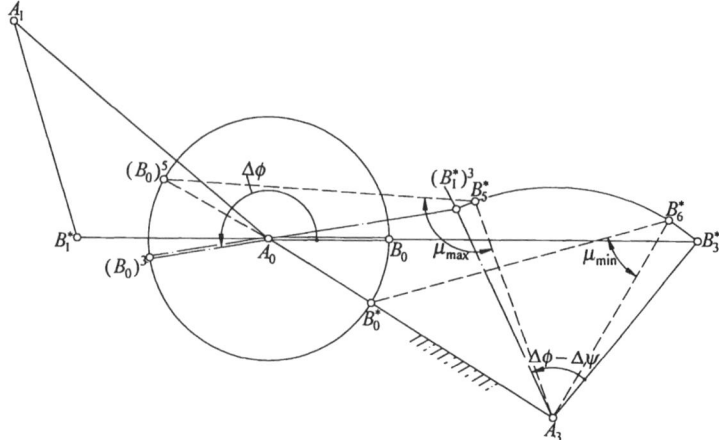

Fig. 7.46. The two positions of the double-crank in Fig. 7.45 correspond to the inversion of two extreme positions of a crank-rocker.

its rocker $A_3 B_3^*$ rotates to the position $A_3(B_1^*)^3$ is $\Delta\phi - \Delta\psi$. Hence synthesizing a drag-link between two unity velocity ratio positions and at the same time coordinating the angle pair $\Delta\phi : \Delta\psi$, is equivalent to synthesizing a crank-rocker between two dead-centre positions and at the same time coordinating the angle pair $\Delta\phi : (\Delta\phi - \Delta\psi)$. Hence all equations in Section 7.7.5 can be used as formulae. It can be seen from Figs. 7.45 and 7.46 that the maximum transmission angle $\mu_{\max}$ and minimum transmission angle $\mu_{\min}$ of the inverted crank-rocker are exactly the same as the $\mu_{\max}$ and $\mu_{\min}$ of the original drag-link. Hence in the synthesis of the inverted crank-rocker, the optimization procedure mentioned in Section 7.7.5 (c) can still be applied to find the drag-link with an optimum transmission angle.

Obviously, if we had assumed $\psi_1' = 1$, $\psi_3' = 1$ in equation (7.56), we could get the same synthesis equations.

Let us return to (a) First kind of problem. Consider first a so-called *central crank-rocker* as shown in Fig. 7.47. In order to avoid confusion with the points of a drag-link, we denote here the points by $\overline{A}_0, \overline{B}_0, \overline{A}, \overline{B}$, and the link lengths $\overline{A_0 A}, \overline{B_0 B}, \overline{AB}, \overline{A_0 B_0}$ by $\overline{a}, \overline{b}, \overline{c}, \overline{f}$ respectively, and $\sphericalangle \overline{A_1 A_0 A_3} = \overline{\phi}_3$, maximum transmission angle by $\overline{\mu}_{\max}$ and minimum transmission angle by $\overline{\mu}_{\min}$. The name central crank-rocker indicates that in both outer dead-centre and inner dead-centre positions the common lines of crank and coupler coincide. In other words, a comparison with Fig. 7.34 shows that $\overline{A}_3, \overline{A}_1, \overline{B}_3, \overline{B}_1$ are now collinear, and $\overline{\phi}_3 = 180°$. It is easy to show that:

Synthesis of function generators

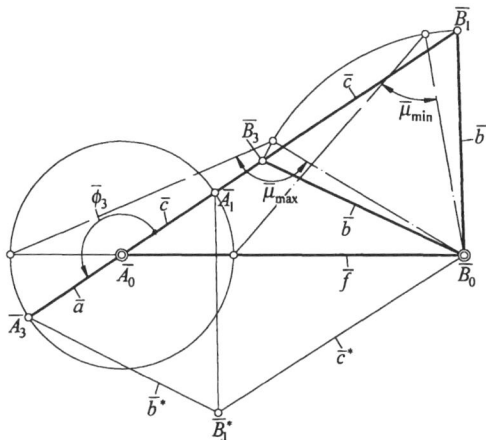

Fig. 7.47. First kind of problem of synthesizing a double-crank in Section 7.7.7 (a) corresponds to an inversion of central crank-rocker.

$$\bar{a}^2 + \bar{f}^2 = \bar{b}^2 + \bar{c}^2 \tag{7.86}$$

Equation (7.86) is the necessary and sufficient condition for a crank-rocker to become a central crank-rocker. From this condition we get

$$\cos \bar{\mu}_{max} = - \cos \bar{\mu}_{min}$$

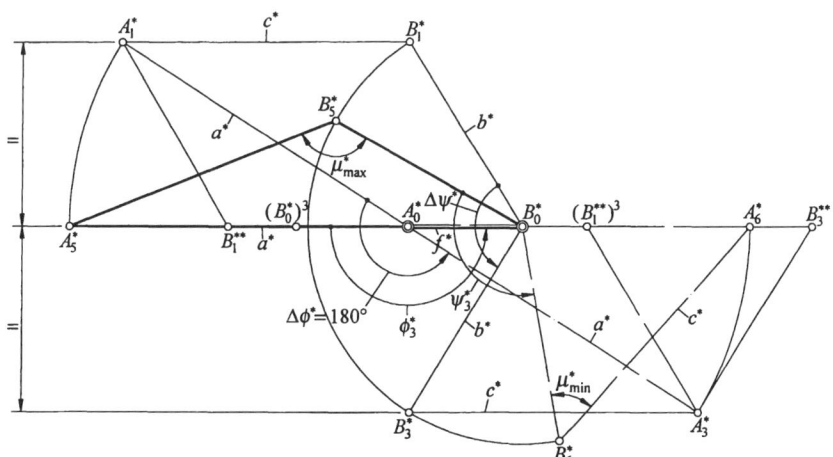

Fig. 7.48. An explanation of Fig. 7.47 according to Fig. 7.46.

or
$$\bar{\mu}_{max} - 90° = 90° - \bar{\mu}_{min} \tag{7.87}$$

Comparing equations (7.86) with (7.82) and (7.87) with (7.79), we see that the drag-link synthesized in (a) is in fact an inversion of a central crank-rocker.

This problem can also be observed from the point of view of synthesizing the first kind of drag-links in (a). In Fig. 7.48, in order to avoid confusion, all points and angles are earmarked with a (*). Suppose $A_0^* A_5^* B_5^* B_0^*$ and $A_0^* A_6^* B_6^* B_0^*$ are the two positions of the synthesized drag-link of the first kind according to Fig. 7.41, where $\mu_{max}^* - 90° = 90° - \mu_{min}^*$, and as $A_0^* A_5^*$ rotates through $\phi_3^* (=180°)$ to $A_0^* A_6^*$, $B_0^* B_5^*$ rotates through ψ_3^* to $B_0^* B_6^*$. Now put this mechanism in the two unity velocity ratio positions $A_0^* A_1^* B_1^* B_0^*$ and $A_0^* A_3^* B_3^* B_0^*$, where $A_1^* B_1^*$ or $A_3^* B_3^*$ is parallel to $A_0^* B_0^*$. It can easily be shown that $A_1^* B_1^*$ and $A_3^* B_3^*$ are equally distant away from $A_0^* B_0^*$, and B_1^*, A_0^*, B_3^* are collinear, and A_1^*, A_0^*, A_3^* are also collinear. These two positions are just those shown in Fig. 7.45 for synthesizing a drag-link of the 2nd kind. Construct as before parallelograms $B_0^* B_1^* A_1^* B_1^{**}$ and $B_0^* B_3^* A_3^* B_3^{**}$, and rotate $A_0^* A_1^* B_1^{**} B_0^*$ about A_0^* through $\Delta\phi^* (=180°)$ to the position A_0^* $A_3^*(B_1^{**})^3 (B_0^*)^3$ until $A_0^* A_1^*$ coincides with $A_0^* A_3^*$. Then $(B_0^*)^3, A_0^*, B_0^*, (B_1^{**})^3, B_3^{**}$ are collinear. The crank-rocker as an inversion of the original drag-link is of course a central crank-rocker.

Compare Fig. 7.48 with Fig. 7.47. Draw in Fig. 7.47 the parallelograms $\bar{A}_1 \bar{B}_1 \bar{B}_0 \bar{B}_1^*$ and $\bar{A}_3 \bar{B}_3 \bar{B}_0 \bar{B}_1^*$. It can be seen that if we made $a^* = \bar{f}, f^* = \bar{a}, b^* = \bar{b}$, $c^* = \bar{c}$, we would certainly have $\mu_{max}^* = \bar{\mu}_{max}$, and $\mu_{min}^* = \bar{\mu}_{min}$. This is the method described by (Tsai, 1983a). Please note that, although in Fig. 7.48 both ϕ_3^* and $\Delta\phi^*$ are 180°, ψ_3^* and $\Delta\psi^*$ are not the same. ψ_3^* can be calculated from equation (7.85), while $\Delta\psi^*$ is

$$\Delta\psi^* = 2\cos^{-1}\frac{f^*}{b^*}$$

Hence if $\Delta\phi^* = 180°$, and the value of $\Delta\psi^*$ are prescribed, the position of B_1^* in Fig. 7.48 is determined. In other words, b^* is determined by f^*. The positions of the three points A_0^*, B_0^*, B_1^* in the figure are all determined. The locus of A_1^* is the straight line passing through B_1^* and parallel to $A_0^* B_0^*$. From equation (7.83) we have

$$\cos\mu_{min}^* = \frac{a^* f^*}{b^* c^*} \tag{7.88}$$

It can be seen from equation (7.88) that, since f^*, b^* are determined, the value of μ_{min}^* is determined by a^* and c^*. From Fig. 7.48 it is clear that a^* is always larger than c^*. Therefore the maximum μ_{min}^* occurs when $\cos\mu_{min}^*$ is a minimum. This

Synthesis of function generators

happens when a^*/c^* is a minimum, or when a^* and c^* are equal, or when both a^* and c^* approach infinity. In other words, maximum μ^*_{min} happens when the drag-link becomes a turning-block linkage.

7.8 Spacing of precision points

(a) Chebyshev (or translated as Tschebychew) minimax theorem
This principle is sometimes introduced in textbooks of engineering mathematics. We shall describe it briefly.

Theorem 27: Let $f(x)$ be a given polynomial. It is required to approximate $f(x)$ by another polynomial $p_n(x)$ of nth degree within the interval $-1 \le x \le 1$. The criterion of the minimax theorem is: $p_n(x)$ should be a polynomial such that the absolute value of the maximum error $|e(x)|=|f(x)-p_n(x)|$ becomes a minimum.

To explain this theorem, we proceed in two steps.
(1) The case $f(x) \equiv 0$
This polynomial with leading coefficient 1 should be such that $\max|p_n(x)|$ is a minimum. We assert that $p_n(x)$ is a polynomial with alternative maximum $(+M)$ values and minimum $(-M)$ values at $n+1$ points.

Proof: Suppose $g_n(x)$ were another polynomial of nth degree with leading coefficient 1 whose extreme value within the interval $-1 \le x \le 1$ is smaller than the extreme value $|M|$ of $p_n(x)$. Then the function $p_n(x) - g_n(x)$ would be alternatively positive and negative at $n+1$ points. In other words, it would have n roots. However, as $p_n(x) - g_n(x)$ is at most an equation of $(n-1)^{th}$ degree, this result is impossible unless $p_n(x) \equiv g_n(x)$.

The Chebyshev polynomial is

$$T_n(x) = \cos n\theta$$

where $x = \cos\theta$, and $-1 \le x \le 1$. By means of the identity

$$\cos(n+1)\theta + \cos(n-1)\theta \equiv 2\cos\theta\cos n\theta$$

we can get a recurrent formula

$$T_{n+1}(x) = 2xT_n(x) - T_{n-1}(x)$$

Hence a series of Chebyshev polynomials can be obtained:

$$T_0(x) = 1$$
$$T_1(x) = 1$$
$$T_2(x) = 2x^2 - 1$$
$$T_3(x) = 4x^3 - 3x$$
$$T_4(x) = 8x^4 - 8x^2 + 1$$
$$T_5(x) = 16x^5 - 20x^3 + 5x$$
$$\dots\dots\dots\dots\dots$$
$$T_n(x) = 2^{n-1}x^n + \cdots$$

Therefore

$$p_n(x) = 2^{1-n} T_n(x)$$

The extreme value M is

$$M = 2^{1-n}$$

appearing at $n+1$ points of x values

$$x = \cos\frac{r\pi}{n}, r = 0, 1, 2, \cdots, n.$$

The n zero points are spaced at

$$x = \cos\left(r + \frac{1}{2}\right)\frac{\pi}{n}, r = 0, 1, 2, \cdots, n-1.$$

Fig 7.49 shows the $p_n(x)$ curve for $n = 8$.

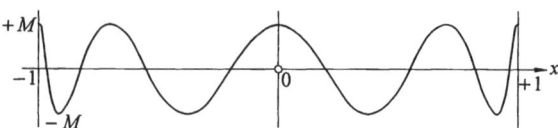

Fig. 7.49. $p_n(x)$-curve for $n = 8$.

(2) General case of $f_n(x)$

Let $f_n(x)$ be a given polynomial of nth degree with leading coefficient a_n. It is required to approximate $f_n(x)$ by a polynomial $q_{n-1}(x)$ of $(n-1)^{th}$ (or even lower) degree. Obviously if $q_{n-1}(x)$ should be a polynomial closest to $f_n(x)$ of a lower degree, their difference has to be minimum. Hence we can let

$$f_n(x) - q_{n-1}(x) = a_n p_n(x) = a_n 2^{1-n} T_n(x)$$

and then find $q_{n-1}(x)$.

Example: Find $q_2(x)$ which is closest to $f_3(x) = x^3 + x^2$
Solution: Since $a_n = 1$, $2^{1-n} = 2^{1-3} = 1/4$, we have

$$f_3(x) - q_2(x) = x^3 + x^2 - q_2(x) = \frac{1}{4}T_3(x) = x^3 - \frac{3}{4}x$$

$q_2(x) = x^2 + 3x/4$ is the polynomial of second degree which is closest to $x^3 + x^2$ in the interval $-1 \le x \le 1$. Fig. 7.50 shows the original curve $f_3(x) = x^3 + x^2$ and the quadratic curve $q_2(x) = x^2 + 3x/4$.

(b) Chebyshev spacing of precision points

Synthesis of function generators

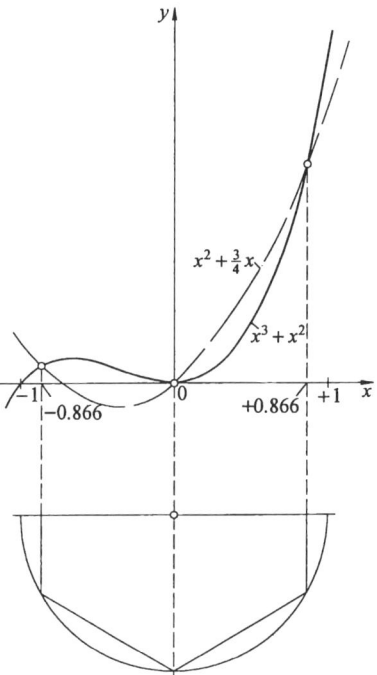

Fig. 7.50. The best quadratic curve $x^2 + 3x/4$ to approximate x^3+x^2.

Let us go back to Fig. 7.4. Suppose a given function is shown as the solid curve. The range of variation of ϕ is also given. The question is how to select the points P_1, P_2, ... within this range, such that the functional relationship generated by the synthesized function generator is as close as possible to the given functional relationship. In Fig. 7.51 a given functional relationship is shown. The procedure starts by drawing a half-circle with the range of variation of ϕ as the diameter. Suppose for instance the number of required precision points is 4, then draw a half inscribed octagon within the half circle. Projecting the four vertices of the octagon onto the given $\psi(\phi)$ curve will locate the four precision points P_1, P_2, P_3, P_4.

The reason for this construction is based on the characteristics of the Chebyshev polynomial $T_n(x) = \cos n\theta$. For $n = 4$, $T_4(x) = 8x^4 - 8x^2 +1$, and $T_4(x) = 0$ occurs at $x = \cos\theta = +0.9239, +0.3827, -0.3827, -0.9239$ (or $\theta = 22.5°, 67.5°, 112.5°, 157.5°$), just like the case $n = 3$ as shown in Fig.7.50. The error of the function generator thus synthesized based on such precision points is comparatively small.

Similarly, if five precision points are required, the inscribed half polygon within the half-circle in Fig. 7.51 should be a half-decagon, and the precision points are obtained by projecting the vertices of the half-decagon onto the given $\psi(\phi)$ curve,

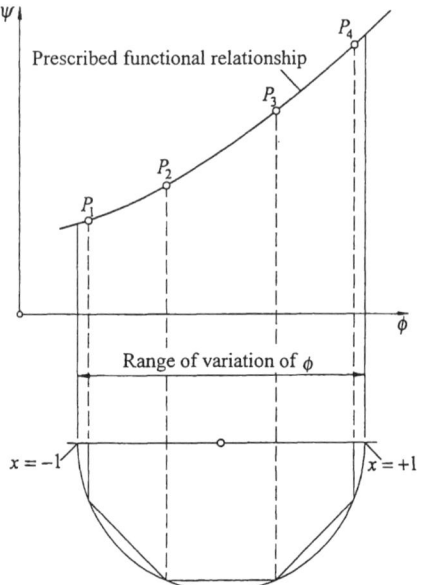

Fig. 7.51. Chebyshev spacing of precision points.

etc. This procedure of obtaining precision points is in general called *Chebyshev spacing*.

However, as the relationship $\psi(\phi)$ generated by a four-bar linkage is not a polynomial, (Freuderstein, 1959a) modified the Chebyshev spacing to further reduce the structural error. This modification applies also to spacing of precision points for path generation. Readers are referred to this article for further reference.

As mentioned above, the traditional Chebsyshev spacing takes into consideration the spacing of points only along the ϕ (or x) -axis, but not on the property of the curve or the function $\psi(\phi)$ (or $y = y(x)$) itself. (Tsai & Chin, 1980) proposed three modifications by performing Chebyshev spacing along (*a*) curvilinear length of the functional curve, (*b*) curvature of the functional curve, (*c*) line interal of the curvature of the functional curve.

7.9 Synthesis of geared five-bar function generators

(*a*) General concepts

A geared five-bar is a sort of mechanism containing links and gears. Although the number of links is five (gears being not counted as links), the degrees of freedom remain at 1 due to gear transmissions. Fig. 7.52 shows a variety of geared five-bar mechanisms.

Synthesis of function generators

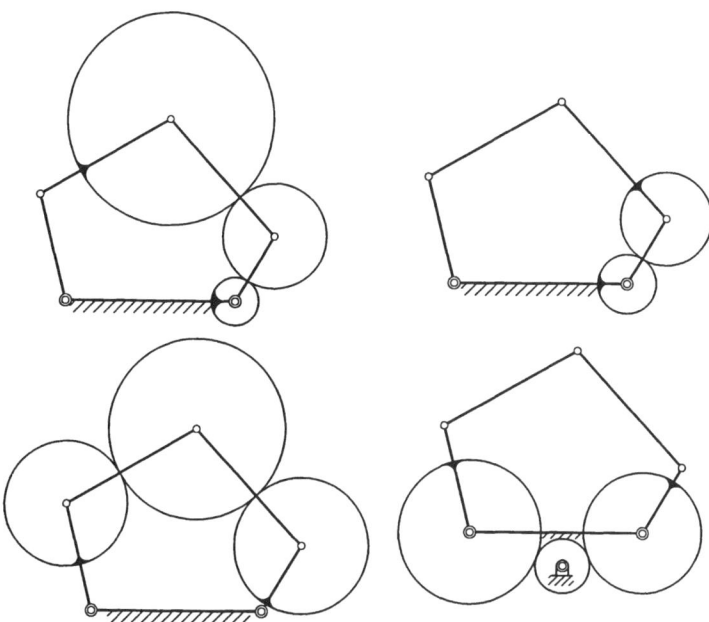

Fig. 7.52. Various kinds of geared five-bar mechanisms.

Methods of synthesizing planar geared five-bar function generators include the matrix method (Suh & Radcliffe, 1967), vector loop equations (Kramer & Sandor, 1970) and the method of coordinate transformation (Oleksa & Tesar, 1971). In general, geared five-bars may generate more kinds of functions than four-bar linkages, or even functions which are difficult to generate by four-bars (Oleksa & Tesar, 1971). In the following we shall introduce two kinds of synthesis methods. The first is the relative pole method that we have been consistently using. The second is a method corresponding to that depicted in Section 3.6.6. In both of these methods the length of the coupler does not appear as an unknown. Please refer to (Lin & Chiang, 1992).

Fig. 7.53 shows a geared five-bar function generator. The initial position angles of the input link a and the output link b are denoted respectively by ϕ_0 and ψ_0 as before. Gears with A_0, B_0 and B as centres are denoted by z_a, z_e and z_c respectively. z_a and link a are bound together, and z_c and c are bound together. z_e is not bound either to b or to f. Let the numbers of teeth of z_a, z_e, z_c be N_a, N_e, N_c respectively. The train value from z_a to z_e is

$$i_{ae} = N_a / N_e$$

and the train value from z_e to z_c is

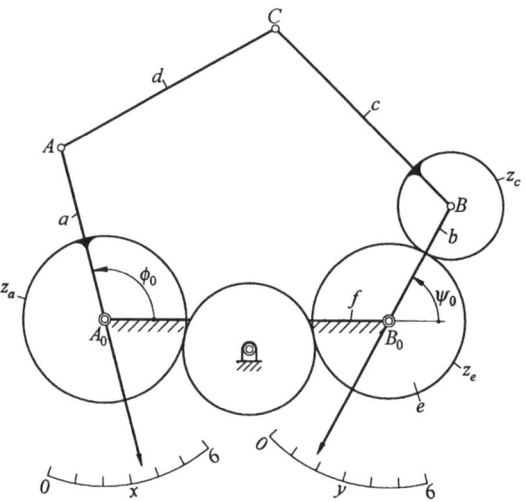

Fig. 7.53. Geared five-bar function generator.

$$i_{ec} = -N_e / N_c$$

To coordinate an angle-pair $\phi_j : \psi_j$, imagine that a is fixed and link d is removed, as shown in Fig. 7.54. Rotate link f through an angle $-\phi_j$ as shown in Fig. 7.55. (Please note that the angle ϕ_j assumed in Fig. 7.55 is clockwise, hence $-\phi_j$ is counterclockwise.) Next rotate link b relative to link f through an angle ψ_j (the angle ψ_j assumed in Fig. 7.55 is also clockwise). The displacement matrix $[\mathbb{D}_j]_{ca}$ of c relative to a can be built as the multiplication of three matrices

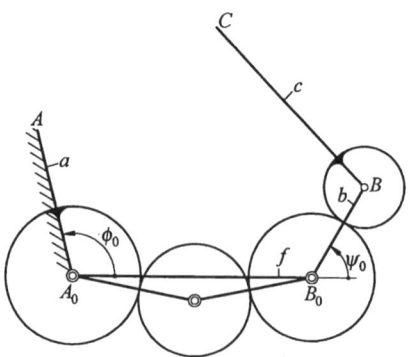

Fig. 7.54. Inversion of the mechanism in Fig. 7.53.

Synthesis of function generators

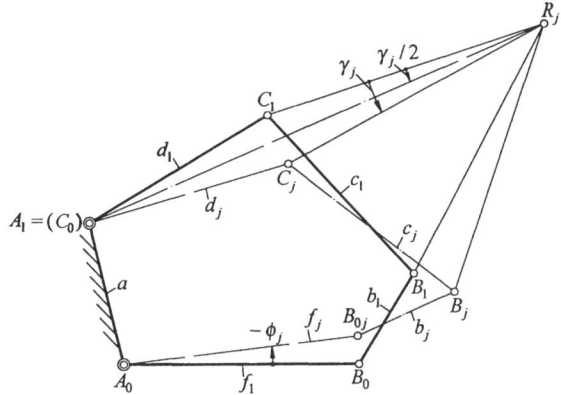

Fig. 7.55. Relative pole R_j for the motion of c relative to a.

$$[\mathbb{D}_j]_{ca} = [\mathbb{D}_j]_{fa}[\mathbb{D}_j]_{bf}[\mathbb{D}_j]_{cb} \qquad (7.89)$$

$[\mathbb{D}_j]_{cb}$ is the relative displacement matrix for the motion of body c relative to body b. Since the angle of rotation now of c relative to b is an angle of $\beta_j = i_{ae}i_{ec}\phi_j - i_{ec}\psi_j$ about B, we have

$$[\mathbb{D}_j]_{cb} = \begin{bmatrix} C_{\beta j} & -S_{\beta j} & x_B(1-C_{\beta j})+y_B S_{\beta j} \\ S_{\beta j} & C_{\beta j} & -x_B S_{\beta j}+y_B(1-C_{\beta_j}) \\ 0 & 0 & 1 \end{bmatrix} \qquad (7.90)$$

where $S_{\beta j} = \sin\beta_j$, $C_{\beta j} = \cos\beta_j$. This displacement matrix can be obtained by replacing γ_j, x_{Pj}, y_{Pj}; by β_j, x_B, y_B respectively in equation (A17.1). $[\mathbb{D}_j]_{bf}$ is the relative displacement matrix for the motion of body b relative to body f. The angle of rotation of b with respect to f about B_0 is ψ_j, hence

$$[\mathbb{D}_j]_{bf} = \begin{bmatrix} C_{\psi j} & -S_{\psi j} & x_{B0}(1-C_{\psi j})+y_{B0}S_{\psi j} \\ S_{\psi j} & C_{\psi j} & -x_{B0}S_{\psi j}+y_{B0}(1-C_{\psi j}) \\ 0 & 0 & 1 \end{bmatrix} \qquad (7.91)$$

$[\mathbb{D}_j]_{fa}$ is the relative displacement matrix for the motion of f relative to a. The angle of rotation of f with respect to a about A_0 is $-\phi_j$, hence

$$[\mathbb{D}_j]_{fa} = \begin{bmatrix} C_{\phi j} & S_{\phi j} & x_{A0}(1-C_{\phi j})-y_{A0}S_{\phi j} \\ -S_{\phi j} & C_{\phi j} & x_{A0}S_{\phi j}+y_{A0}(1-C_{\phi j}) \\ 0 & 0 & 1 \end{bmatrix} \qquad (7.92)$$

If A_0 is taken as the coordinate origin, then $x_{A0}=0, y_{A0}=0$. The matrix $[\mathbb{D}_j]_{fa}$ becomes one of pure rotation. Furthermore, according to the sequence of the matrices in equation (7.89), (x_B, y_B) in equation (7.90) and (x_{B0}, y_{B0}) in equation (7.91) are the coordinates of B and B_0 respectively before rotation. Combining the three equations (7.90), (7.91) and (7.92) into equation (7.89), we can get $[\mathbb{D}_j]_{ca}$.

(b) Relative pole method

As shown in Fig. 7.55, the displacement of c relative to a is the displacement of c from c_1 to c_j. After having found the displacement matrix $[\mathbb{D}_j]_{ca}$, the relative angle of rotation γ_j and the position of the relative pole $R_j(=R_{1j})$ can be found from equations (A17.2) and (A17.3).

For a problem of four precision points, three angle pairs $\phi_2:\psi_2$; $\phi_3:\psi_3$; $\phi_4:\psi_4$ are to be coordinated. Let $j=2, 3, 4$, then the relative circle-point curve and relative centre-point curve should be able to be determined. Now apply the principle of equation (3.1) to Fig. 7.55. Suppose a pair of relative circle-point C_1 (point on coupler c) and relative centre point $C_0(=A_1)$ are known, then

$$\sphericalangle A_1R_jC_1 = \gamma_j/2 \tag{7.93}$$

Equation (7.93) can be formulated either by vectors or complex numbers. The advantage of these mathematical means is that no matter where the coordinate origin lies, the formulated equations are always valid. Suppose the position vectors of the points A_1, R_j, C_1 in Fig. 7.55 are respectively $\mathbf{a}, \mathbf{R}_j, \mathbf{c}$. Equation (7.93) can be written as

$$\frac{\mathbf{c}-\mathbf{R}_j}{\mathbf{a}-\mathbf{R}_j} = h_j \exp(-i\gamma_j/2) \tag{7.94}$$

where h_j is a scalar, and $i=\sqrt{-1}$. Dividing equation (7.94) by its conjutate complex gives

$$\frac{\mathbf{c}-\mathbf{R}_j}{\mathbf{a}-\mathbf{R}_j}\frac{\overline{\mathbf{a}}-\overline{\mathbf{R}}_j}{\overline{\mathbf{c}}-\overline{\mathbf{R}}_j} = \varepsilon_j \tag{7.95}$$

where $\varepsilon_j = \exp(-i\gamma_j)$, being a unit vector. Rearranging equation (7.95), we get

$$\mathbf{c}(\overline{\mathbf{a}}-\overline{\mathbf{R}}_j) - \overline{\mathbf{c}}\varepsilon_j(\mathbf{a}-\mathbf{R}_j) = \overline{\mathbf{a}}\mathbf{R}_j - \varepsilon_j\mathbf{a}\overline{\mathbf{R}}_j + (\varepsilon_j - 1)\mathbf{R}_j\overline{\mathbf{R}}_j \tag{7.96}$$

Setting $j=2, 3, 4$ in equation (7.96), we get three simultaneous linear equations in the two unknowns $\mathbf{c}, \overline{\mathbf{c}}$. The determinant of the coefficients should vanish, or

$$m_{1234}: \begin{vmatrix} \overline{\mathbf{a}}-\overline{\mathbf{R}}_2, & -\varepsilon_2(\mathbf{a}-\mathbf{R}_2), & \overline{\mathbf{a}}\mathbf{R}_2-\varepsilon_2\mathbf{a}\overline{\mathbf{R}}_2+(\varepsilon_2-1)\mathbf{R}_2\overline{\mathbf{R}}_2 \\ \overline{\mathbf{a}}-\overline{\mathbf{R}}_3, & -\varepsilon_3(\mathbf{a}-\mathbf{R}_3), & \overline{\mathbf{a}}\mathbf{R}_3-\varepsilon_3\mathbf{a}\overline{\mathbf{R}}_3+(\varepsilon_3-1)\mathbf{R}_3\overline{\mathbf{R}}_3 \\ \overline{\mathbf{a}}-\overline{\mathbf{R}}_4, & -\varepsilon_4(\mathbf{a}-\mathbf{R}_4), & \overline{\mathbf{a}}\mathbf{R}_4-\varepsilon_4\mathbf{a}\overline{\mathbf{R}}_4+(\varepsilon_4-1)\mathbf{R}_4\overline{\mathbf{R}}_4 \end{vmatrix} = 0 \tag{7.97}$$

Synthesis of function generators

Equation (7.97) is the equation of the relative centre-point curve for the motion of c relative to a, or of the locus of A_1. In this equation $\mathbf{a}(\bar{\mathbf{a}})$ is the only unknown. Similarly, by rearranging equation (7.96) to separate the two unknowns $\mathbf{a}, \bar{\mathbf{a}}$, we can get three simultaneous linear equations in the two unknowns $\mathbf{a}, \bar{\mathbf{a}}$. Setting the determinant of the coefficients of these equations equal to zero, we can get the equation of the relative circle-point curve, or of the locus of C_1. The expansion of equation (7.97) is listed in Appendix 18.

(c) Method by analogy with equations in Section 3.6.6

In Section 3.6.6, we have already shown how to derive equations of centre-point curve and of circle-point curve for four finitely separated position in body guidance problem. However, the derivation there was based on the displacement matrix of equation (3.53). In general cases, the displacement matrix $[\mathbb{D}_j]_{ca}$ in equation (7.89) is not of the form of the displacement matrix in equation (3.53), because the coordinates of the two point E_1, E_j as those mentioned in Section 3.4.2 are not known yet. Let the displacement matrix in the present case be

$$[\mathbb{D}_j]_{ca} = \begin{bmatrix} e_{11j} & e_{12j} & e_{13j} \\ e_{21j} & e_{22j} & e_{23j} \\ 0 & 0 & 1 \end{bmatrix}$$

Hence in place of equation (3.94) we have

$$\left.\begin{array}{l} G_2 x_{A1} + S_2 y_{A1} + T_2 = 0 \\ G_3 x_{A1} + S_3 y_{A1} + T_3 = 0 \\ G_4 x_{A1} + S_4 y_{A1} + T_4 = 0 \end{array}\right\} \quad (7.98)$$

and in place of equation (3.95) we have

$$\left.\begin{array}{l} G_j = (e_{11j} - 1)x_{C1} + e_{12j} y_{C1} + e_{13j} \\ S_j = e_{21j} x_{C1} + (e_{22j} - 1)y_{C1} + e_{23j} \\ T_j = -(e_{11j} e_{13j} + e_{21j} e_{23j})x_{C1} \\ \quad - (e_{12j} e_{13j} + e_{22j} e_{23j})y_{C1} - (e_{13j}^3 + e_{23j}^2)/2 \end{array}\right\} \quad (7.99)$$

Set the determinant of the coefficients in equation (7.98) equal to zero, like the form of equation (3.96)

$$k_{C1234}: \begin{vmatrix} G_2 & S_2 & T_2 \\ G_3 & S_3 & T_3 \\ G_4 & S_4 & T_4 \end{vmatrix} = 0 \quad (7.100)$$

However, equation (7.100) is not the same as equation (3.96). It is derived from equation (7.99). It represents the relative circle-point curve for the motion of c relative to a, or the locus k_{C1234} of the point C_1.

Choose a point C_1 as a relative circle-point on the relative circle-point curve, then the corresponding relative centre-point A_1 can be found according to the method mentioned in Section 3.4.3, or from R_{12}, γ_{12} and R_{13}, γ_{13} according to Fig. 7.55.

If it is required to coordinate four angular displacement pairs $\phi_2 : \psi_2$; $\phi_3 : \psi_3$; $\phi_4 : \psi_4$; $\phi_5 : \psi_5$, the procedure is similar to that mentioned before to find two

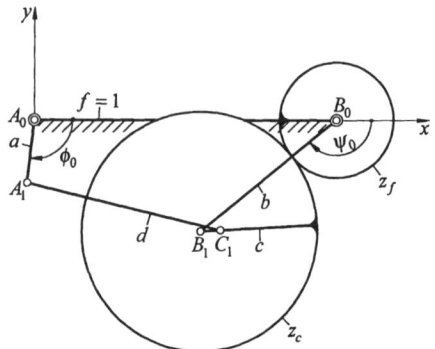

Fig. 7.56. Example: synthesized function generator.

relative circle-point curves k_{C1234} and k_{C1235}, and then to locate their intersections, or the relative Burmester circle-points, and then the relative Burmester centre-points. The function generator can thus be constructed.

Example: In the geared five-bar function generator as shown in Fig. 7.56, the train value of the gear pair z_f, z_c is $i_{fc} = -0.5$. Assume $\overline{A_0 B_0} = f = 1$, and take A_0 as the origin. Prescribed coordinates are: $A_0 (0,0)$, $B_0 (1,0)$, $B (0.5503, -0.3728)$. It is required to coordinate four angle-pairs:

$$\phi_2 = -20.0°, \quad \psi_2 = -10.0°$$
$$\phi_3 = -34.5°, \quad \psi_3 = -18.0°$$
$$\phi_4 = -38.0°, \quad \psi_4 = -20.0°$$
$$\phi_5 = -55.0°, \quad \psi_5 = -30.0°$$

The problem is solved and two relative Burmester pairs are found, namely, A_1 (−1.1362, 0.5254), C_1(−0.3388, 0.7747) and A_1(−0.0205, −0.2057), C_1(0.6111, −0.3677). It is obvious that the link length proportions of the mechanism synthesized according to the first pair of A_1, C_1 values are not satisfoctory. The mechanism shown in Fig. 7.56 is the one based on the second pair of A_1, C_1 values.

Exercises

7.1 According to the description in Section 6.2, in the figure shown, if the points A_0, B_0, E_1, E_2 are prescribed, and it is required to coordinate $\phi_{12} : E_1 \to E_2$;

Synthesis of function generators

and $\psi_{12} : E_1 \rightarrow E_2$, there are ∞^1 possible solutions. If the synthesis procedure is carried out by using the relative pole R_{12} for coordinating $\phi_{12} : \psi_{12}$ as that described in Section 7.2.1, what is this choice?

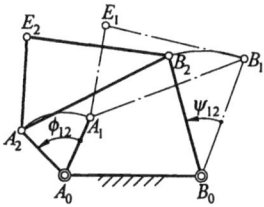

7.2 Construct a four-bar function generator, so that as the driving link A_0A rotates through an angle of 120°, the driven link B_0B rotates through an angle of −90°.

7.3 Construct a slider-crank, so that as the crank A_0A rotates through an angle of 40°, the joint B (the centre of the slider) moves through a distance 20.

7.4 As a continuation of Exercise 7.3, the point B should move through a further distance 20 as the crank A_0A rotates through a further angle of 30°.

7.5 Design a slider-crank mechanism to produce a quick return motion with a time ratio 3:5. The stroke of the slider is 50.

7.6 Construct a four-bar function generator to coordinate the following angular displacements; $\phi_{12} : \psi_{12} = 120° : -90°$; $\phi_{13} : \psi_{13} = 150° : -110°$.

7.7 Construct a four-bar function generator so that as the crank A_0A rotates through 90° and 120°, the link B_0B rotates through −60° and −90° respectively.

7.8 Construct a four-bar function generator to coordinate the following angular displacements: $\phi_{12} : \psi_{12} = 60° : 30°$; $\phi_{13} : \psi_{13} = 90° : 60°$.

7.9 It is known that, in synthesizing a four-bar function generator to coordinate three angular displacement pairs $\phi_{12} : \psi_{12}$; $\phi_{13} : \psi_{13}$; $\phi_{14} : \psi_{14}$, the locus of A_1 is the relative centre-point curve $k_{A1(1234)}$, and the locus of B_1 is the relative circle-point curve $k_{B1(1234)}$. Show that as A_1 is taken at R_{12}, the corresponding point B_1 is at the intersection $(\pi_{12})_b$ of $R_{13}(R_{23})_b$ and $R_{14}(R_{24})_b$, while as A_1 is taken at $(R_{23})_a$, the corresponding B_1 is at the intersection $(\pi_{23})_b$ of $R_{12}R_{13}$ and $(R_{24})_b(R_{34})_b$.

(Please refer to Alt, 1936 and Beyer, 1953, p.101.)

7.10 Design a four-bar function generator to generate the function $y = \sin^2 x$. Assume $45° \leq x \leq 70°$, and the scale factors are:

$$M_\phi = \Delta\phi / \Delta x = 1.5 \, \text{rad} / \text{radian of } x$$

$$M_\psi = \Delta\psi / \Delta y = (\pi/4) / \text{radian of } y$$

Choose two pairs of angular displacements $\phi_2 : \psi_2 ; \phi_3 : \psi_3$.

7.11 Synthesize a four-bar function generator to generate the function $y = \log_{10} x$. Let x be represented by ϕ, and y by ψ. Assume $1 \leq x \leq 10$, and the scale factors are:

$$M_\phi = \frac{\Delta\phi}{\Delta x} = 10°/ \text{unit of } x$$

$$M_\psi = \frac{\Delta\psi}{\Delta y} = -60°/ \text{unit of } y$$

Take four points on the original functional curve: $x = 1.3, 3.8, 7.2, 9.7$ to compute the angles: $\phi_{12}, \phi_{13}, \phi_{14}; \psi_{12}, \psi_{13}, \psi_{14}$. After having synthesized the four-bar mechanism,
(a) calculate the error, and draw the error curve;
(b) find the values of the transmission angle μ at $x = 1$ and $x = 10$.

7.12 Synthesize a four-bar linkage to coordinate

$$\psi' = d\psi / d\phi = 0.45$$
$$\psi'' = d^2\psi / d\phi^2 = 0.35$$

There should be ∞^2 solutions.

7.13 Determine the expression of the diameter d_c of the Carter-Hall circle, in terms of ψ', ψ'', ψ''' and $\overline{A_0 B_0} = f$.

Hint: Consider the relative motion of a relative to b in Fig. 7.15, i.e. imagine that b were fixed. By using equation (3.116) and the relation

$$\frac{1}{l} = \frac{1}{3}\left(\frac{1}{\delta_{ab}} + \frac{1}{\rho_a}\right)$$

where δ_{ab} is the inflection circle diameter for the motion of a relative to b, and the following equation of d_c derived by Hall (Hall, 1957):

$$d_c = m \cos \lambda$$

the expression of d_c in equation (7.8) can be derived. Note that the equation of ρ_a is in equation (4.28), while the relations of r_{A0}, δ_{ab} and λ are:

$$r_{A0} = \frac{\psi'}{1-\psi'}f \qquad [(7.5)]$$

$$\delta_{ab} = \frac{\psi'}{(1-\psi')^2|\sin\lambda|}f$$

Synthesis of function generators

$$\tan \lambda = -\frac{\psi'(1-\psi')}{\psi''} \qquad [(7.7)]$$

7.14 Using the method described in Section 7.6.2, construct a four-bar function generator to match $\psi' = +0.40$, $\psi'' = +0.35$, $\psi''' = -0.30$.

7.15 Construct a four-bar function generator to match the following differential coefficients.

$$\psi' = 0.453 , \ \psi'' = 0.349 , \ \psi''' = -0.806$$

(a) by the method mentioned in Section 7.3.4.
(b) by the method mentioned in Section 7.6.2.

7.16 Similar to the example shown in Fig. 4.6, in which $\psi' = -1$, $\psi'' = -0.57296$, $\psi''' = 0$ are prescribed, but solve by using the method of Section 7.6.2.

7.17 Show that in synthesizing a four-bar function generator
(a) to match prescribed ψ'
(b) to match prescribed ψ', ψ'' the number of possible $\begin{cases} \infty^3 \\ \infty^2 \\ \infty^1 \\ 1 \text{ or } 3 \end{cases}$
(c) to match prescribed ψ', ψ'', ψ''' solutions is
(d) to match prescribed $\psi', \psi'', \psi''', \psi''''$

7.18 Let a four-bar linkage be given: $f = 1$, $a = 0.4$, $c = 0.7$, $b = 0.6$, $\phi_0 = \measuredangle B_0 A_0 A = 45°$.
(a) Calculate ψ''' by means of the expression of ψ''' found in Exercise 2.2.
(b) Find the diameter d_d of the circle k_d according to the geometrical relation shown in Fig. 7.23, and then calculate ψ''' by means of equation (7.16).

7.19 Synthesize a four-bar function generator to generate the function $y = \log_{10} x$. Let x be represented by ϕ, and y by ψ. Assume $2 \leq x \leq 9$, and M_ϕ and M_ψ are the same as those in Exercise 7.11. Use higher order synthesis, and take $x_m = 5$.
(a) Calculate the error, and draw the error curve.
(b) Determine the transmission angles at $x = 2$ and $x = 9$.
(c) Compare the result with the example shown in Fig. 7.24.

7.20 Design a four-bar function generator to generate the function $y = e^x$. Let x be represented by ϕ, and y by ψ. Assume $0 \leq x \leq 2$. The total range of variation of ϕ is 90°, and the total range of variation of ψ is $-60°$. Use higher order synthesis, and take $x_m = 1.2$. Moreover, take $-45° \leq \theta \leq -5°$.

7.21 In Fig. 2.5, $A_0 A$ is the driver. For the link $G_0 G$, what is the transmission angle?

7.22 In Fig. 2.9, a is the driver. For the link b, what is the transmission angle?

7.23 In the type $(P_1 P_2 - P_3 P_4)$ synthesis of a function generator, the $\psi(\phi)$ relation is as that shown in Fig. 7.33. Suppose ψ'_1, ψ'_3 are both prescribed, but only ϕ_3 is prescribed in the angle-pair $\phi_3 : \psi_3$ to be coordinated, while ψ_3 is left free. Moreover, ϕ_0 is prescribed. How can this sort of problem be solved (please refer to Barkan & Tuohy, 1959)?

7.24 Design a crank-rocker mechanism to coordinate $\phi_3 : \psi_3 = 170° : 50°$ (please

refer to Fig. 7.34).
(a) Find the solution of maximum μ_{min}, and determine the values of ϕ_0, μ_{min} and the lengths of the four bars.
(b) If it is required that μ_{min} should not be less than 45°, and hence either ϕ_3 or ψ_3 has to be modified, which one should be modified?

7.25 Design a crank-rocker mechanism to coordinate $\phi_3:\psi_3 = 165°:45°$ (please refer to Fig. 7.34).
(a) Find the solution of maximum μ_{min}, and determine the values of ϕ_0, μ_{min} and the lengths of the four bars.
(b) If it is required that μ_{min} should not be less than 45°, and hence either ϕ_3 or ψ_3 has to be modified, which one should be modified?

7.26 Design a crank-rocker mechanism to coordinate $\phi_3:\psi_3 = 216.5°:85°$ (please refer to Fig. 7.34). Find the solution of maximum μ_{min}, and determine the values of ϕ_0, μ_{min} and the lengths of the four bars.

7.27 Discuss the case $\phi_3 = 180°$ in synthesizing a crank-rocker (symbols are as those shown in Fig. 7.34). Explain that for a central crank-rocker, the transmission angle does not have to be optimized.

7.28 A family of crank-rockers is to be synthesized for the angles $\phi_3 = 220°$, $\psi_3 = 36°$ (symbols are as those shown in Fig. 7.34). Find the solution of maximum μ_{min}, and determine the values of ϕ_0, μ_{min} and the lengths of the four bars.

7.29 Synthesize double-rockers between the points $P_7(P_8)$ and $P_3(P_4)$ in Fig. 7.39(a) or the points $P_{1\Delta}$ ($P_{2\Delta}$) and $P_3(P_4)$ in the following figure. For prescribed values of ϕ_3, ψ_3, let $1/\psi_1' = 0$, $1/\psi_3' = 0$. Take ψ_0 as the parameter. Find the equations of k_{B1} and k_{A1} and those of the lengths of the links. (Please refer to Chiang, 1986b.)

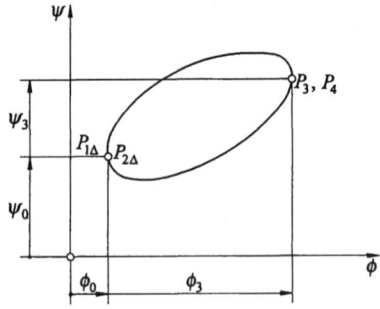

7.30 Find the family of double-rockers to coordinate the angles $\phi_3 = 84°$, $\psi_3 = 62°$ (please refer to the figure of Exercise 7.29).

7.31 Synthesize a double-crank between the two positions shown in Fig. 7.41.

Synthesis of function generators

Assume $\Delta\mu = 61°$ in equation (7.79), and $\psi_2 = 110°$.

7.32 Synthesize a double-crank between the two positions shown in Fig. 7.41. Assume $\Delta\mu = 60°$ in equation (7.79), and $\psi_2 = 110°$.

7.33 Synthesize a double-crank between two positions of unity velocity ratio, as that shown in Fig. 7.45.
(a) Assume $\Delta\phi = 188°$, $\Delta\psi = 123°$
(b) Assume $\Delta\phi = 189°$, $\Delta\psi = 124°$

7.34 Determine $T_8(x)$, and approximate the functions $f(x) = 3x^8 + 5x^3$ and $F(x) = 2x^8 + 5x^3$ by a polynomial of 7th degree.

7.35 Determine $T_7(x)$, and approximate the function $f(x) = 4x^7 + 5x^2$ by a polynomial of 6th degree.

PART III

8
Harmonic analysis of four-link Mechanisms

8.1 General concepts

Harmonic analysis is a means often used in engineering, to analyze a periodic function by expanding it into a Fourier series, i.e. into series of sine and cosine functions of the independent variable. In other words, if a periodic function $f(\phi)$ is continuous within the interval $0 \leq \phi \leq 2\pi$, it should be able to expand it into a Fourier series:

$$f(\phi) = a_0 + a_1 \cos\phi + a_2 \cos 2\phi + \cdots \cdots \\ + b_1 \sin\phi + b_2 \sin 2\phi + \cdots \cdots \quad (8.1)$$

The coefficents $a_0, a_1, \ldots b_1, b_2$, are all constants, being equal to

$$\begin{aligned} a_0 &= \frac{1}{2\pi} \int_{-\pi}^{\pi} f(\phi) \mathrm{d}\phi \\ a_{m(m\neq 0)} &= \frac{1}{\pi} \int_{-\pi}^{\pi} f(\phi) \cos m\phi \mathrm{d}\phi \\ b_m &= \frac{1}{\pi} \int_{-\pi}^{\pi} f(\phi) \sin m\psi \mathrm{d}\phi \\ c_m &= \sqrt{a_m^2 + b_m^2} \end{aligned} \quad (8.2\ a,b,c,d)$$

c_m is the amplitude of the mth harmonic. In order to avoid resonance, the amplitude should attenuate rapidly or even be eliminated as far as possible.

For a four-link mechanism, since its motion is periodic, $f(\phi)$ can be the motion quantity of the output link, the slider, or of the coupler, including displacement, velocity, acceleration and kinetic energy. Works on harmonic analyses of planar mechanisms are numerous, including the earlier work of (Biezeno & Grammel, 1953) on central slider-crank, those of (Meyer zur Capellen, 1956b, 1957, 1958) on swinging-block linkage and offset slider-crank, and those of (Freudenstein, 1959) and (Freudenstein & Mohan, 1961) on four-bar linkages, and those of (Meyer zur Capellen, 1959, 1975) on kinetic energy of slider crank and of swinging block linkage. According to the report of (Rankers, 1960), up to 1959 harmonic analyses on approximately 25 types of mechanisms had been completed at the Technische Hochschule Aachen, as the one worked by (Meyer zur capellen, 1960). In the following we shall introduce some representative literature to illustrate the technique.

8.2 Harmonic analysis of central slider-crank mechanism (Biezeno & Grammel, 1953)

The name *central slider-crank* implies that the straight path of the centre B of the slider joint passes through the centre A_0 of the crank, as shown in Fig. 8.1.

(a) Harmonic analysis of the displacement of the slider

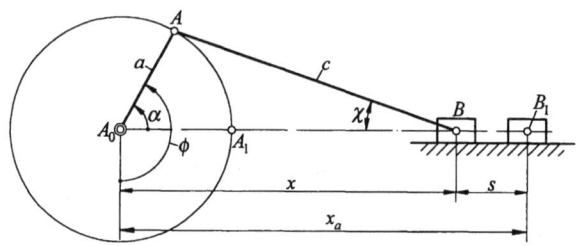

Fig. 8.1.   Central slider-crank mechanism.

In Fig. 8.1, let $\chi = \angle A_0BA$, $\alpha = \phi - 90°$, and $x = \overline{A_0B}$, then

$$\left.\begin{array}{l} x = a\cos\alpha + c\cos\chi \\ a\sin\alpha = c\sin\chi \end{array}\right\} \quad (8.3\ a,b)$$

Eliminating χ from the two equations of (8.3 a,b), and setting $\varepsilon = a/c$, we get

$$\frac{x}{a} = \cos\alpha + \frac{1}{\varepsilon}(1 - \varepsilon^2 \sin^2\alpha)^{1/2} \quad (8.4)$$

The second term on the right hand side of equation (8.4) can be expanded as a power series of ε:

$$\frac{x}{a} = \cos\alpha + \frac{1}{\varepsilon} - \frac{\varepsilon}{2}\sin^2\alpha - \frac{\varepsilon^3}{8}\sin^4\alpha - \frac{\varepsilon^5}{16}\sin^6\alpha - \cdots \quad (8.5)$$

By means of the following formulae

$$\left.\begin{array}{l} \sin^2\alpha = \dfrac{1}{2}(1 - \cos 2\alpha) \\ \sin^4\alpha = \dfrac{1}{8}(3 - 4\cos 2\alpha + \cos 4\alpha) \\ \sin^6\alpha = \dfrac{1}{32}(10 - 15\cos 2\alpha + 6\cos 4\alpha - \cos 6\alpha) \\ \cdots\cdots \end{array}\right\} \quad (8.6)$$

equation (8.5) can be rewritten as

Harmonic analysis of four-link mechanisms

$$\frac{x}{a} = A_0 + A_1 \cos\alpha + A_2 \cos 2\alpha + A_4 \cos 4\alpha + A_6 \cos 6\alpha + \cdots \tag{8.7}$$

where all coefficients are power series of ε:

$$\left.\begin{aligned}
A_0 &= \frac{1}{\varepsilon} - \frac{1}{4}\varepsilon - \frac{3}{64}\varepsilon^3 - \frac{5}{256}\varepsilon^5 - \cdots \\
A_1 &= 1 \\
A_2 &= \left(\varepsilon + \frac{1}{4}\varepsilon^3 + \frac{15}{128}\varepsilon^5 + \cdots\right)\left(\frac{1}{4}\right) \\
A_4 &= \left(\frac{1}{4}\varepsilon^3 + \frac{3}{16}\varepsilon^5 + \cdots\right)\left(-\frac{1}{16}\right) \\
A_6 &= \left(\frac{9}{128}\varepsilon^5 + \cdots\right)\left(\frac{1}{36}\right)
\end{aligned}\right\} \tag{8.8}$$

In order to be compared with the Fourier expansion of the output angle of other mechanisms, equation (8.7) can also be written in form of sine series:

$$\frac{x}{a} = A_0 + A_1 \sum_{m=1}^{\infty} \frac{A_m}{A_1} \sin(m\alpha + 90°) \tag{8.9}$$

(b) Harmonic analyses of coupler angle χ and coupler angular velocity $\dot\chi$

Equation (8.3b) can be rewritten as

$$\cos\chi = (1 - \varepsilon^2 \sin^2 \alpha)^{1/2} = 1 - \frac{\varepsilon^2}{2}\sin^2\alpha - \frac{\varepsilon^4}{8}\sin^4\alpha - \frac{\varepsilon^6}{16}\sin^6\alpha - \cdots \tag{8.10}$$

Differentiating equation (8.10) with respect to time, and dividing it by equation (8.3b), $\sin\chi = \varepsilon\sin\alpha$, we get

$$\dot\chi = \dot\alpha\left(\varepsilon\cos\alpha + \frac{1}{2}\varepsilon^3 \sin^2\alpha\cos\alpha + \frac{3}{8}\varepsilon^5 \sin^4\alpha\cos\alpha + \cdots\right) \tag{8.11}$$

Substituting equations (8.6) into equation (8.11), and using the formula

$$\cos 2k\alpha \cos\alpha = \frac{1}{2}\cos(2k+1)\alpha + \frac{1}{2}\cos(2k-1)\alpha$$

we get

$$\dot\chi = \dot\alpha\,\varepsilon(C_1 \cos\alpha - \frac{1}{3}C_3 \cos 3\alpha + \frac{1}{5}C_5 \cos 5\alpha - + \cdots) \tag{8.12}$$

where the coefficients are

$$\left.\begin{array}{l} C_1 = 1 + \dfrac{1}{8}\varepsilon^2 + \dfrac{3}{64}\varepsilon^4 + \cdots \\ C_3 = \dfrac{3}{8}\varepsilon^2 + \dfrac{27}{128}\varepsilon^4 + \cdots \\ C_5 = \dfrac{15}{128}\varepsilon^4 + \cdots \\ \cdots \cdots \end{array}\right\}$$

For the harmonic analysis of the angular acceleration $\ddot{\chi}$ (please refer to equation (8.53b)), and those of the inertia forces and inertia moments of the coupler please refer to (Biezeno & Grammel, 1953).

8.3 Harmonic analysis of the output angle of a four-bar linkage (Freudenstein, 1959b)

In Fig. 8.2, the output angle ψ as a function of the input angle ϕ can be derived from the relation

$$\psi = 180^\circ - (\psi_s + \psi_u)$$

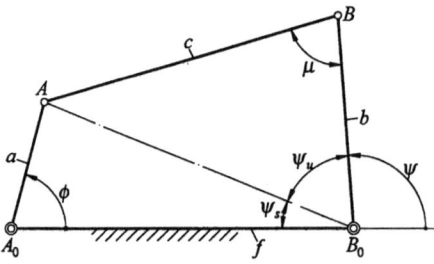

Fig. 8.2. - A general four-bar linkage.

Therefore doing the harmonic analysis of ψ is equivalent to doing the harmonic analyses of ψ_s and ψ_u. We shall proceed separately.

(a) The part ψ_s

We get from geometric relation that

$$\psi_s = \tan^{-1} \frac{\lambda \sin\phi}{1 - \lambda \cos\phi} \tag{8.13}$$

where $\lambda = a/f$. In the case where $f = 1$, then $\lambda = a$. From Fig. 8.2 it is clear that ψ_s is an odd function of ϕ (if ϕ changes sign only, ψ_s also changes sign only). Hence we may assume

Harmonic analysis of four-link mechanisms

$$\psi_s = \sum_{m=1}^{\infty} b_m \sin m\phi \tag{8.14}$$

where

$$b_m = \frac{1}{\pi} \int_{-\pi}^{\pi} \psi_s \sin m\phi \, d\phi \tag{8.15}$$

By means of integration by parts, equation (8.15) can be written as

$$b_m = -\frac{1}{\pi m} \int_{-\pi}^{\pi} \psi_s \, d(\cos m\phi)$$

$$= \frac{1}{\pi m} \int_{-\pi}^{\pi} \frac{d\psi_s}{d\phi} \cos m\phi \, d\phi \tag{8.16}$$

The expression of $d\psi_s/d\phi$ in equation (8.16) can be found by differentiating equation (8.13) with respect to ϕ, or

$$\frac{d\psi_s}{d\phi} = \frac{\lambda(\cos\phi - \lambda)}{1 - 2\lambda\cos\phi + \lambda^2} \tag{8.17}$$

Substituting equation (8.17) into equation (8.16) gives

$$b_m = \frac{\lambda}{\pi m} \int_{-\pi}^{\pi} \frac{\cos\phi - \lambda}{1 - 2\lambda\cos\phi + \lambda^2} \cos m\phi \, d\phi \tag{8.18}$$

The following expansion is already known

$$\frac{\cos\phi - \lambda}{1 - 2\lambda\cos\phi + \lambda^2} = \sum_{n=1}^{\infty} \lambda^{n-1} \cos n\phi \quad (\lambda^2 < 1) \tag{8.19}$$

Substituting equation (8.19) into equation (8.18) and because of

$$\int_{-\pi}^{\pi} \cos m\phi \cos n\phi \, d\phi \begin{cases} = \pi, \text{if } m = n \\ = 0, \text{if } m \neq n \end{cases}$$

we get

$$b_m = \frac{\lambda^m}{m}$$

Hence

$$\psi_s = \sum_{m=1}^{\infty} \frac{\lambda^m}{m} \sin m\phi \tag{8.20}$$

(b) The part ψ_u

Similarly, we get from the geometric relation in Fig. 8.2

$$\psi_u = \tan^{-1} \frac{\delta \sin \mu}{1 - \delta \cos \mu} \qquad (8.21)$$

where $\delta = c/b$. The relationship between the angle μ and the input angle ϕ is

$$\cos \mu = \sigma_1 + \sigma_2 \cos \phi \qquad (8.22)$$

where

$$\left.\begin{array}{l} \sigma_1 = \dfrac{b^2 + c^2 - a^2 - f^2}{2bc} = \dfrac{b^2 + c^2 - a^2 - 1}{2bc} \\ \sigma_2 = \dfrac{af}{bc} = \dfrac{a}{bc} \end{array}\right\} \text{(for } f=1\text{)} \qquad (8.23\ a,b)$$

From Fig. 8.2 it can be seen that ψ_u is an even function of ϕ (if ϕ changes sign only, ψ_u remains unchanged). Hence we may assume

$$\psi_u = \sum_{m=0}^{\infty} a_m \cos m\phi \qquad (8.24)$$

where

$$a_0 = \frac{1}{2\pi} \int_{-\pi}^{\pi} \psi_u \, d\phi \qquad (8.25)$$

and

$$a_{m(m \neq 0)} = \frac{1}{\pi} \int_{-\pi}^{\pi} \psi_u \cos m\phi \, d\phi \qquad (8.26)$$

By means of integration by parts again, equation (8.26) can be written as

$$a_{m(m \neq 0)} = -\frac{1}{\pi m} \int_{-\pi}^{\pi} \frac{d\psi_u}{d\phi} \sin m\phi \, d\phi \qquad (8.27)$$

The expression of $d\psi_u/d\phi$ in equation (8.27) can be found by differentiating equation (8.21) with respect to ϕ, or

$$\frac{d\psi_u}{d\phi} = \frac{\delta(\cos \mu - \delta)}{1 - 2\delta \cos \mu + \delta^2} \cdot \frac{\sigma_2 \sin \phi}{\sin \mu}$$

$$= \left[-\frac{1}{2} + \frac{1 - \delta^2}{2(1 - 2\delta \cos \mu + \delta^2)} \right] \frac{\sigma_2 \sin \phi}{\sin \mu}$$

Because of

$$\frac{\sigma_2}{1 - 2\delta \cos \mu + \delta^2} = \frac{\lambda}{\delta(1 - 2\lambda \cos \phi + \lambda^2)}$$

this can be written as

Harmonic analysis of four-link mechanisms

$$\frac{d\psi_u}{d\phi} = F_\phi G_\phi \tag{8.28}$$

where

$$F_\phi = -\frac{1}{2}\sigma_2 \sin\phi + \frac{\lambda(1-\delta^2)\sin\phi}{2\delta(1-2\lambda\cos\phi+\lambda^2)} \tag{8.29}$$

$$G_\phi = \frac{1}{\sin\mu} = \frac{1}{[1-(\sigma_1+\sigma_2\cos\phi)^2]^{1/2}} \tag{8.30}$$

Therefore equation (8.27) can be written as

$$a_{m(m\neq 0)} = -\frac{1}{\pi m}\int_{-\pi}^{\pi} F_\phi G_\phi \sin m\phi \, d\phi \tag{8.31}$$

A comparison between equations (8.31) and (8.2c) shows that $-ma_m$ is just equal to the mth coefficient f_m of the Fourier expansion of $F_\phi G_\phi = \sum_{m=1}^{\infty} f_m \sin m\phi$ ($F_\phi G_\phi$ is an odd function of ϕ). Hence simply by expanding $F_\phi G_\phi$ into a Fourier series, we can find a_m (please note that the trigonomitric function in the integrand of right hand side of equation (8.31) is now $\sin m\phi$, not $\cos m\phi$ as in equation (8.2b)). Both F_ϕ in equation (8.29) and G_ϕ in equation (8.30) can be expanded into series, as will be shown in the following.

In equation (8.29), the expansion of the following fraction part is also known

$$\frac{\sin\phi}{1-2\lambda\cos\phi+\lambda^2} = \sum_{m=1}^{\infty}(\lambda)^{m-1}\sin m\phi \quad (\lambda^2 < 1) \tag{8.32}$$

Therefore equation (8.29) can be written as

$$F_\phi = -\frac{1}{2}\sigma_2 \sin\phi + \frac{1-\delta^2}{2\delta}\sum_{m=1}^{\infty}\lambda^m \sin m\phi \tag{8.33}$$

Next, the term G_ϕ in equation (8.30) can be expanded by means of Maclaurin series (binomial theorem):

$$(1-x^2)^{-1/2} = 1 + \frac{1}{2}x^2 + \frac{1\cdot 3}{2\cdot 4}x^4 + \frac{1\cdot 3\cdot 5}{2\cdot 4\cdot 6}x^6 + \cdots$$

(where $x = \sigma_1 + \sigma_2 \cos\phi$), and because of the expansion of $\cos^m\phi$ into a series containing§ $\cos m\phi$, $\cos(m-2)\phi$, $\cos(m-4)\phi$, ... in the form

---

§ If m is even, then $\cos^{2k}\phi = \frac{1}{2^{2k-1}}\left\{\sum_{n=0}^{k-1}\binom{2k}{n}\cos[(2k-2n)\phi] + \frac{1}{2}\binom{2k}{k}\right\}$.

If m is odd, then $\cos^{2k+1}\phi = \frac{1}{2^{2k}}\sum_{n=0}^{k}\binom{2k+1}{n}\cos[(2k+1-2n)\phi]$.

$$G_\phi = \sum_{k=0}^{\infty} C_k \cos k\phi \quad (k = \text{even}) \tag{8.34}$$

Combining equations (8.33) and (8.34), we get

$$F_\phi G_\phi = \left[-\frac{1}{2}\sigma_2 \sin\phi + \frac{1-\delta^2}{2\delta} \sum_{m=1}^{\infty} \lambda^m \sin m\phi \right] \sum_{k=0}^{\infty} C_k \cos k\phi \tag{8.35}$$

Using the formula

$$2\sin X \cos Y = \sin(X+Y) + \sin(X-Y)$$

we can write the first term of equation (8.35) as

$$-\frac{1}{2}\sigma_2 \sin\phi \sum_{k=0}^{\infty} C_k \cos k\phi = -\frac{\sigma_2}{4} \sum_{k=0}^{\infty} C_k [\sin(1+k)\phi + \sin(1-k)\phi]$$

$$= -\frac{\sigma_2}{4} \left[\sum_{k=1}^{\infty} C_{k-1} \sin k\phi + C_0 \sin\phi - \sum_{k=1}^{\infty} C_{k+1} \sin k\phi \right]$$

$$= -\frac{\sigma_2}{4} \left[C_0 \sin\phi + \sum_{k=1}^{\infty} (C_{k-1} - C_{k+1}) \sin k\phi \right]$$

and equation (8.35) can then be written as

$$F_\phi G_\phi = -\frac{1}{4}\sigma_2 C_0 \sin\phi - \sum_{m=1}^{\infty} \left[\frac{\sigma_2}{4}(C_{m-1} + C_{m+1}) + \frac{1-\delta^2}{2\delta}(2C_0\lambda^m + D_m + E_m) \right] \sin m\phi \tag{8.36}$$

where

$$\left. \begin{array}{l} D_m = \sum_{n=1}^{m-1} \lambda^n C_{m-n} \quad (m \geq 2); (D_1 = 0) \\ E_m = \sum_{n=1}^{\infty} (-\lambda^n C_{m+n} + \lambda^{m+n} C_n) \end{array} \right\} \tag{8.37}$$

(c) The complete Fourier series of ψ
combining equations (8.20), (8.24) and (8.36), we have

$$180° - \psi = \psi_s + \psi_u$$

$$= a_0 + \sum_{m=1}^{\infty} (b_m \sin m\phi + a_m \cos m\phi) \tag{8.38}$$

In equation (8.38), a_0 is a constant, which need not be considered. a_m is just equal to the coefficient of the term $\sin m\phi$ in equation (8.36) divided by $(-m)$, as has been mentioned before. The result is:

Harmonic analysis of four-link mechanisms

$$\left.\begin{array}{l} b_m = \dfrac{\lambda^m}{m} \\[4pt] a_1 = \dfrac{\sigma_2}{4}(2C_0 + C_2) + \dfrac{1-\delta^2}{2\delta}(2C_0\lambda + E_1) \\[4pt] a_{m(m=even)} = 0 \\[4pt] a_{m(m\neq 1)} = \left[\dfrac{\sigma_2}{4}(C_{m-1}+C_{m+1}) + \dfrac{1-\delta^2}{2\delta}(2C_0\lambda^m + D_m + E_m)\right]\Big/m \end{array}\right\} \quad (8.39)$$

In a special case, if $\sigma_1 = 0$, then equation (8.34) is expanded into

$$G_\phi = C_0 + C_2 \cos 2\phi + C_4 \cos 4\phi + C_6 \cos 6\phi + \cdots \quad (8.40)$$

where

$$\left.\begin{array}{l} C_0 = 1 + \dfrac{1}{4}\sigma_2^2 + \dfrac{9}{64}\sigma_2^4 + \dfrac{25}{256}\sigma_2^6 + \left(\dfrac{35}{128}\right)^2 \sigma_2^8 + \cdots \\[6pt] C_2 = \dfrac{1}{4}\sigma_2^2 + \dfrac{3}{16}\sigma_2^4 + \dfrac{75}{512}\sigma_2^6 + \dfrac{35\cdot 56}{(128)^2}\sigma_2^8 + \cdots \\[6pt] C_4 = \dfrac{3}{64}\sigma_2^4 + \dfrac{15}{256}\sigma_2^6 + \dfrac{35\cdot 28}{(128)^2}\sigma_2^8 + \cdots \\[6pt] C_6 = \dfrac{5}{512}\sigma_2^6 + \dfrac{35}{16\cdot 128}\sigma_2^8 + \cdots \\[6pt] C_8 = \dfrac{35}{(128)^2}\sigma_2^8 + \cdots \end{array}\right\} \quad (8.41)$$

(*d*) Applications of harmanic analysis of four-bar linkages

From equations (8.40) and (8.41) it can be seen that, for high speed running four-bar mechanisms, the link proportions should be chosen as follows:

(1) $c/b = \delta = 1$; this eliminates a portion of the cosine series.

(2) $b^2 + c^2 - a^2 - f^2 = 0$, i.e. $\sigma_1 = 0$; this eliminates another portion of the cosine series, as shown in equations (8.40), (8.41).

(3) $a/f = \lambda$ to be as small as possible; this attenuates the sine series.

Dimenaions of crank-rockers satisfying these proportions can be determined as

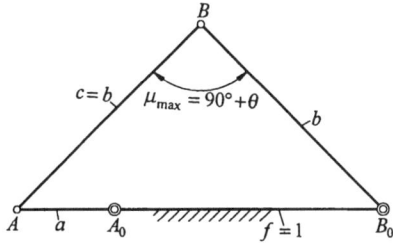

Fig. 8.3.

follows: Please refer to Fig. 8.3. Assume $f = 1$. When the transmission angle μ is a maximum, $\mu_{max} = 90° + \theta$, and θ is the maximum pressure angle. (Please note that this θ is not the pressure angle θ in Figs. 7.19 and 7.20.) We have

and
$$\left. \begin{array}{l} a = \lambda = \tan\dfrac{\theta}{2} \\[6pt] b = c = \dfrac{1}{\sqrt{2}\cos\dfrac{\theta}{2}} \\[10pt] \sigma_2 = \dfrac{a}{b^2} = \sin\theta \end{array} \right\} \qquad (8.42)$$

Hence the pressure angle θ is the sole parameter. Thus the requirements of good transmission and minimum overtones are both satisfied. In fact the convergence of the series is quite rapid, as can be seen by observing the magnitude of C_8 in equations (8.41). With the dimensions thus determined, equations (8.39) can be reduced to

$$\left. \begin{array}{l} b_m = \dfrac{[\tan(\theta/2)]^m}{m} \\[8pt] a_1 = \dfrac{\sin\theta}{4}(2C_0 + C_2) \\[8pt] a_{m(even)} = 0 \\[8pt] a_{m(odd) \atop m\neq 1} = \dfrac{\sin\theta}{4m}(C_{m-1} + C_{m+1}) \end{array} \right\} \qquad (8.43)$$

Equation (8.38) can further be combined into

$$180° - \psi = a_0 + \sum_{m=1}^{\infty} c_m \sin(m\phi + \phi_m)$$

$$= a_0 + c_1 \sum_{m=1}^{\infty} \dfrac{c_m}{c_1} \sin(m\phi + \phi_m) \qquad (8.44)$$

where

$$c_m = \sqrt{a_m^2 + b_m^2}$$
$$\phi_m = \cos^{-1}(a_m / c_m)$$

Example: (taken from Freudenstein, 1959) Assume the maximum pressure angle to be $\theta = 30°$. Then $\sigma_2 = \sin\theta = 1/2$, $f = 1$, $a = 0.2679$, $b = c = 0.7320$, $\lambda = a/f = 0.2679$. Now equation (8.38) becomes

$$180° - \psi = a_0 + \sum_{m=1}^{\infty} \dfrac{(0.2679)^m}{m} \sin m\phi + \dfrac{1}{8}C_0 \cos\phi + \dfrac{1}{8}\sum_{m=1}^{\infty}(C_{m-1} + C_{m+1})\dfrac{\cos m\phi}{m}$$

Harmonic analysis of four-link mechanisms

The coefficients are calculated: $C_0 = 1.073107$, $C_2 = 0.076975$, $C_4 = 0.004079$, $C_6 = 2.19 \times 10^{-4}$, $C_8 = 0.08 \times 10^{-4}$. Since $\sigma_1 = 0$, equations (8.42) are valid. The first six coefficients a_m, b_m and c_m are calculated

m	a_m	b_m	$c_m = (a_m^2 + b_m^2)^{1/2}$
1	0.277899	0.26795	0.386038
2	0	0.035898	0.035898
3	0.003377	0.006413	0.007248
4	0	0.001288	0.001288
5	0.000107	0.000276	0.000296
6	0	0.000062	0.000062

Example: (taken from Freudenstein & Mohan, 1961) Assume $\theta = 25°$. Again by equations (8.42), $\sigma_2 = \sin\theta = 0.422618$, $f = 1$, $a = 0.221695$, $b = c = 0.724275$, $\lambda = a/f = 0.221695$. The coefficients thus calculated are: $C_0 = 1.04977$, $C_2 = 0.051589$, $C_4 = 0.00189$, $C_6 = 0.000073$, $C_8 = 0.0000022$. Again by equations (8.43) the first six coefficients a_m, b_m and c_m are calculated

m	a_m	b_m	$c_m = (a_m^2 + b_m^2)^{1/2}$
1	0.227277	0.221695	0.317496
2	0	0.024574	0.024574
3	0.001883	0.003632	0.004091
4	0	0.000604	0.000604
5	0.000041	0.000107	0.000115
6	0	0.000020	0.000020

8.4 Harmonic analysis of inverted slider-cranks

There are two kinds of inversions of the slider-crank. The one is the swinging-block linkage, or the so-called oscillating cylinder, and the other is the turning block linkage. The following description is taken from (Meyer zur Capellen, 1956b). For simplicity reasons, we use complex numbers. In Fig. 8.4(a), take B_0 as the origin, and $\overline{B_0 A_0} = +1$. Let the length $\overline{B_0 A}$ be denoted by h. Assume A_0A is the driving crank rotating with a constant angular velocity $\dot\phi$. The vector equation is now

$$he^{i\gamma} = 1 + ae^{i\phi}$$

Taking logrithmus on both sides, we have

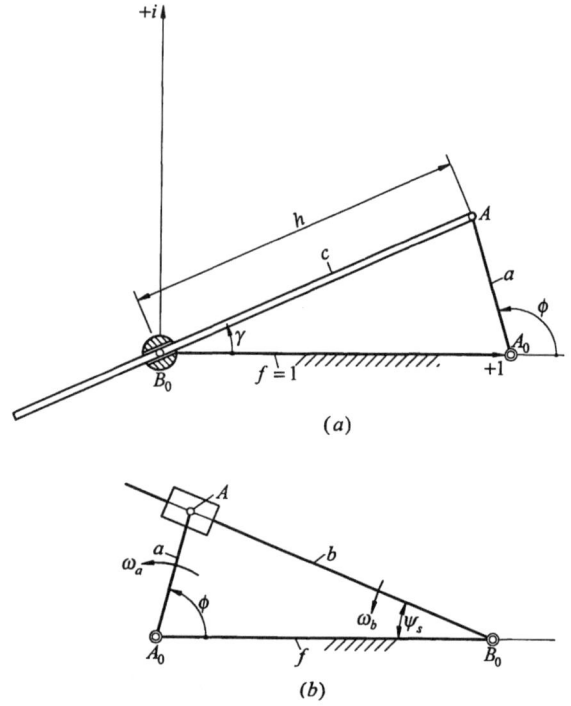

Fig. 8.4. (a)(b) Two types of swinging block linkage.

$$\ln h + i\gamma = \ln(1 + ae^{i\phi}) \tag{8.45}$$

(a) Swinging block linkage

Assume now $a < f$. The coupler c can only swing, but not make complete revolutions. The right hand side of equation (8.45) can be expanded into a power series:

$$\ln(1 + ae^{i\phi}) = ae^{i\phi} - \frac{(ae^{i\phi})^2}{2} + \frac{(ae^{i\phi})^3}{3} - \frac{(ae^{i\phi})^4}{4} + - \cdots \tag{8.46}$$

Substituting the formula

$$(ae^{i\phi})^n = a^n(\cos n\phi + i\sin n\phi)$$

into equation (8.46), and comparing it with the left hand side of equation (8.45) and equating both imaginary parts, we get

$$\gamma = \sum_{n=1}^{\infty} B_n \sin n\phi, \quad B_n = \frac{a^n}{n} \tag{8.47}$$

Harmonic analysis of four-link mechanisms

Differentiating equation (8.47) with respect to ϕ, we get the angular velocity ratio

$$\frac{\omega_c}{\omega_a} = \frac{d\gamma}{d\phi} = \sum_{n=1}^{\infty} A_n \cos n\phi, \quad A_n = a^n \tag{8.48}$$

The mechanisms shown in Figs. 8.4(b) and 8.4(a) are in fact the same, being the well-known quick-return mechanism used in a shaper. The angle ϕ's in both Figs. 8.4(a) and 8.4(b) are supplementary, but the angles γ and ψ_s in both figures are the same. The mechanism shown in Fig. 8.4(b) is just the partial mechanism of that shown in Fig. 8.2. Noting that $\lambda = a/f = a$ here, and combining equations (8.17) and (8.19), we get equation (8.48).

(b) Turning block linkage

For a linkage such as that shown in 8.4 (a), if $a > f$, the coupler c is capable of making complete revolutions. Equation (8.45) can be written as

$$\ln h + i\gamma = \ln\left[ae^{i\phi}\left(1+\frac{1}{ae^{i\phi}}\right)\right] = \ln ae^{i\phi} + \ln\left(1+\frac{1}{ae^{i\phi}}\right)$$

$$= \ln a + i\phi + \left(\frac{1}{ae^{i\phi}} - \frac{1}{2(ae^{i\phi})^2} + \frac{1}{3(ae^{i\phi})^3} - + \cdots\right) \tag{8.49}$$

With the formula

$$\frac{1}{(ae^{i\phi})^n} = a^{-n}e^{-in\phi} = \frac{1}{a^n}(\cos n\phi - i\sin n\phi)$$

substituted into the right hand side of equation (8.49) and comparing this with the imaginary part of its left hand side, we get

$$\gamma = \phi + \sum_{n=1}^{\infty} \overline{B}_n \sin n\phi, \quad \overline{B}_n = -\frac{1}{na^n} \tag{8.50}$$

Similarly, differentiating equation (8.50) with respect to ϕ yields the velocity ratio

$$\frac{\omega_c}{\omega_a} = \frac{d\gamma}{d\phi} = 1 + \sum_{n=1}^{\infty} \overline{A}_n \cos n\phi, \quad \overline{A}_n = -\frac{1}{\lambda^n}$$

The linkage shown in Fig. 8.5 is also a turning-block linkage. Please note that this linkage is not quite the same as the case $a > f$ shown in Fig. 8.4(b). As far as the mechanism is concerned, both are indeed the same, but the driving link a in Fig. 8.5 corresponds to the link b in Fig. 8.4(b), therefore the input and output angles in both figures are exchanged. The harmonic analysis of the motion quantities of the linkage in Fig. 8.5 can proceed according to (Meyer zur Capellen, 1957) as follows. In the figure we have

$$\psi = \phi + \chi \tag{8.51}$$

where $\chi = \angle B_0BA_0$. Differentiating equation (8.51) with respect to ϕ, we get

$$i_{ba} = \frac{\omega_b}{\omega_a} = \frac{d\psi}{d\phi} = 1 + \frac{d\chi}{d\phi} \tag{8.52}$$

Please note that the $d\chi/d\phi$ in equation (8.52) is just the $\dot{\chi}/\dot{\alpha}$ in equation (8.12),

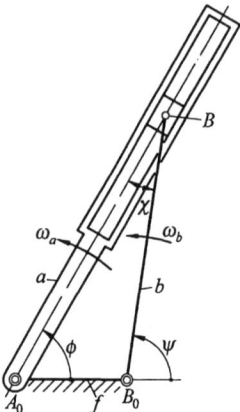

Fig. 8.5.   Turning block linkage.

hence

$$\frac{\omega_b}{\omega_a} = 1 + \varepsilon(C_1 \cos\phi - \frac{1}{3}C_3 \cos 3\phi + \frac{1}{5}C_5 \cos 5\phi - + \cdots) \qquad (8.53a)$$

The coefficients C_1, C_2, ... in equation (8.53a) are those in equation (8.12), but the ε here is equal to $\varepsilon = f/b$ of Fig. 8.5.

Differentiating equation (8.53a) again with respect to time and assuming ω_a = constant, we get

$$\frac{\dot\omega_b}{\omega_a^2} = \varepsilon(-C_1 \sin\phi + C_3 \sin 3\phi - C_5 \sin 5\phi + - \cdots) \qquad (8.53b)$$

8.5 Harmonic analysis of the rotation energy of the coupler of an offset slider-crank (Meyer zur Capellen, 1959)

For an offset slider-crank such as that shown in Fig. 8.6, it would be very complicated to expand the displacement function of the centre point B of the slider into a Fourier series (please refer to Mayer zur Capeller et al., 1958). However, it is rather simple to find the Fourier expansion of the rotation energy of the coupler c. Let the moment of inertia of mass of the coupler c in Fig.8.6 with respect to its centre of mass be denoted by Θ_c, and the angular velocity of c be $\omega_c = \dot\beta$, then its rotation energy is

$$E = \frac{1}{2}\Theta_c \omega_c^2$$

Harmonic analysis of four-link mechanisms

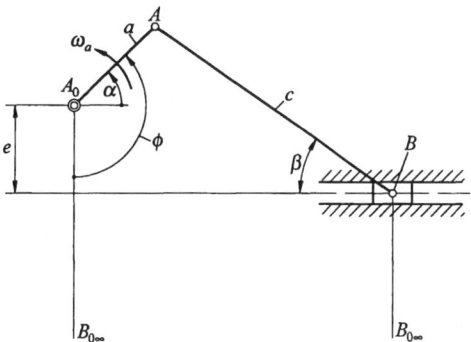

Fig. 8.6.   Offset slider-crank mechanism.

Please note that E is not the total kinetic energy of c. If ω_c is replaced by the velocity ratio $i_{ca} = \omega_c / \omega_a$, or $\omega_c = i_{ca}\omega_a$, then

$$E = \frac{1}{2}\Theta_c \omega_a^2 i_{ca}^2 \tag{8.54}$$

where i_{ca}^2 is the only variable, hence i_{ca}^2 is the only term for which harmonic analysis is needed. Let $\varepsilon = a/c$, $v = e/c$. We can write, according to Fig. 8.6

$$c \sin\beta = e + a \sin\alpha$$

or

$$\sin\beta = v + \varepsilon \sin\alpha \tag{8.55}$$

Differentiating equation (8.55) with respect to α gives

$$i_{ca} = \frac{d\beta}{d\alpha} = \varepsilon \frac{\cos\alpha}{\cos\beta}$$

Hence

$$i_{ca}^2 = \varepsilon^2 \frac{\cos^2\alpha}{\cos^2\beta} \tag{8.56}$$

The problem now is to find the Fourier coefficients of $\cos^2\alpha / \cos^2\beta$. Rewrite equation (8.56) in the form

$$i_{ca}^2 = \frac{\varepsilon^2}{2}\frac{1+\cos 2\alpha}{\cos^2\beta} = \frac{\varepsilon^2}{2}\left(\frac{1}{\cos^2\beta} + \frac{\cos 2\alpha}{\cos^2\beta}\right) \tag{8.57}$$

Equation (8.57) may be dealt with in two parts.

(a) The part $\dfrac{1}{\cos^2\beta}$

$$\frac{1}{\cos^2\beta} = \frac{1}{1-\sin^2\beta} = \frac{1}{1-(v+\varepsilon\sin\alpha)^2}$$

$$= \frac{1/2}{1+v+\varepsilon\sin\alpha} + \frac{1/2}{1-v-\varepsilon\sin\alpha} \tag{8.58}$$

Equation (8.58) may again be dealt with in two further parts.

(i) The part $\dfrac{1/2}{1+v+\varepsilon\sin\alpha}$

In order to transform this expression into the form of equation (8.19), multiply both of its numerator and denominator by a factor μ_1 which is still unknown

$$\frac{\mu_1/2}{\mu_1(1+v)+\mu_1\varepsilon\sin\alpha} = \frac{\mu_1/2}{1-2\lambda_1\cos\alpha^* + \lambda_1^2} \tag{8.59}$$

where

$$\left.\begin{array}{l}\lambda_1 = \dfrac{(1+v)-\sqrt{(1+v)^2-\varepsilon^2}}{-\varepsilon} \\[6pt] \mu_1 = -\dfrac{2\lambda_1}{\varepsilon} \\[6pt] \alpha^* = \dfrac{\pi}{2}-\alpha\end{array}\right\} \tag{8.60}$$

Equation (8.59) can further be transformed into terms containing expressions of the form of equation (8.19) and then expanded according to the latter

$$\frac{\mu_1/2}{1-2\lambda_1\cos\alpha^*+\lambda_1^2} = \frac{\mu_1/2}{(1-\lambda_1^2)}\left[1+\frac{2\lambda_1(\cos\alpha^*-\lambda_1)}{1-2\lambda_1\cos\alpha^*+\lambda_1^2}\right]$$

$$= \frac{\mu_1/2}{1-\lambda_1^2} + \frac{\mu_1}{1-\lambda_1^2}\sum_{n=1}^{\infty}\lambda_1^n\cos n\alpha^*$$

$$= -\frac{\lambda_1}{\varepsilon(1-\lambda_1^2)}\left[1+2\sum_{n=1}^{\infty}\lambda_1^n\cos n\alpha^*\right] \tag{8.61}$$

(ii) The part $\dfrac{1/2}{1-v-\varepsilon\sin\alpha}$

Similar to (i), this part can also be transformed into

$$\frac{\mu_2/2}{\mu_2(1-v)-\mu_2\varepsilon\sin\alpha} = \frac{\mu_2/2}{1-2\lambda_2\cos\alpha^{**}+\lambda_2^2} \tag{8.62}$$

where

Harmonic analysis of four-link mechanisms

$$\left. \begin{array}{l} \lambda_2 = \dfrac{(1-v)-\sqrt{(1-v)^2-\varepsilon^2}}{-\varepsilon} \\ \mu_2 = -\dfrac{2\lambda_2}{\varepsilon} \\ \alpha^{**} = \dfrac{\pi}{2}+\alpha \end{array} \right\} \qquad (8.63)$$

Equation (8.62) can also be transformed into terms containing expressions of the form of equation (8.19), and be written according to equation (8.61)

$$\frac{\mu_2/2}{1-2\lambda_2\cos\alpha^{**}+\lambda_2^2} = -\frac{\lambda_2}{\varepsilon(1-\lambda_2^2)}\left[1+2\sum_{n=1}^{\infty}\lambda_2^n\cos n\alpha^{**}\right] \qquad (8.64)$$

(*iii*) The complete part of $\dfrac{1}{\cos^2\beta}$

From the above definitions of α^* and α^{**} we have

$$\left. \begin{array}{l} \left. \begin{array}{l} \cos 2n\alpha^* \\ \cos 2n\alpha^{**} \end{array} \right\} = \cos(n\pi \mp 2n\alpha) = (-1)^n\cos 2n\alpha \\ \left. \begin{array}{l} \cos(2n-1)\alpha^* \\ \cos(2n-1)\alpha^{**} \end{array} \right\} = \cos\left[\dfrac{2n-1}{2}\pi \mp (2n-1)\alpha\right] = \mp(-1)^n\sin(2n-1)\alpha \end{array} \right\} \qquad (8.65)$$

Adding equations (8.61) and (8.64), and then simplifying according to equations (8.65) we get

$$\frac{1}{\cos^2\beta} = A_0 + \sum_{n=1}^{\infty} A_{2n}\cos 2n\alpha + \sum_{n=1}^{\infty} B_{2n-1}\sin(2n-1)\alpha \qquad (8.66)$$

where

$$\left. \begin{array}{l} A_0 = -\dfrac{1}{\varepsilon}\left(\dfrac{\lambda_1}{1-\lambda_1^2}+\dfrac{\lambda_2}{1-\lambda_2^2}\right) = \dfrac{1}{2}(w_1+w_2) \\ A_{2n} = (-1)^n(w_1\lambda_1^{2n}+w_2\lambda_2^{2n}) \\ B_{2n-1} = (-1)^{n+1}(w_1\lambda_1^{2n-1}-w_2\lambda_2^{2n-1}) \\ w_1 = -\dfrac{2\lambda_1}{\varepsilon(1-\lambda_1^2)} = \dfrac{1}{\sqrt{(1+v)^2-\varepsilon^2}}, w_2 = -\dfrac{2\lambda_2}{\varepsilon(1-\lambda_2^2)} = \dfrac{1}{\sqrt{(1-v)^2-\varepsilon^2}} \end{array} \right\} \qquad (8.67)$$

(*b*) The part $\dfrac{\cos 2\alpha}{\cos^2\beta}$

The expansion of this part can be found by multiplying the expansion of $1/\cos^2\beta$ obtained in (*a*) by $\cos 2\alpha$, and by noting the following relationships

$$\cos 2n\alpha \cos 2\alpha = \frac{1}{2}[\cos 2(n+1)\alpha + \cos 2(n-1)\alpha]$$

$$\sin(2n-1)\alpha \cos 2\alpha = \frac{1}{2}[\sin(2n+1)\alpha + \sin(2n-3)\alpha]$$

or

$$\frac{\cos 2\alpha}{\cos^2 \beta} = \frac{A_2}{2} + \left(A_0 + \frac{A_4}{2}\right)\cos 2\alpha + \frac{A_2 + A_6}{2}\cos 4\alpha$$
$$+ \frac{A_4 + A_8}{2}\cos 6\alpha + \frac{A_6 + A_{10}}{2}\cos 8\alpha + \cdots + \frac{-B_1 + B_3}{2}\sin \alpha$$
$$+ \frac{B_1 + B_5}{2}\sin 3\alpha + \frac{B_3 + B_7}{2}\sin 5\alpha + \frac{B_5 + B_9}{2}\sin 7\alpha + \cdots \quad (8.68)$$

(c) Total rotation energy of the coupler
Combining equations (8.54), (8.56), (8.57), (8.66) and (8.68), we get

$$E = \frac{1}{4}\Theta_c \omega_a^2 \varepsilon^2 [A_0 + A_2 \cos 2\alpha + A_4 \cos 4\alpha + A_6 \cos 6\alpha + A_8 \cos 8\alpha + \cdots$$
$$+ B_1 \sin \alpha + B_3 \sin 3\alpha + B_5 \sin 5\alpha + B_7 \sin 7\alpha + \cdots$$
$$+ \frac{A_2}{2} + \left(A_0 + \frac{A_4}{2}\right)\cos 2\alpha + \frac{A_2 + A_6}{2}\cos 4\alpha + \frac{A_4 + A_8}{2}\cos 6\alpha + \frac{A_6 + A_{10}}{2}\cos 8\alpha + \cdots$$
$$+ \frac{-B_1 + B_3}{2}\sin \alpha + \frac{B_1 + B_5}{2}\sin 3\alpha + \frac{B_3 + B_7}{2}\sin 5\alpha + \frac{B_5 + B_9}{2}\sin 7\alpha + \cdots]$$
$$= \frac{1}{4}\Theta_c \omega_a^2 \varepsilon^2 \left[A_0^* + \sum_{n=1}^{\infty} A_{2n}^* \cos 2n\alpha + \sum_{n=1}^{\infty} B_{2n-1}^* \sin(2n-1)\alpha\right] \quad (8.69)$$

where

$$\left. \begin{array}{l} A_0^* = A_0 + \dfrac{A_2}{2} \\[4pt] A_2^* = A_2 + A_0 + \dfrac{A_4}{2} \\[4pt] A_{2n(n\neq 1)}^* = A_{2n} + \dfrac{A_{2(n-1)} + A_{2(n+1)}}{2} \\[4pt] B_1^* = \dfrac{B_1 + B_3}{2} \\[4pt] B_{2n-1(n\neq 1)}^* = B_{2n-1} + \dfrac{B_{2n-3} + B_{2n+1}}{2} \end{array} \right\}$$

8.6 Harmonic analysis of the kinetic energy of the inverted slider-crank (Meyer zur Capellen & Thünker, 1975)

The two inverted slider-cranks shown in Figs. 8.7(a)(b) are similar to those shown in Figs. 8.4(a), 8.4(b), 8.5. As mentioned before, the driving link in Fig. 8.7(b)

Harmonic analysis of four-link mechanisms

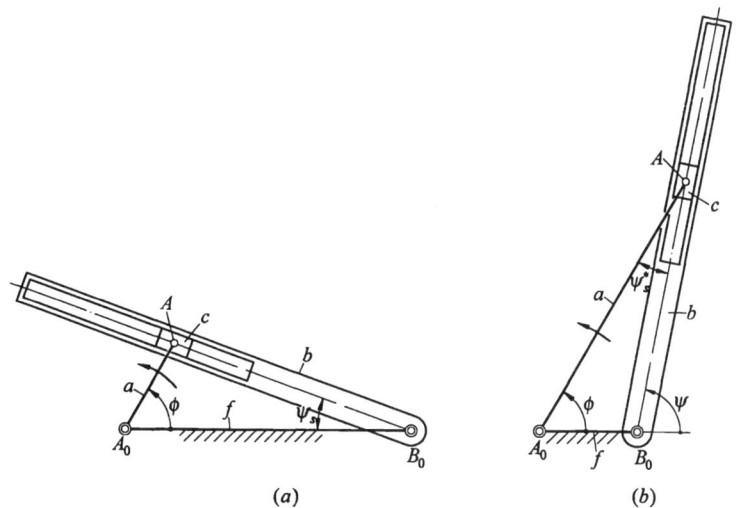

Fig. 8.7.   (a)(b) Two inversions of slider-crank.

and that in Fig. 8.5 are not the same. What we are going to derive, is the harmonic analysis of the kinetic energy E of such mechanisms. Now E consists of four parts. Let the moment of inertia of mass of the crank a with respect to A_0 be denoted by Θ_a, that of link b with respect to B_0 by Θ_b, that of the slider c with respect to its mass centre A by Θ_c, the mass of c by m_c, angular velocity of link a by ω_a, angular velocity ratio ω_b/ω_a by i_{ba}, and the linear velocity of A by v_A, then the kinetic energy of the mechanism is

$$E = \frac{1}{2}\Theta_a \omega_a^2 + \frac{1}{2}\Theta_b (i_{ba}\omega_a)^2 + \frac{1}{2}\Theta_c (i_{ba}\omega_a)^2 + \frac{1}{2}m_c v_A^2$$
$$= \frac{1}{2}\omega_a^2[\Theta_a + m_c a^2 + (\Theta_b + \Theta_c) i_{ba}^2] \qquad (8.70)$$

In equation (8.70) the only variable is i_{ba}^2. Hence i_{ba}^2 is the only term for which harmonic analysis is needed. As the expressions of i_{ba} are not the same in Figs. 8.7(a) and 8.7(b), they have to be treated separately.

(a) Swinging block linkage (Fig. 8.7(a))

The present i_{ba} is equal to $d\psi_s/d\phi$ in Fig. 8.2, or equation (8.17), or

$$i_{ba} = \frac{\lambda(\cos\phi - \lambda)}{1 - 2\lambda\cos\phi + \lambda^2}$$

Let

then
$$F(\phi) = 1 + 2i_{ba} = \frac{1-\lambda^2}{1-2\lambda\cos\phi+\lambda^2}$$

$$i_{ba}^2 = \frac{1-2F+F^2}{4} \tag{8.71}$$

The Fourier expansion of $F(\phi)$ itself can be obtained by means of equation (8.61), or

$$F = 1 + 2\sum_{n=1}^{\infty}\lambda^n\cos n\phi \tag{8.72}$$

The expression of F^2 is

$$F^2 = \left(\frac{1+\lambda^2}{1-\lambda^2}\right)F + 2\lambda\frac{(1+\lambda^2)\cos\phi-2\lambda}{(1-2\lambda\cos\phi+\lambda^2)^2} \tag{8.73}$$

The Fourier expansion of the second term on the right hand side of equation (8.73) can be obtaind by means of equation (8.32). Differentiating both sides of equation (8.32) with respect to ϕ yields

$$\lambda\frac{(1+\lambda^2)\cos\phi-2\lambda}{(1-2\lambda\cos\phi+\lambda^2)^2} = \sum_{m=1}^{\infty}\lambda^m m\cos m\phi \tag{8.74}$$

Combining equations (8.70), (8.71), (8.73) and (8.74) we get

$$\begin{aligned}
E &= a_0 + \sum_{n=1}^{\infty}a_n\cos n\phi \\
a_0 &= \frac{\omega_a^2}{2}\left[\Theta_a + m_c a^2 + \frac{\lambda^2}{2(1-\lambda^2)}(\Theta_b+\Theta_c)\right] \\
a_n &= \frac{(\Theta_b+\Theta_c)\omega_a^2}{4}\left(n+\frac{3\lambda^2-1}{1-\lambda^2}\right)\lambda^n
\end{aligned}$$

(b) Turning block linkage (Fig. 8.7(b))

It is now

$$\psi = \phi + \psi_s^*$$

Therefore

$$i_{ba}^* = 1 + \frac{d\psi_s^*}{d\phi} = \frac{1-\lambda^*\cos\phi}{1-2\lambda^*\cos\phi+\lambda^{*2}}$$

The symbol λ^* here is not the λ in equation (8.17), but $\lambda^* = f/a$, because $f < a$. Let

$$F^*(\phi) = 2i_{ba}^* - 1 = \frac{1-\lambda^{*2}}{1-2\lambda^*\cos\phi + \lambda^{*2}}$$

We have then

$$i_{ba}^{*2} = \frac{1+2F^* + F^{*2}}{4} \tag{8.75}$$

It can be seen that the form of F^* is the same as that of F in (a). All we have to do is to replace λ by λ^*, hence equation (8.72) can be used as a formula for the expansion of F^*. The expansion of F^{*2} can also be obtained by means of equations (8.73) and (8.74). Combining equations (8.70), (8.75) and the expansions of F^* and F^{*2}, we get

$$\left.\begin{aligned}E^* &= a_0^* + \sum_{n=1}^{\infty} a_n^* \cos n\phi \\ a_0^* &= \frac{\omega_a^2}{2}\left[\Theta_a + m_c a^2 + \frac{2-\lambda^{*2}}{2(1-\lambda^{*2})}(\Theta_b + \Theta_c)\right] \\ a_n^* &= \frac{(\Theta_b+\Theta_c)\omega_a^2}{4}\left(n + \frac{3-\lambda^{*2}}{1-\lambda^{*2}}\right)\lambda^{*n}\end{aligned}\right\}$$

APPENDIX 1
Homogeneous coordinates and circular points

A1.1 Homogeneous coordinates

Suppose the rectangular coordinates of a certain point on a plane are (x,y). Let Z be any real number, zero or non-zero. In addition, let

$$x = \frac{X}{Z}, \quad y = \frac{Y}{Z} \tag{A1.1}$$

and represent (x,y) by (X,Y,Z). For example, a point ($x = 1, y = 2$) can be represented by (1, 2, 1), or (2, 4, 2), or (6, 12, 6), or ..., etc.

In this way any polynomial containing (x,y) can always be transformed into a homogeneous polynomial containing (X,Y,Z). For example:

$$x^3y + 3x^2y - 2xy + 5y = 0$$

can be transformed into

$$X^3Y + 3X^2YZ - 2XYZ^2 + 5YZ^3 = 0$$

which is a homogeneous quartic equation in (X,Y,Z), and (X, Y, Z) are called the *homogeneous coordinates* of the point.

A1.2 Circular points

Every circle is a quadratic curve. Two circles should intersect in 2×2=4 points. If two circles are drawn, they can at most intersect in two real points. In other words, if the equations of the two circles are solved simultaneously, at most two sets of real roots can be found. What are the other two sets of imaginary roots? Assume two circles are given:

$$\left.\begin{array}{l} x^2 + y^2 = 2 \\ (x-2)^2 + y^2 = 2 \end{array}\right\}$$

To solve these simultaneous equations to find the intersections, transform them first into homogeneous equations

$$X + Y = 2Z^2$$
$$(X - 2Z)^2 + Y^2 = 2Z^2$$

The four sets of solutions are (1, 1, 1), (1, –1, 1), (1, i, 0), (1, –i, 0). The last two points are denoted by I, J, or I (1, i, 0), J (1, –i, 0), generally called *circular points*, being common to all circles.

Another explanation is as follows: Let the equation of a certain circle be

$$X^2 + Y^2 + 2gXZ + 2hYZ + eZ^2 = 0$$

where g, h, e are constants. This circle intersects the line at infinity in two points which are determined by the following two equations:

$$X^2 + Y^2 = 0$$
$$Z = 0$$

The two points are I (1, i, 0), J (1, –i, 0). This means that every circle intersects the line at infinity in the two points I, J.

Using homogeneus coordinates can avoid the ambiguity of writing I (∞, $i\infty$), J (∞, –$i\infty$).

An algebraic curve (please refer to Section A2.1) $F(x, y) = 0$ containing factors of $x^2 + y^2$ in its highest term is called a *circular curve*. A cubic, quartic, ... passing through I, J are called a circular cubic, circular quartic, ... respectively, etc. Of course there is no circular quadric (or a circular conic), because a circular quadric is a circle.

If a circular curve has a double point at I, it must also have a double point at J, and is called a bicircular curve. If a circular curve has a triple root at I, J respectively, it is called a tricircular curve, etc.

APPENDIX 2
On some topics regarding plane algebraic curves

The materials in the following sections, being taken substantially from (Primiose, 1955), (Walker, 1953), are basic concepts frequently used in investigating a plane algebraic curve. For simplicity purposes, no proof will be given to some formulae. Readers are referred to these two books or other books dealing with plane algebraic curves.

A2.1 Tangent of an algebraic curve

A curve is called a plane algebraic curve if it can be expressed by an equation

$$F(x, y) = 0 \qquad (A2.1)$$

where $F(x, y)$ is a polynomial of nth degree in the two variables x, y. In the following we shall assume that $F(x, y)$ is not factorizable, or that $F(x, y) = 0$ is non-degenerate. Let $P_1(x_1, y_1)$ be a point on this curve, or $F(x_1, y_1) = 0$. Differentiating equation (A2.1) gives

$$dF = \frac{\partial F}{\partial x} dx + \frac{\partial F}{\partial y} dy = 0$$

hence

$$\left(\frac{dy}{dx}\right)_1 = -\left(\frac{\partial F}{\partial x}\right)_1 \bigg/ \left(\frac{\partial F}{\partial y}\right)_1 \qquad (A2.2)$$

where $\left(\frac{dy}{dx}\right)_1$, $\left(\frac{\partial F}{\partial x}\right)_1$, $\left(\frac{\partial F}{\partial y}\right)_1$ represent the corresponding values of that differential coefficients at (x_1, y_1) respectively. The slope $(dy/dx)_1$ of the curve at the point $P_1(x_1, y_1)$ can then be found by equation (A2.2). The equation of the tangent of the curve at $P_1(x_1, y_1)$ is

$$(x - x_1)\left(\frac{\partial F}{\partial x}\right)_1 + (y - y_1)\left(\frac{\partial F}{\partial y}\right)_1 = 0 \qquad (A2.3)$$

A2.2 Double points

Assume as before that $P_1(x_1, y_1)$ is a point on the curve $F(x, y) = 0$, and that $F = 0$ is a polynomial of nth degree with respect to x, y. If (x, y) are replaced respectively by (x_1+x, y_1+y) in $F(x, y) = 0$, the latter becomes an equation representing the curve but with P_1 as the origin and two new axes parallel to the original axes. By Taylor theorem we have

$$F(x_1 + x, y_1 + y) = F(x_1, y_1) + \left[x\left(\frac{\partial F}{\partial x}\right)_1 + y\left(\frac{\partial F}{\partial x}\right)_1 \right]$$
$$+ \frac{1}{2!}\left[x^2\left(\frac{\partial^2 F}{\partial x^2}\right)_1 + 2xy\left(\frac{\partial^2 F}{\partial x \partial y}\right)_1 + y^2\left(\frac{\partial^2 F}{\partial y^2}\right)_1 \right] + \cdots +$$
$$+ \frac{1}{n!}\left[x^n\left(\frac{\partial^n F}{\partial x^n}\right)_1 + \cdots + y^n\left(\frac{\partial^n F}{\partial y^n}\right)_1 \right] \quad (A2.4)$$

In equation (A2.4), $F(x_1, y_1) = 0$. Please note that, as mentioned above, (x, y) in equation (A2.4) are the new coordinates of a point with respect to the axes passing through the point P_1. The intersections of a straight line $y = mx$ passing through P_1 and the curve are given by the roots of the equation obtained by substituting $y = mx$ into equation (A2.4)

$$x\left[\left(\frac{\partial F}{\partial x}\right)_1 + m\left(\frac{\partial F}{\partial y}\right)_1 \right]$$
$$+ \frac{x^2}{2!}\left[\left(\frac{\partial^2 F}{\partial x^2}\right)_1 + 2m\left(\frac{\partial^2 F}{\partial x \partial y}\right)_1 + m^2\left(\frac{\partial^2 F}{\partial y^2}\right)_1 \right]$$
$$+ \frac{x^3}{3!}\left[\left(\frac{\partial^3 F}{\partial x^3}\right)_1 + \cdots \right] + \cdots +$$
$$+ \frac{x^n}{n!}\left[\left(\frac{\partial^n F}{\partial x^n}\right)_1 + \cdots + \left(\frac{\partial^n P}{\partial y^n}\right)_1 \right] = 0 \quad (A2.5)$$

In case P_1 is a double-point, the line $y = mx$ always intersects the curve at least in two points whatever m may be. Therefore the coefficient of x in equation (A2.5) must vanish, or

$$\left.\begin{array}{l}\left(\dfrac{\partial F}{\partial x}\right)_1 = 0 \\[4pt] \left(\dfrac{\partial F}{\partial y}\right)_1 = 0\end{array}\right\} \quad (A2.6)$$

Equations (A2.6) are the necessary and sufficient conditions that P_1 becomes a double-point. It is clear that under such conditions, it is not possible to find the slope of the curve at the point P_1 from equation (A2.2). However, we see from equation

(A2.5) that a straight line $y = mx$, the m value of which satisfies the equation

$$\left(\frac{\partial^2 F}{\partial x^2}\right)_1 + 2m\left(\frac{\partial^2 F}{\partial x \partial y}\right)_1 + m^2\left(\frac{\partial^2 F}{\partial y^2}\right)_1 = 0 \tag{A2.7}$$

intersects the curve in three points, and such lines are the two tangents of the curve at P_1. In other words, the equation of the two tangents with respect to the original coordinate system is

$$(x-x_1)^2\left(\frac{\partial^2 F}{\partial x^2}\right)_1 + 2(x-x_1)(y-y_1)\left(\frac{\partial^2 F}{\partial x \partial y}\right)_1 + (y-y_1)^2\left(\frac{\partial^2 F}{\partial y^2}\right)_1 = 0 \tag{A2.8}$$

If the roots of equation (A2.7) are two different real numbers, P_1 is a *crunode*, as that shown in Fig. A2.1(*a*). If the root of equation (A2.7) is a double root, P_1 is a *cusp*, as that shown in Fig. A2.1(*b*). Both crunode and cusp are called *nodes*. If the roots of equation (A2.7) are two imaginary roots, then P_1 is an *isolated point*, as shown in Fig. A2.1(*c*).

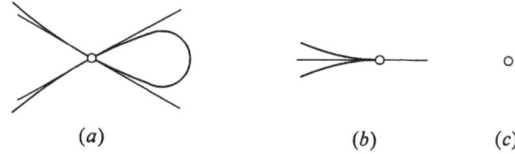

Fig. A2.1. (*a*) Crunode. (*b*) Cusp. (*c*) Isolated point.

Example: Find the double-point of $F(x, y) = x^4 - 2y^3 - 2x^2 + 3y^2 = 0$ and their nature.

Solution: $\partial F/\partial x = 4x^2(x^2-1)$, $\partial F/\partial x = 0$ occurs at $x = 0, 1, -1$.
$\partial F/\partial y = -6y(y-1)$, $\partial F/\partial y = 0$ occurs at $y = 0, 1$.

(0, 0), (1, 1), (−1, 1) are three double-points on the curve $F = 0$. [(0,1), (1,0), (−1,0) are not points on the curve].

$\partial^2 F/\partial x^2 = 12x^2 - 4$, $\partial^2 F/(\partial x \partial y) = 0$, $\partial^2 F/\partial y^2 = -12y + 6$

The equation of the two tangents at (0, 0) is $-4x^2 + 6y^2 = 0$, or the two lines $\sqrt{2}x \pm \sqrt{3}y = 0$. The equation of the two tangents at (1,1) is $2(x-1) \pm \sqrt{3}(y-1) = 0$, and that of the two tangents at (−1, 1) is $2(x+1) \pm \sqrt{3}(y-1) = 0$. All three points are crunodes.

Example: Find the double-points of $F(x, y) = xy^2 + x^2 - 3x + 2y + 3 = 0$ and their nature.

Solution: $\left.\begin{array}{l}\partial F/\partial x = y^2 + 2x - 3\\ \partial F/\partial y = 2xy + 2\end{array}\right\}$ $\partial F/\partial x = \partial F/\partial y = 0$ occurs at $(1, -1)$ and $(-1/2, 2)$. $(1, -1)$ is a point on $F = 0$. [$(-1/2, 2)$ is not a point on $F = 0$.]

$$\partial^2 F/\partial x^2 = 2, \quad \partial^2 F/(\partial x \partial y) = 2y, \quad \partial^2 F/\partial y^2 = 2x$$

Hence the equation of the two tangents at $(1, -1)$ is

$$2(x-1)^2 - 4(x-1)(y+1) + 2(y+1)^2 = 0$$

or

$$[(x-1) - (y+1)]^2 = 0$$

Therefore the two tangents of $F = 0$ at $(1, -1)$ coincide, and $(1, -1)$ is a cusp.

For a curve passing through the origin, it is to be noticed that:

(a) the *order* of the multiple point of the curve $F(x, y) = 0$ at the origin $(0, 0)$ is equal to the degree of the lowest term of $F(x, y)$. For example, in the first example in the present section, there is a double-point at the origin;

(b) the equation of the tangents at the origin is obtained by setting the lowest term equal to zero. Thus in the first example in the present section, the equation of the tangents at the origin is $2x^2 - 3y^2 = 0$, or $\sqrt{2}x \pm \sqrt{3}y = 0$, while in the second example in the present section, we transform $F(x, y)$ into an equation with $(1, -1)$ as the new origin by replacing (x, y) by $(x+1, y-1)$ and get $xy^2 + (x-y)^2 = 0$. From $(x-y)^2 = 0$ we know that this cubic has a cusp at the new origin $(1, -1)$.

A2.3 Asymptotes

The asymptote of a plane curve is the tangent of the curve at infinity. Let the equation of a plane curve of nth degree be

$$F(x, y) = F_n(x, y) + F_{n-1}(x, y) + \cdots + a_n = 0 \tag{A2.9}$$

where $F_r(x, y)$ is the term of homogeneous r^{th} degree with respect to x, y.

(a) In general, the direction of the asymptote of the curve can be determined by

$$F_n(x, y) = 0$$

In other words, setting the highest term of $F(x, y)$ equal to zero will give the direction of the asymptote. (Please note that this is not the equation of the asymptote.) However, there are exceptions. Please refer to the later Section A2.4 (c).

(b) Let us find the intersections of a straight line

$$y = mx + c$$

and the curve $F(x, y) = 0$. m and c are two constants of the straight line, and m is the slope of the line. After substituting $y = mx + c$ into equation (A2.9) and rearranging, suppose the latter becomes

$$f(x) = a_0 x^n + a_1 x^{n-1} + \cdots + a_n = 0 \tag{A2.10}$$

$f(x) = 0$ is an equation containing n roots, which correspond to the n intersections. Let these be denoted by (x_1, y_1), (x_2, y_2), $\cdots (x_n, y_n)$. Now let m, c vary, so that (x_1, y_1) and (x_2, y_2) approach each other indefinitely to become one point, then $x_1(=x_2)$ should be a double root of $f(x) = 0$. In other words, x_1 is a root of

$$f'(x) = n a_0 x^{n-1} + (n-1) a_1 x^{n-2} + \cdots + a_{n-1} = 0$$

If $y = mx + c$ is an asymptote of $F(x, y) = 0$, it should be tangent to the curve at infinity. Hence, in order that $x_1 (\to \infty)$ becomes a double root of $f(x) = 0$, we should have

$$\left.\begin{array}{c} a_0 = 0 \\ a_1 = 0 \end{array}\right\} \tag{A2.11}$$

The two equations in (A2.11) serve to determine the values of m, c, or the eqation of the asymptote.

Example: Find the equation of the asymptote of the curve $(x^2 - y^2)(2x - y) - xy = 0$.

Solution: Substituting $y = mx + c$ into the above equation, we get the coefficients of x^3, x^2

$$a_0 = m^3 - 2m^2 - m + 2$$
$$a_1 = (3m^2 - 4m - 1)c - m$$

Setting a_0, a_1 equal to zero and solving these equations for (m, c), we get $(1, -1/2)$; $(-1, -1/6)$; $(2, 2/3)$. The three asymptotes of this curve are: $2x - 2y - 1 = 0$, $6x + 6y + 1 = 0$, $6x - 3y + 2 = 0$.

If a curve has r asymptotes in the m direction, or if equation (A2.10) has an r-ple root at infinity, then of course the first r roots of equation (A2.10) are zero. To find the value of c, the $(r+1)^{th}$ coefficient should be set to zero, or $a_r = 0$, to get it.

This method may not work in special cases. For a more comprehensive method, please refer to Section A2.4(c).

A2.4 Formulae of a curve in homogeneous coordinates

It is often expedient to use homogeneous coodinates in curve equations. Assume in the following that

$$F(x, y, z) = 0 \tag{A2.12}$$

is a given curve. For simplicity reasons, the symbol (x, y, z) is used directly to represent the homogeneous coordinates, with the understanding that $F(x, y, 1) = 0$ is the original curve equation $F(x, y) = 0$.

(a) Equation of the tangent

Suppose $P_1(x_1, y_1, z_1)$ is a point on the curve $F(x, y, z) = 0$ of nth degree. P_2 (x_2, y_2, z_2) is another point on the plane. A point on the line P_1P_2 can be represented by $P(x_1 + \lambda x_2, y_1 + \lambda y_2, z_1 + \lambda z_2)$. The intersection of P_1P_2 with $F(x, y, z) = 0$ is determined by

$$F(x_1 + \lambda x_2, y_1 + \lambda y_2, z_1 + \lambda z_2) = 0$$

Taking the Taylor expansion of this equation gives

$$F(x_1,y_1,z_1) + \lambda \left[x_2 \left(\frac{\partial F}{\partial x}\right)_1 + y_2 \left(\frac{\partial F}{\partial y}\right)_1 + z_2 \left(\frac{\partial F}{\partial z}\right)_1 \right]$$
$$+ \frac{\lambda^2}{2!} \left[x_2^2 \left(\frac{\partial^2 F}{\partial x^2}\right)_1 + y_2^2 \left(\frac{\partial^2 F}{\partial y^2}\right)_1 + z_2^2 \left(\frac{\partial^2 F}{\partial z^2}\right)_1 + \right.$$
$$\left. + 2y_2 z_2 \left(\frac{\partial^2 F}{\partial y \partial z}\right)_1 + 2z_2 x_2 \left(\frac{\partial^2 F}{\partial z \partial x}\right)_1 + 2x_2 y_2 \left(\frac{\partial^2 F}{\partial x \partial y}\right)_1 \right]$$
$$+ \cdots + \frac{\lambda^n}{n!} \left[x_2^n \left(\frac{\partial^n F}{\partial x^n}\right)_1 + \cdots \right] = 0 \quad \text{(A2.13)}$$

According to the assumption, $F(x_1, y_1, z_1) = 0$. The n roots of λ values determined by equation (A2.13) determine the n intersections of P_1P_2 with $F = 0$, including $\lambda = 0$, or the point P_1. Assume $(\partial F/\partial x)_1$, $(\partial F/\partial y)_1$, $(\partial F/\partial z)_1$ do not vanish identically, then equation (A2.13) has a double root $\lambda = 0$, 0 only when the point P_2 lies on the straight line

$$x \left(\frac{\partial F}{\partial x}\right)_1 + y \left(\frac{\partial F}{\partial y}\right)_1 + z \left(\frac{\partial F}{\partial z}\right)_1 = 0 \quad \text{(A2.14)}$$

Therefore equation (A2.14) is the equation of the tangent of $F = 0$ at P_1.

A *rational curve* is a curve such that x and y can be expressed respectively as rational functions of a parameter t. Suppose the parametric representation of a curve is $x(t), y(t), z(t)$, then the equation of the tangent of this curve at t is

$$\begin{vmatrix} x & y & z \\ x(t) & y(t) & z(t) \\ x'(t) & y'(t) & z'(t) \end{vmatrix} = 0 \quad \text{(A2.15)}$$

where the symbol (') indicates differentiation with respect to t. To prove that equation (A2.15) is the equation of the tangent, assume two points t_1, t_2 on the curve. The equation of the line $t_1 t_2$ is

APPENDIX 2

$$\begin{vmatrix} X & Y & Z \\ x(t_1) & y(t_1) & z(t_1) \\ x(t_2) & y(t_2) & z(t_2) \end{vmatrix} = 0 \qquad (A2.16)$$

where (X, Y, Z) represents a point on the line $t_1 t_2$, or the variables. Rewrite equation (A2.16) as

$$\begin{vmatrix} X & Y & Z \\ x(t_1) & y(t_1) & z(t_1) \\ \dfrac{x(t_2)-x(t_1)}{t_2-t_1} & \dfrac{y(t_2)-y(t_1)}{t_2-t_1} & \dfrac{z(t_2)-z(t_1)}{t_2-t_1} \end{vmatrix} = 0 \qquad (A2.17)$$

As t_2 and t_1 approach each other indefinitely, equation (A2.17) becomes equation (A2.15).

(b) Double-point

If
$$\left(\frac{\partial F}{\partial x}\right)_1 = \left(\frac{\partial F}{\partial y}\right)_1 = \left(\frac{\partial F}{\partial z}\right)_1 = 0 \qquad (A2.18)$$

at P_1, then every line passing through P_1 intersects $F = 0$ in two points. Therefore equations (A2.18) are the conditions that P_1 becomes a double-point.

A *non-degenerate* curve of nth degree can have at most $(n-1)(n-2)/2$ double points. An r-ple point has to be counted as $r(r-1)/2$ double points. Hence a non-degenerate cubic can have one or no double point, and a non-degenerate quartic can have at most three double points.

The number of double points of a curve of nth degree is of course less than the maximum number $(n-1)(n-2)/2$. The number of shortage is called *deficiency* or *genus*. (Please note that deficiency refers only to curves with nodes, cusps and ordinary multiple points.)

If the second partial differential coefficients $(\partial^2 F/\partial x^2)_1$, $(\partial^2 F/(\partial y \partial z))_1,\ldots$ at the double point P_1 do not vanish identically, and if a point (x, y, z) on a straight line satisfies

$$x^2 \left(\frac{\partial^2 F}{\partial x^2}\right)_1 + y^2 \left(\frac{\partial^2 F}{\partial y^2}\right)_1 + z^2 \left(\frac{\partial^2 F}{\partial z^2}\right)_1$$
$$+ 2yz \left(\frac{\partial^2 F}{\partial y \partial z}\right)_1 + 2zx \left(\frac{\partial^2 F}{\partial z \partial x}\right)_1 + 2xy \left(\frac{\partial^2 F}{\partial x \partial y}\right)_1 = 0 \qquad (A2.19)$$

from equation (A2.13) we know that this line intersects $F(x, y, z) = 0$ in three points. Hence equation (A2.19) is the equation of the tangent at P_1. (Please note that this quadratic equation is not a conic, but two straight lines, i.e. the two tangents at P_1.)

Example: Find the double points and their nature of the curve $F(x,y,z) = x^2 y^2 + 4x^2 z^2 - 9y^2 z^2 = 0$.

Solution:

$$\frac{\partial F}{\partial x} = 2xy^2 + 8xz^2 = 2x(y^2 + 4z^2), \frac{\partial F}{\partial x} = 0 \text{ occurs at } x = 0, y = \pm 2iz.$$

$$\frac{\partial F}{\partial y} = 2x^2y - 18yz^2 = 2y(x^2 - 9z^2), \frac{\partial F}{\partial x} = 0 \text{ occurs at } y = 0, x = \pm 3z.$$

$$\frac{\partial F}{\partial z} = 8x^2z - 18y^2z = 2z(4x^2 - 9y^2), \frac{\partial F}{\partial z} = 0 \text{ occurs at } z = 0, 2x = \pm 3y.$$

Hence there are three double points: (0, 0, 1); (0, 1, 0); (1, 0, 0).

$$\frac{\partial^2 F}{\partial x^2} = 2(y^2 + 4z^2), \frac{\partial^2 F}{\partial y^2} = 2(x^2 - 9z^2), \frac{\partial^2 F}{\partial z^2} = 2(4x^2 - 9y^2),$$

$$\frac{\partial^2 F}{\partial y \partial z} = -36yz, \frac{\partial^2 F}{\partial z \partial x} = 16zx, \frac{\partial^2 F}{\partial x \partial y} = 4xy$$

Hence the equation of the tangent at (0, 0, 1) is $(2x + 3y)(2x - 3y) = 0$, this point being a node. The equation of the tangent at (0, 1, 0) is $(x + 3z)(x - 3z) = 0$, this point being a node at infinity. The equation of the tangent at (1, 0, 0) is $(y + 2iz)(y - 2iz) = 0$, the point being an isolated point at infinity.

(c) Asymptotes

Let the equation of a plane curve of nth degree be

$$F(x,y,z) = F_n(x,y) + zF_{n-1}(x,y) + \cdots + a_n z^n = 0 \tag{A2.20}$$

Equation (A2.20) is comparable with equation (A2.9). The point at infinity of $F = 0$ is determined by

$$F_n(x,y) = 0, z = 0$$

This is why we asserted in Section A2.3(a) that the direction of the asymptote can be obtained by setting the highest term of $F(x, y)$ equal to zero.

Now we can use equation (A2.14) or (A2.19), the equation of the tangent, to find the asymptote of a homogeneous equation. There are three cases:

(i) $F_n(x, y)$ contains a single factor $(ax + by)$, then there is an asymptote of $F(x, y, z) = 0$ in the direction $ax + by = 0$.

Example: Consider again the example in Section A2.3(b). The homogeneous equation of the curve is $F(x, y, z) = (x^2 - y^2)(2x - y) - xyz = 0$. The three points at infinity of this curve are (1, 1, 0), (1, −1, 0), (1, 2, 0).

$$\frac{\partial F}{\partial x} = 6x^2 - 2xy - 2y^2 - yz$$

$$\frac{\partial F}{\partial y} = -x^2 - 4xy + 3y^2 - xz$$

$$\frac{\partial F}{\partial z} = -xy$$

Substituting these expressions into equation (A2.14) respectively, we get equations of the three asymptotes $2x-2y-z=0$, $6x+6y+z=0$, $6x-3y+2z=0$, which are identical with the results obtained before.

(ii) $F_n(x, y)$ contains repeated factor $ax+by$, and $F_{n-1}(x, y)$ also contains a factor $ax+by$, then $F(x, y; z) = 0$ has a pair of parallel asymptotes in the direction $ax+by=0$.

Example: Find the asymptotes of $F(x, y, z) = y(x+y)^2 - y(x+y)z + z^3 = 0$.

Solution: The points at infinity of this curve are $(1, 0, 0)$; $(1, -1, 0)$; $(1, -1, 0)$. The tangent at $(1, 0, 0)$ can still be found by equation (A2.14), or $y = 0$. This is the asymptote at $(1, 0, 0)$. At the point $(1, -1, 0)$, since this is a double point, hence $(\partial F / \partial x)_1 = (\partial F / \partial y)_1 = (\partial F / \partial z)_1 = 0$, as expected by equation (A2.18). Equation (A2.19) can now be applied to find the tangent at $(1, -1, 0)$:

$$\frac{\partial^2 F}{\partial x^2} = 2y, \quad \left(\frac{\partial^2 F}{\partial x^2}\right)_1 = -2$$

$$\frac{\partial^2 F}{\partial y^2} = 4x + 6y - 2z, \quad \left(\frac{\partial^2 F}{\partial y^2}\right)_1 = -2$$

$$\frac{\partial^2 F}{\partial z^2} = 6z, \quad \left(\frac{\partial^2 F}{\partial z^2}\right)_1 = 0$$

$$2\frac{\partial^2 F}{\partial y \partial z} = -2x - 4y, \quad 2\left(\frac{\partial^2 F}{\partial y \partial z}\right)_1 = 2$$

$$2\frac{\partial^2 F}{\partial z \partial x} = -2y, \quad 2\left(\frac{\partial^2 F}{\partial z \partial x}\right)_1 = 2$$

$$2\frac{\partial F}{\partial x \partial y} = 4x + 8y - 2z, \quad 2\left(\frac{\partial^2 F}{\partial x \partial y}\right)_1 = -4$$

Hence equation (A2.19) becomes now $(x + y)(x + y - 1) = 0$. The asymptotes at $(1, -1, 0)$, $(1, -1, 0)$ become two parallel lines $x + y = 0$ and $x + y - z = 0$.

Example: Find the asymptotes of $F(x, y, z) = 2x^2y^2 + xyz^2 + yz^3 + 3xz^3 - 3z^4 = 0$.

Solution: The points at infinity on this curve are: a double root at each of the points $(0, 1, 0)$ and $(1, 0, 0)$. We find as before the equation of the tangent at $(0, 1, 0)$ to be $x^2 = 0$ which means this is a cusp. Similarly, the equation of the tangent at $(1, 0, 0)$ is $y^2 = 0$, which also means this is a cusp.

(iii) $F_n(x, y)$ contains repeated factor $ax + by$, but $F_{n-1}(x, y)$ does not contain the factor $ax + by$. Then $F(x, y, z) = 0$ has no asymptote in the direction $ax + by = 0$.

Example: Find the asymptotes of $F(x, y, z) = (x-y)(x+y)^2 - (3xy+y^2)z = 0$.

Solution: The points at infinity of this curve are $(1, 1, 0)$; $(1, -1, 0)$; $(1, -1, 0)$. According to equation (A2.14) the tangent at $(1, 1, 0)$ is $x - y - z = 0$; this being the asymptote at this point. $(1, -1, 0)$ is a double point. The tangent at $(1, -1, 0)$ is $z = 0$, hence the tangent at this point is the line at infinity, or the line at infinity is the asymptote at $(1, -1, 0)$. In other words, there is no asymptote in the direction $x + y = 0$.

A2.5 Foci and singular foci

The *class* of an algebraic curve is the number of tangents that can be drawn from a point to the curve outside of the curve (please refer to Section A2.7). For instance if the class of a certain curve is m, then m tangents can be drawn from a point to the curve outside of the curve. For a curve of class m, m tangents can be drawn from the circular point I to the curve, and also m tangents can be drawn from J. (Assume for the time being that the curve does not pass through I, J.) These two sets of m tangents intersect altogether in m^2 points. These intersections are called *foci* of the curve. For instance $x+iy = a+ib$ is a tangent passing through the point I, then $x-iy = a-ib$ must be a tangent passing through the point J. The intersection (a,b) of these two tangents is a real focus. Hence this curve possesses altogether m real and m^2-m imaginary foci.

Consider the straight line $x + iy = a + ib$. The slope of this straight line is i, and the negative reciprocal of i is also i. Hence this line is perpendicular to itself. Such a line is called an *isotropic line*. Similarly, $x - iy = a - ib$ is also an isotropic line.

If the curve passes through I, J, the tangents at I, J are its asymptotes. To distinguish the intersections of such asymptotes at I, J from the ordinary foci mentioned above, these are called *singular foci*. For a curve with a k-ple point at each of I, J, there are k^2 singular foci, among which k foci are real.

By the above definition, the two real foci of an ellipse are the two commonly known foci of the ellipse. A paradox may arise: How could the intersections of the tangents of an ellipse appear inside the ellipse? We applied concepts from real geometry to imaginary geometry, but the results thus derived may not be interpreted by real geometry. This is also the case with isotropic lines mentioned before. As a rule we know that the slopes of two mutually perpendicular lines are negative reciprocals, but the isotropic lines cannot be interpreted by real geometry.

A2.6 Line coordinates

Just like in point coordinates, where a point can be represented by the coordinates (x, y), a line can be represented by the coordinates $[l, m]$[§] in Cartesian coordinates. The equation of this line is written as $lx + my + 1 = 0$. Thus, just like a curve, or the locus of a set of points (x, y) that can be represented by $f(x, y) = 0$, similarly, an envelope, or the *locus* of a set of straight lines $[l, m]$ can be represented by $\phi(l, m) = 0$.

Consider for example $\phi(l,m) = l - m^2 = 0$. In this equation $\phi(l, m) = 0$, two m's can be found for one assumed value of l. All lines in this set are tangent to a parabola $y^2 = 4x$, as shown in Fig. A2.2.

If the function is $\phi(l, m) = al + bm + c = 0 (c \neq 0)$, this equation then represents a set of straight lines passing through $(a/c, b/c)$, hence can also be called an envelope

---

[§] Please do not confuse these symbols with the symbols l, m used in Sections 3.7.1 and 3.7.2.

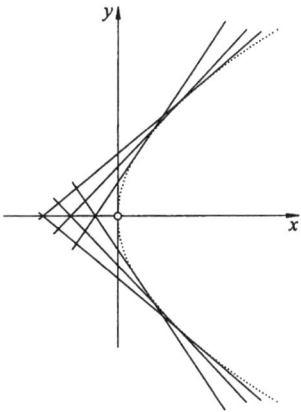

Fig. A2.2. Curve of line equation $l - m^2 = 0$.

equation of the point $(a/c, b/c)$, although this set of enveloping lines becomes a pencil of lines.

The straight line equation $lx + my + 1 = 0$ cannot be used to represent lines passing through the origin. Hence, by analogy with homogeneous coordinates for point coordinates (please refer to Section A2.4), line coordinates can also be expressed in homogeneous coordinates $[l, m, n]$, and the line is $lx + my + n = 0$.

If homogeneous coordinates are used both for point coordinates and for line coordinates, then the (envelope) equation of a point (x_1, y_1, z_1) is $lx_1+my_1+nz_1 = 0$, and the equation of a line $[l_1, m_1, n_1]$ is $l_1x+m_1y+n_1z = 0$. Hence the equation of the origin is $n = 0$, and the equation of the point at infinity (or $z_1 = 0$) is $lx_1+my_1 = 0$.

Let us see how can we transform the point coordinate equation of a curve to a line coordinate equation. For a rational curve (i.e. x, y can be expressed respectively as rational functions of a parameter t), suppose the homogeneous coordinates x, y, z of a point are expressed as functions of a parameter t. According to equation (A2.15) the equation of the tangent at the point t is

$$\begin{vmatrix} x & y & z \\ x(t) & y(t) & z(t) \\ x'(t) & y'(t) & z'(t) \end{vmatrix} = 0 \qquad [(A2.15)]$$

Expanding this equation, we get the line coordinates of the tangent:

$$\lambda l = y(t)z'(t) - y'(t)z(t)$$
$$\lambda m = z(t)x'(t) - z'(t)x(t)$$
$$\lambda n = x(t)y'(t) - x'(t)y(t)$$

where λ is a constant factor. Eliminating λ, t will result in the equation of the envelope of the original curve, or the line coordinate equation.

Example: As mentioned before, the point coordinate equation of a parabola is $y^2 - 4x = 0$, or in homogeneous coordinates $y^2 - 4xz = 0$. In a parametric equation this is $x : y : z = t^2 : 2t : 1$. Hence the equation of the tangent is

$$\begin{vmatrix} x & y & z \\ t^2 & 2t & 1 \\ 2t & 2 & 0 \end{vmatrix} = 0$$

or

$\lambda l = -2$
$\lambda m = 2t$
$\lambda n = -2t^2$

or $l/1 = m/(-t) = n/t^2$, or $l/n = 1/t^2$, $m/n = -1/t$. In homogeneous coordinates (l, m, n), this is $ln - m^2 = 0$, but in non-homogeneous coordinates (l,m), this becomes $l - m^2 = 0$.

conversely, to find the point coordinate equation from the given line coordinate equation $l - m^2 = 0$, transform it first into homogeneous form $ln - m^2 = 0$. In a parametric equation in terms of the parameter τ this is $l = \tau^2$, $m = \tau$, $n = 1$. The *dual* of equation (A2.15), or the point coordinate equation is

$$\begin{vmatrix} l & m & n \\ \tau^2 & \tau & 1 \\ 2\tau & 1 & 0 \end{vmatrix} = 0$$

From this we get $x/(-1) = y/(2\tau) = z/(-\tau^2)$, hence $x/z = 1/\tau^2$, $y/z = -2/\tau$. The homogeneous point coordinate equation is $y^2 - 4xz = 0$, and the non-homogeneous point coordinate equation is $y^2 - 4x = 0$.

A straight line can be denoted by its normal distance d from the origin and the

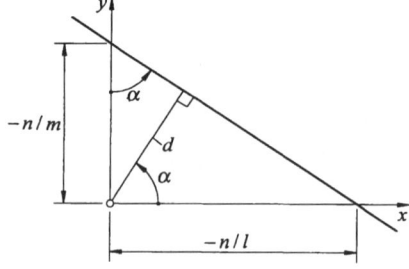

Fig. A2.3. Coordinates (d, α) of a line.

APPENDIX 2

inclination α of this normal as shown in Fig. A2.3. For a line with homogeneous equation $lx + my + nz = 0$, we have $\tan \alpha = m/l$. Let $L = \sqrt{l^2 + m^2}$, then $\sin \alpha = m/L$, $\cos \alpha = l/L$, $d = -n/L$.

A2.7 Duality

The statements "On a plane, a point is determined by two straight lines" and "On a plane, a straight line is determined by two points" are mutually *dual*. The words "straight lines" in the former statement is the *dual* of "points" in the latter statement, while "point" in the former is the *dual* of "line" in the latter. This is called *duality*. As is well-known, the Pappus theorem in projective geometry says that there are three points lying on one line. The *dual* of Pappus theorem says that there are three lines intersecting in one point.

The *dual* of the statement "An algebraic locus is a curve with the polynomial $f(x, y) = 0$ or homogeneous polynomial $f(x, y, z) = 0$ as its point coordinate equation. If the degree of the polynomial is r, the locus is said to have an order r." is "An algebraic envelope is a curve with the polynomial $\phi(l, m) = 0$ or homogeneous polynomial $\phi(l, m, n) = 0$ as its line coordinate equation. If the degree of the polynomial is k, the envelope is said to have a class k."

According to this, the problem of "three collinear points" discussed in Section 3.3.2 and the problem of "three concurrent lines" discussed in Section 3.3.3 may be considered as mutually *dual*.

APPENDIX 3
Equation of q_1 - curve (3.76) in rectangular coordinates

This is an equation of the locus of $A_1(x_1, y_1)$. The point P_1 is taken as the origin, poletangent P_1T_1 as the x-axis, and polenormal P_1N_1 as the y-axis. The complete equation (3.76) is:

$$q_1: U_1(x_1^2 + y_1^2)^4 + U_2(x_1^2 + y_1^2)^3 + [V_1(x_1^2 + y_1^2)^3 + U_3(x_1^2 + y_1^2)^2]$$
$$+ [U_0(x_1^2 + y_1^2)^3 + V_2(x_1^2 + y_1^2)^2 + U_4(x_1^2 + y_1^2)]$$
$$+ W_1(x_1^2 + y_1^2)^2 + V_3(x_1^2 + y_1^2) + U_5 + s^6(x_1^2 + y_1^2)^2$$
$$+ W_2(x_1^2 + y_1^2) + V_4 + W_3 + s^6\delta_1^2 y_1^2 = 0 \qquad (A3.1)$$

In equation (A3.1), the subscripts of the symbols $U_1, \ldots V_3, \ldots W_2,\ldots$ indicate the respective degrees of that factor. For instance, V_3 is a factor of 3rd degree in x_1, y_1.

$$U_1 = 2(-sU + \delta_2 V - \delta_1 y_1)$$
$$U_2 = [s^2 + 2\delta_2 s \sin(v - \sigma)](x_1^2 + y_1^2) + 4sU(3\delta_1 y_1 - \delta_2 V + 2sU)$$
$$+ (\delta_1^2 y_1^2 - \delta_2^2 V^2)$$
$$V_1 = 2[s^2(-3\delta_2 y_1 + \delta_2 V - 4sU) - s\delta_2 \sin(v - \sigma)(\delta_2 V + 2sU)]$$
$$U_3 = -2sU[4sU(sU + 3\delta_1 y_1) + 3\delta_1^2 y_1^2]$$
$$U_0 = 2s^4 + \delta_2 s \sin(v - \sigma)[2s^2 - s\delta_2 \sin(v - \sigma)]$$
$$V_2 = 3s^2[4sU(sU + 2\delta_1 y_1) + \delta_1^2 y_1^2]$$
$$U_4 = 4\delta_1 s^2 U^2 y_1(4sU + 3\delta_1 y_1)$$
$$W_1 = -6s^4(sU + \delta_1 y_1)$$
$$V_3 = -12s^3 \delta_1 U y_1(2sU + \delta_1 y_1)$$
$$U_5 = -8\delta_1^2 s^3 U^3 y_1^2$$
$$W_2 = 3s^4 \delta_1 y_1(4sU + \delta_1 y_1)$$
$$V_4 = 12s^4 \delta_1^2 U^2 y_1^2$$
$$W_3 = -2\delta_1 s^4 y_1[s^2(x_1^2 + y_1^2) + 3s\delta_1 U y_1]$$
$$U = x_1 \cos\sigma + y_1 \sin\sigma$$
$$V = y_1 \cos v - x_1 \sin v$$

APPENDIX 4
Expressions of the 9 terms in equation (3.96)

Having adopted the coordinate system in Fig. 3.27, we can write the terms of G_j, S_j, $T_j (j = 2, 3, 4)$ in equation (3.96) in the following forms, where each of the coefficients G_{3z}, S_{3z} and T_{3x}, T_{3y}, T_{3z} in G_3, S_3, T_3 are given in two forms, the second form being in terms of $s = \overline{P_{12} P_{23}}$ as shown in Fig. 3.27.

$$G_2 = G_{2x} x_{A1} + G_{2y} y_{A1} + G_{2z}$$

where

$$G_{2x} = -\text{vers}\,\gamma_{12}$$
$$G_{2y} = -\sin\gamma_{12}$$
$$G_{2z} = 0$$
$$S_2 = S_{2x} x_{A1} + S_{2y} y_{A1} + S_{2z}$$
$$S_{2x} = \sin\gamma_{12}$$
$$S_{2y} = -\text{vers}\,\gamma_{12}$$
$$S_{2z} = 0$$
$$T_2 = T_{2x} x_{A_1} + T_{2y} y_{A1} + T_{2z}$$
$$T_{2x} = 0$$
$$T_{2y} = 0$$
$$T_{2z} = 0$$
$$G_3 = G_{3x} x_{A1} + G_{3y} y_{A1} + G_{3z}$$
$$G_{3x} = -\text{vers}\,\gamma_{13}$$
$$G_{3y} = -\sin\gamma_{13}$$
$$G_{3z} = x_{P13}\text{vers}\,\gamma_{13} + y_{P13}\sin\gamma_{13} = s\,\text{vers}\,\gamma_{23}$$
$$S_3 = S_{3x} x_{A1} + S_{3y} y_{A1} + S_{3z}$$
$$S_{3x} = \sin\gamma_{13}$$
$$S_{3y} = -\text{vers}\,\gamma_{13}$$
$$S_{3z} = -x_{P13}\sin\gamma_{13} + y_{P13}\text{vers}\,\gamma_{13} = -s\sin\gamma_{23}$$

$$T_3 = T_{3x} x_{A1} + T_{3y} y_{A1} + T_{3z}$$
$$T_{3x} = x_{P13} \text{vers}\gamma_{13} - y_{P13} \sin\gamma_{13} = s(-\cos\gamma_{13} + \cos\gamma_{12})$$
$$T_{3y} = x_{P13} \sin\gamma_{13} + y_{P13} \text{vers}\gamma_{13} = s(\sin\gamma_{13} - \sin\gamma_{12})$$
$$T_{3z} = -(x_{P13}^2 + y_{P13}^2)\text{vers}\gamma_{13} = -s^2 \text{vers}\gamma_{23}$$
$$G_4 = G_{4x} x_{A1} + G_{4y} y_{A1} + G_{4z}$$
$$G_{4x} = -\text{vers}\gamma_{14}$$
$$G_{4y} = -\sin\gamma_{14}$$
$$G_{4z} = x_{P14} \text{vers}\gamma_{14} + y_{P14} \sin\gamma_{14}$$
$$S_4 = S_{4x} x_{A1} + S_{4y} y_{A1} + S_{4z}$$
$$S_{4x} = \sin\gamma_{14}$$
$$S_{4y} = -\text{vers}\gamma_{14}$$
$$S_{4z} = -x_{P14} \sin\gamma_{14} + y_{P14} \text{vers}\gamma_{14}$$
$$T_4 = T_{4x} x_{A1} + T_{4y} y_{A1} + T_{4z}$$
$$T_{4x} = x_{P14} \text{vers}\gamma_{14} - y_{P14} \sin\gamma_{14}$$
$$T_{4y} = x_{P14} \sin\gamma_{14} + y_{P14} \text{vers}\gamma_{14}$$
$$T_{4z} = -(x_{P14}^2 + y_{P14}^2)\text{vers}\gamma_{14}$$

APPENDIX 5
Coefficients in equations (3.97), (3.99)

The coefficients in equation (3.97), the expanded form of equation (3.96), are listed as follows. For brevity reasons, each 3×3 determinant will be represented only by one row, and j stands for 2, 3, 4. For example

$$d_1 = \begin{vmatrix} G_{2y} & S_{2y} & T_{2y} \\ G_{3y} & S_{3y} & T_{3y} \\ G_{4y} & S_{4y} & T_{4y} \end{vmatrix} = \begin{vmatrix} G_{jy} & S_{jy} & T_{jy} \end{vmatrix} \qquad (j = 2, 3, 4)$$

$$d_2 = \begin{vmatrix} G_{jx} & S_{jy} & T_{jy} \end{vmatrix} + \begin{vmatrix} G_{jy} & S_{jx} & T_{jy} \end{vmatrix} + \begin{vmatrix} G_{jy} & S_{jy} & T_{jx} \end{vmatrix}$$

$$d_3 = \begin{vmatrix} G_{jx} & S_{jx} & T_{jy} \end{vmatrix} + \begin{vmatrix} G_{jx} & S_{jy} & T_{jx} \end{vmatrix} + \begin{vmatrix} G_{jy} & S_{jx} & T_{jx} \end{vmatrix} = d_1$$

$$d_4 = \begin{vmatrix} G_{jx} & S_{jx} & T_{jx} \end{vmatrix} = d_2$$

$$d_5 = \begin{vmatrix} G_{jy} & S_{jz} & T_{jz} \end{vmatrix} + \begin{vmatrix} G_{jz} & S_{jy} & T_{jz} \end{vmatrix} + \begin{vmatrix} G_{jz} & S_{jz} & T_{jy} \end{vmatrix}$$

$$d_6 = \begin{vmatrix} G_{jx} & S_{jz} & T_{jz} \end{vmatrix} + \begin{vmatrix} G_{jz} & S_{jx} & T_{jz} \end{vmatrix} + \begin{vmatrix} G_{jz} & S_{jz} & T_{jx} \end{vmatrix}$$

$$d_7 = \begin{vmatrix} G_{jy} & S_{jy} & T_{jz} \end{vmatrix} + \begin{vmatrix} G_{jy} & S_{jz} & T_{jy} \end{vmatrix} + \begin{vmatrix} G_{jz} & S_{jy} & T_{jy} \end{vmatrix}$$

$$d_8 = \begin{vmatrix} G_{jx} & S_{jy} & T_{jz} \end{vmatrix} + \begin{vmatrix} G_{jx} & S_{jz} & T_{jy} \end{vmatrix} + \begin{vmatrix} G_{jy} & S_{jx} & T_{jz} \end{vmatrix}$$
$$+ \begin{vmatrix} G_{jz} & S_{jx} & T_{jy} \end{vmatrix} + \begin{vmatrix} G_{jy} & S_{jz} & T_{jx} \end{vmatrix} + \begin{vmatrix} G_{jz} & S_{jy} & T_{jx} \end{vmatrix}$$

$$d_9 = \begin{vmatrix} G_{jx} & S_{jx} & T_{jz} \end{vmatrix} + \begin{vmatrix} G_{jx} & S_{jz} & T_{jx} \end{vmatrix} + \begin{vmatrix} G_{jz} & S_{jx} & T_{jx} \end{vmatrix}$$

Since $d_3 = d_1$, $d_4 = d_2$, equation (3.97) is simplified so as not to contain d_3, d_4.
The coefficients in equation (3.99) are

$$\left. \begin{aligned} D_2 &= d_1 \sin\theta_{A1} + d_2 \cos\theta_{A1} \\ D_1 &= d_7 \sin^2\theta_{A1} + d_8 \sin\theta_{A1}\cos\theta_{A1} + d_9 \cos^2\theta_A \\ D_0 &= d_5 \sin\theta_{A1} + d_6 \cos\theta_{A1} \end{aligned} \right\} \qquad (A5.1)$$

APPENDIX 6
Coefficients in equation (3.102)

Having adopted the coordinate system in Fig. 3.27, the terms of L_j, Q_j, H_j ($j = 2, 3, 4$) in equation (3.102) can be written in the following terms, where each of the coefficients L_{3z}, Q_{3z} and H_{3x}, H_{3y}, H_{3z} in L_3, Q_3, H_3 are given in two forms, the second form being in terms of $s = \overline{P_{12}P_{23}}$ as shown in Fig. 3.27.

$$L_2 = L_{2x}x_0 + L_{2y}y_0 + L_{2z}$$
$$L_{2x} = -\text{vers}\gamma_{12}$$
$$L_{2y} = \sin\gamma_{12}$$
$$L_{2z} = 0$$
$$Q_2 = Q_{2x}x_0 + Q_{2y}y_0 + Q_{2z}$$
$$Q_{2x} = -\sin\gamma_{12}$$
$$Q_{2y} = -\text{vers}\gamma_{12}$$
$$Q_{2z} = 0$$
$$H_2 = H_{2x}x_0 + H_{2y}y_0 + H_{2z}$$
$$H_{2x} = 0$$
$$H_{2y} = 0$$
$$H_{2z} = 0$$
$$L_3 = L_{3x}x_0 + L_{3y}y_0 + L_{3z}$$
$$L_{3x} = -\text{vers}\gamma_{13}$$
$$L_{3y} = \sin\gamma_{13}$$
$$L_{3z} = x_{P13}\text{vers}\gamma_{13} - y_{P13}\sin\gamma_{13} = s(\cos\gamma_{12} - \cos\gamma_{13})$$
$$Q_3 = Q_{3x}x_0 + Q_{3y}y_0 + Q_{3z}$$
$$Q_{3x} = -\sin\gamma_{13}$$
$$Q_{3y} = -\text{vers}\gamma_{13}$$
$$Q_{3z} = x_{P_{13}}\sin\gamma_{13} + y_{P13}\text{vers}\gamma_{13} = s(\sin\gamma_{13} - \sin\gamma_{12})$$

APPENDIX 6

$$H_3 = H_{3x}x_0 + H_{3y}y_0 + H_{3z}$$
$$H_{3x} = x_{P13}\operatorname{vers}\gamma_{13} + y_{P13}\sin\gamma_{13} = s\operatorname{vers}\gamma_{23}$$
$$H_{3y} = -x_{P13}\sin\gamma_{13} + y_{P13}\operatorname{vers}\gamma_{13} = -s\sin\gamma_{23}$$
$$H_{3z} = -(x_{P13}^2 + y_{P13}^2)\operatorname{vers}\gamma_{13} = -s^2\operatorname{vers}\gamma_{23}$$
$$L_4 = L_{4x}x_0 + L_{4y}y_0 + L_{4z}$$
$$L_{4x} = -\operatorname{vers}\gamma_{14}$$
$$L_{4y} = \sin\gamma_{14}$$
$$L_{4z} = x_{P14}\operatorname{vers}\gamma_{14} - y_{P14}\sin\gamma_{14}$$
$$Q_4 = Q_{4x}x_0 + Q_{4y}y_0 + Q_{4z}$$
$$Q_{4x} = -\sin\gamma_{14}$$
$$Q_{4y} = -\operatorname{vers}\gamma_{14}$$
$$Q_{4z} = x_{P14}\sin\gamma_{14} + y_{P14}\operatorname{vers}\gamma_{14}$$
$$H_4 = H_{4x}x_0 + H_{4y}y_0 + H_{4z}$$
$$H_{4x} = x_{P14}\operatorname{vers}\gamma_{14} + y_{P14}\sin\gamma_{14}$$
$$H_{4y} = -x_{P14}\sin\gamma_{14} + y_{P14}\operatorname{vers}\gamma_{14}$$
$$H_{4z} = -(x_{P14}^2 + y_{P14}^2)\operatorname{vers}\gamma_{14}$$

APPENDIX 7
Coefficients in equations (3.103), (3.104)

The coefficients in equation (3.103), the expanded form of equation (3.102), are listed as follows:

$$\left.\begin{array}{l} m_1 = \begin{vmatrix} L_{2y} & Q_{2y} & H_{2y} \\ L_{3y} & Q_{3y} & H_{3y} \\ L_{4y} & Q_{4y} & H_{4y} \end{vmatrix} = \begin{vmatrix} L_{jy} & Q_{jy} & H_{jy} \end{vmatrix} \quad (j=2,3,4) \\ m_2 = \begin{vmatrix} L_{jx} & Q_{jy} & H_{jy} \end{vmatrix} + \begin{vmatrix} L_{jy} & Q_{jx} & H_{jy} \end{vmatrix} + \begin{vmatrix} L_{jy} & Q_{jy} & H_{jx} \end{vmatrix} \\ m_3 = \begin{vmatrix} L_{jx} & Q_{jx} & H_{jy} \end{vmatrix} + \begin{vmatrix} L_{jx} & Q_{jy} & H_{jx} \end{vmatrix} + \begin{vmatrix} L_{jy} & Q_{jx} & H_{jx} \end{vmatrix} = m_1 \\ m_4 = \begin{vmatrix} L_{jx} & Q_{jx} & H_{jx} \end{vmatrix} = m_2 \\ m_5 = \begin{vmatrix} L_{jy} & Q_{jz} & H_{jz} \end{vmatrix} + \begin{vmatrix} L_{jz} & Q_{jy} & H_{jz} \end{vmatrix} + \begin{vmatrix} L_{jz} & Q_{jz} & H_{jy} \end{vmatrix} \\ m_6 = \begin{vmatrix} L_{jx} & Q_{jz} & H_{jz} \end{vmatrix} + \begin{vmatrix} L_{jz} & Q_{jx} & H_{jz} \end{vmatrix} + \begin{vmatrix} L_{jz} & Q_{jz} & H_{jx} \end{vmatrix} \\ m_7 = \begin{vmatrix} L_{jy} & Q_{jy} & H_{jz} \end{vmatrix} + \begin{vmatrix} L_{jy} & Q_{jz} & H_{jy} \end{vmatrix} + \begin{vmatrix} L_{jz} & Q_{jy} & H_{jy} \end{vmatrix} \\ m_8 = \begin{vmatrix} L_{jx} & Q_{jy} & H_{jz} \end{vmatrix} + \begin{vmatrix} L_{jx} & Q_{jz} & H_{jy} \end{vmatrix} + \begin{vmatrix} L_{jy} & Q_{jx} & H_{jz} \end{vmatrix} \\ \quad + \begin{vmatrix} L_{jz} & Q_{jx} & H_{jy} \end{vmatrix} + \begin{vmatrix} L_{jy} & Q_{jz} & H_{jx} \end{vmatrix} + \begin{vmatrix} L_{jz} & Q_{jy} & H_{jx} \end{vmatrix} \\ m_9 = \begin{vmatrix} L_{jx} & Q_{jx} & H_{jz} \end{vmatrix} + \begin{vmatrix} L_{jx} & Q_{jz} & H_{jx} \end{vmatrix} + \begin{vmatrix} L_{jz} & Q_{jx} & H_{jx} \end{vmatrix} \end{array}\right\} \quad (A7.1)$$

Since $m_3 = m_1$, $m_4 = m_2$, equation (3.103) is simplified so as not to contain m_3, m_4.

The coefficients in equation (3.104) are

$$\left.\begin{array}{l} M_2 = m_1 \sin\theta_0 + m_2 \cos\theta_0 \\ M_1 = m_7 \sin^2\theta_0 + m_8 \sin\theta_0 \cos\theta_0 + m_9 \cos^2\theta_0 \\ M_0 = m_5 \sin\theta_0 + m_6 \cos\theta_0 \end{array}\right\} \quad (A7.2)$$

APPENDIX 8
Coefficients in equation (3.143)

For brevity reasons, C_θ stands for $\cos\theta_0$. Equation (3.141) is written as

$$\frac{1}{C_\theta^2}\begin{vmatrix} M_2/C_\theta & M_0/C_\theta \\ E_2/C_\theta & E_0/C_\theta \end{vmatrix}^2 + \begin{vmatrix} M_2/C_\theta & M_1/C_\theta^2 \\ E_2/C_\theta & E_1/C_\theta^2 \end{vmatrix}\begin{vmatrix} M_0/C_\theta & M_1/C_\theta^2 \\ E_0/C_\theta & E_1/C_\theta^2 \end{vmatrix} = 0 \quad [(3.141)]$$

Let $\tau = \tan\theta_0$, and we get from equations (A7.2) in Appendix 7

$$\left.\begin{aligned} M_2/C_\theta &= m_1\tau + m_2 \\ M_1/C_\theta^2 &= m_7\tau^2 + m_8\tau + m_9 \\ M_0/C_\theta &= m_5\tau + m_6 \end{aligned}\right\} \quad (A8.1)$$

$$\left.\begin{aligned} E_2/C_\theta &= e_1\tau + e_2 \\ E_1/C_\theta^2 &= e_7\tau^2 + e_8\tau + e_9 \\ E_0/C_\theta &= e_5\tau + e_6 \end{aligned}\right\} \quad (A8.2)$$

Values of $e_1, e_2, \ldots$ can be calculated by analogy with $m_1, m_2, \ldots$ of equations (A7.1) in Appendix 7. Use the symbol

$$\Phi_{ij} = \begin{vmatrix} m_i & m_j \\ e_i & e_j \end{vmatrix} = m_i e_j - m_j e_i$$

and calculate the values of the following 26 2×2 determinants:

Subscript	5	6	7	8	9
1	Φ_{15}	Φ_{16}	Φ_{17}	Φ_{18}	Φ_{19}
2	Φ_{25}	Φ_{26}	Φ_{27}	Φ_{28}	Φ_{29}
5	—	—	Φ_{57}	Φ_{58}	Φ_{59}
6	—	—	Φ_{67}	Φ_{68}	Φ_{69}

Use the following symbols

$$\Omega_1 = \Phi_{15} + \Phi_{26}$$
$$\Omega_2 = \Phi_{16} + \Phi_{25}$$
$$\Omega_3 = \Phi_{18} + \Phi_{27}$$
$$\Omega_4 = \Phi_{19} + \Phi_{28}$$
$$\Omega_5 = \Phi_{58} + \Phi_{67}$$
$$\Omega_6 = \Phi_{59} + \Phi_{68}$$

The coefficients in equation (3.142) are

$$\left.\begin{aligned}g_6 &= \Phi_{15}^2 + \Phi_{17}\Phi_{57}\\g_5 &= 2\Phi_{15}\Omega_2 + \Phi_{17}\Omega_5 + \Phi_{57}\Omega_3\\g_4 &= \Phi_{15}^2 + \Omega_2^2 + 2\Phi_{15}\Phi_{26} + \Phi_{17}\Omega_6 + \Omega_3\Omega_5 + \Phi_{57}\Omega_4\\g_3 &= 2\Omega_1\Omega_2 + \Phi_{17}\Phi_{69} + \Omega_3\Omega_6 + \Omega_4\Omega_5 + \Phi_{29}\Phi_{57}\\g_2 &= \Omega_2^2 + 2\Phi_{15}\Phi_{26} + \Phi_{26}^2 + \Phi_{69}\Omega_3 + \Omega_4\Omega_6 + \Phi_{29}\Omega_5\\g_1 &= 2\Phi_{26}\Omega_2 + \Phi_{69}\Omega_4 + \Phi_{29}\Omega_6\end{aligned}\right\} \quad (A8.3)$$

Let $\tau_1 = \tan(-\gamma_{12}/2)$. The coefficients in equation (3.143) are

$$\left.\begin{aligned}g_6 &= \Phi_{15}^2 + \Phi_{17}\Phi_{57}\\g_{03} &= g_6\tau_1 + g_5\\g_{02} &= g_6\tau_1^2 + g_5\tau_1 + g_4\\g_{01} &= g_6\tau_1^3 + g_5\tau_1^2 + g_4\tau_1 + g_3\\g_{00} &= g_6\tau_1^4 + g_5\tau_1^3 + g_4\tau_1^2 + g_3\tau_1 + g_2\end{aligned}\right\} \quad (A8.4)$$

The value of p_0 corresponding to a certain real root τ_0 ($= \tan\theta_0$) of equation (3.143), or the expanded form of equation (3.144), is

$$p_0 = -\frac{\Phi_{15}\tau_0^2 + \Omega_2\tau_0 + \Phi_{26}}{(\Phi_{17}\tau_0^3 + \Omega_3\tau_0^2 + \Omega_4\tau_0 + \Phi_{29})\cos\theta_0} \quad (A8.5)$$

Please note that, in the calculation process of equation (A8.5), both numerator and denominator are frequently small differences between two large values, hence taking insufficient significant figures could result in incorrect answers. The best way of checking the results is to substitute the value of θ_0 ($= \tan^{-1}\tau_0$) obtained into equations (A7.2) to calculate M_2, M_1, M_0; E_2, E_1, E_0, and to solve separately the two quadratic equations in (3.139). The value of p_0 that is identical in both quadratic equations is the right answer.

APPENDIX 9
The Frost equation of radius of curvature in bipolar coordinates

A9.1 Derivation of Frost (1880-81) equation

In Figs. A9.1 and A9.2, S_1, S_2 are two fixed points. $F(r_1, r_2) = 0$ is the equation of a curve in the two independent variables r_1, r_2, and r_1, r_2 are the distances from a point P on the curve to S_1, S_2. In the following, the symbols for the partial differential coefficients are:

$$F_1 = \frac{\partial F}{\partial r_1}, F_2 = \frac{\partial F}{\partial r_2}, F_{11} = \frac{\partial^2 F}{\partial r_1^2}, F_{22} = \frac{\partial^2 F}{\partial r_2^2}, F_{12} = \frac{\partial^2 F}{\partial r_1 \partial r_2}.$$

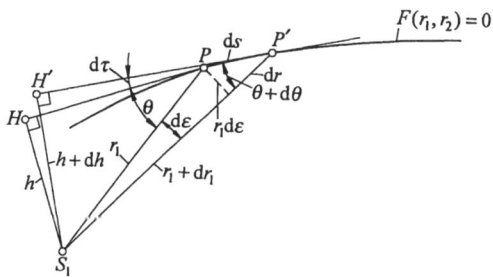

Fig. A9.1. For derivation of Frost equation.

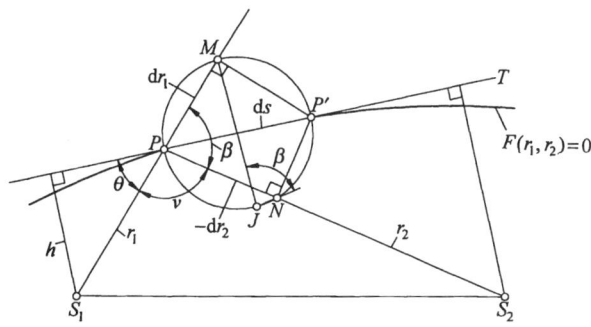

Fig. A9.2. For derivation of Frost equation.

In Fig. A9.1, P is a point on the curve $F(r_1, r_2) = 0$, and P' is another point on the curve but close to P. $\overline{PP'}$ is an infinitesimal distance ds on the curve. Draw tangents PH; $P'H'$ of the curve at P and P'. From S_1 draw perpendiculars S_1H, S_1H' to PH and $P'H'$ respectively. Let $\overline{S_1H} = h$, $\angle S_1PH = \theta$, $d\tau$ the angle between PH and $P'H'$ and $\angle P'S_1P = d\varepsilon$. We have

$$h = r_1 \sin\theta$$

Hence

$$dh = r_1 \cos\theta\, d\theta + \sin\theta\, dr_1 = r_1 \cos\theta (d\tau - d\varepsilon) + \sin\theta\, dr_1$$

Since $\cos\theta \approx dr_1/ds$, $\sin\theta \approx r_1 d\varepsilon/ds$, we have

$$dh = \frac{r_1 dr_1}{ds}(d\tau - d\varepsilon) + \frac{r_1 d\varepsilon}{ds} dr_1 = \frac{r_1 dr_1}{ds} d\tau$$

The radius of curvature is $\rho = ds/d\tau$. Hence

$$\frac{r_1 dr_1}{\rho} = dh = \sin\theta\, dr_1 + r_1 d(\sin\theta)$$

or

$$\frac{1}{\rho} = \frac{\sin\theta}{r_1} + \frac{d(\sin\theta)}{dr_1} \tag{A9.1}$$

Take differential of $F(r_1, r_2) = 0$

$$dF = F_1 dr_1 + F_2 dr_2 = 0$$

Let

$$\sigma = \frac{dr_1}{F_2} = -\frac{dr_2}{F_1}$$

Fig. A9.2 is an enlarged view of the portion PP' in Fig. A9.1. Draw perpendiculars $P'M$, $P'N$ respectively to S_1P, S_2P from P'. Then $\overline{PM} = dr_1$, $\overline{PN} = -dr_2$. Draw a circle passing through the four points P, M, P', N. The diameter of this circle is $\overline{MJ} = \overline{PP'} = ds$. Further let $\angle NPM = \beta$, and

$$\begin{aligned} V_1 &= F_1 - F_2 \cos\beta \\ V_2 &= F_2 - F_1 \cos\beta \\ E^2 &= F_1 V_1 + F_2 V_2 \end{aligned} \tag{A9.2}$$

From equations (A9.1) and (A9.2) we get

APPENDIX 9

$$\frac{\sigma E^2}{\rho} = \frac{\sigma}{\rho}(F_1 V_1 + F_2 V_2)$$
$$= \frac{\sigma F_2 V_2}{\rho} + \frac{\sigma F_1 V_1}{\rho} \tag{A9.3}$$

The two terms in equation (A9.3) have to be found separately. It can be seen from Fig. A9.2 that

$$(ds)^2 \sin^2 \beta = \overline{MN}^2 = (dr_1)^2 + (dr_2)^2 + 2dr_1 dr_2 \cos\beta \quad (\text{The last term is } (+)$$
because $dr_2 < 0$.)

Hence $$[(ds)^2 - (dr_1)^2]\sin^2 \beta = (\cos\beta\, dr_1 + dr_2)^2$$

and

$$\sin^2 \theta = \frac{[(ds)^2 - (dr_1)^2]/\sigma^2}{(ds)^2/\sigma^2} = \frac{(F_2 \cos\beta - F_1)^2}{F_2^2 + F_1^2 - 2F_1 F_2 \cos\beta}$$
$$= \frac{(F_1 - F_2 \cos\beta)^2}{F_1^2 + F_2^2 - 2F_1 F_2 \cos\beta} = \frac{V_1^2}{E^2}$$

or $\sin\theta = \pm\dfrac{V_1}{E}$. Take now

$$\sin\theta = +\frac{V_1}{E} \tag{A9.4}$$

Substituting equation (A9.4) into equation (A9.1), we get

$$\left.\begin{array}{l}
\dfrac{\sigma F_2}{\rho} = \sigma F_2\left(\dfrac{\sin\theta}{r_1} + \dfrac{d(\sin\theta)}{dr_1}\right) = \dfrac{\sigma F_2}{r_1}\dfrac{V_1}{E} + \dfrac{EdV_1 - V_1 dE}{E^2} \\[2mm]
\text{Similarly, it can be proved that} \\[2mm]
\dfrac{\sigma F_1}{\rho} = \dfrac{\sigma F_1 V_2}{r_2 E} - \dfrac{EdV_2 - V_2 dE}{E^2}
\end{array}\right\} \tag{A9.5}$$

Substituting the two equations in (A9.5) into equation (A9.3), we get

$$\frac{\sigma E^2}{\rho} = \sigma\left(\frac{F_2}{r_1} + \frac{F_1}{r_2}\right)\frac{V_1 V_2}{E} + \frac{V_2 dV_1 - V_1 dV_2}{E} \tag{A9.6}$$

The last term in equation (A9.6) can be found as follows:

$$V_2 dV_1 - V_1 dV_2 = (F_2 dF_1 - F_1 dF_2)\sin^2\beta + (F_1^2 - F_2^2)d(\cos\beta) \tag{A9.7}$$

but
$$dF_1 = F_{11}dr_1 + F_{12}dr_2 = \sigma(F_2F_{11} - F_1F_{12})$$
$$dF_2 = F_{12}dr_1 + F_{22}dr_2 = \sigma(F_2F_{12} - F_1F_{22})$$

Hence
$$F_2 dF_1 - F_1 dF_2 = \sigma(F_2^2 F_{11} - 2F_1F_2 F_{12} + F_1^2 F_{22}) \qquad (A9.8)$$

Furthermore
$$r_1^2 + r_2^2 + 2r_1r_2\cos\beta = \overline{S_1S_2}^2$$

Differentiating this equation gives
$$2r_1 dr_1 + 2r_2 dr_2 + 2r_2\cos\beta dr_1 + 2r_1\cos\beta dr_2 + 2r_1r_2 d(\cos\beta) = 0$$

Hence
$$r_1 r_2 d(\cos\beta) = -\sigma(r_1V_2 - r_2V_1)$$

or
$$d(\cos\beta) = \sigma\left(\frac{V_1}{r_1} - \frac{V_2}{r_2}\right) \qquad (A9.9)$$

Substituting equations (A9.8), (A9.9) into equation (A9.7) and then into equation (A9.6), and taking the following two relations into consideration
$$V_1^2 = E^2 - F_2^2 \sin^2\beta$$
$$V_2^2 = E^2 - F_1^2 \sin^2\beta$$

we get finally
$$\frac{E^3}{\rho} = (F_2^2 F_{11} - 2F_1F_2F_{12} + F_1^2 F_{22})\sin^2\beta$$
$$+ E^2\left(\frac{F_1}{r_1} + \frac{F_2}{r_2}\right) - \left(\frac{F_2}{r_1} + \frac{F_1}{r_2}\right) F_1 F_2 \sin^2\beta \qquad (A9.10)$$

where E can be calculated from the following relation
$$E^2 = F_1^2 + F_2^2 - 2F_1F_2\cos\beta \qquad (A9.11)$$

This means that if the values of F_1, F_2, F_{11}, F_{12}, F_{22} and β can be found for the curve at a certain point P, the value of the radius of curvature ρ can then be found from equation (A9.10).

APPENDIX 10
Coefficients in equation (6.9) of k_{A0} –curve

Use again notations as in Appendix 5 to represent each 3×3 determinant by only one row, and j stands for 2, 3, 4.

In the following equations:

$$A_{1j} = -x_1 + x_j \cos\phi_{1j} + y_j \sin\phi_{1j}$$
$$B_{1j} = -y_1 - x_j \sin\phi_{1j} + y_j \cos\phi_{1j}$$
$$C_{1j} = (x_1^2 + y_1^2 - x_j^2 - y_j^2)/2$$

$k_1 = k_3 = \left|-\sin\phi_{1j}, \text{vers}\phi_{1j}, x_j \sin\phi_{1j} + y_j \text{vers}\phi_{1j}\right| + \left|-\sin\phi_{1j}, B_{1j}, -\text{vers}\phi_{1j}\right|$

$k_2 = k_4 = \left|-\sin\phi_{1j}, \text{vers}\phi_{1j}, x_j \text{vers}\phi_{1j} - y_j \sin\phi_{1j}\right| + \left|A_{1j}, \sin\phi_{1j}, -\text{vers}\phi_{1j}\right|$

$k_5 = \left|-\sin\phi_{1j}, B_{1j}, C_{1j}\right| + \left|A_{1j}, \text{vers}\phi_{1j}, C_{1j}\right| + \left|A_{1j}, B_{1j}, x_j \sin\phi_{1j} + y_j \text{vers}\phi_{1j}\right|$

$k_6 = \left|\text{vers}\phi_{1j}, B_{1j}, C_{1j}\right| + \left|A_{1j}, \sin\phi_{1j}, C_{1j}\right| + \left|A_{1j}, B_{1j}, x_j \text{vers}\phi_{1j} - y_j \sin\phi_{1j}\right|$

$k_7 = \left|-\sin\phi_{1j}, \text{vers}\phi_{1j}, C_{1j}\right| + \left|A_{1j}, \text{vers}\phi_{1j}, x_j \sin\phi_{1j} + y_j \text{vers}\phi_{1j}\right|$
$\quad + \left|-\sin\phi_{1j}, B_{1j}, x_j \sin\phi_{1j} + y_j \text{vers}\phi_{1j}\right| + \left|A_{1j}, B_{1j}, -\text{vers}\phi_{1j}\right|$

$k_8 = \left|\text{vers}\phi_{1j}, B_{1j}, x_j \sin\phi_{1j} + y_j \text{vers}\phi_{1j}\right| + \left|A_{1j}, \sin\phi_{1j}, x_j \sin\phi_{1j} + y_j \text{vers}\phi_{1j}\right|$
$\quad + \left|-\sin\phi_{1j}, B_{1j}, x_j \text{vers}\phi_{1j} - y_j \sin\phi_{1j}\right| + \left|A_{1j}, \text{vers}\phi_{1j}, x_j \text{vers}\phi_{1j} - y_j \sin\phi_{1j}\right|$

$k_9 = \left|\text{vers}\phi_{1j}, \sin\phi_{1j}, C_{1j}\right| + \left|\text{vers}\phi_{1j}, B_{1j}, x_j \text{vers}\phi_{1j} - y_j \sin\phi_{1j}\right|$
$\quad + \left|A_{1j}, \sin\phi_{1j}, x_j \text{vers}\phi_{1j} - y_j \sin\phi_{1j}\right| + \left|A_{1j}, B_{1j}, -\text{vers}\phi_{1j}\right|$

$k_{10} = \left|A_{1j}, B_{1j}, C_{1j}\right|$

The coefficients in equation (6.12) are

$$K_2 = k_1' \sin\theta_0 + k_2' \cos\theta_0$$
$$K_1 = k_7' \sin^2\theta_0 + k_8' \sin\theta_0 \cos\theta_0 + k_9' \cos^2\theta_0$$
$$K_0 = k_5' \sin\theta_0 + k_6' \cos\theta_0$$

APPENDIX 11
Selection of γ_2 in equation (6.15)

The three minors in the determinant of equation (6.15) are represented by the following symbols:

$$\Delta_2 = \begin{vmatrix} \lambda_3-1 & \delta_3 \\ \lambda_4-1 & \delta_4 \end{vmatrix}, \Delta_3 = \begin{vmatrix} \lambda_4-1 & \delta_4 \\ \lambda_2-1 & \delta_2 \end{vmatrix}, \Delta_4 = \begin{vmatrix} \lambda_2-1 & \delta_2 \\ \lambda_3-1 & \delta_3 \end{vmatrix}$$

Δ_2, Δ_3, Δ_4 are all vectors. Join these three vectors in series, then join the tail of Δ_2 with the head of Δ_4, and call it Δ_1, as shown by the solid lines in Fig. A11.1, or

$$\Delta_2 + \Delta_3 + \Delta_4 = \Delta_1 \tag{A11.1}$$

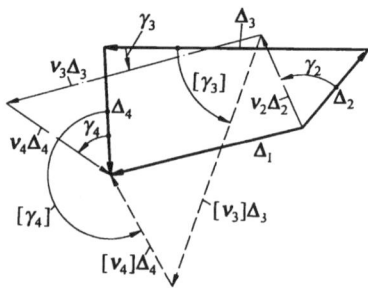

Fig. A11.1.   A virtual four-bar linkage $\Delta_2+\Delta_3+\Delta_4=\Delta_1$.

Expand equation (6.15) in the following form

$$-(\Delta_2 + \Delta_3 + \Delta_4) + \nu_2\Delta_2 + \nu_3\Delta_3 + \nu_4\Delta_4 = 0$$

or
$$\nu_2\Delta_2 + \nu_3\Delta_3 + \nu_4\Delta_4 = \Delta_1 \tag{A11.2}$$

Equation (A11.2) is called the *compatibility equation*. It says that in order for equation (6.14) to have a solution, equation (A11.2) has to be satisfied. As can be seen from Fig. A11.1, equation (A11.2) indicates that the four vectors in equation (A11.1) may be considered as a four-bar linkage (please note that this is a virtual four-bar linkage, not to be confused with the four-bar in Fig. 6.14), and equation (A11.2) corresponds to the case in which Δ_1 is considered fixed and the linkage is moved to the chain-line position. This means that when Δ_2, Δ_3, Δ_4 are given, γ_3, γ_4

APPENDIX 11

are uniquely determined for an assumed γ_2. However, there are two sets of solutions, the one being γ_3, γ_4, the other being $[\gamma_3]$, $[\gamma_4]$, as shown by the dotted lines in Fig. A11.1.

There is a certain limit for the assignment of γ_2, as shown by the regions I, II (or II′), III, IV (or IV′) in Figs. A11.2(a)(b). In these positions, $|\Delta_3|$ and $|\Delta_4|$ become collinear, and $|\Delta_2|$ reaches its dead-centre position. If such positions exist, the virtual four-bar becomes a triangle. In position I it is

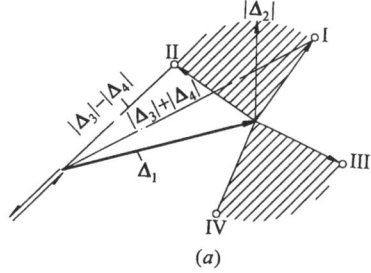

(a)

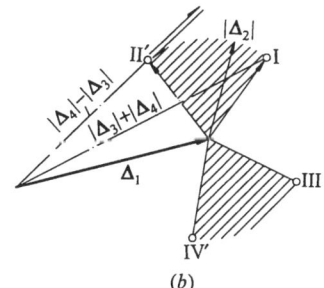

(b)

Fig. A11.2. (a) Dead-centre position of the virtual four-bar, $|\Delta_4|$ being the shortest link.
(b) Dead-centre position of the virtual four-bar, $|\Delta_3|$ being the shortest link.

$$|\Delta_1| + |\Delta_2| > |\Delta_3| + |\Delta_4| \tag{A11.3}$$

and in position II it is

$$|\Delta_3| - |\Delta_4| > \left||\Delta_1| - |\Delta_2|\right|, \text{ if } |\Delta_3| > |\Delta_4| \tag{A11.4}$$

while in position II′ it is

$$\left||\Delta_4|-|\Delta_3|\right| > \left||\Delta_1|-|\Delta_2|\right|, \text{ if } |\Delta_4|>|\Delta_3| \qquad (A11.5)$$

We can conclude as follows:

(a) If the three inequalities (A11.3), (A11.4), (A11.5) do not exist, the limits I, II (or II′), III, IV (or IV′) in Figs. A11.2(a)(b) do not exist at all. γ_2 can be arbitrarily assumed.

In the following cases, γ_2 should be so selected that $\nu_2\Delta_2 = e^{i\gamma_2}\Delta_2$ shall not fall beyond the crosshatched region shown in the figures.

(b) If the inequality (A11.3) exists, and (A11.4) or (A11.5) also exists, then all the limits I, II (or II′), III, IV (or IV′) in Figs. A11.2 (a)(b) exist.

(c) If the inequality (A11.3) does not exist, but (A11.4) or (A11.5) exists, then the limits I and III in Figs. A11.2 (a)(b) combine together.

(d) If the inequality (A11.3) exists, but (A11.4) or (A11.5) does not exist, then the limits II (or II′) and IV (or IV′) in Figs. A11.2(a)(b) combine together.

APPENDIX 12
Equations of the major and minor axes of an ellipse

A12.1 Equation of an allipse

An ellipse belongs to one of the conic sections, or one of the quadratic curves. Fig. A12.1 shows a general ellipse, the equation of which can be written as

$$ax^2 + 2hxy + by^2 + 2gx + 2fy + 1 = 0 \qquad (A12.1)$$

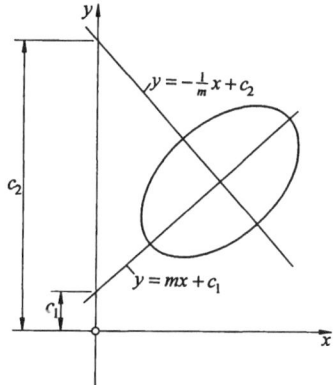

Fig. A12.1. A general ellipse.

Apart from the constant 1 in equation (A12.1), there are altogether five coefficients. Therefore an ellipse on a plane can be determined by five points. Please note that equation (A12.1) may also be a hyperbola or parabola, but there will be no problem in mechanism applications. Because if five points are given, it can easily be observed whether the curve passing through the five points is an ellipse.

After having selected the five points, substituting the coordinates (x_1, y_1), (x_2, y_2), ..., (x_5, y_5) into equation (A12.1) will give five linear equations in the five unknowns a, h, b, g, f, which can easily be solved.

A12.2 Major and minor axes of the ellipse

With the equation of an ellipse known, its major and minor axes can be found as follows:

$$p = \frac{a-b}{h}$$
$$q = \frac{h(g^2 - f^2) - (a-b)fg}{h(h^2 - ab)}$$
$$u = \frac{2h(ag - fh) - (a-b)(af + gh)}{h(h^2 - ab)}$$
$$v = \frac{2h(hg - bf) - (a-b)(bg + fh)}{h(h^2 - ab)}$$
(A12.2)

The major and minor axes of the ellipse are

$$y = mx + c_1$$
$$y = -\frac{1}{m}x + c_2$$
(A12.3a,b)

where

$$c_1 = \frac{-v + \sqrt{v^2 - 4q}}{2}$$
$$c_2 = \frac{-v - \sqrt{v^2 - 4q}}{2}$$
$$m = \frac{u + pc_1}{c_2 - c_1}$$

Having found the major and minor axes of a general ellipse, we can determine the four vertices of the ellipse by the intersections of these axes and the ellipse. Hence the lengths $2a_e$ of the major axis and $2b_e$ of the minor axis can also be determined (Fig. A12.2). The parametric equations of an ellipse with its major and minor axes as the coordinate axes are

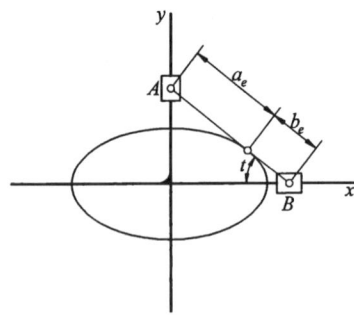

Fig. A12.2. Major axis $2a_e$ and minor axis $2b_e$ of an ellipse.

APPENDIX 12

$$\left. \begin{array}{l} x = a_e \cos t \\ y = b_e \sin t \end{array} \right\} \tag{A12.4}$$

where t is the parameter. The illustration shown in Fig. A12.2 is just the well-known technique of generating an ellipse by means of a double-slider mechanism. t is the inclination of the link AB with respect to the x-axis.

In case the constant term in equation (A12.1) is not equal to 1, the number of independent coefficients is still five. The values of p, q, u, v depend only on the ratios among these five coefficients, not on themselves. Equations (A12.2) are still valid, even if the constant term is zero.

APPENDIX 13
Derivation of equation (7.8) --- diameter of Carter-Hall circle

Fig. A13.1 shows an illustration substantially similar to Fig. 7.15. For the motion of a relative to b, i.e. for the motion by considering b as the fixed link and a as the coupler, R_{12} is the velocity pole of a. Build $\sphericalangle TR_{12}A_0 = \sigma_0 = -\lambda$ according to

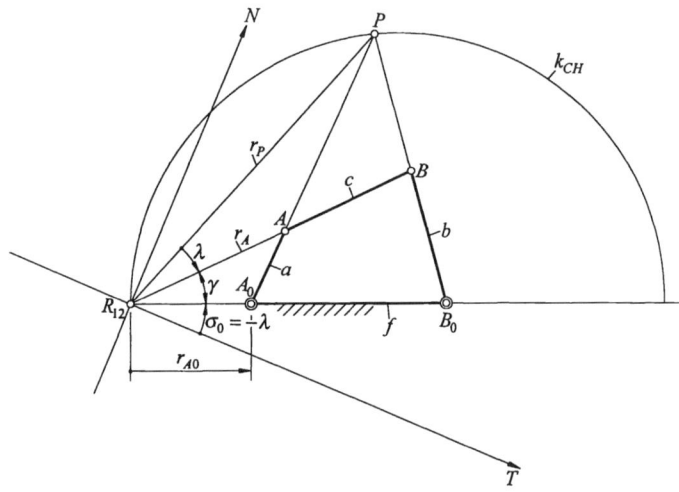

Fig. A13.1. For the derivation of diameter of k_{CH}.

the Bobillier theorem, then $R_{12}T$ is the poletangent. Let $r_{A0} = \overline{R_{12}A_0}$, $r_A = \overline{R_{12}A}$, $r_P = \overline{R_{12}P}$, $\gamma = \sphericalangle A_0R_{12}A$, and δ_{ab} be the inflection circle diameter of this relative motion. We can write according to equation (3.116)

$$\frac{1}{r} = \frac{1}{l\sin\sigma} + \frac{1}{m\cos\sigma} \qquad (A13.1)$$

(r, σ) in equation (A13.1) are the polar coordinates of a certain point on the relative circling-point curve with respect to the polar line $R_{12}T$, and m, l are the two constants of this curve. Since both A_0 and A are points on this relative circling-point curve, we have

$$\frac{1}{r_{A0}} = \frac{1}{l\sin\sigma_0} + \frac{1}{m\cos\sigma_0} \tag{A13.2}$$

$$\frac{1}{r_A} = \frac{1}{l\sin(\sigma_0+\gamma)} + \frac{1}{m\cos(\sigma_0+\gamma)} \tag{A13.3}$$

Eliminating l from equations (A13.2) and (A13.3), we get

$$r_A \sin\sigma_0 - r_{A0}\sin(\sigma_0+\gamma) + \frac{r_A r_{A0}}{m}\frac{\sin\gamma}{\cos\sigma_0 \cos(\sigma_0+\gamma)} = 0 \tag{A13.4}$$

From the relationship of areas

$$\Delta R_{12} AP + \Delta R_{12} A_0 A - \Delta R_{12} A_0 P = 0$$

we get

$$r_A r_P \sin\sigma_0 - r_{A0} r_P \sin(\sigma_0+\gamma) + r_A r_{A0} \sin\gamma = 0$$

Dividing by r_P gives

$$r_A \sin\sigma_0 - r_{A0}\sin(\sigma_0+\gamma) + \frac{r_A r_{A0}}{r_P}\sin\gamma = 0 \tag{A13.5}$$

A comprison between equations (A13.4) and (A13.5) shows that

$$r_P = m\cos\sigma_0 \cos(\sigma_0+\gamma) \tag{A13.6}$$

Equation (A13.6) is the equation of Carter-Hall circle, hence its diameter is

$$d_c = m\cos\sigma_0$$
$$= m\cos\lambda \tag{A13.7}$$

The l in equation (A13.1) can be found as follows. From the two equations (3.126) we get

$$\frac{1}{l} = \frac{1}{3}\left(\frac{1}{\delta_{ab}} + \frac{1}{\rho_a}\right) \tag{A13.8}$$

Substituting equations (A13.7), (A13.8) into equation (A13.2), and changing back $\sigma_0 = -\lambda$, we get

$$\frac{1}{d_c} = \frac{1}{r_{A0}} + \frac{1}{3\sin\lambda}\left(\frac{1}{\delta_{ab}} + \frac{1}{\rho_a}\right) \tag{A13.9}$$

Applying Euler-Savary equation to the two points A_0, B_0 gives

$$\frac{1}{r_{A0}} - \frac{1}{r_{A0}+f} = \frac{1}{\delta_{ab}\sin\sigma_0}$$

or

$$(\delta_{ab}\sin\sigma_0 - r_{A0})f = r_{A0}^2 \qquad (A13.10)$$

Substituting equation (7.5)

$$r_{A0} = \frac{\psi'}{1-\psi'}f \qquad [(7.5)]$$

into equation (A13.10) gives

$$\delta_{ab} = \frac{\psi'}{(1-\psi')^2}\frac{f}{\sin\sigma_0} \qquad (A13.11)$$

Furthermore, from equation (7.7)

$$\cot\lambda = -\frac{\psi''}{\psi'(1-\psi')} \qquad [(7.7)]$$

we get

$$\sin^2\lambda = \frac{\psi'^2(1-\psi')}{\psi'^2(1-\psi')^2 + \psi''^2} \qquad (A13.12)$$

We already have equation (4.28) as an expression of ρ_a in terms of ψ', ψ'', ψ'''

$$\rho_a = \frac{[\psi'^2(1-\psi')^2 + \psi''^2]^{3/2}}{[\psi'^2(1-\psi') + 2\psi''^2 - \psi'\psi'''](1-\psi')^3}f \qquad [(4.28)]$$

Substituting equations (7.5), (A13.11) and (4.28) into equation (A13.9), and replacing $\sin^2\lambda$ by equation (A13.12), we obtain finally equation (7.8).

APPENDIX 14
Program for fourth-order synthesis of four-bar function generators

//---
Assume $\psi' = 0.453450$, $\psi'' = 0.349066$, $\psi''' = -0.806133$; $\theta = -41.5°$.

```
#include <stdio.h>
#include <stdlib.h>
#include <string.h>
#include <math.h>
#pragma hdrstop
int main(int argc, char **argv)
{
   double Ps1,Ps2,Ps3,Tta;
   double Lda,Psm,Eta,Afa,Fim;
   double a,c,f,dd,K;
   //----------------------------------- Input data ----------------------------------
   Ps1 = 0.453450;
   Ps2 = 0.349066;
   Ps3 = - 0.806133;
   Tta = - 41.5*M_PI/180;
   //---------------------------------- Main program --------------------------------
   K = Ps1*(1-Ps1);
   dd = 3*(K*K+Ps2*Ps2)/(K*(1-2*Ps1)+Ps3);           // (7.16)
   Lda = atan(-K/Ps2);                               // (7.17)
   Psm = atan(sin(Tta)*cos(Tta)/(-(1-Ps1)/dd-cos(Tta)*cos(Tta)));
                                                     // (7.18)
   Eta = atan(sin(Tta)*cos(Tta)/(-Ps1/dd+cos(Tta)*cos(Tta)));   // (7.19)
   Afa = Lda+Tta;                                    // (7.20)
   a = Ps1*sin(Afa)/sin(Afa-Eta);
   if (a<0){
      Eta = Eta-M_PI;
      a = Ps1*sin(Afa)/sin(Afa-Eta);
   }
   f = (a*sin(Afa-Eta)-sin(Afa))/sin(Afa+Psm);       // (7.21)
   if (f<0){
      Psm = Psm+M_PI;
      f = (a*sin(Afa-Eta)-sin(Afa))/sin(Afa+Psm);
```

```
}
Fim = Psm+Eta;                                              // (7.22)
c = fabs((-a*sin(Fim)+sin(Psm))/sin(Lda+Tta+Psm));          // (7.23)
//----------------------------------- Results ----------------------------------

print f (" dd   = %f \n",dd);
print f (" Lda = %f \n",Lda*180/M_PI);
print f (" a    = %f \n",a);
print f (" c    = %f \n",c);
print f (" f    = %f \n",f);
print f (" Psm = %f \n",Psm*180/M_PI);
print f (" Eta = %f \n",Eta*180/M_PI);
print f (" Fim = %f \n",Fim*180/M_PI);
return 0;
}
//-------------------------------------------------------------------------------------
```

APPENDIX 15
Program for fourth order synthesis of slider-crank function generators

```
//--------------------------------------------------------------------------
```
Assume $s' = ds/d\phi = 29.46464$ mm, $s'' = d^2s/d\phi^2 = -39.54568$ mm, $s''' = -59.28341$ mm, and $\theta = 60°$.

```c
#include <stdio.h>
#include <stdlib.h>
#include <string.h>
#include <math.h>
//--------------------------------------------------------------------------
int main(int argc, char **argv)
{
   double S1,S2,S3,Tta;
   double Lda,Eta;
   double a,c,dd;
   double Bx,By;
   //----------------------------------- input data -----------------------------------
   S1 = 29.46464;
   S2 = -39.54568;
   S3 = -59.28341;
   Tta = 60*M_PI/180;
   //-------------------------------- main prognam --------------------------------
   dd = 3*(S1*S1+S2*S2)/(S1+S3);
   Lda = atan(-S1/S2);
   Eta = atan(sin(Tta)*cos(Tta)/(-S1/dd+cos(Tta)*cos(Tta)));
   a = S1*sin(Lda+Tta)/sin(Lda+Tta-Eta);
   if (a<0){
      Eta = Eta+M_PI;
      a = S1*sin(Lda+Tta)/sin(Lda+Tta-Eta);
   }
   c = (-a*sin(Eta)-dd*sin(Tta)*cos(Tta))/sin(Lda+Tta);
   Bx = dd*sin(Tta)*cos(Tta);
   By = a*cos(Eta)+c*cos(Lda+Tta);
   //----------------------------------- Results -----------------------------------
   print f (" dd   = %f \n",dd);
   print f (" Lda = %f \n",Lda*180/M_PI);
   print f (" a     = %f \n",a);
```

```
    print f (" c     = %f \n",c);
    print f (" B = (%f,%f) \n",Bx,By);
    return 0;
}
```

APPENDIX 16
Algebraic equations for synthesis of function generators

A16.1 Four precision points (P_1–P_2–P_3–P_4)

Three angular displacement pairs $\phi_2:\psi_2$; $\phi_3:\psi_3$; $\phi_4:\psi_4$ are to be corrdinated. Calculate first the values of r_{aj}, r_{bj} by equations (7.29a,b)

$$\left. \begin{array}{l} r_{aj} = \dfrac{\sin\dfrac{\psi_j}{2}}{\sin\dfrac{\phi_j - \psi_j}{2}} f \\[2ex] r_{bj} = \dfrac{\sin\dfrac{\phi_j}{2}}{\sin\dfrac{\phi_j - \psi_j}{2}} f \end{array} \right\} \quad (j = 2, 3, 4) \qquad [(7.29a,b)]$$

For simplicity reasons, assume $f = 1$.
The following symbols are adopted:

$$s_{1j} = r_{aj}$$
$$s_{2j} = \sin\frac{\phi_j - \psi_j}{2}$$
$$s_{3j} = r_{bj}$$
$$s_{4j} = \cos\frac{\phi_j - \psi_j}{2}$$
$$s_{5j} = \sin\frac{\phi_j}{2}$$
$$s_{6j} = \cos\frac{\phi_j}{2}$$
$$s_{7j} = \sin\frac{\psi_j}{2}$$
$$s_{8j} = \cos\frac{\psi_j}{2}$$

and $c_{18j} = s_{1j} s_{8j}$, $c_{36j} = s_{3j} s_{6j}$, etc.

Equation (7.36), being of the form

$$a_2 a^2 + (a_1 \cos\phi_0)a + a_0 = 0$$

is now written as

$$(u_1 t + u_2)a^2 + [(u_7 t^2 + u_8 t + u_9)\cos\phi_0]a + (u_5 t + u_6) = 0$$

where

$$\left. \begin{aligned} u_1 &= |c_{36j}, s_{4j}, s_{2j}| \\ u_2 &= |c_{35j}, s_{4j}, s_{2j}| \\ u_5 &= -|c_{36j}, c_{18j}, c_{17j}| \\ u_6 &= -|c_{35j}, c_{18j}, c_{17j}| \\ u_7 &= |c_{36j}, c_{18j}, s_{4j}| + |c_{36j}, s_{2j}, c_{17j}| \\ u_8 &= |c_{36j}, c_{18j}, s_{2j}| + |c_{35j}, c_{18j}, s_{4j}| - |c_{36j}, s_{4j}, c_{17j}| + |c_{35j}, s_{2j}, c_{17j}| \\ u_9 &= |c_{35j}, c_{18j}, s_{2j}| - |c_{35j}, s_{4j}, c_{17j}| \end{aligned} \right\} \quad \text{(A16.1)}$$

The solved values of (ϕ_0, a), transformed into $t_0 = \tan \phi_0$, $X_0 = 1/(a\cos\phi_0)$, are then substituted into the following equations to calculate Δ, Δ_Y, Δ_q:

$$\Delta = (w_1 t_0 + w_2) X_0 + (w_3 t_0^2 + w_4 t_0 + w_5)$$
$$\Delta_Y = w_6 X_0^2 + (w_7 t_0 + w_8) X_0 + (w_9 t_0^2 + w_{10})$$
$$\Delta_q = w_{11} t_0^2 + (w_{12} X_0 + w_{13}) t_0 + (w_{14} X_0 + w_{15})$$

where

$$\left. \begin{aligned} w_1 &= -|c_{36j}, c_{18j}, 1| \\ w_2 &= -|c_{35j}, c_{18j}, 1| \\ w_3 &= |c_{36j}, s_{2j}, 1| \\ w_4 &= -|c_{36j}, s_{4j}, 1| + |c_{35j}, s_{2j}, 1| \\ w_5 &= -|c_{35j}, s_{4j}, 1| \\ w_6 &= -|c_{17j}, c_{18j}, 1| \\ w_7 &= |c_{17j}, s_{2j}, 1| + |s_{4j}, c_{18j}, 1| \\ w_8 &= -|c_{17j}, s_{4j}, 1| + |s_{2j}, c_{18j}, 1| \\ w_9 &= -|s_{4j}, s_{2j}, 1| \\ w_{10} &= |s_{2j}, s_{4j}, 1| \\ w_{11} &= -|c_{36j}, s_{4j}, 1| \\ w_{12} &= |c_{36j}, c_{17j}, 1| \\ w_{13} &= -|c_{36j}, s_{2j}, 1| - |c_{35j}, s_{4j}, 1| \\ w_{14} &= |c_{35j}, c_{17j}, 1| \\ w_{15} &= -|c_{35j}, s_{2j}, 1| \end{aligned} \right\} \quad \text{(A16.2)}$$

APPENDIX 16

$$Y_0 = \frac{1}{b_0 \cos\psi_0} = \frac{\Delta_Y}{\Delta}$$

$$q_0 = \tan\psi_0 = \frac{\Delta_q}{\Delta}$$
(A16.3)

Values of ψ_0 and Y_0 can readily be calculated from Δ, Δ_Y, Δ_q. Please note that two ψ_0-values can be found from one q_0, and only the correct one should be chosen. The value of b should be positive.

A16.2 Five precision points (P_1–P_2–P_3–P_4–P_5)

Four angle-pairs $\phi_2:\psi_2$; $\phi_3:\psi_3$; $\phi_4:\psi_4$; $\phi_5:\psi_5$ are to be coordinated. Equation (7.42) can be expanded by analogy with equation (3.141) and by referring to Appendix 8. Use the symbol $\Delta_{ij} = \begin{vmatrix} u_i & u_j \\ v_i & v_j \end{vmatrix} = u_i v_j - u_j v_i$ and calculate the values of the following 16 2×2 determinants:

Subscript	5	6	7	8	9
1	Δ_{15}	Δ_{16}	Δ_{17}	Δ_{18}	Δ_{19}
2	Δ_{25}	Δ_{26}	Δ_{27}	Δ_{28}	Δ_{29}
5	—	—	Δ_{57}	Δ_{58}	Δ_{59}
6	—	—	Δ_{67}	Δ_{68}	Δ_{69}

Use the following symbols

$$\Lambda_1 = \Delta_{15} + \Delta_{26}$$
$$\Lambda_2 = \Delta_{16} + \Delta_{25}$$
$$\Lambda_3 = \Delta_{18} + \Delta_{27}$$
$$\Lambda_4 = \Delta_{19} + \Delta_{28}$$
$$\Lambda_5 = \Delta_{58} + \Delta_{67}$$
$$\Lambda_6 = \Delta_{59} + \Delta_{68}$$

The first four coefficients in equation (7.43) are

$$f_6 = \Delta_{15}^2 + \Delta_{17}\Delta_{57}$$
$$f_5 = 2\Delta_{15}\Lambda_2 + \Delta_{17}\Lambda_5 + \Delta_{57}\Lambda_3$$
$$f_4 = \Delta_{15}^2 + \Lambda_2^2 + 2\Delta_{15}\Delta_{26} + \Delta_{17}\Lambda_6 + \Lambda_3\Lambda_5 + \Delta_{57}\Lambda_4$$
$$f_3 = 2\Lambda_1\Lambda_2 + \Delta_{17}\Delta_{69} + \Lambda_3\Lambda_6 + \Lambda_4\Lambda_5 + \Delta_{29}\Delta_{57}$$

Let

$$t_1 = \tan(-\phi_2/2)$$
$$t_2 = \tan(-\phi_3/2)$$
$$t_3 = \tan[-(\phi_2+\phi_3)/2]$$
$$t_{10} = t_1 + t_2 + t_3$$
$$t_{20} = t_1 t_2 + t_2 t_3 + t_3 t_1$$
$$t_{30} = t_1 t_2 t_3$$

The coefficients in equation (7.44) are

$$\left. \begin{array}{l} f_{20} = t_{10} f_6 + f_5 \\ f_{10} = (t_{10}^2 - t_{20}) f_6 + t_{10} f_5 + f_4 \\ f_{00} = [t_{10}(t_{10}^2 - 2t_{20}) + t_{30}] f_6 + (t_{10}^2 - t_{20}) f_5 + t_{10} f_4 + f_3 \end{array} \right\} \quad (A16.4)$$

The value of a corresponding to a certain real root t_0 ($= \tan \phi_0$) of equation (7.44), or the expanded form of equation (7.45) is

$$a = -\frac{\Delta_{15} t_0^2 + \Lambda_2 t_0 + \Delta_{26}}{(\Delta_{17} t_0^3 + \Lambda_3 t_0^2 + \Lambda_4 t_0 + \Delta_{29}) \cos\phi_0} \quad (A16.5)$$

APPENDIX 17
Relations between displacement matrix, rotation angle and pole coordinates

If the coordinates (x_{Pj}, y_{Pj}) of the pole and the angle of rotation γ_j are given, the displacement matrix (equation (3.53)) is

$$[\mathbb{E}_j] = \begin{bmatrix} C_{\gamma j}, & -S_{\gamma j}, & x_{Pj}(1-C_{\gamma j}) + y_{Pj}S_{\gamma j} \\ S_{\gamma j}, & C_{\gamma j}, & -x_{Pj}S_{\gamma j} + y_{Pj}(1-C_{\gamma j}) \\ 0, & 0, & 1 \end{bmatrix} \quad (A17.1)$$

where $S_{\gamma j} = \sin \gamma_j$, $C_{\gamma j} = \cos \gamma_j$

Conversely, if the displacement matrix $[\mathbb{E}_j]$ is given first

$$[\mathbb{E}_j] = \begin{bmatrix} e_{11}, & e_{12}, & e_{13} \\ e_{21}, & e_{22}, & e_{23} \\ 0, & 0, & 1 \end{bmatrix}$$

the angle of rotation γ_j is then

$$\gamma_j = \text{SGN}(e_{21}) \cos^{-1} e_{11} \quad (A17.2)$$

where SGN (e_{21}) represents the sign of e_{21}, and the coordinates of the pole are

$$x_{Pj} = \frac{W_2}{W_1}, \quad y_{Pj} = \frac{W_3}{W_1} \quad (A17.3)$$

where

$$W_1 = \begin{vmatrix} e_{11} - 1 & e_{12} \\ e_{21} & e_{22} - 1 \end{vmatrix}, \quad W_2 = \begin{vmatrix} e_{12} & e_{13} \\ e_{22} - 1 & e_{23} \end{vmatrix}, \quad W_3 = \begin{vmatrix} e_{13} & e_{11} - 1 \\ e_{23} & e_{21} \end{vmatrix}$$

APPENDIX 18
Expansion of equation (7.97)

This expansion is cited from (Chiang & Chen, 1987). In order to facilitate computer programming, equation (7.97) is written as

$$m_{1234}: \left|\bar{\mathbf{a}}\mathbf{g}_{j1} + \mathbf{h}_{j1}, \mathbf{a}\mathbf{g}_{j2} + \mathbf{h}_{j2}, \bar{\mathbf{a}}\mathbf{g}_{j3} + \mathbf{a}\mathbf{g}_{j4} + \mathbf{h}_{j3}\right| = 0, \quad (j = 2, 3, 4) \quad (A18.1)$$

where $\mathbf{g}_{ji}$ and $\mathbf{h}_{ji}$ are all complex numbers. Assume that on the plane of complex coordinates, the origin is taken at R_{12}, then $\mathbf{R}_2 = 0$. This relative centre-point curve can then be transformed into a quadratic equation in the unknown a.

$$m_{1234}: \mathbf{a}_2 a^2 + \mathbf{a}_1 a + \mathbf{a}_0 = 0 \qquad (A18.2)$$

where

$$\mathbf{a}_2 = \cos\phi_0(t\mathbf{u}_1 + \mathbf{u}_2)$$
$$\mathbf{a}_1 = \cos^2\phi_0(t^2\mathbf{u}_5 + t\mathbf{u}_6 + \mathbf{u}_7)$$
$$\mathbf{a}_0 = \cos\phi_0(t\mathbf{u}_3 + \mathbf{u}_4)$$
$$t = \tan\phi_0$$

and

$$\mathbf{u}_1 = -i\left|g_{j1}, g_{j2}, g_{j3} + g_{j4}\right|$$

$$\mathbf{u}_2 = \left|g_{j1}, g_{j2}, g_{j3} - g_{j4}\right|$$

$$\mathbf{u}_3 = i\begin{vmatrix} -g_{11} & g_{12} & 0 \\ h_{21} & h_{22} & h_{23} \\ h_{31} & h_{32} & h_{33} \end{vmatrix}$$

$$\mathbf{u}_4 = \begin{vmatrix} g_{11} & g_{12} & 0 \\ h_{21} & h_{22} & h_{23} \\ h_{31} & h_{32} & h_{33} \end{vmatrix}$$

$$\mathbf{u}_5 = \begin{vmatrix} g_{11} & g_{12} & 0 \\ -h_{21} & h_{22} & h_{23} \\ g_{31} & g_{32} & -g_{33} - g_{34} \end{vmatrix} + \begin{vmatrix} g_{11} & g_{12} & 0 \\ g_{21} & g_{22} & -g_{23} - g_{24} \\ -h_{31} & h_{32} & h_{33} \end{vmatrix}$$

APPENDIX 18

$$\mathbf{u}_6 = i \left\{ \begin{vmatrix} -g_{11} & g_{12} & 0 \\ h_{21} & h_{22} & h_{23} \\ g_{31} & g_{32} & g_{33} - g_{34} \end{vmatrix} - \begin{vmatrix} g_{11} & g_{12} & 0 \\ h_{21} & h_{22} & h_{23} \\ g_{31} & -g_{32} & g_{33} + g_{34} \end{vmatrix} \right.$$

$$\left. + \begin{vmatrix} -g_{11} & g_{12} & 0 \\ g_{21} & g_{22} & g_{23} - g_{24} \\ h_{31} & h_{32} & h_{34} \end{vmatrix} - \begin{vmatrix} g_{11} & g_{12} & 0 \\ g_{21} & -g_{22} & g_{23} + g_{24} \\ h_{31} & h_{32} & h_{33} \end{vmatrix} \right\}$$

$$\mathbf{u}_7 = \begin{vmatrix} g_{11} & g_{12} & 0 \\ h_{21} & h_{22} & h_{23} \\ g_{31} & g_{32} & g_{33} - g_{34} \end{vmatrix} + \begin{vmatrix} g_{11} & g_{12} & 0 \\ g_{21} & g_{22} & g_{23} - g_{24} \\ h_{31} & h_{32} & h_{33} \end{vmatrix}$$

REFERENCES

Alt, H. (1921). Zur Synthese der ebenen Mechanismen. *Zeitschrift für angewandte Mathematik und Mechanik* **1**, 373-98.
Alt, H. (1925). Über die Totlagen des Gelenkvierecks. *Zeitschrift für angewandte Mathematik und Mechanik* **5**, 337-46.
Alt, H. (1932a). Koppelgetriebe als Rastgetriebe. *VDI-Z.* **76**, 456-62; 533-37.
Alt, H. (1932b). Zur Geometrie der Koppelrastgetriebe. *Ingenieur-Archiv* **3**, 394-411.
Alt, H. (1932c). Der Übertragungswinkel und seine Bedeutung für das Konstruieren periodischer Getriebe. *Werkstattstechnik* **26**, 61-4.
Alt, H. (1936). Beziehungen zwischen Punkten der Mittelpunkt- und der Kreispunktkurve. *Maschinenbau/Der Betrieb, Beilage Getriebetechnik* **4**, 407-9.
Alt, H. (1944). Die Kardanlagen von Getriebegliedern und die Krümmung der Polkurven. *Ingenieur-Archiv* **14**, 319-31.
Antuma, H. J. (1972). Symmetrical Curves Generated by Multi-Bar Mechanisms. In *Conf. on Mechanisms 1972*, 44-59. I. Mech. E. London.
Antuma, H. J. (1978). Triangular Nomograms for Symmetrical Coupler Curves. *Mechanism and Machine Theory* **13**, 251-68.
Ball, R. (1849). Notes on Applied Mechanics. *Proc. Roy. Irish Acad.* Ser. II, 243-5.
Barkan, P. & Tuohy, E. J.(1959). Synthesis of the Four-Bar Linkage to Match Prescribed Velocity Ratios. *Trans. ASME, J. Eng. Ind.* **81B**, 169-77.
Beyer, R. (1931). *Technische Kinematik.* Leipzig: Barth.
Beyer, R. (1938a). Anwendung von Schmiegungskegelschnitten in der Getriebetechnik. *Maschinenbau/Der Betrieb, Beilage Getriebetechnik* **6**, 253-8.
Beyer, R. (1938b). Koppelkurven mit drei Spitzen und spezielle Koppelkurven-Büschel. *VDI-Z.* **82**, 124.
Beyer, R. (1939). Zur synthese ebener und räumlicher Kurbeltriebe. *VDI-Forschungshefte* Nr. 394, Berlin.
Beyer, R. (1950). Profilfräser mit logarithmisch-spiralförmig hinterdrehten Zähnen. *Werkstatt und Betrieb* **83**, 270-3.
Beyer, R. (1953). *Kinematische Getriebesynthese.* Springer-Verlag, Berlin.
Biezeno, C. B. & Grammal, R. (1953). *Engineering Dynamics*, Vol. 4, *Internal Combustion Engine*. 2nd ed. English Translation (1954), Blackie and Son, London.
Bobillier, E. (1870). *Cours de géométrie.* 12th ed., p. 232.
Bottema, O. (1961). On the Instantaneous Invariants of the Motion of a Rigid Plane System. In *Proceedings of the International Conference for Teachers of Mechanisms*, 159-64. Yale University. Show String Press, New Haven, New

REFERENCES

England.
Burmester, L. (1876). Über die Geradführung durch das Kurbelgetriebe. *Civilingenieur* **22**, 597-606.
Burmester, L. (1877a). Über die Geradführung durch das Kurbelgetriebe. *Civilingenieur* **23**, 227-50.
Burmester, L. (1877b). Über die Geradführung durch das Kurbelgetriebe. *Civilingenieur* **23**, 319-42.
Carter, W. J. (1957). Kinematic Analysis and Synthesis Using Collineation-Axis Equations. *Trans. ASME* **79**, 1305-12.
Cayley, A. (1876). On Three-Bar Motion. *Proc. London Math. Soc.* **7**, 142.
Chang, C. F. & Hwang, W. M. (1994). Kinematic Synthesis of Watt-I Mechanisms Generating Closed Coupler Curves With Up to Four Cusps. *Mechanism and Machine Theory* **24**, 501-11.
Chebyshev, P.L. (1879). Sur les parallélogrammes composés de trois éléments quelconques. *Mémoires de l'Académie des Sciences de Saint-Pétersbourg* **36**, suppl. 3.
Chen, F. Y. (1969a). An Analytical Method for Synthesizing the Four-Bar Crank-Rocker Mechanism. *Trans. ASME, J. Eng. Ind.* **91B**, 45-54.
Chen, F. Y. (1969b). Position Synthesis of the Double-Rocker Mechanism. *J. Mechanisms* **4**, 303-10.
Chiang, C. H. (1970). Simplified Graphical Acceleration Analysis of Four-Bar Linkages. *J. Mechanisms* **5**, 549-62.
Chiang, C. H. (1971). Einfache Verfahren zur Ruckermittlung eines Gelenkvierecks und zur Maßsynthese des Funktionsgetriebes. *Feinwerktechnik* **75**, 306-13.
Chiang, C. H. (1973). Maßsynthese des viergelenkigen Funktionsgetriebes zum Anpassen bis zur vierten Ableitung einer gegebenen Funktion. *Feinwerktechnik+micronic* **77**, 215-8.
Chiang, C. H. (1986a). Design of Spherical and Planar Crank-Rockers and Double-rockers as Function Generators-I. *Mechanism and Machine Theory* **21**, 287-96.
Chiang, C. H. (1986b). Design of Spherical and Planar Crank-Rockers and Double-Rockers as Function Generators-II. *Mechanism and Machine Theory* **21**, 297-305.
Chiang, C. H. & Chen, J. S. (1987). An Algebraic Treatment of Burmester Points by Means of Three Basic Poles. *Mechanism and Machine Theory* **22**, 47-53.
Chiang, C. H., Pennestri, E. and Chung, W. Y. (1989). On a Technique for Higher Order Synthesis of Four-Bar Function Generators. *Mechanism and Machine Theory* **24**, 195-205.
Dijksman, E. A. (1971a). Six-Bar Cognates of Watt's Form. *Trans. ASME, J. Eng. Ind.* **93B**, 183-90.
Dijksman, E. A. (1971b). A Strong Relationship Between New and Old Inversion Mechanisms. *Trans. ASME, J. Eng. Ind.* **93B**, 334-9.
Dijksman, E. A. (1976). *Motion Geometry of Mechanisms*. Cambridge University Press.

Erdman, A. G. & Sandor, G. N. (1991). *Mechanism Design, Analysis and Synthesis*, Vol. I. 2nd ed. Prentice Hall, Englewood Cliffs, New Jersey.

Filemon, E. (1972). Useful Ranges of Centerpoint Curves for Design of Crank-and-Rocker Linkages. *Mechanism and Machine Theory* **7**, 47-53.

Freudenstein, F. (1955). Approximate Synthesis of Four-Bar Linkages. *Trans. ASME* **77**, 853-61.

Freudenstein, F. (1959a). Structural Error Analysis in Plane Kinematic Synthesis. *Trans. ASME, J. Eng. Ind.* **81B**, 15-22.

Freudenstein, F. (1959b). Harmonic Analysis of Crank-and-Rocker Mechanisms With Applications. *Trans. ASME, J. Applied Mechanics* **26**, 673-5.

Freudenstein, F. (1960). The Cardan Positions of a Plane. In *Trans. Sixth Conf. Mechanisms*. Purdue University, Lafayette, Indiana, 129-33.

Freudenstein, F. (1965). Higher Path-Curvature Analysis in Plane Kinematics. *Trans. ASME, J. Eng. Ind.* **87B**, 184-90.

Freudenstein, F. (1983). Discussion on (Tsai, 1983b). *Trans. ASME, J. Mechanisms, Transmissions, and Automation in Design* **105**, 691.

Freudenstein, F. & Mohan, K. (1961). When Linkages Need Harmonic Analysis. *Product Engineering*, March 6, 1961, 47-50.

Freudenstein, F. & Primrose, E. J. F. (1972). The Classical Transmission-Angle Problem. In *Conference on Mechanisms 1972*, 105-10. I. Mech. E., London.

Freudenstein, F. & Sandor, G. N. (1959). Synthesis of Path-Generating Mechanisms by Means of a Programmed Digital Computer. *Trans. ASME, J. Eng. Ind.* **81B**, 159-68.

Freudenstein, F. & Sandor, G. N. (1961). On the Burmester Points of a Plane. *Trans. ASME, J. Applied Mechanics* **28**, 41-9. (Discussion 433-75).

Frost, P. (1880-81). General Expression for the Radius of Curvature in Dipolar Coordinates. *The Messenger of Mathematics* **10**, 18-20.

Grübler, M. (1884). Über die Krümmungsmittelpunkte der Polbahnen. *Z. für Mathematik und Physik* **29**, 212-21.

Grübler, M. (1889). Die Krümmungsradien der Polbahnen. *Z. für Mathematik und Physik* **34**, 305-10.

Grübler, M. (1892). Über die Kreisungspunkte einer complan bewegten Ebene. *Z. für Mathematik und Physik* **37**, 35-36.

Gustavson, R. E. (1967). Computer-Designed Car Window Linkage. *Mechanical Engineering*, 45-51.

Hain, K. (1941a). Koppelkurven mit Spitzen an Gelenkvierecken. *Feinmechanik u. Präzision* **49**, 61-3.

Hain, K. (1941b). Koppelkurven mit Spitzen und ihre getriebetechnische Anwendung. *Maschinenbau/Der Betrieb, Beilage Getriebetechnik* **9**, 313-6.

Hain, K. (1942). Punktlagen-, Kurbelwinkel- und Schwingenwinkelzuordnungen. *Maschinenbau/Betrieb, Beilage Getriebetechnik* **10**, 218-21.

Hain, K. (1944). Punktlagen- und Winkelzuordnungen an Kurble und Schwinge von Gelenkvierecken. *Maschinenbau/Der Betrieb, Beilage Getriebetechnik* **12**, 253-6.

REFERENCES

Hain, K. (1955). Die Analyse und Synthese der achtgliedrigen Gelenkgetriebe. *VDI-Berichte* **5**, 81-93.

Hain, K. (1957). How to Apply Drag-Link Mechanisms in the Synthesis of Mechanisms. In *Trans. Fourth Mechanism Conference*, Purdue University, Lafayette, Indiana, 66-75.

Hain, K. (1961). *Angewandte Getriebelehre*. VDI-Verlag, Düsseldorf.

Hall, A. S. (1957). Discussion on (Carter, 1957). *Trans. ASME* **79**, 1312.

Hall, A. S. (1958). Inflection Circle and Polode Curvature. In *Trans. Fifth Conf. on Mechanisms*. Purdue University, Lafayette, Indiana, 207-31.

Hart, H. (1883). Quaternion Proof of the Triple Generation of Three-Bar Motion. *Messenger of Math.* **12**, 32.

Hartmann, W. (1893). Ein neues Verfahren zur Aufsuchung des Krümmungskreises. *VDI-Z.* **37**, 95-102.

Hirschhorn, J. (1962). *Kinematics and Dynamics of Plane Mechanisms*. McGraw-Hill Book Company, New York.

Hunt, K. H. (1978). *Kinematic Geometry of Mechanisms*. Oxford University Press.

Hunt, K. H. & Fichter, E. F. (1981). Equations for Four-Bar Line-Envelopes. *Trans. ASME, J. Mechanical Design* **103**, 743-9.

Hwang, W. M. & Chang, C. F. (1989). Synthesis of Watt-I Mechanism Generating Closed Coupler Curves With Up to Three Cusps. *J. Chinese Soc. Mech. Engrs.* **10**, 211-8.

de Jonge, A. E. R. (1942). What Is Wrong With "Kinematics" and "Mechanisms". *Mechanical Engineering* **64**, 273-8 and 747-51.

de Jonge, A. E. R. (1943). A Brief Account of Modern Kinematics. *Trans. ASME* **65**, 663-83.

Keller, R. E. (1965). Sketching Rules for the Curves of Burmester Mechanism Synthesis. *Trans. ASME, J. Eng. Ind.* **87B**, 155-60.

Kracke, J. (1970). Annährung von Funktionen durch viergliedrige Kurbelgetriebe. *VDI-Berichte* Nr. **140**, 75-82.

Kraemer, O. (1959). *Getriebelehre*. Verlag G. Braun, Karlsruhe.

Kramer, S. N. & Sandor, G. N. (1970). Finite Kinematic Synthesis of Cycloidal-Crank Mechanism for Function Generation. *Trans. ASME, J. Eng. Ind.* **92B**, 531-6.

Kraus, R. (1935). Lagenreduktion als Hilfsmittel zur Synthese ebener Kurbeltriebe. *Maschinenbau/Der Betrieb, Beilage Getriebetechnik* **3**, 637-9.

Kraus, R. (1952). Wertigkeitsbilanz und ihre Anwendung auf eine Geradführung für Meßgeräte. *Feinwerktechnik* **56**, 57-63.

Kraus, R. (1954). *Getriebelehre* Bd. I. VEB Verlag Technik, Berlin.

Kraus, R. (1956). *Getriebelehre* Bd. III. VEB Verlag Technik, Berlin.

Lakshminarayana, K. (1971). On the Carter-Hall Circle and Its Application. *J. Mechanisms* **6**, 517-32.

Lichtenheldt, W. (1936). Die Koppelkurvenfräsmaschine. *Werkstattstechnik* **30**, 266-9.

Lin, C. K. & Chiang, C. H. (1992). Synthesis of Planar and Spherical Geared Five-Bar

Function Generators by the Polode Method. *Mechanism and Machine Theory* **27**, 131-41.

Mayer, A. E. (1938). Koppelkurven mit drei Spitzen und spezielle Koppelkurven-Büschel. *Z. Math. u. Physik* **43**, 389-445.

Meyer zur Capellen, W. (1949). Die Bahn des Momentanpols und die Kardanlage. *Ingenieur-Archiv* **17**, 308-16.

Meyer zur Capellen, W. (1956a). Der Zykloidenlenker und seine Weiterentwicklung. *Konstruktion* **8**, 510-8.

Meyer zur Capellen, W. (1956b). Harmonische Analyse bei der Kurbelschleife. *Z. angew. Math. Mech.* **36**, 151-2.

Meyer zur Capellen, W. (1957). Die Kurbelschleife zweiter Art. *Werkstatt und Betrieb* **90**, 306-8.

Meyer zur Capellen, W. und Mitarbeiter. (1958). Bewegungsverhältnisse an der geschränkten Schubkurbel. *Forschungsberichte des Landes Nordrhein-Westfalen*, Nr. 449. Westdeutscher Verlag, Köln und Opladen.

Meyer zur Capellen, W. (1959). Die Harmonischen der Rotationsenergie bei der Schubkurbel und verwandte Fourier-Reihen. *Z. für angw. Math. und Physik* **11**, 207-18.

Meyer zur Capellen, W. (1960). Harmonische Analyse an Kurbeltrieben. *Konstruktion* **12**, 38-41.

Meyer zur Capellen, W. & Thünker, N. (1975). Die kinetische Energie der ebenen Kurbelschleifen un ihre harmonische Analyse. *Mechanism and Machine Theory* **10**, 147-54.

Modler, K. -H. (1972). Reihenfolge der homologen Punkte. *Maschinenbautechnik* **21**, 258-65.

Müller, R. (1889a). Über die Doppelpunkte der Koppelkurve. *Z. Math. Physik* **34**, 303-5.

Müller, R. (1889b). Über die Doppelpunkte der Koppelkurve. *Z. Math. Physik* **34**, 372-5.

Müller, R. (1891a). Über die Gestaltung der Koppelkurven für besondere Fälle des Kurbelgetriebes. *Z. Math. Physik* **36**, 11-20.

Müller, R. (1891b). Über die Doppelpunkte der Koppelkurve. *Z. Math. Physik* **36**, 65-70.

Müller, R. (1892). Über die Bewegung eines starren ebenen Systems durch fünf unendlich benachbarte Lagen. *Z. Math. Physik* **37**, 129-50.

Müller, R. (1897). Beiträge zur Theorie des ebenen Gelenkvierecks. *Z. Math. Physik* **42**, 247-71.

Müller, R. (1903). Über einige Kurven, die mit der Theorie des ebenen Gelenkvierecks im Zusammenhang stehen. *Z. Math. Physik* **48**, 224-48.

Oleksa, S. A. & Tesar, D. (1971). Multiply Separated Position Design of the Geared Five-Bar Function Generator. *Trans. ASME, J. Eng. Ind.* **93B**, 74-84.

Pflieger-Haertel, H. (1944). Abgewandelte Kurbelgetriebe und der Satz von Roberts. *Maschinenbau/Der Betrieb, Beilage Getriebetechnik* **12**, 197-9.

REFERENCES

Primrose, E. J. F. (1955). *Plane Algebraic Curves*. MACMILLAN & CO LTD, London.

Rauh, K. (1951). *Praktische Getriebelehre*. Bd. 1, 2nd ed. 65-121. Springer-Verlag, Berlin.

Rauh, K., Marks, H, Bündgens, W. & Otto, K. (1938). *Praktische Getriebetechnik*, #2 Kardanbewegung und Koppelbewegung. Neudruck 1948, Verlag W. Girardet, Essen.

Rankers, H. (1960). Harmonische Analyse und Maß-Synthese. *Konstruktion* **12**, 41.

Rischen, K. -A. (1962). Über die achtfache Erzeugung der Koppelkurven der zweiten Koppelebene. *Konstruktion* **14**, 381-5.

Roberts, S. (1875). Three Bar Motion in Plane Space. *Proceedings, London Math. Soc.* **7**, 14-23.

Rosenauer, N. (1938). Koppeltriebe ohne Gelenkvierecke. *Maschinenbau/Der Betrieb, Beilage Getriebetechnik* **6**, 147-9.

Roth, B. (1965). On the Multiple Generation of Coupler-Curves. *Trans. ASME, J. Eng. Ind.* **87B**, 177-83.

Roth, B. & Yang, A. T. (1977). Application of Instantaneous Invariants to the Analysis and Synthesis of Mechanisms. *Trans. ASME, J. Eng. Ind.* **99B**, 97-103.

Sieker, K. -H. (1948). Ermittlung von Gelenkvierecken aus den Krümmungshalbmessern der Polbahnen und deren Änderungen. *Die Technik* **3**, 170-4.

Sieker, K. -H. (1956). Zur algebraischen Maßsynthese ebener Kurbelgetriebe. *Ingenieur-Archiv* **24**, 188-215, 233-57.

Sieker, K. -H. & Beyer, R. (1943). Größt-und Kleinstwerte der Polbahnkrümmungs halbmesser eines Gelenkviereckes und ihre getriebesynthetische Anwendung. *Maschinenbau/der Betrieb, Beilage Getriebetechnik* **11**, 425-8.

Soni, A. H. (1970). Coupler Cognate Mechanisms of Certain Parallelogram Forms of Watt's Six-link Mechanism. *J. Mechanisms* **5**, 203-15.

Soni, A. H. (1971a). Coupler Congnates of Eight-Link Mechanisms With Ternary and Quaternary Links - Part 1. *Trans. ASME, J. Eng. Ind.* **93B**, 294-8.

Soni, A. H. (1971b). Coupler Cognates of Eight-Link Mechanisms With Ternary Links and Double Joints - Part 2. *Trans. ASME, J. Eng. Ind.* **93B**, 299-304.

Suh, C. H. & Radcliffe, C. W. (1967). Synthesis of Plane Linkages With Use of the Displacement Matrix. *Trans. ASME, J. Eng. Ind.* **89B**, 206-14.

Tesar, D. (1967). The Generalized Concept of Three Multiply Separated Positions in Coplanar Motion. *J. Mechanisms* **2**, 461-74.

Tsai, L. W. (1983a). Design of Drag-Link Mechanisms With Optimum Transmission Angle. *Trans. ASME, J. Mechanisms, Transmissions, and Automation in design* **105**, 254-8.

Tsai, L. W. (1983b). Design of Drag-Link Mechanisms With Minimax

Transmission Angle Deviation. *Trans. ASME, J. Mechanisms, Trannsmissions and Automation in Design* **105**, 686-91.

Tsai, Y. C. & Chin, P. C. (1986). Modified Chebyshev Spacing Methods for Kinematic Synthesis. *J. Chinese Soc. Mech. Engineers* **7**, 335-44.

Veldkamp, G. R. (1963). *Curvature Theory in Plane Kinematics.* J. B. Wolters, Groningen.

Veldkamp, G. R. (1967). Some Remarks on Higher Curvature Theory. *Trans. ASME, J. Eng. Ind.* **89B**, 84-6.

Volmer, J. & Jensen, P. W. (1962). Charts give best dimensions for Four-bar power linkages. *Product Engineering*, November 12, 1962, 71-76.

Waldron, K. J. (1976). Elimination of the Branch Problem in Graphical Burmester Mechanism Synthesis for Four Finitely Separated Positions. *Trans. ASME, J. Eng. Ind.* **98B**, 176-82.

Waldron, K. J. & Strong, R. T. (1978). Improved Solutions of the Branch and Order Problems of Burmester Linkage Synthesis. *Mechanism and Machine Theory*, **13**, 199-207.

Walker, R. (1953). *Cartesian and Projective Geometry.* Edward Arnold & Co. London.

Wu, L. I. & Yan, H. S. (1987). Counterexamples in Transmission Angles. In *Proc. 10th Appld. Mechanism Conf.*, Vol. 1, Session 2B, New Orleans, Louisiana.

Yan, H. S. (1992). A Methodology for Creative Mechanism Design. *Mechanism and Machine Theory* **27**, 235-42.

Yan, H. S. & Harary, F. (1987). On the Maximum Value of the Maximum Degree of Kinematic Chains. *Trans. ASME, J. Mechanisms, Transmissions, and Automation in Design* **109**, 487-90.

Yan, H. S. & Hwang, Y. W. (1990). Number Synthesis of Kinematic Chains Based on Permutation Groups. *Mathematical and Computer Modeling* **13**, 29-42.

NAME INDEX

Alt, H., 51, 78, 94, 100, 104, 112, 147, 260, 319, 325, 375
Antuma, H. J., 287, 292
Ball, R., 157
Barkan, P., 377
Beyer, R., *iii*, 52, 66, 67, 117, 122, 148, 199, 202, 214, 248, 267, 268, 271, 275, 280, 284, 319, 375
Biezeno, C. B., 384
Bobillier, E., 18
Bottema, O., 185
Burmester, L., 113
Bündgens, W., 147
Carter, W. J., 324
Cayley, A., 254
Chang, D. F., 285
Chebyshev, P. L., 252
Chen, F., Y., 337
Chen, J. S., 172, 452
Chiang, C. H., 43, 172, 328, 347, 369, 378, 452
Chin, P. C., 368
Chung, W. Y., 328
De La Hire, P., 9
Dijksman, E. A., 27, 164, 259, 298
Erdman, A. G., 172
Fichter, E. F., 30
Filemon, E., 121
Freudenstein, F., 39, 147, 172, 242, 295, 299, 337, 351, 361, 368, 386, 392, 393
Frost, P., 199, 429
Grammel, R., 384
Grübler, M., 194, 195
Gustavson, R. E., 238
Hain, K., 51, 227, 233, 283, 357, 359

Hall, A. S., 194, 196, 198, 324
Harary, F., 51
Hart, H., 254
Hartmann, W., 5
Hirschhorn, J., 72, 74, 195, 197
Hunt, K. H., 27, 30, 93
Hwang, W. M., 285
Hwang, Y. W., 51
Jensen, P. W., 351
de Jonge, A. E. R., *iii*
Keller, R. E., 119
Kracke, J., 309, 310
Kraemer, O., 55, 56
Kramer, S. N., 369
Kraus, R., *iii*, 115, 141, 217, 221, 225, 250
Lakshminarayana, K., 328
Lichtenheldt, W., 268
Lin, C. K., 369
Marks, H., 147
Mayer, A. E., 284
Meyer zur Capellen, W., 147, 285, 287, 383, 393, 395, 396, 400
Modler, K. -H., 121
Mohan, K., 393
Müller, R., 58, 148, 163, 262, 264, 267
Oleksa, S. A., 369
Otto, K., 147
Pennestrì, E., 328
Pflieger-Haertel, H., 254
Primrose, E. J. F., 25, 351, 407
Radcilffe, C. W., 82, 337, 369
Rankers, H., 383
Rauh, K., 147, 259, 285
Rischen, K. -A., 259
Roberts, S., 246, 252
Rosenauer, N., 42

Roth, B., 191, 259
Sandor, G. N., 172, 242, 369
Sieker, K. -H., 199, 202, 209, 214, 337,
Soni, A. H., 259
Strong, R. T., 121
Suh, C. H., 82, 337, 369
Tesar, D., 166, 369
Thünker, N., 400
Tsai, L. W., 359, 361

Tsai, Y. C., 368
Tuohy, E. J., 377
Veldkamp, G. R., 185, 304
Volmer, J., 351
Waldron, K. J., 121
Walker, R., 407
Wu, L. I., 327
Yan, H. S., 51, 327
Yang, A. T., 191

SUBJECT INDEX

A_0 corresponding to A_1, A_2, A_3, 84
A_1 corresponding to A_0, 85
acceleration analysis, 43
    of complex mechanisms, 47
acceleration pole, 12
adjustable dwell mechanism, 100, 122, 152
algebraic envelope, 419
angle of rotation, 82, 451
angular velocity ratio, 320
Antuma triangular nomogram, 287
Archimedes spiral, 275
asymptote, 33, 410, 414
axis of symmetry, 286
balancing of number of coordinates, 217, 220-229
Ball curve, 159
Ball point, 22, 156, 194, 298
basic pole, 172, 319
beak cusp, 147, 267
bicircular curve, 406
bipolar coordinates, 199, 429
bitangent, 33, 37
Bobillier construction, 17-23
Bobillier theorem, 18, 197
break-ups
    of centre-point curve and circle-point curve, 121
    of circling-point and centering-point curves, 149
Bresse's circles, 11
Burmester centres, 161
Burmester curves, 113
Burmester point-pair, 163
Burmester points, 160, 163
    Generalized, 304
        of a given four-bar linkage, 165

relative, 341
canonical system, 189, 191, 192
Cardan circle pair, 21, 22, 33, 61, 91, 147, 149, 151, 275, 305, 369
Cardan position, 147, 148
cardinal line, 77
cardinal point, 64, 74, 77
cardioid, 298, 302
cardioid motion, 298, 302
Carter-Hall circle, 323
centre-point, 63
centre-point curve, 113
    break up of, 121
centering-point curve, 137
    construction of, 139
central crank-rocker, 362
central slider-crank mechanism, 155, 384
central swinging block linkage, 156
centro, 3
charactcristic numbcr, 297
Chebyshev minimax theorem, 365
Chebyshev polynomial, 365
Chebyshev spacing of precision points, 366
circle-point, 63
circle-point curve, 113, 119
    break up of, 124
circling-and centering-point curve method, 210
circling-point curve, 113, 119
    construction of, 139
    Müller equation of, 136
circular cubic, 119, 127, 149, 406
circular curve, 119, 406
circular point, 405
circular quartic, 406
class, 33, 248, 416

closed curve, 354
cognate linkages, 254, 287
    Cayley diagram, 254
    of a slider-crank, 257
collineation axis, 18, 148
compatibility equation, 434
conic section point curve, 269, 271
constrained kinematic chain, 51
contingence angle, 135, 145
coordinate system,
    selection of, 88
coordinations of five point-positions
    with four crank rotations, 245
coordinations of four point-positions
    with three crank rotations – k_{A0} –
    curve, 236, 243
coordinations of three point-positions
    with two crank rotations, 234
coordination of two point-positions
    with one crank rotation, 231
coordinations of two point-positions
    with one pair of crank rotations, 232
correlated lines, 74
correlated points, 63
correlated positions, 73
coupler, 3
coupler cognates, 259
coupler curve(=coupler point curve),
coupler dwell mechanism, 20, 67
coupler point, 19
coupler point curve, 246
    algebraic equation of, 246
    bicursal and unicursal, 248
    nodes of, 249
    singular foci of, 252
    symmetrical, 285-292
        by six-bar linkage, 292
    with two cusps, 280
    with three cusps, 284
crank-rocker,
    synthesis of, 346
cubic of stationary curvature, 133
curve cognates, 259
curve of equal radii of curvature of
the first kind (the ρ- curve), 94
curve of equal radii of curvature of
    the second kind (the q_1-courve), 100
cusp, 60, 93, 144, 266, 409
cusp circle, 27, 93
dimensional synthesis, 51
displacement matrix, 83
    based on pole coordintes, 87
double point, 33, 408
double slider-crank mechanism, 61
double-crank, 357
    synthesis of, 357
double-rocker, 354
    synthesis of, 354
drag-link mechanism, 357
dual, 33, 73, 419
duality, 419
dwell mechanism, 20, 30, 75, 257, 259
    by means of osculating ellipse, 276, 279
dyad, 20, 232, 241, 257, 269, 295
ellipse,
    generation of, 268
    tangent to a given curve at five
        points, 279
    tangent to a given curve at four
        points, 275
envelope, 28
envelope equation, 416
equivalent instantaneous four-bar, 50
error,
    of a function generator, 325
Euler-Savary equation, 13-17
evolute, 295
excess, 164
fifth order synthesis, 333
five infinitesimally separated positions, 163
fixed polode, 4
focus, 416
four finitely separated positions, 113
four homologous lines passing
    through a point, 124
four homologous points on a line, 124,

SUBJECT INDEX

130
four infinitesimally separated positions, 133
fourth order synthesis, 323
Freudenstein equation, 39
function cognates, 259
function generator, 309
   coordination of a pair of angular and linear displacements, 316
   order type synthesis, 319
     simplified geometrical method, 328
   coordination of finitely separated angular displacements,
     algebraic method, 337
     geometrical method, 313
   geared five-bar, 368
geared five-bar, 259
generating curve, 28
Grübler-Hall equation, 194
graph theory, 51
Grashof double-rocker, 354
harmonic analysis, 383
   of central slider-crank, 384
   of kinetic energy of inverted slider-crank, 400
   of offest slider-crank, 396
   of output angle of four-bar linkage, 386
   of swinging block, 394
   of turning block linkage, 395
Hartmann construction, 5, 13, 20
higher order path curvature, 295
higher order synthesis, 319
homogeneous coordinate equation, 31
homogeneous coordinates, 405
homologous lines, 74
homologous points, 63
homologous positions, 73
inflection circle, 7, 15, 34
   diameter of, 8
inflection pole, 8, 34
inner dead centre position 149, 361
inner frame position, 152, 325
instantaneous centre of velocity, 3, 56

instantaneous invariants, 185-194, 210
intermediate cases, 166, 343
inverted slider-crank, 393
involute, 302
isolated point, 267, 409, 414
isotropic line, 267, 416
k_{A0} - curve, 236
kinematic inversion, 4, 27
Lagrange multiplier, 351
λ_1-curve, 298
line at infinity, 33, 406
line coordinates, 31, 416
line coordinate equation, 31, 416
linearization, 310
locus, 4
logarithmic spiral, 275
loop equation, 172, 244, 369
major axis and minor axis of an ellipse, 437
minimax theorem, 365
mirror image, 64, 100
moving polode, 4, 58, 160, 266
Muller's equation of circling-point curve, 136
node, 249, 265, 266, 409
non-Grashof double-rocker, 354
non-parallel equal crank linkage, 61
number synthesis, 51
offset slider-crank, 396
opposite-pole quadrilateral, 115
optimization of transmission angle, 328, 351
order type synthesis, 319
order, 211
order, 410, 419
orthogonal velocity, 42
oscillating cylinder, 393
osculating curve 76, 273, 275, 295
osculating ellipse, 273
outer dead centre position 151, 361
outer frame posistion 152, 325
overtones, 392
path generating mechanism, 223, 231
path generation, 231

vector loop method, 242
plane algebraic curve, 407
point-position reduction method, 225
pole, 3
pole changing velocity, 4, 93
pole triangle, 64
polenormal, 5
poletangent, 5
polode, 4
   fixed, 4
   fixed, equation of, 58
   maximum and minimum redii of curvature of, 202
   moving, 4
   moving, equation of, 58
   relative, 195
polode method, 194, 210
position vector, 83, 372
potency, 104
precision point, 311
pressure angle, 392
q_1 - curve, 100
q_M - curve, 112
quadratic transformation, 23
quartic of derivative curvature, 298
R^1- curve, 81, 86
R_M- curve, 79, 86
   as a special case of coupler curve, 260
relative Burmester centre point 341
relative centre-point curve, 339
relative circle-point curve, 340
relative pole, 313
relative polodes, 195
required speeification, 217
resonance, 383
return circle, 27, 90
return pole, 27, 152
ρ - curve, 94
ρ_M - curve, 100
Roberts-Chebyshev theorem, 252-259, 287
rotation coefficients, 244
rotation, 83

second acceleration, 191
second angular acceleration, 191, 328
second order synthesis, 321
Sieker-Beyer equation, 199
singular focus, 138, 252, 416
singular point, 10
slider-crank, 316
   fourth order synthesis of, 332
source linkage, 259
source mechanism, 259
stretch and rotation, 259, 296, 303
structural error, 312, 368
swinging block linkage, 394
synthesis, 51
   type, 51
   number, 51
   dimensional, 51
   Releaux, 51
   Grübler, 51
tangential circle, 9, 193
   diameter of, 10
Thales circle, 282
third order synthesis, 322, 329
three finitely separated positions, 62, 82
   algebraic method, 82
   geometrical method, 62
three homologous lines passing through one point, 73
three homologous points on a line, 69
three infinitesimally separated positions, 89
   algebraic method, 93
   geometrical method, 89
transition curve, 262-268
translation, 83
transmission angle, 75, 325
   optimization of, 328, 351
tricircular curve, 406
tricircular sextic, 80, 248, 260
turning block linkage, 365, 395
two finitely separated positions, 53
two infinitesimally separated positions, 55
type synthesis, 51

SUBJECT INDEX

undulation point, 157
unit vector, 243
valence, 217
vector loop equation, 244, 369
velocity analysis, 40
    of 8-bar linkage, 42
velocity pole, 3, 11, 56, 144, 280, 290, 291
velocity pole as a moving point, 144